AF248981

Ductile Iron Handbook

Copyright © 1992

Revised 1993

American Foundrymen's Society, Inc.
Des Plaines, Illinois

All rights reserved. This book, or parts thereof may not be reproduced without permission of the publisher.

The American Foundrymen's Society, Inc., as a body, is not responsible for the statements and opinions advanced in this publication. Nothing contained in any publication of the American Foundrymen's Society, Inc. is to be construed as granting any right, by implication or otherwise, for manufacture, sale, or use in connection with any method, apparatus or product covered by letters patent, nor as insuring anyone against liability for infringement of letters patent.

ISBN 0-87433-124-2

Cover photo of "Hardenable Ductile Iron" courtesy of the research foundry of Intermet Foundries, Inc.

Printed in the U.S.A.

Foreword

The American Foundrymen's Society, Inc. would like to extend its appreciation to Miller and Company for making available for publication the material contained in their *Ductile Iron Compendium*. Numerous micrographs along with tables and graphs from the *Compendium* were incorporated in the *Ductile Iron Handbook*.

On behalf of AFS, I would also like to thank William A. Henning and Jerry Mercer for their technical review of the manuscript. In addition to their many critical comments and changes, their "Editors' Notes" provide valuable perspective to the discussions of controversial or varied foundry practices.

The Society also wishes to express its sincere appreciation to the authors and members of the Cast Iron Molten Metal Processing Committee who contributed to the development of this book.

Michael F. Burditt
Technical Editor

Preface

The ductile iron process is now a celebrated 40+ years old. In that time span, the technology has improved substantially, from the choice of raw materials to be melted, to techniques that confirm casting specifications.

To a certain extent, credit for the advances in technology is due the pioneers of the process who were successful in handing down a valuable part of their experience. More recently, that individual communication has continued, but we now benefit from a history of technical developments in ductile iron production that were well recorded in trade or research journals in many languages. This new AFS *Ductile Iron Handbook* was developed to continue the communication process; to provide a comprehensive reference on specific aspects of a growing foundry processes; and to provide a basic bibliography.

Where it was deemed appropriate to alert the reader to additional or, in some cases, contradictory information, we have inserted this information as an "Editors' Note," abbreviated "Ed. N."

The authors for the chapters of this handbook were selected for their known expertise and their established contributions to industry. These are some of the same foundrymen who have either presented papers at AFS technical conferences or Congress technical sessions, or were members of the audience, but in either case, facilitating the transfer of technology between sectors.

It is hoped that, by providing a sound base of technical information on the ductile iron process, the capabilities of foundries and individual foundrymen will be enhanced, permitting them to meet the continuously escalating quality demanded by our casting customers. In the process, we hope to promote and provide for a growing industry whose technological changes will require revisions to this book based on (your!) future technical contributions.

The editors join the Society in thanking the following individuals:

AUTHORS

Al Alagarsamy
Grede Foundries, Inc.

Donald B. Craig
Elkem Metals

Richard B. Gundlach
Climax Research Services

Harvey Henderson
Consultant

Hugh Kind
Foseco, Inc.

Bela V. Kovacs
AFC Technical Center

Richard Kryzanek
Deere & Co.

Carl R. Loper, Jr.
University of Wisconsin—Madison

Robert Neuman
Foseco, Inc.

William L. Powell
Waupaca Foundry

Warren Spear
Nickel Development Institute

Doru M. Stefanescu
University of Alabama

T. Stoecker
Sandy Hill Corp.

Nick Wukovich
Foseco, Inc.

Timothy Zeh
Internet Corp.

COMMITTEE MEMBERS

Michael Barstow *
  Elkem Metals
Robert Bigge
  Briggs & Stratton
James D. Bopp
  Sibley Machine & Foundry
Jim Csonka
  American Alloys
J.E. Foltz, Jr. *
  Intermet Corp.
Jeffrey T. Fowler
  CMI International
Gary Garlough *
  Goulds Pumps, Inc.
George M. Goodrich *
  Taussig Associates
Randy Hunt
  Alabama Ductile Casting Co.
Dan Jarvis
  Waupaca Foundry
David P. Jones
  Deere & Co.

Seymour Katz
  General Motors Corp.
C. Ronald Kern *
  ACIP Co.
Jack Klein *
  SKW Metals and Alloys
Prem P. Mohla
  Globe Metallurgical
Bernardo Morgansteren *
  Ford Motor Co.
James D. Mullins *
  QIT -Fer et Titane, Inc.
Leon Pasternak
  Wells Manufacturing
R. Alan Patrick *
  Buck Co.
William Thomas
  Wagner Castings
Larry B. White
  Grede Vassar, Inc.
James E. Woods
  Hickman, Williams & Co.

---

* Authors

The *AFS Ductile Iron Handbook* could never have been completed without the dedicated effort of the individual authors as well as other members of AFS Cast Iron Molten Metal Processing Committee.

William A. Henning
  Miller and Company
Jerry Mercer
  Pickands Mather & Co.

Editors-in-Chief

In memory of
Jerry Longest Mercer,
1921-91,
A Foundryman.

Contents

Contents

Contents

Contents

— Chapter 12: Influence of Section Size on Microstructure and Mechanical Properties

— Chapter 13: Heat Treatment

— Chapter 14: Ductile Iron Casting Defects

Contents

— **Chapter 15: Ductile Iron Process Control**

— **Chapter 16: Welding**

— **Appendix**

— **Index** 270

1

Theory of Solidification and Graphite Growth in Ductile Iron

D. M. Stefanescu

University of Alabama
Tuscaloosa, Alabama

■ INTRODUCTION

Understanding the mechanisms involved in the formation of spheroidal graphite during the solidification of cast iron has been a coveted goal of cast iron metallurgists for more than four decades. In the first paper ever to deal with spheroidal graphite in cast iron, written by Morrogh and Williams[1] in 1947, a first theory was proposed. This theory, as well as many of the following ones, while being a tribute to the imagination of the authors, suffered from a lack of understanding of crystal growth mechanisms, knowledge that was unavailable at that time. Thus, their value is rather limited. Since that time, a good number of monographs have tackled this subject. (The most recent ones are cited in references 2, 3, and 4.)

This chapter will attempt to cover most of the theories related to the crystallization of spheroidal graphite. The evolution of the scientific understanding of this phenomenon will be explained and, hopefully, the reader will be able to form his or her own opinion on the subject.

Cast iron is a binary Fe-C or a multicomponent Fe-C-X alloy that is rich in carbon and exhibits a considerable amount of eutectic in the solid state. The high carbon content of cast iron has remarkable consequences on the structure and properties of both the liquid and the solid material. To understand the mechanism of the solidification of ductile iron, it is first necessary to discuss the structure of liquid iron-carbon alloys, the structure of graphite, and the fundamentals of nucleation and growth in cast iron. Then it will be possible to present various theories that try to explain the occurrence of spheroidal graphite.

■ STRUCTURE OF LIQUID IRON-CARBON ALLOYS

Observations from X-ray, neutron diffraction, and sound velocity measurements on liquid binary iron-carbon alloys at temperatures approximately 20°C (68°F) above the liquidus indicate that, for up to 1.8% C, the distance between nearest iron neighbors, r_1, as well as the number of nearest neighbors, N_1, in the first coordination sphere increases (Fig. 1-1). Above 1.8% C, the distance remains constant, while the number of nearest neighbors continues to grow.[5] Above 3.5% and up to 5.5% C, both the distance and the

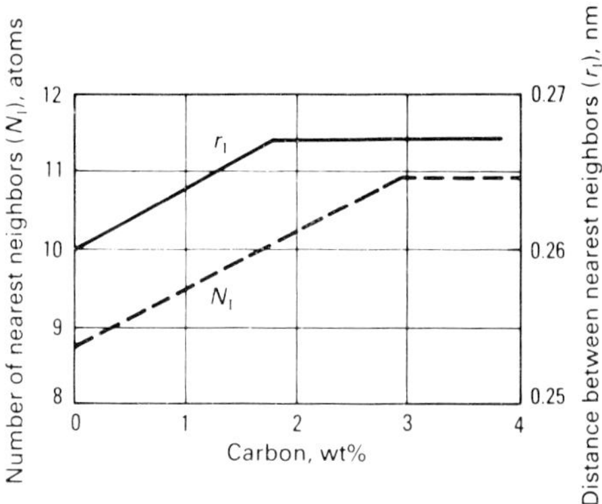

Fig. 1-1. Variation of distance between nearest neighbors (r_l) and the number of nearest neighbors (N_l) as a function of the percentage of carbon in the iron-carbon alloy.[5]

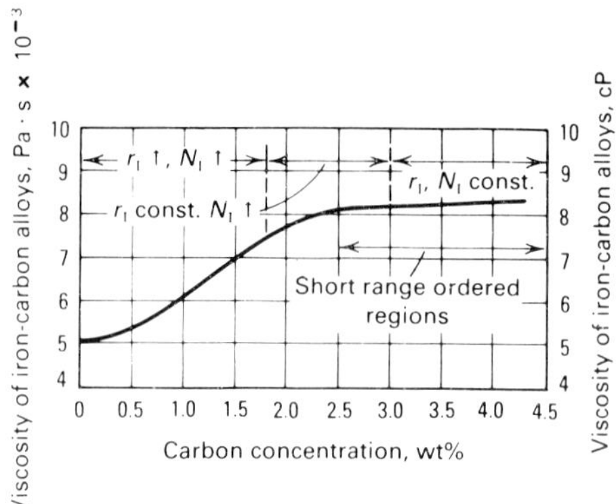

Fig. 1-2. Viscosity of iron-carbon alloys as a function of carbon concentration at a temperature ≈20°C above liquidus.[6]

number of neighbors remain constant. Above 3.5% C, short-range order regions, rich in carbon, exist in the melt. This means that the melt becomes more dense with the addition of carbon. A maximum packing density is reached at 3% concentration. The excess carbon forms carbon-rich regions (nonhomogeneities) in the melt.

Viscosity measurements (Fig. 1-2) show a correlation between viscosity and percentage of carbon.[6] This correlation can be further explained in terms of increased viscosity as the interatomic distance becomes smaller.

Liquid iron-carbon alloys with low carbon content (<3.5% C)—steels and cast irons poor in carbon—are microscopically homogeneous. Liquid iron-carbon alloys with high carbon (>3.5% C)—cast irons rich in carbon—are colloidally dispersed systems with microgroups of carbon in liquid solution. The nature of these microgroups is not clear. It is hypothesized that they are either Fe_3C clusters[7] or C_n clusters[5,8] (where n equals the number of carbon atom clusters). The size of the C_n clusters is considered to be in the range of 1–20 µm, and it increases with the carbon equivalent, lower silicon content, and lower holding time and temperature. It is to be expected that the carbon-rich configurations existing in molten iron-carbon alloys are in dynamic equilibrium, and that they diffuse within the melt.

STRUCTURE OF SPHEROIDAL GRAPHITE

The graphite phase in cast iron is a faceted crystal bounded by low index planes. For graphite crystallizing from an iron-carbon melt, the normally observed bounding

planes are [0001] and [$10\bar{1}0$], as shown in Figure 1-3a. The crystallographic structure of graphite and the possible growth directions, A and C, are shown in Figure 1-3b. Because unstable growth occurs on the [$10\bar{1}0$] planes, the edges of the platelike graphite crystals are not well defined. Graphite growing out of liquid iron-carbon alloys has a layer-type structure, with strong covalent bonds (4.19 x 10^5 to 5 x 10^5 J/mol) between atoms in the same layer. There is a trielectronic bond of each atom with its neighbors, while the fourth electron is common for the layer, giving the metallic properties of graphite. Weak molecular forces exist between layers (4.19 x 10^3 to 8.37 x 10^3 J/mol). The prism plane is a high-energy plane at which impurities adsorb preferentially. Strength and hardness are higher in the C direction of the graphite crystal.

Complete destruction of the graphite structure occurs only at about 4000C (7232F). This explains the presence of some graphite aggregates in molten iron, even at temperatures considerably higher than the liquidus temperature.

Depending on the chemical composition and on the temperature gradient/growth rate ratio (G/R), or on the cooling rate (G·R), a variety of graphite shapes can solidify as part of the austenite-graphite eutectic or as a primary phase. Basically, they are as follows:

- flake (actually plate) graphite (FG);
- compacted/vermicular graphite (CG);
- coral graphite;
- spheroidal (nodular) graphite (SG).

A schematic of these graphite types, showing the traces of the [$10\bar{1}0$] planes, is given in Figure 1-4.

There is no unanimity of views as to what is the growth pattern of spheroidal graphite, nor on the causes and mechanisms of this growth. Before discussing the various

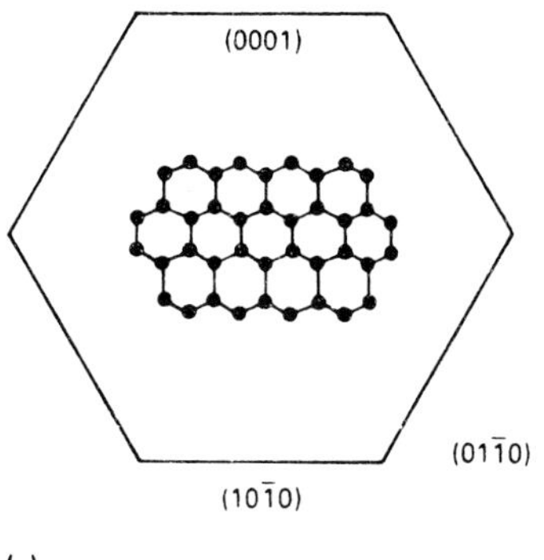
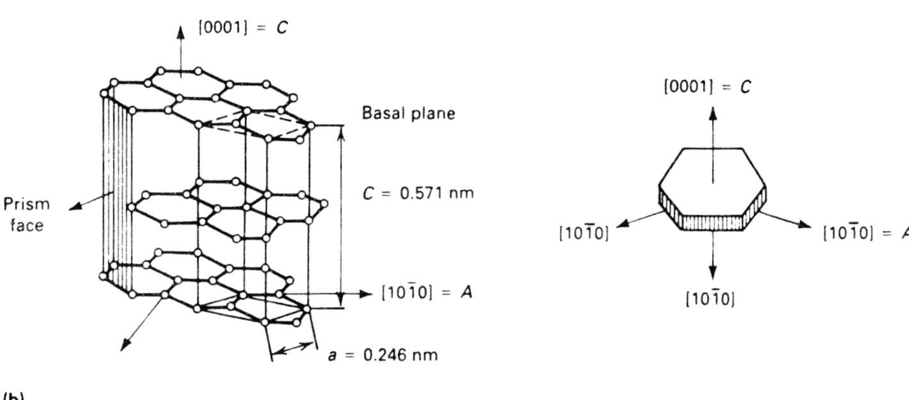

Fig. 1-3. Crystalline structure of graphite: (a) Crystal of graphite bounded by [0001] and [10$\bar{1}$0] type planes. The hexagonal structure of the atoms within the [0001] plane is shown relative to the bounding [10$\bar{1}$0] faces; (b) hexagonal structure of graphite showing the unit cell (heavy lines).[2]

theories proposed for the mechanisms of spheroidal graphite formation in cast iron, it is necessary to review some of the models suggested for the crystalline growth of spheroidal graphite.

GROWTH BY SCREW DISLOCATION MODEL

Hillert and Lindblom[9] suggested that graphite spheroids grow by a screw dislocation mechanism. Cerium or magnesium atoms attach to the carbon atoms at the growing edge of the close-packed plane, thus producing the disturbances required for the development of new screw dislocations. As shown in Figure 1-5, the spirals will grow into one another, then will branch, and growth will become spherulitic.

GROWTH BY INTERFACE BREAKDOWN MODEL

Oldfield, Geering and Tiller[10] proposed a model for spheroi-dal graphite growth based on the breakdown of the interface. The process starts from a polyhedral crystal growing with a stable faceted interface that becomes unstable when a critical size is exceeded. A protuberance that forms on any of the faces of the unstable crystal can grow at the same velocity as the corner, but it will lag by a distance, d, behind the corner (Fig. 1-6). As growth continues, the shape of the graphite particle increasingly approximates to a spherical form. The growing instabilities are dendritic in nature and grow by a screw dislocation mechanism.

CIRCUMFERENTIAL GROWTH MODELS

As will be discussed later in this chapter, it is conceivable that the tendency of the liquid/graphite system to minimize its free energy can result in the imposition of forces on the growing graphite crystal, which can eventually result in curved crystal growth. A first model based on this concept

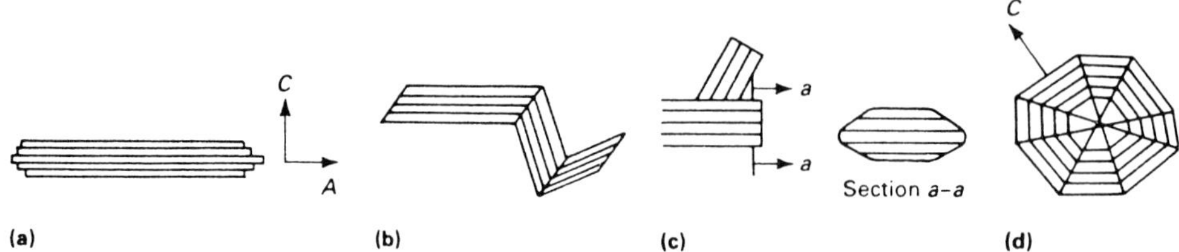

Fig. 1-4. Schematic of graphite types occurring in the austenite-graphite eutectic: (a) flake graphite; (b) compacted/vermicular graphite; (c) coral graphite; (d) spheroidal graphite.

was proposed by Double and Hellawell,[11] who noted that vapor-grown graphite fibers are sometimes closed at their ends by conical caps showing specific apical angles. This observation led to the assumption that the structure of these fibers consists of graphite basal sheets rolled upon themselves into the form of conical helices (Fig. 1-7a). Near the center of the spheroid, the initial growth may begin as a loose combination of conical helices (Fig. 1-7b). The slight misorientation between the numerous radial crystal filaments may produce an uneven multistepped surface with many tilt/twist boundaries (Fig. 1-7c). Such defects will allow one subgrain to grow over the other, creating many active growth sources on the surface of the spheroid, and decreasing the role of single-step dislocation sources.

Sadocha and Gruzleski[12] proposed a simpler model, where it is assumed that spheroidal growth occurs in a circumferential manner from the beginning, as shown in Figure 1-8, by movement of steps around the surface of the spheroid. These steps grow in the A direction by curved crystal growth, with the low energy basal plane of graphite exposed to the liquid. The growing steps run into one another forming boundaries on the surface. From these boundaries new steps can develop and grow over the surface, producing a cabbage-leaf effect.

Fig. 1-5. Growth of graphite spheroids by screw dislocations.[9]

FUNDAMENTALS OF NUCLEATION AND GROWTH IN CAST IRON

NUCLEATION IN CAST IRON

Two different issues must be addressed here. One needs to understand first nucleation of the primary austenite, and then nucleation of the graphite phase, which is either part of the eutectic or is a primary phase.

Nucleation of Primary Austenite

Formation and growth characteristics of austenite dendrites have received considerably less attention from investigators than graphite morphology, simply because dendrites are not readily discernible in the structure. Nevertheless, it has been demonstrated that inoculants that are effective in increasing cell count in flake graphite iron, such as Ca-Ti, Sr, and 0.75%Fe-Si have little effect on the solidification pattern of primary austenite dendrites.[13] A number of elements, as for example Ti, V, and Al, were proven to generally increase the number of dendrites and their length as compared with the base iron. For Ti and V, this was attributed to the formation of carbides, nitrides and carbonitrides, which then act as substrates for austenite solidification. The influence of Al was not explained. Other elements, such as Ce and B, although being instrumental in increasing the number and length of dendrites, were not considered to be nucleants for austenite. It was suggested that their effect is based on the fact that they restrict the growth of the eutectic cells, which results in larger undercooling for eutectic solidification. This means that more time (larger temperature interval) is allowed for the nucleation and growth of austenite.[13]

Nucleation of the Austenite-Flake Graphite Eutectic

Homogeneous nucleation is improbable in cast iron because typical undercoolings are much smaller than

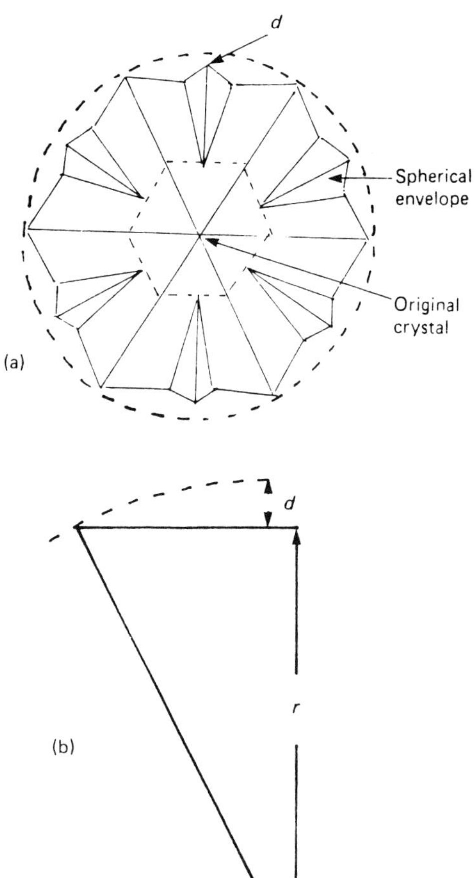

Fig. 1-6. Growth of graphite spheroids by interface breakdown: (a) growth of protuberances from the polyhedral crystal; (b) the spherical envelope is a distance, *d*, ahead of the growing crystallite.[10]

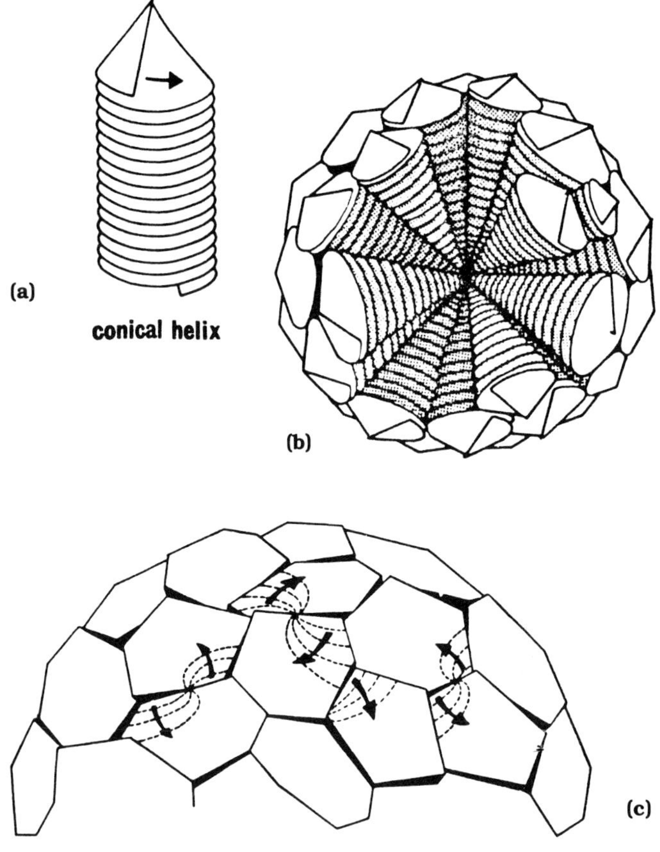

Fig. 1-7. Growth of graphite spheroids from conical helix crystals: (a) conical-helix structure of graphite basal sheet; (b) nucleus of a spheroid composed of numerous close-packed, conical helices growing from a common center; (c) tilt/twist boundaries between individual crystal segments on the surface of a graphite spheroid.[11]

required by the theory of homogeneous nucleation (1–10C/ 35–50F, as opposed to 230C/446F). Nevertheless, some C_n clusters (as discussed earlier in this section) or undissolved graphite may act as heterogeneous nuclei for the solidification of the graphite-austenite eutectic. When microstructures of castings produced from two separate melt charges of similar composition—but containing different amounts of white iron and gray iron—are compared, it appears reasonable to assume that some type of residual graphite serves as nuclei during solidification.[14] Further, additions of graphite to the melt enhance nucleation, since the eutectic cell count is increased while the chilling tendency decreases.[15]

A rather wide variety of compounds have been claimed to serve as nuclei for flake graphite cast iron, including oxides (for example, silicon dioxide) or silicates, sulfides,

nitrides (boron nitride), carbides (for example, Al_4C_3), and intermetallic compounds.[14] In addition, a number of metals, such as sodium, potassium, calcium, strontium, barium, and yttrium and the lanthanides, act as inoculants in flake graphite iron and can therefore be considered to play a significant role in the heterogeneous nucleation of the austenite-flake graphite eutectic.[14, 16-18] High silicon concentrations in the melt can also contribute to the heterogeneous nucleation of graphite.

The two best supported theories on the heterogeneous nucleation of the flake graphite eutectic are: 1) the nucleation of graphite on silicon dioxide particles[19]; and 2) the nucleation of graphite on saltlike carbides.[16] Saltlike carbides are carbides containing the ion, C_2^{2-} such as CaC_2, SrC_2, BaC_2, YC_2, and LaC_2. These theories are described in some detail in reference 20. The influence of cooling rate

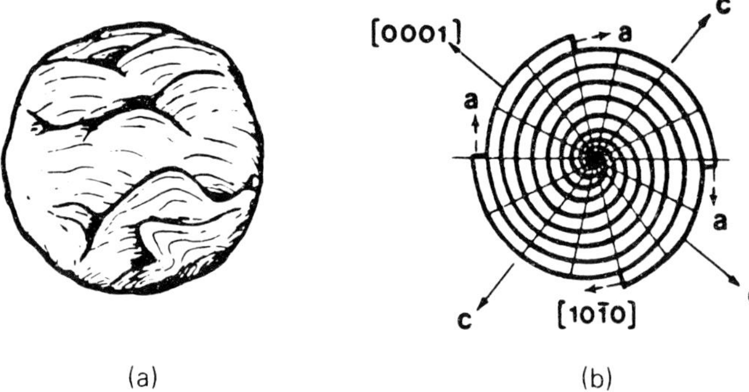

(a)   (b)

Fig. 1-8. Circumferential growth of graphite spheroids: (a) surface showing leaf effect; (b) diametral section showing growth in the *a* direction.[12]

and undercooling on nucleation is also discussed in the same reference.

Nucleation of the Austenite-Spheroidal Graphite Eutectic

Because most of the inoculants used in flake graphite iron treatment are also effective for ductile irons, it is reasonable to assume that similar mechanisms and substrates are active in the nucleation of both the flake graphite-austenite and spheroidal graphite-austenite eutectics. Extensive transmission electron microscopy and scanning electron microscopy studies have been done to identify the composition of graphite nuclei in ductile irons, with a rather wide variety of results. Numerous types of compounds have been found in the middle of graphite spheroids; therefore, it was hypothesized that they could act as nuclei. Some of these compounds are as follows:

1) $3MgO \cdot 2SiO_2 \cdot 2H_2O$ (chrystobalite)[21]
2) $xMgO \cdot yAl_2O_3 \cdot zSiO_2$
3) $xMgO \cdot ySiO_2 \cdot xMgO \cdot ySiO_2 \cdot zMgS$[22]
4) MgS[23]
5) $Te + Mn + S$[24]
6) lanthanide sulfides.[25]

A rather extensive work on what appeared to be substrates for spheroidal graphite in chilled ductile iron[26] concluded that the substrates are duplex sulfide-oxide inclusions with a diameter of 1 µm. The core is probably made of Ca-Mg or Ca-Mg-Sr sulfides, while the outer shell is made of complex Mg-Al-Si-Ti oxides with a spinel structure. The x-ray diffraction data showed that the first few graphite layers, adjacent to the oxide, had a dilated lattice (0.264 nm instead of 0.246 nm). It was suggested that the spacing within the graphite layers decreases away from the oxide until unconstrained spacing is reached. Dislocations were frequently observed in the matrix, and it was suggested that these were generated to relieve some of the elastic strain in the graphite layers adjacent to the oxide.

These observations allow for a theory of nucleation of spheroidal graphite similar to the catalytic theory of nucleation of flake graphite on silicon dioxide. It can be assumed that the nucleation process begins with the formation of complex sulfides that serve as nuclei for complex oxides, which, in turn, serve as nuclei for spheroidal graphite.[20]

Nucleation of Primary Graphite

It is reasonable to assume that nuclei that are active in the solidification of eutectic graphite also serve as nuclei for hypereutectic primary graphite. Nevertheless, it must be noted that inoculation does not seem to increase significantly the number of eutectic cells in hypereutectic flake graphite iron. On the other hand, inoculation is quite effective in hypoeutectic ductile iron.

GROWTH KINETICS OF THE EUTECTIC IN CAST IRON

The degree and type of eutectic growth that occurs in cast iron can be determined by using tools such as growth rate curves to locate coupled zone regions (i.e., regions where the two phases of the eutectic grow together rather than separately), isothermal time-temperature diagrams to gauge susceptibility to carbide formation, and growth rate-composition plots to ascertain parameters that affect both directional and multidirectional solidification.

The Coupled Zone in Cast Iron

The initial step that must be undertaken in the effort to understand solidification of ductile iron must be directed to the comprehension of the coupled growth region of the eutectic. The coupled growth region of the eutectic in cast iron is asymmetric. It is possible to construct a theoretical coupled zone for gray iron from the condition of equal growth rate of the austenite (γ) and graphite (Gr) phases.[27] First one must consider the growth rate curves for γ and Gr in the Fe-C system (Fig. 1-9). For flake graphite iron, the growth rate of austenite, R_γ, and that of graphite along the $[10\bar{1}0]$ direction, R_{Gr} $[10\bar{1}0]$, intersect; therefore, a coupled zone can be constructed. Indeed, Jones and Kurz[28] have determined, experimentally, the transition from a fully eutectic to a eutectic plus dendrite structure in pure Fe-C alloys of eutectic composition, solidifying white or with flake graphite, and have calculated the γ-Fe and Gr-Fe eutectic boundaries (Fig. 1-10).

For ductile iron, where the predominant growth direction is along [0001], the two rates, R_γ and R_{Gr}[0001] do not intersect, which means that coupled growth is impossible. The γ-flake graphite eutectic is a coupled irregular eutectic of the faceted (Gr)/nonfaceted (γ) type. The γ-spheroidal graphite eutectic is a divorced eutectic.

Isothermal Solidification

Solidification studies are usually performed athermally,

because of the high rate of the liquid-solid transformation. Nevertheless, useful information can be extracted from the isothermal time-temperature-transformation diagrams shown in Figure 1-11. It is apparent that ductile iron is more susceptible to carbide formation than flake graphite iron. Graphite precipitates earlier in ductile iron than in flake graphite iron at all undercoolings, although the time interval for complete gray iron solidification is smaller in flake graphite irons. Also, it is evident that, for solidification temperatures below 1090C (2174F), the process starts with formation of austenite even for eutectic composition.

Growth in Directional Solidification

It is quite obvious from the previous discussion that undercooling and composition must be carefully selected to achieve coupled growth of the eutectic in cast iron.

The basic parameters affecting the morphology of the eutectic are the G/V ratio and composition (where G is the temperature gradient at the interface, and V is the growth velocity of the interface). It is possible to achieve a variety of graphite and matrix structures in cast iron when varying G/V and/or the level of impurities, such as magnesium or cerium.[30-32] Argo and Gruzleski[31] have achieved transition from a spheroidal through compacted to flake graphite structure by directional solidification of Mg-containing ductile iron. The results have been presented in terms of the solidification rate and residual Mg concentration (Fig. 1-12). The problem of Mg fading was avoided, in this case, by pouring the Mg-treated liquid metal into a tube in the directional solidification furnace. The thermal gradient in the liquid at the solid-liquid interface was 15 K/cm. Cooling rates used in this study varied from 0.0015 to 0.013°C/s, which is rather on the low side. Consequently, complete structural transition, to include stable/metastable transition, was not obtained.

The complete structural transition from metastable to stable, and for different graphite morphologies, has recently been documented for cast irons of hypoeutectic composition as a function of growth velocity, temperature gradients at the solid/liquid interface, and cerium concentration, by Bandyopadhyay et al.[32] It was found that, while the metastable (white) to stable (gray) transition depends mostly on the G/V ratio, the transition between different graphite shapes (lamellar to vermicular to spheroidal) depends mostly on the cerium concentration (Fig. 1-13).

Based on existing experimental work,[32,33] a sequence of changes in the eutectic morphology of directionally solidified cast iron is proposed in Figure 1-14. As G/V decreases or the composition (e.g., Mg or Ce) increases, the solid-liquid interface changes from planar, to cellular, and then to equiaxed, while graphite remains basically flake (lamellar).[34] Cooperative growth of austenite and graphite occurs. Further change of G/V or of composition brings about formation of an irregular interface, with austenite dendrites

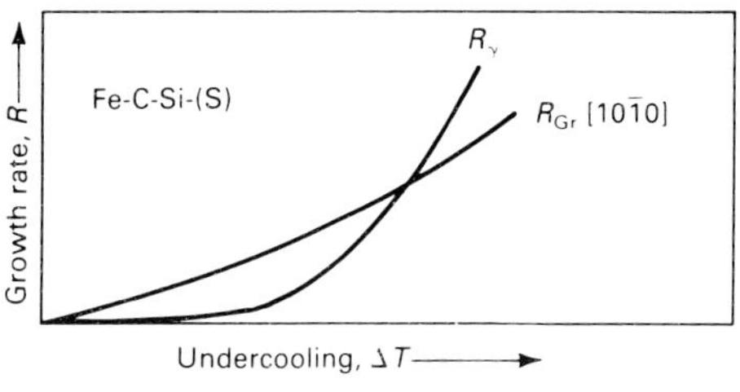

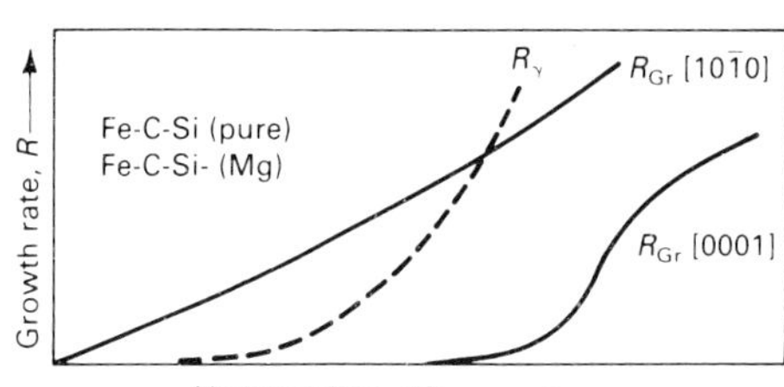

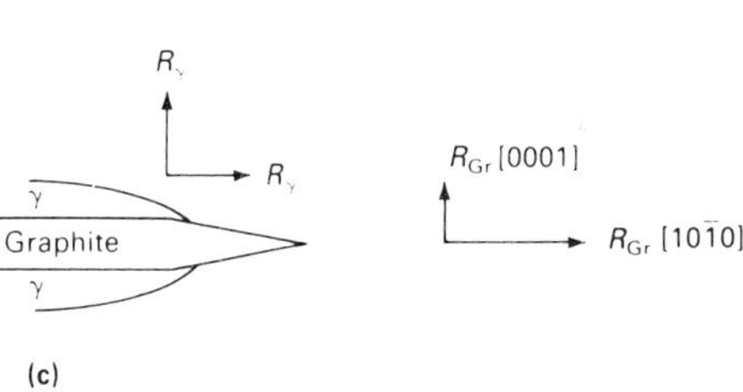

Fig. 1-9. Suggested growth rate curves for austenite and graphite in the Fe-C-Si alloys: (a) austenite-flake graphite eutectic; (b) austenite-spheroidal graphite eutectic; (c) different graphite growth rates in lamellar growth. R_G [$10\bar{1}0$] determines the lengthwise growth of graphite flakes, while R_{Gr} [0001] gives the rate of thickening for the flake.[2]

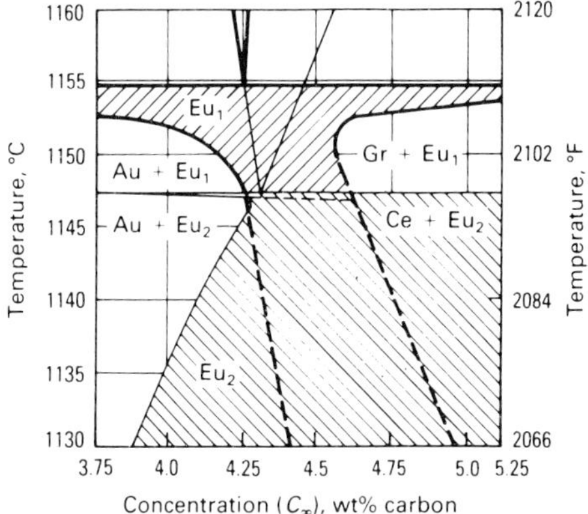

Fig. 1-10. Coupled zone of directionally grown Fe-C eutectic: (Eu_1) stable eutectic; (Eu_2) metastable eutectic; (Au) austenite dendrites; (Gr) primary graphite; (Ce) primary cementite.[28]

7

protruding in the liquid. Graphite becomes compacted and then spheroidal; eutectic growth is divorced.

THEORIES ON THE FORMATION OF SPHEROIDAL GRAPHITE IN CAST IRON

The complexity of the solidification process of the austenite-spheroidal graphite eutectic is reflected in the large number of theories trying to explain the occurrence of spheroidal graphite in cast iron. In the following

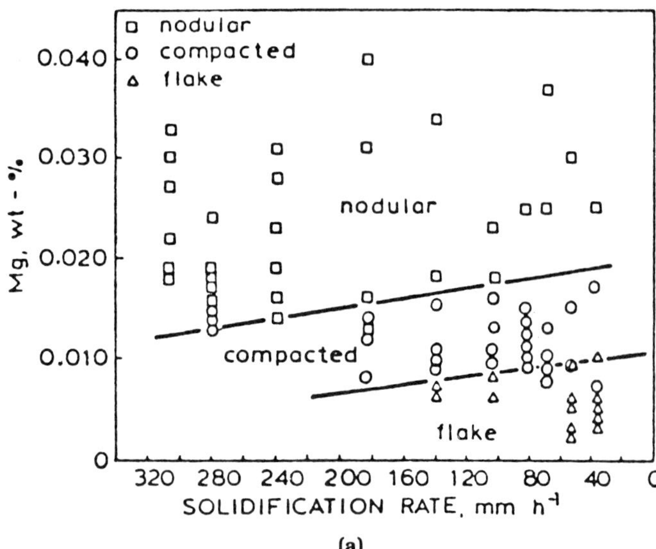

(a)

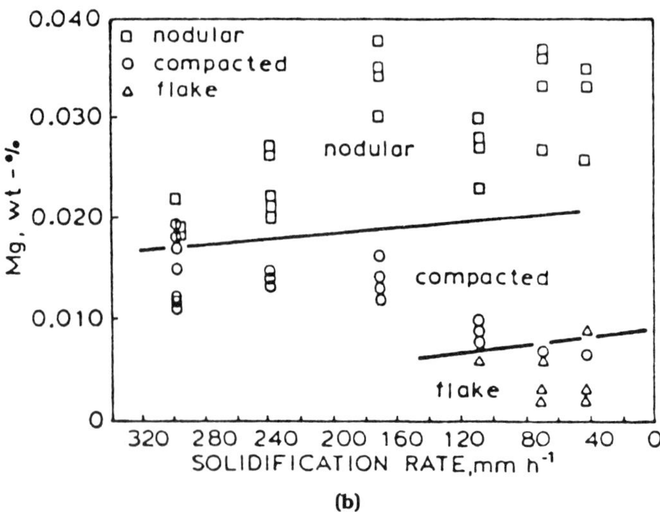

(b)

Fig. 1-12. Graphite morphology as a function of magnesium concentration and solidification rate: (a) hypereutectic iron, CE = 4.51; (b) hypoeutectic iron, CE = 3.98.[31]

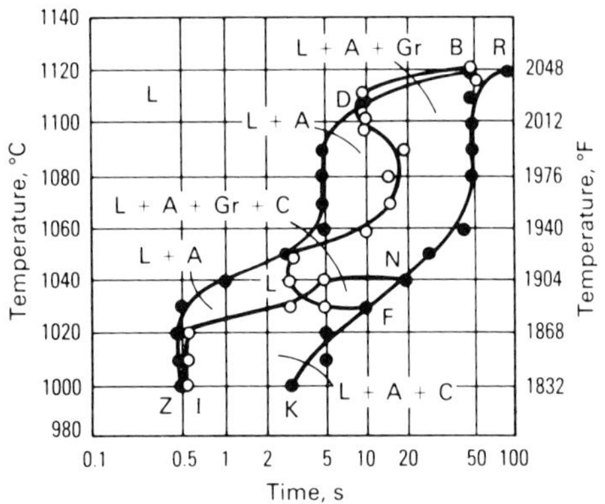

(a)

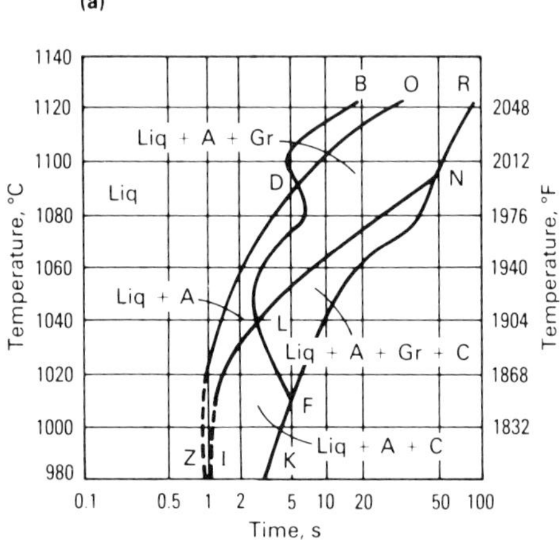

(b)

Fig. 1-11. Time-temperature transformation diagrams for isothermal solidification of eutectic cast iron: (a) flake graphite cast iron; (b) spheroidal graphite cast iron; (L) liquid; (A) austenite; (C) cementite; (Gr) graphite.[29]

paragraphs it will be attempted to critically discuss the most important theories.

SOLID STATE GROWTH THEORY

The first theories on spheroidal graphite formation postulated that it is mostly a solid state process, in which graphite grows in an austenite shell by decomposition of iron carbide (Morrogh, Dunphy, Vastchenko), or by carbon depletion of the supersaturated austenite (Wittmoser, Scheil, De Sy). Although these mechanisms can contribute to the growth of spheroidal graphite, their role is limited. Indeed it was demonstrated that spheroidal graphite can result from the solid solution or from the decomposition of iron carbide. Nevertheless, quenching of partially solidified Mg-treated melts,[35-37] observation of graphite flotation,[38] and segregation of spheroidal graphite particles during centrifuging,[37, 39] have clearly established the fact that spheroidal graphite originates in the melt. Furthermore, according to Loper and Heine,[40] in both hypo- and hypereutectic irons, graphite spheroids occur at temperatures of 1320–1350C (2408–2462F), that is even above the liquidus temperature. This could either be the effect of local carbon supersaturation, or an artifact of the quenching experiments.

GROWTH ON PARTICULAR NUCLEI THEORY

According to De Sy and Vits,[41] spheroidal graphite grows on nuclei of particular shape and/or crystallographic structure. Thus, inclusions that have a cubic lattice, such as MgS, Mg_3N_2, Mg_2Si, and MgO, promote formation of spheroidal graphite. Inclusions with a hexagonal or other lattice, such as SiO_2, SiO, and SiC, promote solidification of flake graphite. Indeed, the occurrence of spheroidal graphite in vacuum-melted Fe-C-Si alloys, without any treatment with Mg or Ce, can be attributed to the absence of SiO_2 from the melt,[42] and, thus, can be construed as a supportive argument for this theory. The weakness of this hypothesis consists in the lack of theoretical or experimental evidence, which would correlate the shape of the grown crystal to that of the nucleus, at late stages of the crystallization process.

GROWTH IN GAS BUBBLES THEORY

Some researchers postulate that, in order for spheroidal graphite to grow, spaces with free surfaces must be present first. Graphite nucleates on the outside of this space and growth occurs inside it.[3, 43] These spaces are gas bubbles resulting from the high vapor pressure of elements such as Mg or Zn.[44] Other gases, such as hydrogen and nitrogen, as possible sites for growth of spheroidal graphite, are suggested by Yamamoto.[45] He demonstrated experimentally that degassed rare earths did not cause graphite spheroidization in cast iron, while hydrogen-saturated rare earths—or simply bubbling hydrogen or nitrogen through the melt—produced spheroidal graphite. Finally, Karsay[3] proposed the following mechanism (Fig. 1-15), in which the gas responsible for spheroidal graphite formation is CO:

- at temperatures approaching the solidification temperature most of the oxygen dissolved in iron precipitates as microscopic SiO_2 inclusions;

- some of these inclusions react with carbon according to the reaction:

$$SiO_2 + 2C = Si + 2CO$$

- graphite nucleates at the surface of the CO gas bubbles;

- the individual crystallites grow along the [0001] direction (C direction) as opposed to flake graphite that grows along the [10$\bar{1}$0] direction (A direction);

- growth continues until the bubble is nearly filled with graphite.

Basically this theory suggests that spheroidal graphite grows in gas bubbles. If the bubble collapses, the result is flake graphite; if it does not collapse, the result is spheroidal graphite. This particular theory has not gained too much credibility, because it relies on two concepts that are not widely accepted. First, it assumes the existence of gas bubbles in liquid iron, for which there is no experimental evidence. Second, it postulates polycrystallinity of graphite, while it is accepted that at least flake graphite growing in cast iron is a single crystal.

GROWTH OF GRAPHITE THROUGH THE AUSTENITIC SHELL

As discussed previously in this section, the γ-spheroidal graphite eutectic is a divorced eutectic. It has been rather widely accepted that the growth of this eutectic begins with the nucleation and growth of graphite in the liquid, followed by early encapsulation of these graphite spheroids in austenite shells (envelopes). A schematic of the process is shown in Figure 1-16. Graphite nucleation and growth deplete the melt of carbon in the vicinity of the graphite; this creates conditions for austenite nucleation and growth around the graphite spheroid.[27] Once the austenite shell is formed, further growth of graphite can occur only by solid diffusion of carbon from the liquid through the austenite.

Calculations of diffusion-controlled growth of graphite through the austenite shell were originally made based on Zener's growth equation for an isolated spherical particle in a matrix of low supersaturation.[46] The following equation was derived:

$$\frac{dr_{Gr}}{dt} = \frac{V_m^{Gr}}{V_m^{\gamma}} D_c^{\gamma} \frac{r_{\gamma}}{r_{Gr}(r_{\gamma} - r_{Gr})} \frac{X^{\gamma/L} - X^{\gamma/Gr}}{X^{Gr} - X^{\gamma/Gr}} \tag{1}$$

where:

dr_{Gr}/dt is the rate of growth of graphite;

V_m^{Gr} and V_m^{γ} are the molar volumes of graphite and austenite, respectively;

D_c^{γ} is the diffusivity of carbon in austenite;

r_{Gr} and r_{γ} are the radii of graphite and austenite, respectively;

$X^{\gamma/L}$, $X^{\gamma/Gr}$ are the molar fraction of austenite at the austenite/liquid, and austenite/graphite boundaries, respectively;

X^{Gr} is the molar fraction of carbon in graphite.

It has also been shown that $r_\gamma = 2.4\ r_{Gr}$.

A slightly different approach, based on a steady-state diffusion model of carbon through the austenite shell[47] also allowed derivations for r_{Gr} and r_γ.

However, recent research has shown that the solidification mechanism of ductile iron is more complicated and that austenite dendrites play a significant role in eutectic solidification.[33,34] Even for hypereutectic irons, the graphite spheroids do not grow in independent austenite envelopes, but rather are associated with austenite dendrites (Fig. 1-17). The eutectic austenite is dendritic and cannot be distinguished from the primary austenite. The sequence of solidification is shown in Figure 1-18. At the eutectic temperature, austenite dendrites and graphite spheroids nucleate independently in the liquid. Limited growth of spheroidal graphite occurs in contact with the liquid. Flotation or convection then determines the collision of spheroidal graphite with the austenite dendrites. Graphite encapsulation in austenite can occur before or immediately after the contact between graphite and austenite dendrites. Further growth of graphite occurs by carbon diffusion through the austenite shell.

At some point it was suggested that the encapsulation of graphite by austenite is the main reason for the spheroidal shape of graphite, since graphite will grow mostly through solid-state diffusion of carbon through austenite shell, and since solid-state growth is isotropic. While this argument is quite reasonable, this theory can not explain the spheroidal shape of graphite before encapsulation. Thus, it is reasonable to conclude that encapsulation, while playing an important role, is only part of the story.

SURFACE ENERGY THEORY

Many theories capitalize on the observation that the graphite/liquid surface energy is higher in ductile iron than in flake graphite iron. Marincek and co-workers[48] used the capillary rise method to determine the surface tension in cast iron melts. Indeed, it was found that a surface tension of 800–1100 dyn/cm was associated with flake graphite, while a value of 1400 dyn/cm was associated with graphite spheroids.

The theories in this group explain spheroidal graphite formation by either simply implying that a sphere will have less free surface energy than a lamella with the same volume above a certain critical interface energy,[49] or by suggesting that the high interface energy will curve the growing crystal in order to decrease the energy/volume ratio, resulting in spheroidal rather than lamellar graphite.[12]

The original model by Buttner, Taylor and Wulff[49] suggested that, under a critical solid/liquid interface energy value, γ^*_{SL}, flake graphite is more stable than spheroidal graphite (Fig. 1-19). ΔG_D is the energy stored inside the graphite spheroid at low angle boundaries. The graphite/melt interface energy is a function of sulfur level. When reactive impurities, such as Mg, remove sulfur from the melt, the graphite/melt interface energy is increased and the spheroidal shape of graphite becomes thermodynamically more favorable.

A more complex theory, which includes critical interface energies for both basal and prism planes of the graphite crystal, was developed by Geilenberg.[50] The main limitation of these theories is the assumption of thermodynamic equilibrium, but there are other complications, too.

It was noted that all factors that prevent occurrence of spheroidal graphite, such as addition of deleterious elements (Sb, Pb, As, etc.), or superheating and/or holding of liquid iron after the magnesium treatment, decrease the value of surface tension. Initially, an increase in the surface tension is observed upon addition of Mg, which is normally explained by the desorption of surface-active elements, such as oxygen and sulfur.[51] Then the surface tension decreases as shown in Figure 1-20a. Nevertheless, other elements that degas (e.g., Al) and desulfurize (e.g., Ca) the melt, do not promote formation of spheroidal graphite, even though they increase the surface tension to 1500–1600 dyn/cm. Furthermore, it was observed that the highest value of the surface tension is associated with a residual Mg of 0.01–0.02%, which normally results in interdendritic flake graphite (Fig. 1-20b). It is well known that spheroidal graphite occurs at residual Mg levels in excess of 0.025–0.03%.

According to Sadocha and Gruzleski,[12] the "natural" graphite morphology in pure Fe-C-Si alloys is a spheroidal one. This shape results from the circumferential growth model of graphite explained earlier in this paper (Fig. 1-8). In the presence of impurities, such, as S or O_2, which decrease the surface tension, the spherical shape is deteriorated into a lamellar one. The major role of spheroidizing elements, such as Mg or Ce, is to act as scavengers and to produce a high interface surface tension, which helps the curved growth of graphite.

SURFACE ADSORPTION THEORY

This model postulates that the change from lamellar to spheroidal graphite occurs because of the change in the ratio between growth on the $[10\bar{1}0]$ face and growth on the [0001] face of graphite.[52] For equilibrium conditions, the Gibbs-Curie-Wulf law states that the crystalline phase with the higher interface energy has a slow rate of growth in the normal direction. Bravais' rule stipulates that the growth

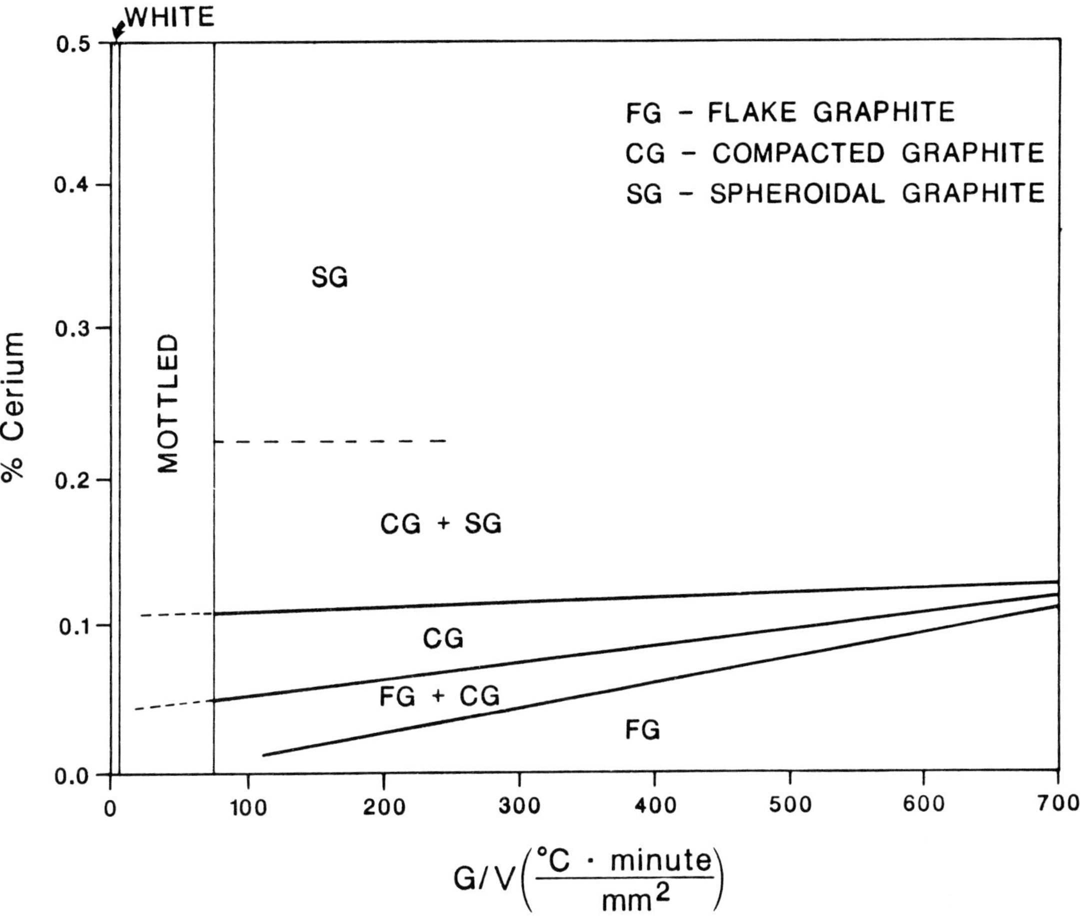

$$G/V\left(\frac{°C \cdot minute}{mm^2}\right)$$

Fig. 1-13. Influence of G/R ratios and percent cerium on structural transitions in cast iron.[32]

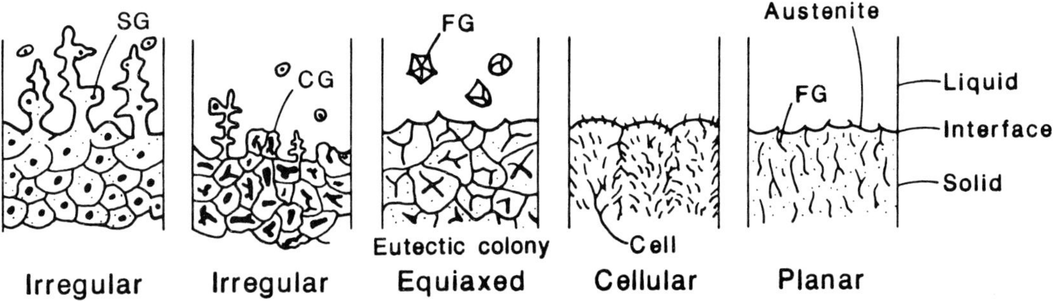

Fig. 1-14. Possible morphologies of the solid-liquid interface and of the graphite in eutectic cast iron.[34]

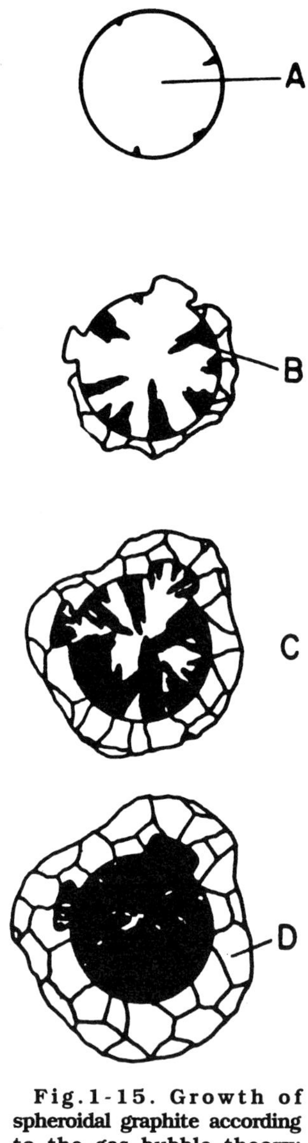

Fig. 1-15. Growth of spheroidal graphite according to the gas bubble theory: (A) gas; (B) graphite single crystal; (C) melt; (D) solid iron.[3]

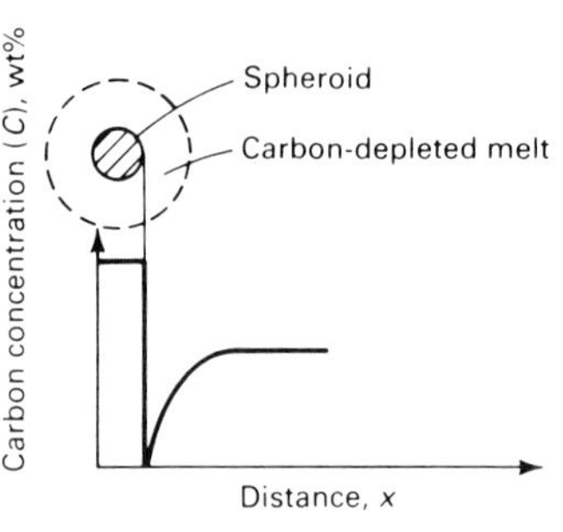

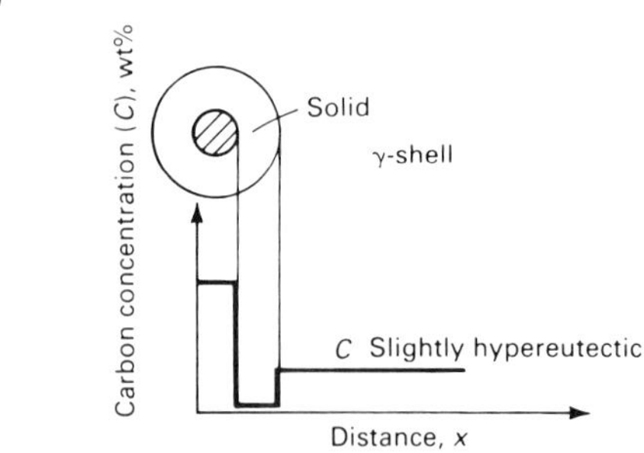

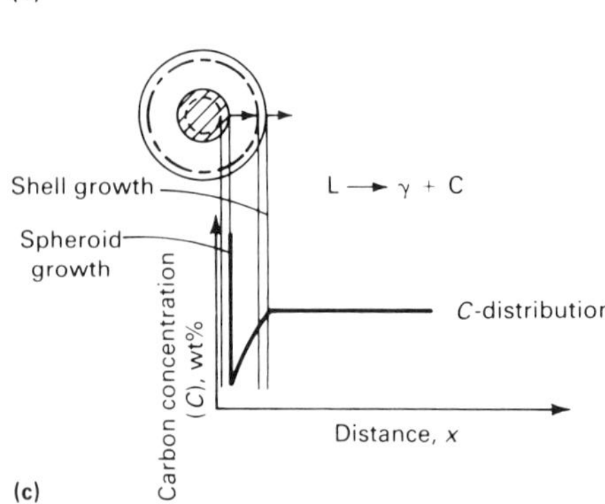

Fig. 1-16. Isothermal growth of a graphite spheroid within an austenite shell and growth of the shell with a smooth interface: (a) growth of spheroidal graphite in contact with the melt; (b) envelopment by austenite; (c) growth of spheroidal graphite within the austenite shell.[27]

rate in the direction normal to a plane is inversely proportional to the density of atoms located on the plane. Accordingly, it follows that, under equilibrium conditions, the crystallographic plane with the highest density of atoms has the lowest interface energy and the minimum growth rate in a direction perpendicular to the plane.

Nevertheless, under the nonequilibrium conditions prevailing during the solidification of cast iron, kinetic considerations become important. Assuming growth by two-dimensional nucleation, the highest rate of growth will be experienced by the face with the higher density of atoms, where the probability for nucleation is higher.

Therefore, in a pure environment, the highest growth rate will be in the [0001] direction of the graphite crystal (i.e., V_B) resulting in the formation of unbranched single crystals (coral graphite), as shown in Figure 1-21b. In a contaminated environment, surface-active elements, such as sulfur or oxygen, are absorbed on the high-energy plane

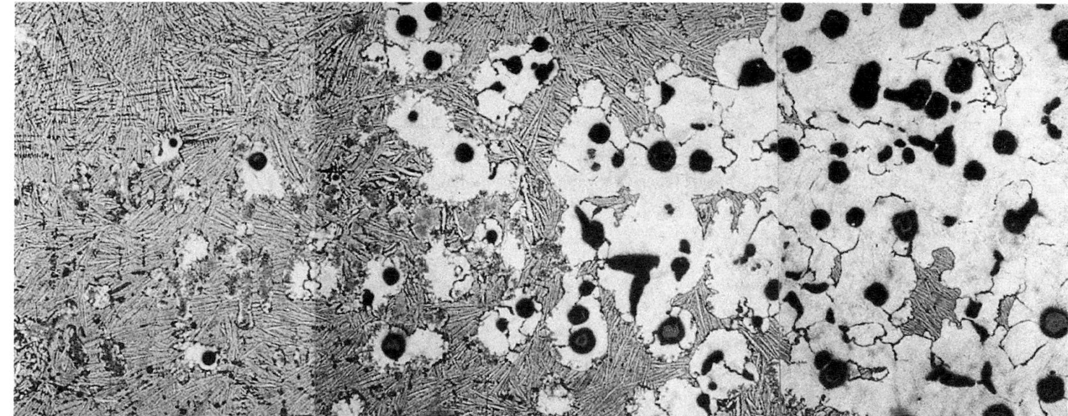

Fig. 1-17. Microstructure of the solid/liquid interface for a directionally solidified, hypereutectic, cerium-treated, spheroidal graphite cast iron.[15]

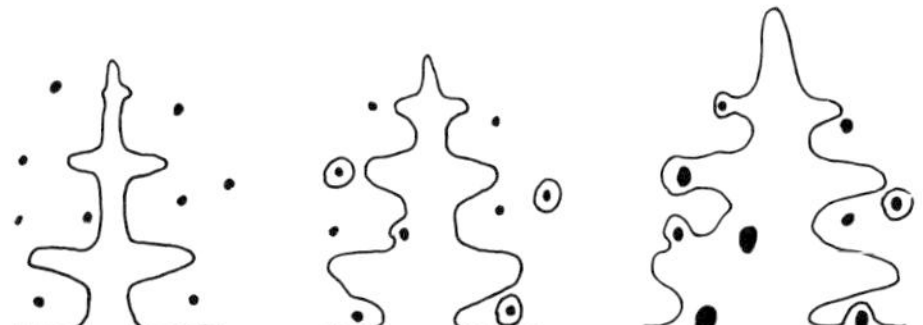

Fig. 1-18. Schematic illustration of the progression of growth of the austenite-spheroidal graphite eutectic.[15]

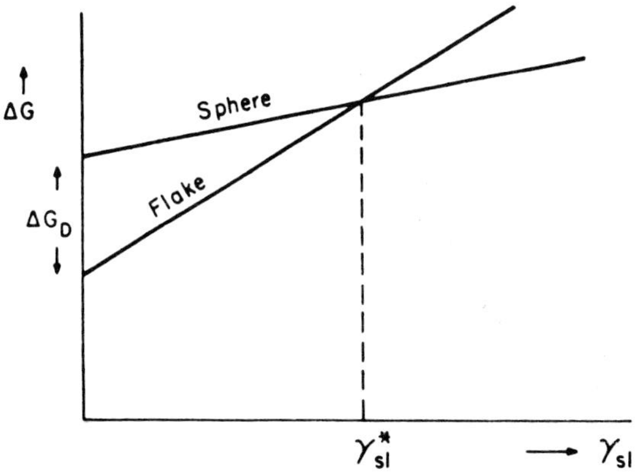

Fig. 1-19. Variation of free energy for crystallization of graphite, ΔG, as a function of solid graphite/melt interface energy γ^*_{SL}.[49]

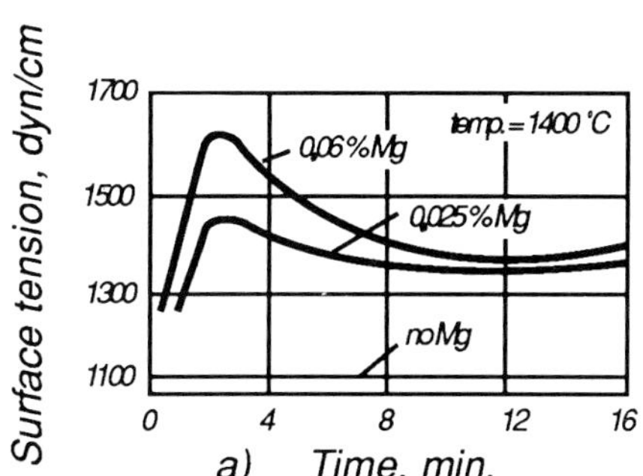

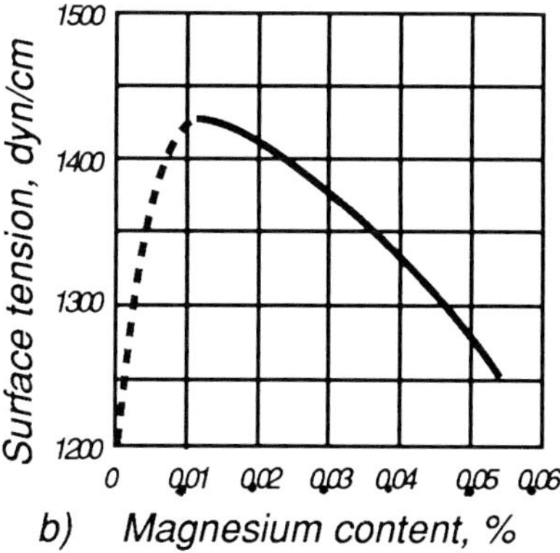

Fig. 1-20. Variation of the surface tension of magnesium-treated iron with holding time at constant temperature (a), and with residual magnesium (b).[51]

$[10\bar{1}0]$, which has fewer satisfied bonds. Subsequently, the $[10\bar{1}0]$ plane face achieves a lower surface energy and higher atomic density than the [0001] face, and growth becomes predominant in the $[10\bar{1}0]$ direction (i.e., V_p), resulting in lamellar (plate) graphite (Fig. 1-21c). Finally, the reactive impurities (such as magnesium, cerium, and lanthanum) in the environment scavenge the melt of surface-active elements (sulfur, diatomic oxygen, lead, antimony, titanium, and so on), after which they also block growth on the $[10\bar{1}0]$ prism face. A polycrystalline spheroidal graphite results (Fig. 1-21a).

DEFECT GROWTH OF GRAPHITE THEORY

Minkoff and Lux[53] attempted to explain spheroidal graphite formation based on the possible growth mechanisms of faceted crystals, such as graphite. Three growth mechanisms are considered: two-dimensional nucleation (Fig. 1-22a), step of a defect (twisted) boundary (Fig. 1-22b), and screw dislocation. The first two mechanisms are governed by exponential laws and apply to the $[10\bar{1}0]$ surface, while the third is governed by a parabolic law and applies to the [0001] surface of the graphite crystal.

When weak, reactive impurities, such as sulfur, are present in the melt, a contaminated environment occurs. These elements change the edge energy of steps, resulting in a relative position change of the growth rates involved, as shown in Figure 1-23a. The curve for growth on the step of a defect boundary, R_{step}, is at a lower undercooling than those for growth by two-dimensional nucleation, R_{2D}, or by screw dislocation, R_{screw}.

In a pure environment, such as an iron-carbon-silicon alloy with no sulfur contamination, the growth rate curves are displaced to higher undercooling (Fig. 1-23b). In a melt of sufficient purity, or when increasing cooling rate, the higher degree of undercooling may allow growth with R_{screw} so that graphite spheroids can form. This has been achieved experimentally for pure nickel-carbon alloys by increasing the cooling rate of the melt, or for ultra pure iron-carbon alloys by cooling slowly in a vacuum.

In an environment with reactive impurities (for example, magnesium), the impurity will react with the surface, and the growth at a step of a twist boundary will be neutralized. Only the curves for R_{2D} and R_{screw} are left, and they are displaced to greater undercoolings (Fig. 1-23c).

Another theory relates graphite shape in cast iron with undercooling (kinetic plus constitutional) during solidification.[2] As shown in Figure 1-24, each graphite form has its own temperature for growth, which is achieved by a specific cooling rate and composition. Steps on surfaces can change graphite morphology from plate to rod. With an increase in undercooling, pyramidal instabilities will occur on the faces of the graphite crystal. At undercoolings of 29–350C (82.7–662F), instabilities occur on the $[10\bar{1}1]$ faces of the pyramid, and it is suggested that graphite spheroids form at these undercoolings. Finally, at large undercoolings of 40°C (104°F), the growth form noted is a pyramidal one, bounded by $[10\bar{1}1]$ faces. These pyramidal crystals are part of the series of imperfect forms observed particularly in thick-wall ductile iron castings.

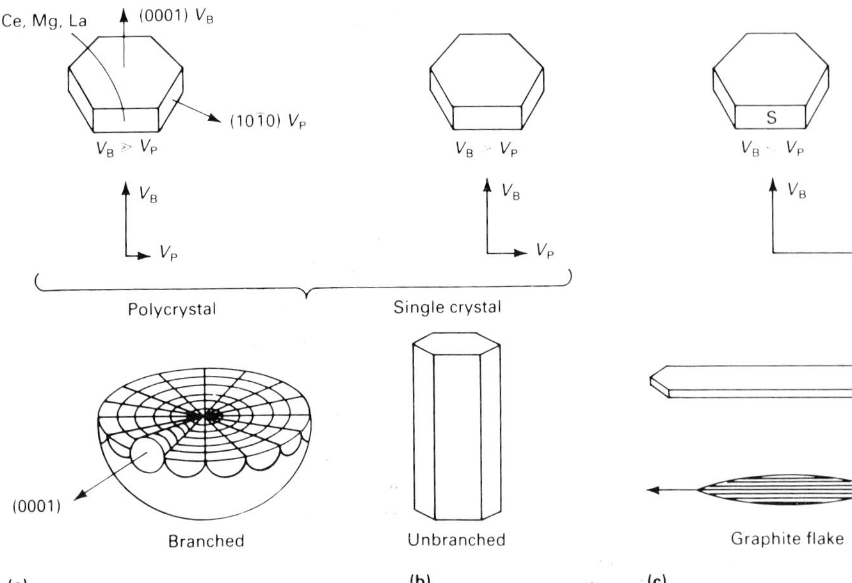

Fig. 1-21. Schematic of the change in the growth velocity of graphite due to adsorbtion of foreign atoms: (a) nodularizer (Mg, Ce, La) added as a reactive impurity; (b) pure environment; (c) environment contaminated with surface-active elements, such as S or O_2 adsorbed on the prism faces.[52]

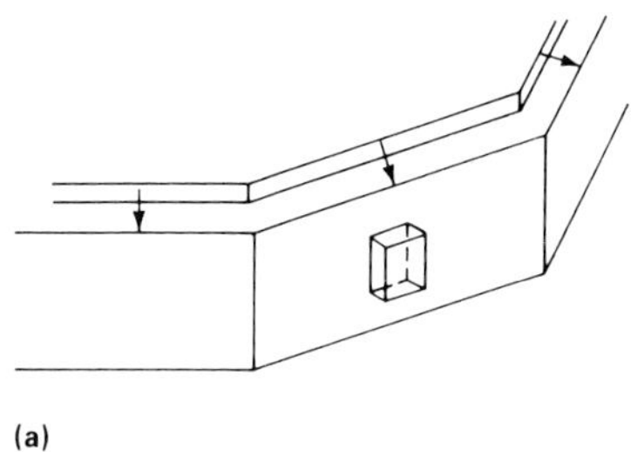

Fig. 1-22. Growth of graphite in the <[1010]> direction: (a) growth by two-dimensional nucleation on [1010] faces illustrates that steps on (0001) surfaces will advance only as far as bounding crystal edges; (b) growth from step to twist boundary illustrates that the [1010] faces grow by nucleation of planes at the step.[53]

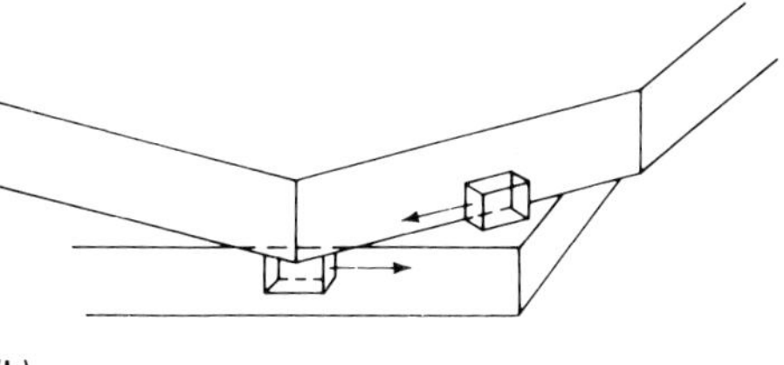

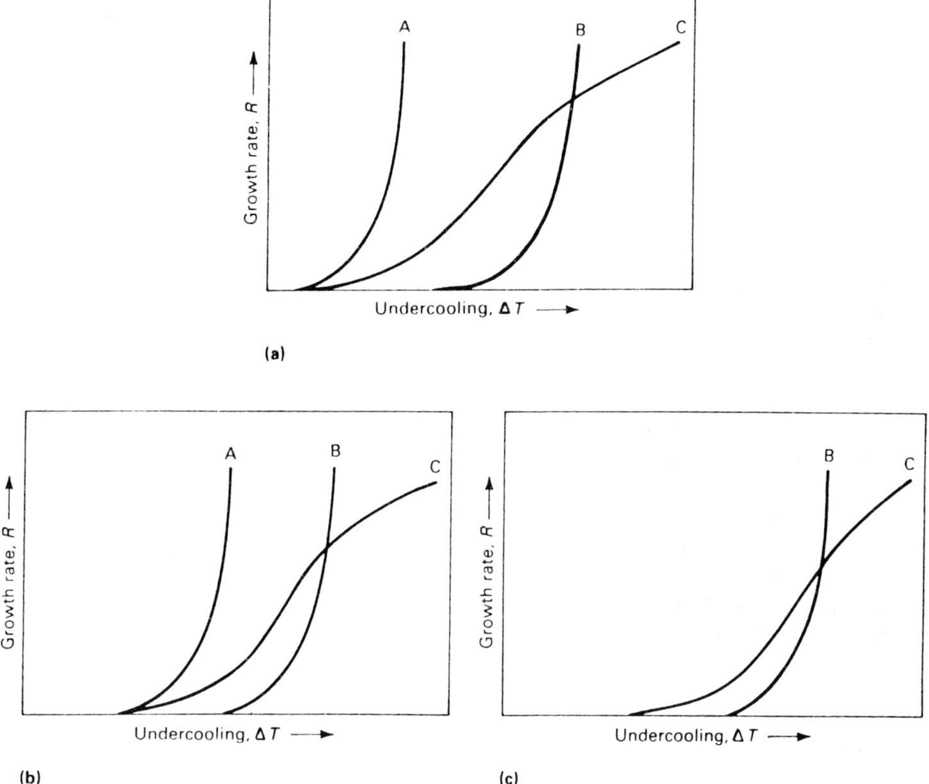

Fig. 1-23. Suggested ΔT–R correlation for [1010] and [0001] crystal faces of graphite growing in various environments: (a) contaminated environment; (b) pure environment. (c) environment with reactive impurities. Where (A) is growth on step of defect boundary, R_{step}; (B) is growth by two-dimensional nucleation, R_{2D}; (C) is growth by screw dislocations, R_{screw}.[53]

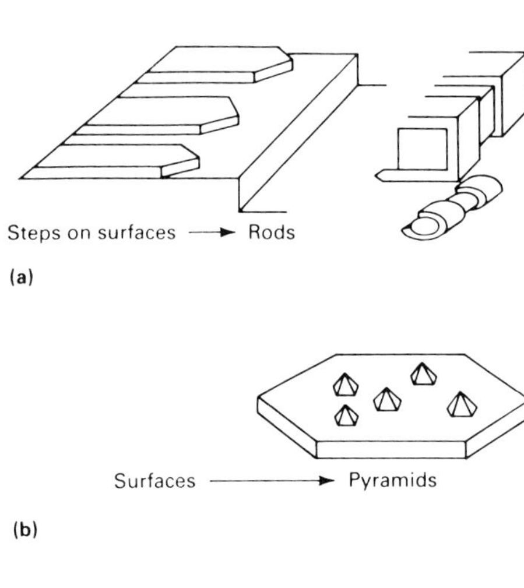

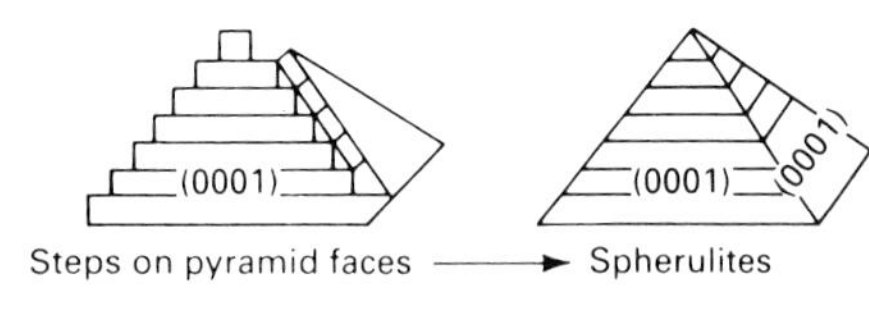

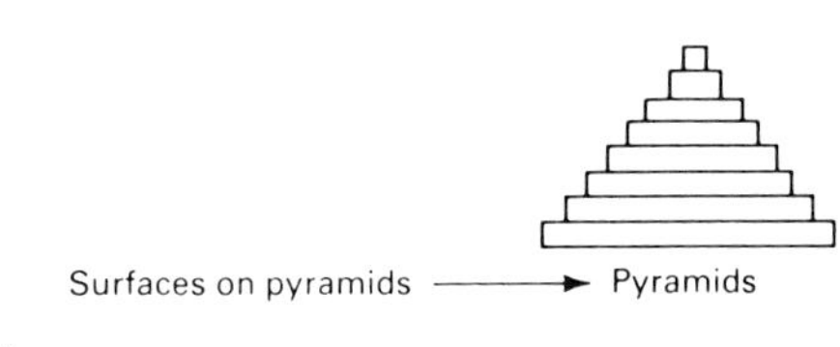

Fig. 1-24. Correlation between the different types of instabilities, growth morphology and undercooling ΔT: (a) $\Delta T = 4°C$; (b) $\Delta T = 9°C$; (c) $\Delta T = 30°C$; (d) $\Delta T = 40°C$.[2]

Imperfect spheroidal graphite shapes are due to incorrect kinetic undercooling and an imbalance of impurities influencing constitutional undercooling. An insufficient amount of impurities influencing kinetic undercooling (for example, magnesium) will favor the formation of intermediate graphite shapes, such as compacted/vermicular, while an insufficient amount of impurities influencing constitutional undercooling (for example, lead) will perturb the surface of the spheroid and promote the formation of protuberances.

MODELS FOR GROWTH OF COMPACTED AND DEGENERATED GRAPHITE

A complete understanding of spheroidal graphite formation requires some explanation for the occurrence of graphite shapes that are not spheroidal, but depart enough from the lamellar structure to attract the attention of the metallurgist preoccu-

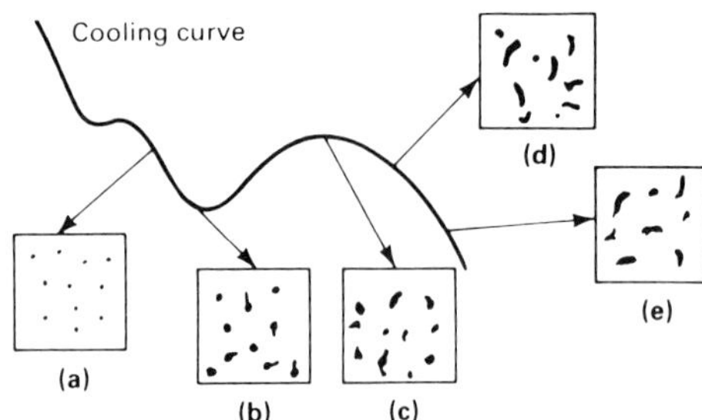

Fig. 1-25. Schematic of the sequence of development of compacted graphite in iron melts: (a) small spheroids; (b) and (c) some spheroids having tails; (d) compacted plus spheroidal graphite; (e) compacted graphite.[20]

pied with spheroidal graphite. Such graphite shapes are compacted graphite, chunky graphite and exploded graphite. The last two are termed also as degenerated spheroidal graphite.

Models for Growth of Compacted/Vermicular Graphite

The sequence of growth of compacted/vermicular graphite during the eutectic transformation is shown schematically in Figure 1-25, based on experimental data on rapidly quenched samples from successive stages during the solidification process.[54] It can be seen that, at the beginning, graphite precipitates as spheroids, which then degenerate during growth and subsequently develop into compacted graphite. Compacted graphite develops as interconnected segments within an austenitic matrix.

It was proposed that the growth of compacted graphite occurs by the twin/tilt of boundaries.[55] The initiation of a twin/tilt is related to the unstable growth of the $[10\overline{1}0]$ interface, which can be induced by the presence of some reactive elements. It was further hypothesized that, when insufficient reactive elements are present, the tilt orientation of twin boundaries can alternate, resulting in typical compacted graphite (Fig. 1-26a). On the other hand, when there is enough impurity in the melt, the tilt orientation is singular, and spheroidal graphite may occur (Fig. 1-26b).

Based on this growth mechanism, compacted graphite can be considered to grow either by transition from flake graphite (Fig. 1-27a) or by degeneration of spheroidal graphite (Fig. 1-27b).[56] The degree of interconnection of the compacted graphite within the eutectic cell is also apparent from the same figure. Based on cooling curve configuration, it is reasonable to assume that the growth rate of compacted graphite is similar to that of spheroidal graphite, which is not surprising because the growth direction is supposedly in the [0001] direction of the graphite crystal.

MODELS FOR DEGENERATED GRAPHITE MORPHOLOGIES

The two most important forms of degenerated spheroidal graphite morphologies are chunky graphite and exploded graphite. Both these morphologies can be highly detrimental to mechanical properties and must be avoided by appropriate control of melt processing.

Chunky Graphite

According to Liu et al.,[57] chunky graphite forms by extensive branching of graphite spheroids (Fig. 1-28). While chunky graphite is highly interconnected and does not consist of broken pieces of graphite spheroids, some branches can be disturbed and fractured by the thermally-induced turbulence of the melt during solidification. The tendency for formation of chunky graphite can be decreased by lowering the carbon equivalent and the silicon content and by increasing the cooling rate when possible.

Exploded Graphite

Exploded graphite seems to be a less severe form of spheroidal graphite degeneration, but very similar in nature, resulting from interface instability. High undercooling in front of the interface results in cellular or dendritic growth

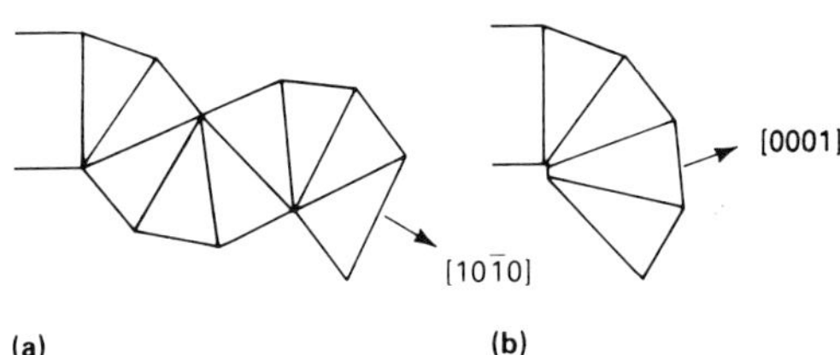

Fig. 1-26. Change of direction of the tilt of the boundary due to the amount of reactive impurity: (a) insufficient spheroidization; (b) sufficient spheroidization.[55]

of the graphite (Fig. 1-29).[58] Branching crystals can have a length considerably greater than the radius of the graphite crystal from which they grow. Overtreatment of hypereutectic irons with rare earth seems to be the main reason for the occurrence of exploded graphite, which is also associated with graphite flotation.

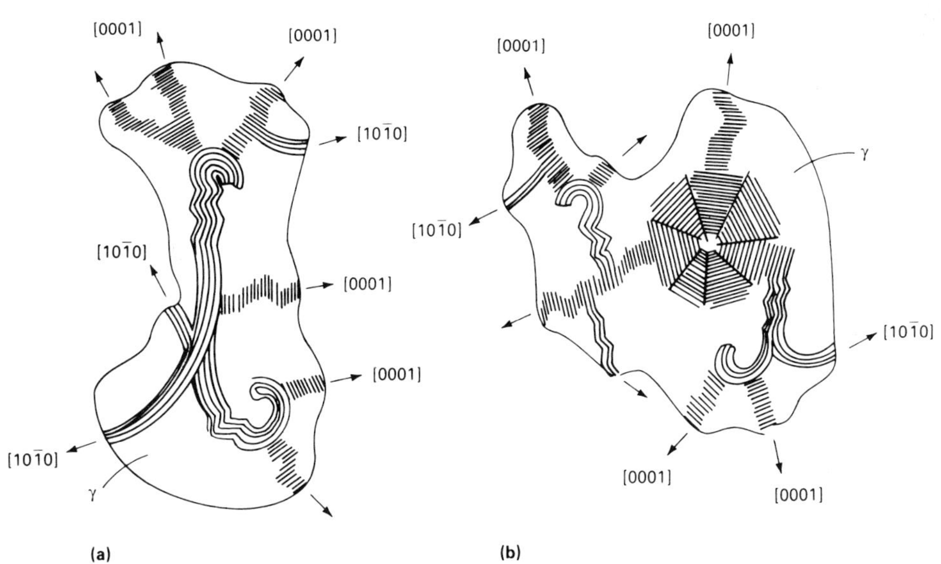

Fig. 1-27. Transition from flake to compacted graphite (a) and from spheroidal to compacted graphite (b), based on the twin/tilt of boundaries growth mechanism.[56]

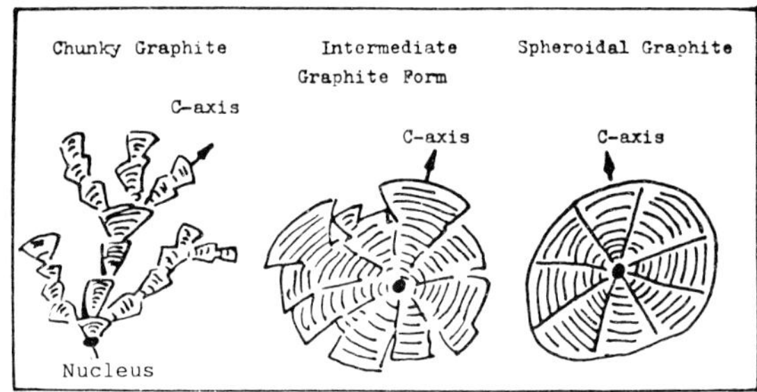

Fig. 1-28. Schematic representation of degeneration of graphite spheroids into chunky graphite.[57]

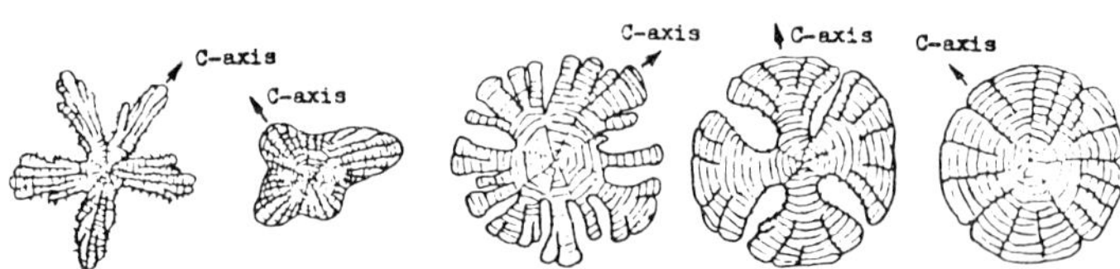

Fig. 1-29. Schematic representation of degeneration of graphite spheroids into exploded graphite.[58]

■ REFERENCES

1. H. Morrogh, W.J. Williams; *J. Iron and Steel Inst.*, London, vol 155, p 321 (1947).

2. I. Minkoff; *The Physical Metallurgy of Cast Iron*, John Wiley & Sons (1983).

3. S.I. Karsay; *Ductile Iron Production Practices*, AFS, Des Plaines, IL (1985).

4. R. Eliott; *Cast Iron Technology*, Butterworth—Heinemann Limited, London (1988).

5. S. Steeb, U. Maier; "Structure of Molten Fe-C Alloys by Means of X-Ray and Neutron Wide Angle Diffraction as Well as Sound Velocity Measurements," *The Metallurgy of Cast Iron*, ed. B. Lux et al., Georgi Publishing, Switzerland, pp 1-9 (1975).

6. W. Krieger, H. Trenkler; *Arch. Eisenhuttenwes.*, vol 42, no. 3, p 175 (1971).

7. L.S. Darken; "Equilibria in Liquid Iron With Carbon and Silicon," *Technical Publication 1163*, American Institute of Mining, Metallurgical, and Petroleum Engineers, p 1 (1940).

8. A.A. Vertman et al.; *Liteinoe Proizvod.*, no. 10 (1964).

9. M. Hillert, Y. Lindblom; *J. Iron and Steel Inst.*, London, vol 176, p 388 (1954).

10. W. Oldfield, G.T. Geering, W.A. Tiller; "Solidification of Spheroidal and Flake Graphite Cast Iron," *Proc. of Solidification of Metals, Iron and Steel Inst.*, Publ. 110, p 256 (1968).

11. D.D. Double, A. Hellawell; "Growth Structure of Various Forms of Graphite" *The Metallurgy of Cast Iron*, ed. B. Lux et al., Georgi Publishing, Switzerland, pp 509–525 (1975).

12. J.P. Sadocha, J.E. Gruzleski; "The Mechanism of Graphite Spheroid Formation in Pure Fe-C-Si Alloys," *The Metallurgy of Cast Iron*, ed. B. Lux et al., Georgi Publishing, Switzerland, pp 443–459 (1975).

13. G.F. Ruff, J.F. Wallace, "Control of Graphite Structure and Its Effect on Mechanical Properties of Gray Iron," *AFS Transactions*, vol 84, p 705 (1976).

14. J.F. Wallace; "Effects of Minor Elements on the Structure of Cast Iron," *AFS Transactions*, vol 83, p 363 (1975).

15. S.L. Liu, C.R. Loper, Jr., S. Shirvani; "Inoculation of Gray Cast Irons With Carbonaceous Materials," *AFS Transactions*, vol 93, p 501 (1985).

16. B. Lux; "Nucleation of Graphite in Fe-C-Si Alloys," *Recent Research on Cast Iron*, ed. H.D. Merchant, Gordon and Breach, p 241 (1968).

17. J.V. Dawson, S. Maitra; "Recent Research on the Inoculation of Cast Iron," *British Foundryman*, p 117 (Apr 1967).

18. D.M. Stefanescu; "Inoculation of Gray Iron with Sodium and Barium," *Giesserei-Prax.*, no. 24, p 429 (1972).

19. W. Weiss; "The Importance of Deoxidation in the Crystallization of Cast Iron," *The Metallurgy of Cast Iron*, ed. B. Lux et al., Georgi Publishing, Switzerland, pp 69–77 (1975).

20. D.M. Stefanescu; "Cast Iron," *Metals Handbook 9th Edition*, ASM International, pp 168–181 (1989).

21. M.B. Zeedijk; "Identification of the Nuclei in Graphite Spheroids by Electron Microscopy," *J. Iron Steel Inst.*, p 737 (Jul 1965).

22. P. Poyet, J. Ponchon; "Contributions to the Study of Graphite Nucleation in Cast Iron for Ingot Molds," *Fonderie*, no. 277, p 183 (1969).

23. J.C. Mercier et al.; "Inclusions in Graphite Spheroids," *Fonderie*, no. 277, p 191 (1969).

24. H. Nieswaag, A.J. Zuithoff; "The Occurrence of Nodules in Cast Iron Containing Small Amounts of Tellurium," Paper presented at the 37th International Foundry Congress, CIATF, Brighton, England (Sep 1970).

25. R.J. Warrick; "Spheroidal Graphite Nuclei in Rare Earth and Magnesium Inoculated Irons," *AFS Transactions*, vol 76, p 722 (1968).

26. M.M. Jacobs, T.J. Law, D.A. Melford, M.J. Stowell; "Basic Processes Controlling the Nucleation of Graphite Nodules in Chill Cast Iron," *Metals Technology*, vol 1, part II, p 490 (1974).

27. B. Lux, F. Mollard, I. Minkoff, "On the Formation of Envelopes Around Graphite in Cast Iron," *The Metallurgy of Cast Iron*, ed. B. Lux et al., Georgi Publishing, Switzerland, pp 371–401 (1975).

28. H. Jones, W. Kurz; "Growth Temperatures and the Limits of Coupled Growth in Unidirectional Solidification of Fe-C Eutectic Alloys," *Metall. Trans. A*, vol 11A, p 1265 (1980).

29. H.D. Merchant, "Solidification of Cast Iron—A Review of Literature," *Recent Research on Cast Iron*, ed. H.D. Merchant, Gordon and Breach, p 1–100 (1968).

30. D.M. Stefanescu, P.A. Curreri, M.R. Fiske; "Microstructural Variations Induced by Gravity Level During Directional Solidification of Near-Eutectic Iron-Carbon Type Alloys," *Metall. Trans. A*, vol 17A, p 1121 (1986).

31. D. Argo, J.E. Gruzleski; *Mat. Sci. Tech.*, vol 10, no. 2, pp 1019–1024 (1986).

32. D.K. Bandyopadhyay, D.M. Stefanescu, I. Minkoff, S.K. Biswal; "Structural Transitions in Directionally Solidified Spheroidal Graphite Cast Iron," *The International Symposium on the Physical Chemistry of Cast Iron*, Tokyo, Mat. Res. Soc. Proc. (1989).

33. A. Rickert, S. Engler; "Solidification Morphology of Cast Irons," *The Physical Metallurgy of Cast Iron*, ed. H. Fredriksson and M. Hillert, Proceedings of the Materials Research Society, North Holland, vol 34, p 165 (1985).

34. D.M. Stefanescu, D.K. Bandyopadhyay; "On the Solidification of Spheroidal Graphite Cast Iron," *The International Symposium on the Physical Chemistry of Cast Iron*, Tokyo, Mat. Res. Soc. Proc. (1989).

35. E. Scheil; *Giess. Techn. Wiss. Beih.*, vol 24, pp 1313–1338 (1959).

36. E. Scheil, J.D. Schobel; *Giess. Techn. Wiss. Beih.*, vol 13, pp 203–213 (1961).

37. J.D. Schobel; "Precipitation of Graphite During the Solidification of Nodular Cast Iron," *Recent Research on Cast Iron*, ed. H.D. Merchant, Gordon and Breach, New York, pp 303–346 (1968).

38. H. Kempers; *Giesserei.*, pp 841–846 (1966).

39. S. Banerjee; *British Foundryman*, pp 344–352 (1965).

40. C.R. Loper, Jr., R.W. Heine; *AFS Transactions*, pp 135–152 (1963).

41. A. De Sy, J. Vidts; *Fonderie Belge*, no. 3, p 41 (1952).

42. M. Maruyama; "The Influence of SiO_2 on the Formation of Spheroidal Graphite in Cast Iron" *30th International Foundry Congress, CIATF*, Prague (1963).

43. S.I. Karsay; "Hypothetical Considerations on the Role of Phase Boundaries During Graphitization," *Recent Research on Cast Iron*, ed. H.D. Merchant, Gordon and Breach, New York, pp 215–224 (1968).

44. S.I. Karsay; *Ductile Iron Production*, Quebec Iron and Titanium Co., p 145 (1970).

45. S. Yamamoto; "Producing Spheroidal Graphite Cast Iron by Suspension of Gas Bubbles in Melts," *AFS Transactions*, pp 217–226 (1957).

46. S.E. Wetterfall, H. Fredriksson, M. Hillert; "Solidification Process of Nodular Cast Iron," *J. Iron Steel Inst.*, p 323 (May 1972).

47. K.C. Su, I. Ohnaka, I. Yamauchi, T. Fukusako; "Computer Simulation of Solidification of Nodular Cast Iron," *The Physical Metallurgy of Cast Iron.*, ed. H. Fredriksson and M. Hillert, Proceedings of the Materials Research Society, North Holland, vol 34, p 181 (1985).

48. B. Marincek et al.; *Giesserei*, vol 40, p 663 (1953); vol 41, p 313 (1954), vol 42, p 121 (1955).

49. F.H. Buttner, H.F. Taylor, J. Wulff; *American Foundryman*, vol 20, no. 49 (1951).

50. H. Geilenberg; "A Critical Review of the Crystallization of Graphite From Metallic Solutions After the Surface Tension Theory," *Recent Research on Cast Iron*, ed. H.D. Merchant, Gordon and Breach, pp 195–210 (1968).

51. C. Cosneanu; *Metalurgia*, no. 10, pp 563–567 (1966).

52. K. Herfurth; "Investigations Into the Influence of Various Additions on the Surface Tension of Liquid Cast Iron With the Aim of Finding Relationships Between the Surface Tension and the Occurrence of Various Forms of Graphite," *Freiberg Forschungs*, vol 105, p 267 (1965).

53. I. Minkoff, B. Lux; "Graphite Growth From Metallic Solution," *The Metallurgy of Cast Iron*, ed. B. Lux et al., Georgi Publishing Co., Switzerland, pp 473–491 (1975).

54. E.N. Pan, K. Ogi, C.R. Loper, Jr.; "Analysis of the Solidification Process of Compacted/Vermicular Graphite Cast Iron," *AFS Transactions*, vol 90, p 509 (1982).

55. P. Zhu, R. Sha, Y. Li; "Effect of Twin/Tilt on the Growth of Graphite," *The Physical Metallurgy of Cast Iron*, ed. H. Fredriksson and M. Hillert, Proceedings of the Materials Research Society, North Holland, vol 34, p 3 (1985).

56. X. Den, P. Zhu, Q. Liu; "Structure and Formation of Vermicular Graphite," *The Physical Metallurgy of Cast Iron*, ed. H. Fredriksson and M. Hillert, Proceedings of the Materials Research Society, North Holland, vol 34, p 141 (1985).

57. P.C. Liu, C.L. Li, D.H. Wu, C.R. Loper, Jr.; "SEM Study of Chunky Graphite in Heavy-Section Ductile Iron," *AFS Transactions*, vol 91, pp 119–126 (1983).

58. G.X. Sun, C.R. Loper, Jr.; "Graphite Flotation in Cast Iron," *AFS Transactions*, vol 91, pp 841–854 (1983).

Engineering Properties, Specifications and Physical Constants of Specific Ductile Irons

James D. Mullins

QIT-Fer et Titane, Inc.
Montreal, Quebec
CANADA

SECTION I: UNALLOYED OR LOW-ALLOYED DUCTILE IRONS

■■■ INTRODUCTION

Designers specifying a particular grade of ductile should not assume that the tensile strength, yield strength, or elongation values for a given part will be above the minimum properties stated in the specification, unless actual test data is available from individual producers. Casting hardness and strength values can be correlated for these individual castings. This chapter will present information useful in determining the interrelationships of the mechanical properties of ductile iron.

A summary of the most commonly used specifications are listed in Table 2-1. The typical chemical composition ranges relating to the five major grades are shown in Table 2-2. Those that are interested in more information are referred to the individual specifications, or the specifying body. The ferritic ductile irons include grades 60-40-18 through 65-45-12. These are the softest of the unalloyed grades, containing up to 30% pearlite in the matrix, with tensile strengths up to 75,000 psi. Grade 60-40-18 is made

TABLE 2-1. SUMMARY OF COMMONLY USED DUCTILE IRON SPECIFICATIONS

SPECIFICATION NUMBER	CLASS OR GRADE	MIN TENSILE STRENGTH PSI	MIN YIELD STRENGTH PSI	ELONGATION IN 2 IN, %	HEAT TREATMENT	OTHER REQUIREMENTS						
	60-40-18	60,000	40,000	18	May be annealed	Chemical composition is subordinate						
	65-45-12	65,000	45,000	12	Usually as cast	to mechanical properties; however,						
	80-55-06	80,000	55,000	6	Usually as cast	the content of any chemical element						
ASTM A536-84	100-70-03	100,000	70,000	3	Usually normalized	may be specified by mutual agreement.						
Mil-i-11466B)	120-90-02	120,000	90,000	2	Quenched and tempered							
						Hardness, Bhn		Microstructure				
	D-4018				May be annealed	170 max		Ferritic				
	D-4512					156-217		Ferritic-pearlitic				
SAEJ434C	D-5506					187-255		Ferritic-pearlitic				
	D-7003				May be normalized	241-302		Pearlitic				
	DQ&T				Quenched and tempered	Range specified		Martensitic				
							Composition, %					
							TC	Si	P	Other	CE	Bhn
ASTM A395-88					Ferritized by annealing	Min 3.0						
ASME SA395	60-40-18	60,000	40,000	18	Bhn 143-187	Max		2.5	0.08			
					To be used in as-cast condition. Hardness	Min 3.0						
ASTM A476-82	80-60-03	80,000	60,000	3	shall be minimum Bhn 201	Max		2.5	0.08			
MIL-i-24137	(REPLACED BY MIL-C-24707/F THAT SPECIFICES ASTM A395 GRADE 60-45-15											
ASTM A445-					Heat treated to	Min 3.0						143
API604	60-40-18	60,000	40,000	18	ferritic structure	Max		2.5	0.08			187
						Min 3.2		1.7				
AMS 5315		60,000	45,000	15	Annealed	Max 4.0		2.5	0.08	0.8		190
						Min 3.2		1.7				202
AMS 5316		80,000	60,000	3		Max 4.0		2.5	0.08	0.8		269
	165	65,000	45,000	10	Annealed							
	180	70,000	55,000	7								
	210	85,000	70,000	5								
	225	89,000	75,000	4								
	255	103,000	87,000	3	Normalized							
	265	107,000	92,000	2	and tempered							
	285	115,000	100,000	1.5								
AGMA 244.02	300	123,000	105,000	1								
	350	143,000	123,000	0.5								
	180	98,000	75,000	7								
	210	105,000	82,000	6								
	255	115,000	90,000	4	Quenched							
	265	120,000	95,000	3.5	and tempered							
	285	130,000	105,000	3								
	300	135,000	110,000	2.5								
	350	158,000	130,000	1								
AWWA C151	Pressure Pipe 60-42-10	60,000	42,000	10	Impact 7 ft lbs.							

TABLE 2-2. TYPICAL CHEMICAL COMPOSITIONS RELATED TO MECHANICAL PROPERTIES OF UNALLOYED AND LOW-ALLOY DUCTILE IRONS

CHEMICAL COMPOSITION	GRADE 60-40-18	GRADE 65-45-12	GRADE 80-55-06	GRADE 100-70-03	GRADE 120-90-02
CARBON	3.50-3.90	3.50-3.90	3.50-3.90	3.50-3.80	3.50-3.80
SILICON	2.20-3.00	2.50-2.80	2.20-2.70	2.20-2.70	2.20-2.70
MANGANESE	.30 MAX	.40 MAX	.20-.50	.60 MAX	.60 MAX
*PHOSPHORUS	.05 MAX	.05 MAX	.05 MAX	.05 MAX	.05 MAX
SULFUR	.015 MAX	.015 MAX	.015 MAX	.015 MAX	.015 MAX
CHROMIUM	.06 MAX	.10 MAX	.10 MAX	.10 MAX	.10 MAX
NICKEL	–	–	–	**	**
COPPER	–	–	.20-.40	.20-.50	.20-.50

* For optimum elongation and impact properties, phosphorus should not exceed .03% and silicon should be kept as low as possible.
** Nickel additions of up to 1% may be used to enhance strength properties

Notes:

1. Sections under 1/4 in. and over 2 in. may require variations to the chemistries listed above (especially carbon and silicon).
2. See chapters on specific chemistry contents and heat treating. Properties will depend upon chemistry and thermal treatments as well as other processing variables.
3. Certain specifications will restrict the chemical limits, and each producer must refer to all applicable specifications, which should be designated by the customer.

either as cast, or may be annealed to obtain maximum ductility. Grade 65-45-12 is generally produced as cast.

The next grade that is classified as a mixed pearlitic-ferritic iron is grade 80-55-06. The microstructure will range from 70% ferrite to 80% pearlite with tensile strength to over 90,000 psi. This grade is produced either as cast or normalized.

Grade 100-70-03 is called the fully pearlitic iron containing a maximum of 20% ferrite, with tensile strengths of over 100,000 psi. While it can be produced as cast, it is typically made by a normalizing heat treatment.

A quenched and tempered iron containing either bainite or tempered martensite is grade 120-90-02. This grade is distinguished from the high strength pearlitic iron by a higher yield-strength-to-tensile-strength ratio. The impact properties also are generally superior to that of pearlitic irons of similar tensile strength.

The most recent addition to the ductile iron family are the Austempered Ductile Irons (ADI). ADI grades have tensile strengths of 125,000–230,000 psi, while still retaining high elongation and toughness characteristics. Additionally, these irons afford exceptional wear resistance and fatigue strength, allowing component weight (and costs)

to be reduced, but with equivalent or improved performance.

The austempered properties are developed by a closely controlled heat treatment that develops a unique matrix of acicular ferrite and high carbon retained austenite. The retained austenite is stable to extremely low temperatures, but will work-harden and locally transform to martensite under suitable stress conditions.

Alternatively, surface stresses can be deliberately imposed prior to service by shotpeening or fillet rolling in order to achieve significant improvements in wear resistance and fatigue life. The new specifications are shown in Table 2-3. The typical chemical ranges to produce ADI, including the recommended maximum variations for each element, are listed in Table 2-4.

The data presented here will give some guidance in choosing suitable maximum design stress levels for casting applications under tensile, compressive, and fatigue stresses. Information from the data tables in this chapter can be used for stress calculation required in casting design. It is hoped that this type of information will be of assistance to foundrymen called on to work with engineers in solving casting design problems.

PROPOSED GRADE	TENSILE STRENGTH	YIELD STRENGTH	PERCENT ELONGATION	CHARPY* (unnotched)	BHN HARDNESS RANGE**
(KSi)	(KSi)	(KSi)		(ft-lb)	
125/80/10	125	80	10	75	269 – 321
150/100/7	150	100	7	60	302 – 363
175/125/4	175	125	4	45	341 – 444
200/155/1	200	155	1	25	388 – 477
230/185/-	230	185	–	--	444 – 555
* CHARPY TESTED AT 72°F (22°C)		** NOT PART OF PROPOSED SPECIFICATION			

TABLE 2-3. ASTM A897-90 ADI SPECIFICATIONS
(Reprinted with permission, from the *Annual Book of ASTM Standards*, copyright, American Society for Testing and Materials.)

Element	Range	Variation for Best Results(±)
C.E.	4.3–4.6	0.10
C	3.4–3.8	0.10
Si	2.4–2.8	0.10
Mn	0.35 MAX	0.05
P	0.03 MAX	—
S	0.015 MAX	—
Cr	0.07 MAX	—
Mg	0.06 MAX	0.005
Cu	0–0.8	Sections less than 3/4 of an inch will usually not require alloying.
Ni	0–2.0	
Mo	0.30 MAX	

TABLE 2-4. TYPICAL CHEMISTRY RANGES FOR ADI PRODUCTION

■ MECHANICAL PROPERTY DEFINITIONS

Following are the definitions of mechanical properties used in subsequent strength-of-materials calculations.

Tensile strength is the load value obtained by dividing the maximum load observed during tensile straining to the breaking point by the cross sectional area, before a load is applied.

Tensile Strength (psi) = Load in Pounds ÷ Original CSA of Specimen

Yield Strength (Proof Stress is the British or European equivalent) is that amount of stress at which a material exhibits specified limiting deviation from the proportionality of stress to strain. In soft steels, it is the amount of stress at which a marked increase in deformation occurs without an increase in the applied loads, as indicated in Figure 2-1.

However, in the case of ductile iron, there is no well-defined marked increase in deformation without an increase in applied load. The stress-strain curve gradually merges into a horizontal straight line without the sudden extension which occurs in soft steels.

For this reason, a method of determining yield strength has been established, setting up definite amounts of permanent elongation as criteria of yield strength. In this "offset method," shown in Figure 2-2, a percentage of a 2 in. gauge length is used to indicate the yield strength on the stress-strain curve. This specified set is laid off on a plot of the stress-strain diagram as an intercept, and a line that intercepts it can then be drawn parallel to the beginning of the stress-strain curve. This line cuts the stress-strain curve at the yield strength.

Proportional Limit (Limit of Proportionality is the English or European equivalent) is defined as the maximum stress that a material is capable of developing without a deviation from the proportional relationship between stress (or load) and strain (or state of deformation).

The proportional limit of ductile iron compared to steel is exceptionally low and approaches approximately 40% of the ultimate tensile value, or a yield strength with a 0.005% offset.

Hardness Consideration for Ductile Iron. Much effort has been expended to establish mathematical relationships between hardness and tensile strength. However, the information developed has been only of academic interest. Nevertheless, some degree of correlation can be made on the basis of established practice in a particular foundry. This involves correlation of tensile strength data and test bar hardness results.

The Brinell Hardness test yields the most consistently accurate results. Rockwell "B" and "K," both of which have ball penetrators, can be used to advantage for determining the hardness of light-section castings where the Brinell test cannot be used. Rockwell "C" can be used to determine the hardness on surfaces that have been hardened. However, hardness values are usually from 5 to 8 Rockwell "C" numbers lower than the actual matrix hardness, due to the fact that diamond point penetrant readings are influenced by the graphite nodules imbedded in the matrix.

Compressive Yield Strength (CYS) is the maximum strength that a metal, subjected to compression, can withstand without a predefined amount of deformation. *Ultimate CYS* is the maximum stress that a brittle material can withstand without fracture when subjected to compression.

Shear Strength has been defined as the intensity of force acting in the plane of the area. Shear strength is equal to shearing force divided by the cross-sectional area.

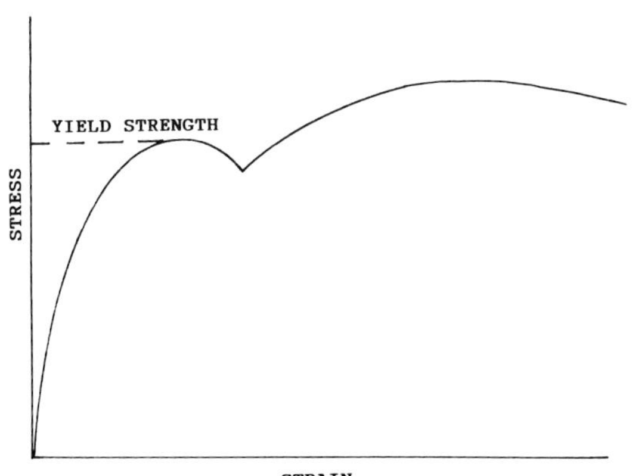

Fig. 2-1. Stress-strain curve for soft steel.

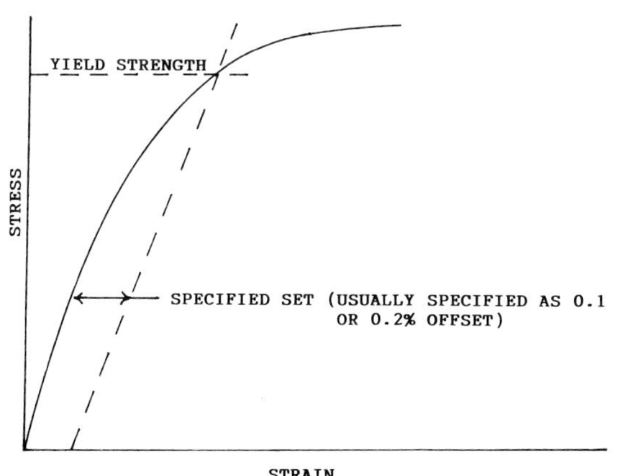

Fig. 2-2. Stress-strain curve for ductile iron.

Modulus of Elasticity is the slope of the elastic portion of the stress-strain curve in mechanical testing. The stress is divided by the unit elongation.

Torsional Strength is the stress within a material resisting twisting.

Modulus of Rigidity is the ratio (in a torsion test) of the unit shear stress to the displacement caused by its per-unit length in the elastic range. This modulus corresponds to the modulus of elasticity in the tension test.

Poisson's Ratio is the ratio of the transverse contraction of a strained test specimen to its longitudinal elongation, essentially a statement of constancy of volume during deformation.

TENSILE YIELD STRENGTH

Many of the mechanical properties of ductile iron are directly related to the 0.1% yield strength value in tension, and it is important that this value should be known accurately in order to calculate other mechanical properties.

Tensile strengths are related to yield strength values, and ratios are given in the data sheets. Such ratios can be used fairly confidently for calculating yield strength values from known tensile strengths in fully ferritic ductile irons, but they should not be regarded as necessarily very reliable for higher strength irons. In such irons, the full tensile strength may not always be realized, and in such cases the ratio will give an inaccurate (low) figure for yield strength. If other mechanical properties are then based on the calculated yield strength figure, low values will also be obtained. It is important, therefore, to know the actual yield strength value for the iron under consideration before applying the various formulas given in the general data sheet.

In ductile cast irons, the limit of proportionality (based on a 0.005% yield strength) is low in relation to the 0.1% yield strength value. The ratio is about 0.75 in grade 60-40-18, and 0.70 in grade 65-45-12 annealed ferritic ductile irons, and about 0.6 in as-cast pearlitic irons. Little data is available for irons with intermediate properties, or for normalized or hardened and tempered irons, but it seems appropriate to use a ratio of 0.6 for these as well as for as-cast pearlitic irons.

The ratio of 0.1% yield strength to tensile strength is higher in ferritic ductile irons than in pearlitic irons, and it is assumed that the ratio decreases uniformly as the strength increases, and the structure changes from ferritic to pearlitic in the intermediate grades of ductile iron. The ratio of yield strength to tensile strength is higher in hardened and tempered irons than in other ductile irons.

Because the stress-strain curve is steeper in the plastic range in pearlitic irons than in ferritic irons, the stress increment required to increase the plastic strain from 0.1 to 0.2%, or from 0.1 to 0.5% is greater in pearlitic irons than in ferritic irons, and this is shown in the yield strength values given in the data sheets.

The stress increments for obtaining 0.2% and 0.5% yield strength values from the 0.1% values for intermediate grades have been assumed to vary uniformly as the strength increases and the structure changes from ferritic to pearlitic. At a stress level of 94,000 psi, it is found that the stress increments are less in hardened and tempered irons than in pearlitic irons, but they increase as the tensile strength of hardened and tempered irons increases.

Austempered ductile irons can have tensile strength levels nearly twice those of conventional ductile irons, and that compare with those of steels. Yield strength values are slightly lower than those of steel. The range of tensile strength is from 125,000–230,000 psi, while yield strengths run from 80,000–185,000 psi.

NOTCHED TENSILE PROPERTIES

Notches increase the tensile strength of annealed ferritic ductile irons but decrease the strength of pearlitic irons. It is thought they will also decrease the strength of hardened and tempered irons, although no direct evidence is available. In ferritic ductile irons the depth of the notch as well as the notch profile are important in determining the increase in strength obtained, as shown by the information given in the general data sheet. In pearlitic and hardened and tempered irons, however, it is thought that the decrease in strength is mainly related to the notch profile and not very dependent on the depth of notch. Irons with intermediate strengths are assumed to have properties in between those of ferritic and pearlitic irons.

COMPRESSIVE STRENGTH

Failure in compression occurs at very high strains, and compressive strength figures for ductile irons are of little use to the designer. Only limit of proportionality and yield strength values are, therefore, quoted. The limit of proportionality is always higher in compression than in tension. For pearlitic and ferritic ductile irons with intermediate strengths, and also for hardened and tempered irons, it is approximately equal to eight-tenths of the 0.1% compressive yield strength. It is assumed that a similar ratio applies for irons. The 0.1% yield strength value is also slightly higher in compression than in tension because the onset of plastic deformation is delayed slightly under compressive loading.

The 0.1% yield strength in compression exceeds that in tension by a greater amount in ferritic irons than in pearlitic irons. The difference between the corresponding 0.5% yield strength values is small, however, and slightly greater in pearlitic irons. The difference between the yield strength values in tension and compression is considered to change uniformly as the strength of intermediate grades of iron increases with the structural change from ferritic to pearlitic. The values for hardened and tempered irons are assumed to be similar to those of pearlitic irons.

SHEAR STRENGTH

Little if any data is available on the shear strength of ductile cast irons. This is mainly because it is very difficult to obtain accurate shear strength values in materials that show some ductility, since in a double shear test, for instance, it is very difficult to avoid bending. It is thought, however, that the shear strength of ductile cast irons will be about nine-tenths of the tensile strength. This is based on early tests carried out on cerium-treated ductile irons, and also on results obtained on malleable cast irons.

TORSIONAL STRENGTH

All ductile irons deform very considerably under torsional shear stresses, and it is not always possible to continue a torsional test to failure. Based mainly on the results obtained on pearlitic ductile cast irons, however, it is thought that the torsional strength will be about nine-tenths of the tensile strength. This ratio was used to calculate the values given in the data sheets.

The limit of proportionality and yield strength values quoted are more reliable and, of course, of much more value to the designer. The limit of proportionality value in torsion is about three-quarters of the 0.1% yield strength value in torsion for all irons. The 0.1% yield strength value in torsion varies between 0.7875 and 0.7 of the 0.1% yield strength value in tension in ferritic and pearlitic irons, respectively, intermediate values of the ratio being obtained in irons with mixed structures. Hardened and tempered irons are assumed to have values similar to those of pearlitic irons. The 0.2% and 0.5% yield strength values in torsion can be obtained from the corresponding values of yield strength in tension by multiplying by the same ratios as used for the 0.1% values.

MODULUS OF ELASTICITY

The modulus of elasticity of pearlitic ductile cast irons is about 1×10^6 psi higher than that of annealed irons, and it is thought it will be about 0.5×10^6 psi higher than that of hardened and tempered irons. Normally a value of 25.5×10^6 psi is appropriate for a pearlitic iron giving values of 25×10^6 and 24.5×10^6 psi for hardened and tempered/annealed irons, respectively. Austempered ductile irons will have the same approximate value as hardened and tempered irons. These values are quoted in the general data sheets.

MODULUS OF RIGIDITY

Ductile cast irons have a value of Poisson's ratio of about of 0.280, and the modulus of rigidity is related to the modulus of elasticity as follows:

$$E = 2G(l + v),$$

where

E = Modulus of elasticity

G = Modulus of rigidity

v = Poisson's ratio.

This modulus of rigidity is, therefore, about 0.39 of the modulus of elasticity. This has been confirmed by mechanical tests and this ratio has been used for calculating values in the data sheets.

POISSON'S RATIO

This value shows little variation in ductile irons and can be taken as 0.280.

FATIGUE STRENGTH (ENDURANCE)

Castings usually fail due to repeated or cyclic stresses rather than impact loading. A failure of this nature occurs at a level below the yield strength value, and is related to the stress level and the number of cycles. The fatigue limit is then the maximum stress that will lead to a failure at a specified number of cycles. The stresses applied in testing are usually completely reversed, alternating between tension and compression. Notches in samples or parts increase stress concentrations in the area of the notch. Notched fatigue strength values will be considerably lower than the unnotched numbers. The fatigue strength reduction factor is a ratio of the unnotched fatigue limit divided by the notched fatigue limit, and shows the sensitivity of a material to notches.

The endurance limit is defined as the maximum stress level below which a material can endure an infinite number stress cycles. This is synonymous with fatigue limit numbers. The endurance ratio is then the un-notched fatigue limit (endurance limit) divided by the tensile strength.

In general, the fatigue limits of ductile irons increase with increasing tensile strengths. In annealed ferritic ductile irons, the unnotched fatigue limit is about 0.5 times the tensile strength. This endurance ratio decreases with increasing tensile strengths to 0.4 for pearlitic, and hardened and tempered irons and is reduced further at strengths above 110,000 psi.

Unlike the conventional ductile irons, the austempered grades show the highest fatigue strengths with irons of the lowest tensile strength. The endurance ratio is approximately 0.4 for the 125,000 psi grade, and drops to 0.3 for the 175,000 psi grade. Increasing section size also reduces the fatigue limit somewhat.

Notched fatigue values of all grades of conventional ductile irons are about 0.6 times the unnotched fatigue limit. Austempered irons are more notch sensitive, and the fatigue limit is reduced to 0.5 times the unnotched fatigue. This would produce a fatigue strength reduction factor of 2.0, whereas the standard grades will range from 1.4–1.7.

A casting's microstructure and surface condition will have quite large effects on fatigue values. Nodularity, carbides, porosity, inclusions, and surface finish are all important. Surface treatments such as shot peening, rolling, and hardening can increase the fatigue limit. Increases of over 50% have been reported in ADI parts that have had subsequent surface treatments. This is due not only to the creation of compressive stresses on the surface, but also to strain induced transformation of the retained high-carbon austenite into martensite.

IMPACT PROPERTIES AND TOUGHNESS

The toughness of a material is a measure of the amount of energy absorbed in fracturing it. One of the methods to measure the impact strength is the Charpy impact test. This test utilizes a standard specimen, either notched or unnotched, that is placed in a fixture and struck by a pendulum. The values obtained are in foot-pounds. The reported original purpose of the Charpy test was to determine the transition temperature at which a material would pass from a ductile to a brittle state.

There is often a considerable spread in data when conducting this test, and relating this data on ductile irons to other materials does not allow accurate comparisons. Also, the application of this data to engineering design and performance is difficult. Even so, some people continue to specify this test, and the data will be discussed here and presented in the data sheets.

Ferritic ductile irons with tensile strengths at below 60,000 psi, and high elongation values of 15% or more, are normally specified when a high degree of toughness is required. Ductility decreases with increasing tensile strength and is at its lowest value with a fully pearlitic structure. Slightly improved impact properties over those of pearlitic grades can be obtained with a tempered martensite structure. Austempered ductile irons have impact properties that compare with ferritic grades and approach those of

wrought steels. Again, these values decrease with increasing tensile strength.

Controlling chemistry and heat treatment to produce the desired microstructure in all grades will give the best impact values. Elements such as silicon and phosphorus must be controlled. Increasing amounts of these elements will not only lower the impact values, but raise the transition temperature. Some specifications limit the silicon to 2.50% maximum (or lower), and phosphorus to 0.03% maximum, for best results.

Ductile to brittle transition can occur over a wide range of temperatures. Transition temperatures increase with increasing amounts of pearlite. Fully annealed irons have much lower transition temperatures than sub-critically annealed irons. Hardened and tempered irons have much lower transition temperatures than pearlitic irons, and can approach those of annealed ferritic irons. Austempered irons have transition temperatures between those of fully ferritic and pearlitic grades. Notched samples have higher transition temperatures than unnotched specimens. The difference can be as much as 120F (49C).

Another test that shows the amount of energy required to fracture a specimen at a specific temperature is the Dynamic Tear (DT) test. This test utilizes a larger specimen than the Charpy bar, and has a pressed-in sharp notch rather than the blunt notch used in the Charpy V-notch test. This test can show the relative quality of an iron, and reflects the chemistry and microstructure. It has better reproducibility than the Charpy test.

A recently developed test is based on fracture mechanics, in which it is assumed that all materials contain flaws. The application of fracture mechanics allows designers to determine the maximum allowable design stress from the largest expected flaw size and the plane strain fracture toughness of the material. The test results show at what temperature the fracture changes from ductile to brittle, and the magnitude of fracture toughness when a ductile fracture is observed. Again, this test utilizes a larger specimen than the Charpy test and has a very sharp notch (fatigue precrack). The critical stress intensity (fracture toughness) is designated as K_{Ic}, which can be calculated from the flaw size and the design stress.

The data from the dynamic tear and fracture toughness tests are more likely to simulate the behavior of a material under actual service conditions than the Charpy test. There is much more information available on these tests and their results, and the reader is encouraged to investigate other sources.

COEFFICIENT OF THERMAL EXPANSION

Most ductile irons have values for thermal expansion that are slightly higher than gray iron and lower than malleable iron and steel. Typical values are given in the data sheets and will vary with matrix and temperature. It is reported that austempered ductile irons have a coefficient of thermal expansion that is 25% higher than that of pearlitic irons.

THERMAL CONDUCTIVITY

Ferritic ductile irons have higher thermal conductivities than pearlitic irons. Hardened and tempered irons have values in between those of ferritic and pearlitic grades. Increasing silicon contents in all grades reduces thermal conductivity. This reduction can be as high as 20% as the silicon content increases from 2.0 to 3.0%. Nickel has a similar effect to that of silicon. Austempered ductile iron has about the same values of thermal conductivity as that of pearlitic irons.

THERMAL AND DIMENSIONAL STABILITY (GROWTH)

Annealed ferritic ductile iron is essentially free from growth up to 1500F (815C), while pearlitic irons will begin to grow due to decomposition of the pearlite into ferrite and graphite at temperatures above 1000F (538C). Pearlite will completely break down at the transformation temperature (approximately 1400F [760C]). Both types will show large amounts of growth above 1500F (815C), with pearlitic structures growing more than the ferritic grades. Pearlitic irons after decomposition and below 1500F (815C) will have up to 0.5% growth. Above 1500F (815C) thin sections may grow an additional 1.0%, while thick sections can grow another 0.5%. Recycling pearlitic irons through the transformation temperature may increase growth as much as 3.0%.

Ductile irons are much more resistant to growth than gray irons. Gray irons can also experience increased growth due to oxidation of the flake graphite. The spheroidal shape of graphite in ductile iron provides a much lower surface area available for oxidation to occur.

Increasing silicon contents up to 4% and more increases the resistance to growth and oxidation. Other elements such as chromium and molybdenum will also inhibit growth.

Since ductile irons have a tendency to grow at elevated temperatures, even without the application of stress, it is recommended to fully anneal, or at least sub-critically anneal parts requiring high temperature dimensional stability. All high silicon ductile irons should be annealed either fully or sub-critically to remove any carbides that may be present, and to reduce the brittleness of the material. The lower strength grades of ADI are stable to 650–750F (343–399C). Increasing the temperature to 1000F (538C) will have a tempering effect on all grades.

THERMAL SHOCK

When all types of cast irons were subjected to alternately heating to 1500F (815C) and water quenching, ductile iron was found to be the most resistant to cracking. Ferritic grades are more resistant to thermal shock than the pearlitic grades.

ELEVATED TEMPERATURE STRENGTH

Pearlite in ductile iron is stable to about 800F (427C). Above this temperature the pearlite slowly begins to spheroidize. Strengths of both pearlitic and ferritic grades drop off quite rapidly over 800F (427C), while elongation values increase rapidly. The elevated temperature tensile strengths for various grades are shown in Table 2-5.

CREEP STRENGTH

In general, the pearlitic grades are somewhat superior below 900F (482C), but the ferritic grades are better above 900F (482C) (Table 2-6). The high silicon and high silicon plus molybdenum grades have good creep resistance. They both will increase the elevated temperature strength, due to the fact that the critical temperature is increased. Molybdenum is more effective than silicon in this regard.

SPECIFIC GRAVITY

The specific gravity of an annealed ductile iron is slightly less than that of a pearlitic iron, and it is assumed that hardened and tempered irons have values in between those of pearlitic and ferritic irons. In irons with intermediate strengths, the value tends to increase as the strength increases, as shown in the general data sheet.

MELTING TEMPERATURE (LIQUIDUS)

The melting range for both the pearlitic and ferritic forms is 2050–2150F (1121–1177C) depending on composition.

■ MAGNETIC PROPERTIES

The magnetic properties of ductile irons produced commercially are determined by the matrix structure. Ferritic ductile irons are magnetically soft and principally used when good magnetic properties are required. Pearlitic irons are magnetically harder and have a higher magnetic loss. The highest magnetic permeabilities are obtained in

Grade	Tensile Strength (PSI) at Temperature					
	200F (93C)	400F (204C)	600F (316C)	800F (427C)	1000F (538C)	1200F (649C)
120-90-02	122,000	120,000	108,000	82,000	45,000	17,000
100-70-03	102,000	100,000	88,000	68,000	41,000	15,000
80-55-06	82,000	82,000	75,000	60,000	36,000	14,500
65-45-12	68,000	68,000	64,000	50,000	29,000	13,500
60-40-18	59,000	59,000	53,000	40,000	25,000	12,500
4%Si + 1%Mo	88,000	—	—	61,000	44,000	19,000

TABLE 2-5. ELEVATED TEMPERATURE TENSILE STRENGTH

TEMPERATURES °F (°C)	LIMITING CREEP STRESS (PSI) FOR 0.0001% PER HOUR MINIMUM RATE	
	PEARLITIC	FERRITIC
800 (427)	18,950	14,000
1000 (538)	1,745	4,025
1200 (649)	448	560

TABLE 2-6. CREEP STRENGTH FOR PEARLITIC AND FERRITIC GRADES AT VARIOUS TEMPERATURES

nickel-free ductile irons with very high silicon contents, greater than those normally used commercially.

The magnetic properties given in the data sheets are for normal, commercially available ductile irons with nickel contents of 0.5–1% and silicon contents of 2–2.5%. Data are given for fully ferritic and fully pearlitic irons. For irons with intermediate structures and properties, it has been assumed that the change in properties will be uniform as the structure changes from pearlitic to ferritic. The data suggest that irons with hardened and tempered structures will have a lower permeability, higher inherent magnetism, coercive force and hysteresis loss than pearlitic grades of ductile iron with similar tensile strengths. No detailed figures for hardened and tempered irons can be given.

ELECTRICAL PROPERTIES

Ductile irons have lower electrical resistivities than flake graphite cast irons, and the primary elements affecting their resistivity are silicon and nickel, both of which increase resistivity. The electrical resistivity of normally available pearlitic ductile irons with 0.5–1% nickel and

2–2.5% silicon is approximately 54×10^6 ohm cm. These values have been assumed in the data sheets, and irons with intermediate structures have been given intermediate values.

TEST SPECIMENS FOR DUCTILE IRONS

Figures 2-3 through 2-6 and (and accompanying Tables 2-7 through 2-9) show the types of test bar molds and machined tensile specimens used in the control of the mechanical properties of ductile irons. Test coupons should be cast in completely baked, oil-bonded core-sand molds or airset molds. The modified keel block (Fig. 2-3) is made in two pieces and glued together. The test bar can then be easily sawed or broken off the feeding section of the casting. This test bar can be readily machined into a tensile test specimen or used as a sonic tensile test specimen.

The reduced section should have a gradual taper from the ends toward the center, with the ends not more than 0.005 in. (0.13 mm) larger in diameter than the center of the standard specimen, and not more than 0.003 in. (0.076 mm) larger in diameter than the center on the small size specimens.

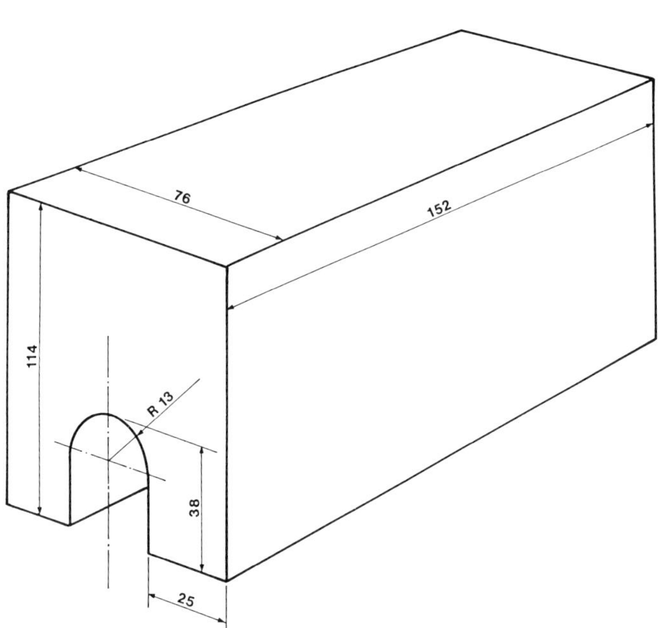

Fig. 2-3. Keel block for test coupons. (Dimensions are in mm, 25 mm = 1 in. Note: The length of the keel block shall be 6 inches.)

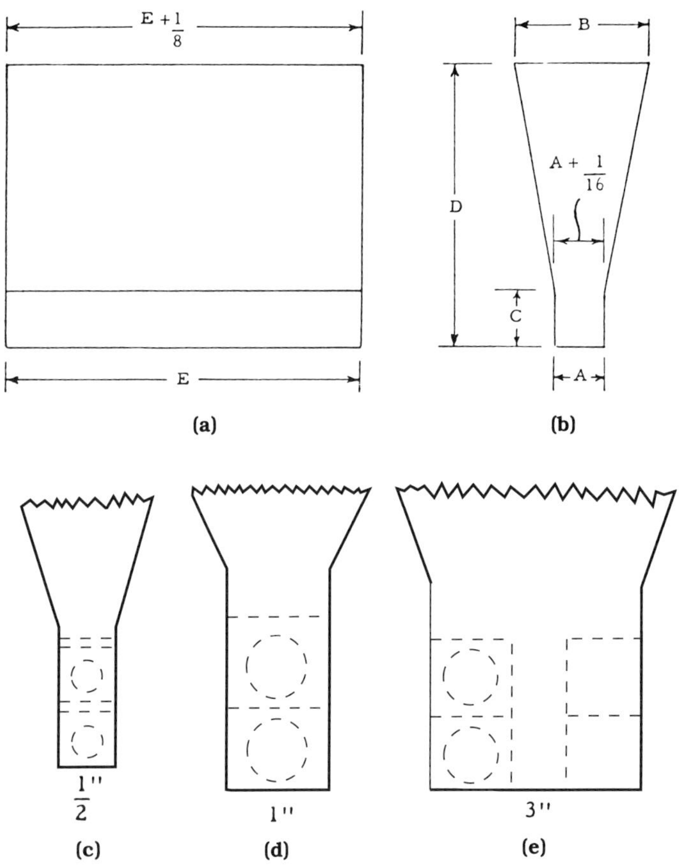

Fig. 2-4a–e. Y-block sizes and location of test specimens.

If desired, on the small size specimens the length of the reduced section may be increased to accommodate an extensometer. However, reference marks for measurement of elongation should nevertheless be spaced at the indicated gauge length.

The gauge length and fillets shall be shown, but the ends may be of any form to fit the holders of the testing machine in such a way that the load shall be axial. If the ends are to be held in grips, it is desirable, if possible, to make the length of the grip section great enough to allow the specimen to extend into the grips a distance equal to two-thirds or more of the length of the grips.

TABLE 2-7. Y-Block Dimensions

DIMENSION	Y BLOCK SIZE (ALL DIMENSIONS IN INCHES)		
	1/2 IN	1 IN	3 IN
A	1/2	1	3
B	1 5/8	2 1/8	5
C	2	3	4
D	4	6	8
E	6 TO 7	6 TO 7	6 TO 7
TO REPRESENT CASTING SECTIONS	< 1/2	1/2 TO 1 1/2	>1 1/2

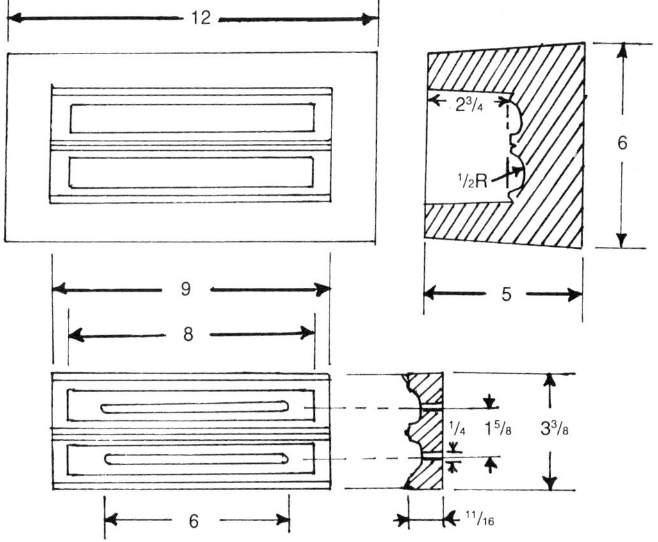

Fig. 2-5. Modified keel block mold for 1 inch round specimens (two-piece mold).

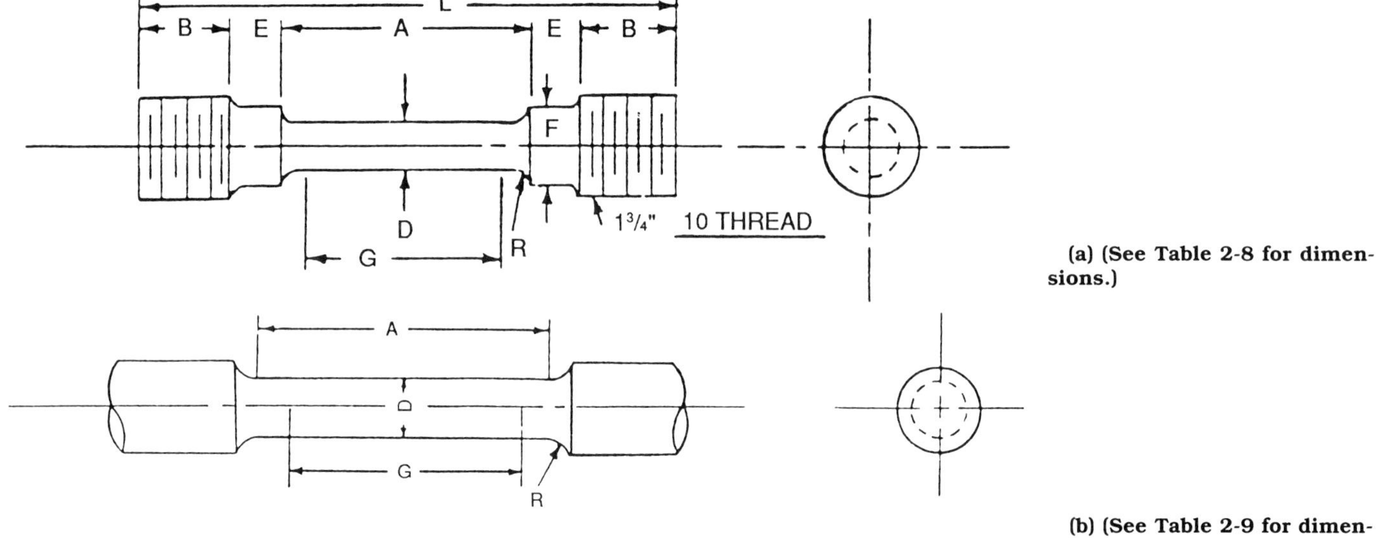

(a) (See Table 2-8 for dimensions.)

(b) (See Table 2-9 for dimensions.)

Fig. 2-6. Threaded tensile specimen (a); machined tensile bar (b).

G = GAGE LENGTH	2.00 + 0.005 INCHES
D = DIAMETER (NOTE -1)	0.500 + 0.010 INCHES
R = RADIUS	3/8 INCH MAXIMUM
A = LENGTH OF REDUCED SECTION (NOTE -2)	2 3/4 INCHES
L = OVERALL LENGTH	8 INCHES
B = LENGTH OF END SECTION (NOTE -3)	1 INCH
C = DIAMETER OF END SECTION	3/4 INCH
E = LENGTH OF SHOULDER	5/8 INCH - APPROXIMATELY
F = DIAMETER OF SHOULDER	5/8 INCH - APPROXIMATELY

TABLE 2-8. THREADED TENSILE SPECIMEN DIMENSIONS

DIMENSIONS	STANDARD SPECIMEN IN. (MM)	SMALL SIZE SPECIMENS PROPORTIONATE TO STANDARD IN. (MM)	
	1/2 ROUND (13 MM)	0.350 ROUND (8.89 MM)	0.250 ROUND (6.35 MM)
G = GAGE LENGTH	2.000 ± 0.005 (50.8 ± 0.13)	1.4 ± 0.005 (35.6 ± 0.13)	1.0 ± 0.005 (25.4 ± 0.13)
D = DIAMETER (NOTE 1)	0.500 ± 0.010 (13 ± 0.25	0.350 ± 0.007 (8.9 ± 0.18)	0.250 ± 0.005 (6.35 ± 0.13)
R = RADIUS OF FILLET	3/8 MIN (9.5, MIN)	3/8 MIN (9.5, MIN)	1/4 MIN (6.35, MIN)
A = LENGTH OF REDUCED SECTION (NOTE 2)	2 1/4 MIN (57.2, MIN)	1 3/4 MIN (44.5, MIN)	1 1/4 MIN (31.8, MIN)

TABLE 2-9. MACHINED TENSILE BAR

SECTION II: GENERAL DATA SHEETS FOR DUCTILE IRONS*

■■■ DUCTILE IRON GRADE 60-40-18

This grade of ductile will have a ferritic matrix usually obtained by annealing. It is specified where ductility and good impact properties are required. While generally annealed, it can be produced in the as-cast condition and can be used to replace ferritic malleable and low-carbon steel. The structure is completely ferritic. It is used in marine applications, valves, fittings, truck and agricultural implements, and automotive steering knuckles.

TYPICAL MECHANICAL PROPERTIES

- Tensile Strength: 60,000 psi
    - proportional limit: 29,000 psi
    - 0.1% yield strength: 40,000 psi
    - 0.2% yield strength: 42,000 psi
    - 0.5% yield strength: 43,000 psi
    - elongation: 18% (minimum) up to 27%
- Compressive Strength
    - proportional limit: 34,000 psi
- Shear Strength (approximately 0.9 × tensile strength): 56,700 psi.
- Torsional Strength (approximately 0.9 × tensile strength): 56,700 psi
    - proportional limit (0.75 × 0.1% yield strength): 22,500 psi
    - 0.1% yield strength: 30,000 psi
    - 0.2% yield strength: 31,000 psi
    - 0.5% yield strength: 32,500 psi
- Modulus of Elasticity
    - tension: 24,500,000 psi
    - compression: 24,500,000 psi
- Modulus of Rigidity (0.39 × modulus of elasticity in tension): 9,600,000 psi
- Poisson's Ratio: 0.280
- Fatigue Limit (unnotched): 27,000–30,000 psi
- Hardness: 130–170 Brinell
- Specific Gravity: 7.1
- Density: 0.245–0.255 lbs/in^3

---

*Typical microstructures are shown in Figures 7a–j.

IMPACT PROPERTIES (CHARPY)

- Ductile to Brittle Transition Temperatures
    - notched: +15F to –60F as tensile increases
    - unnotched: –60F to –130F as tensile increases
- Charpy Impact Values at 72F
    - minimum, notched: 10 ft-lb
    - notched, ductile fracture: 12–16 ft-lb
    - unnotched, ductile fracture: 80–100 ft-lb
    - unnotched, brittle fracture: 2–3 ft-lb

COEFFICIENT OF THERMAL EXPANSION: GRADES 60-40-18 AND 65-45-12

Temperature Range	Mean Coefficient of Expansion
68–212F	6.2×10^6 in./°F/in.
68–392F	6.8×10^6 in./°F/in.
68–572F	7.1×10^6 in./°F/in.
68–752F	7.3×10^6 in./°F/in.
68–932F	7.5×10^6 in./°F/in.
68–1112F	7.6×10^6 in./°F/in.
68–1400F	8.2×10^6 in./°F/in.
68–1600F	8.5×10^6 in./°F/in.

THERMAL CONDUCTIVITY (CAL/SEC/CM^2/°C/CM): GRADES 60-40-18 AND 65-45-12

- 20C (68F): 0.070–0.085
- 100C (212F): 0.085–0.105
- 400C (752F): 0.075–0.085
- 1000C (1832F): 0.0575

MAGNETIC AND ELECTRIC PROPERTIES

- Maximum Magnetic Permeability (at 5700 gauss): 1500–2100 oersteds
- Remanent Magnetism (induction at 10,000 gauss): 5600 gauss
- Coercive Force: 2.0 oersteds
- Hysteresis Loss per Cycle for B (induction at 10,000 gauss): 4–7 ergs/cm^3
- Specific Electrical Resistivity: 50×10^6 ohm cm

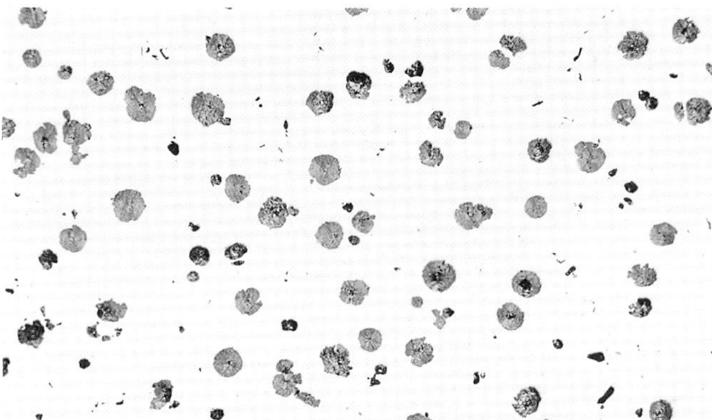

Fig. 2-7a. Ferritic ductile iron, as-cast microstucture: T.C. = 3.70%; Si = 2.55%; Mn = 0.15%; Mg = 0.04%, unetched, ×100.

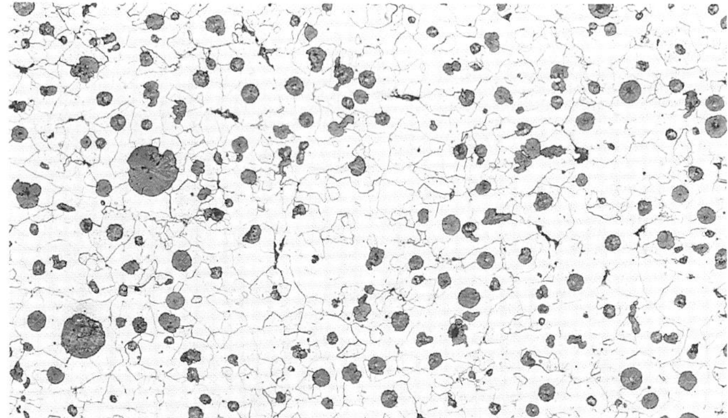

Fig. 2-7b. Ferritic ductile iron, as-cast microstructure: T.S. = 64,400 psi; Y.S. = 44,300 psi; elongation = 21.0%; BHN = 156; 2% nital etched, ×100.

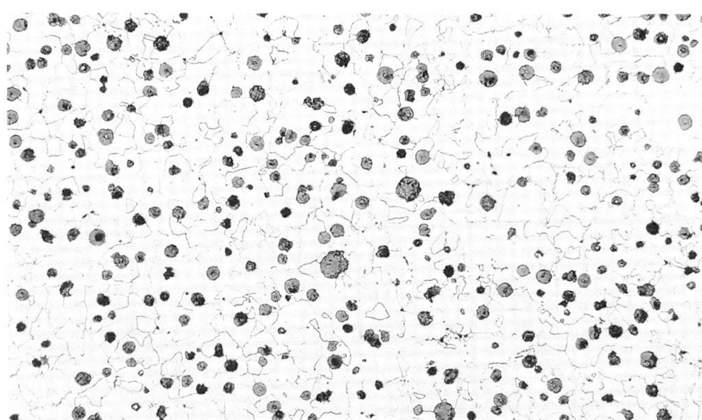

Fig. 2-7c. Grade 60-40-18 (fully ferritized): T.S. = 60,200 psi; Y.S. = 42,100; elongation = 26%; BHN = 139; 2% nital etched, ×50.

Design Stress

- Direct Tension: 20,000 psi
- Direct Compression: 23,500 psi
- Alternating Fatigue: 9500 psi (unnotched)

■■■ DUCTILE IRON GRADE 65-45-12

This grade of ductile will have primarily a ferritic matrix with a relatively small amount of pearlite present. This structure can be produced from a low manganese base iron or heat treatment. In either the as-cast or annealed condition it has good toughness and machinability, and can replace steel or malleable. Its structure is essentially ferritic. It is not as readily flame or induction hardened as are 80-55-06, 100-70-03 or 120-90-02.

This grade is used in pressure castings; valves; connecting rods; compressor and pump bodies; ingot molds; pipe fittings; automotive; agricultural; electrical; railroad; machine; and marine use. It is also used in applications involving severe thermal and mechanical shock and high temperatures.

Typical Mechanical Properties

- Tensile Strength: 65,000 psi
    - proportional limit: 29,000 psi
    - 0.1% yield strength: 41,000 psi
    - 0.2% yield strength: 45,000 psi
    - 0.5% yield strength: 47,000 psi
    - elongation: 12% (minimum) up to 17%
- Compressive Strength
    - proportional limit: 33,500 psi
- Shear Strength (approximately 0.9 × tensile strength): 59,500 psi
- Torsional Strength (approximately 0.9 × tensile strength): 59,500 psi
    - proportional limit (0.75 × 0.1% yield strength): 48,750 psi
    - 0.1% yield strength: 32,000 psi
    - 0.2% yield strength: 33,000 psi
    - 0.5% yield strength: 34,500 psi
- Modulus of Elasticity
    - tension: 24,500,000 psi
    - compression: 24,500,000 psi
- Modulus of Rigidity (0.39 × modulus of elasticity in tension): 9,600,000 psi
- Poisson's Ratio: 0.280
- Fatigue Limit (unnotched) 29,000–31,000 psi

- Hardness: 150–220 Brinell
- Specific Gravity: 7.1
- Density: 0.245–0.255 lbs/in^3

Impact Properties (Charpy)

- Ductile to Brittle Transition Temperatures
        notched: +14 to +86F as tensile increases
        unnotched: –76F to +14F as tensile increases
- Charpy Impact Values at 72F
        notched, typical: 10–14 ft-lb
        notched, fully ductile: 12–16 ft-lb
        unnotched, ductile: 70–100 ft-lb.
        unnotched, completely brittle: 2–3 ft-lb

Coefficient of Thermal Expansion

—approximately the same as 60-40-18

Thermal Conductivity

—approximately the same as 60-40-18

Magnetic and Electric Properties

—approximately the same as 60-40-18

Design Stress

- Direct Tension: 20,000 psi
- Direct Compression: 25,000 psi
- Alternating Fatigue: 9900 psi (unnotched)

▬ DUCTILE IRON GRADE 80-55-06

This grade of ductile iron generally has a matrix structure of ferrite and pearlite. It is usually produced in the as-cast condition, but can be produced by a step-normalizing heat treatment. An intermediate grade of ductile iron, it has a higher tensile strength than the ferritic grade, but with a considerable degree of ductility and impact resistance.

As cast it has an essentially pearlitic structure with good machinability and toughness. It responds readily to flame or induction hardening and may be cast against a chill to produce a carbidic, abrasion resistant surface. It can be welded and can withstand severe stresses.

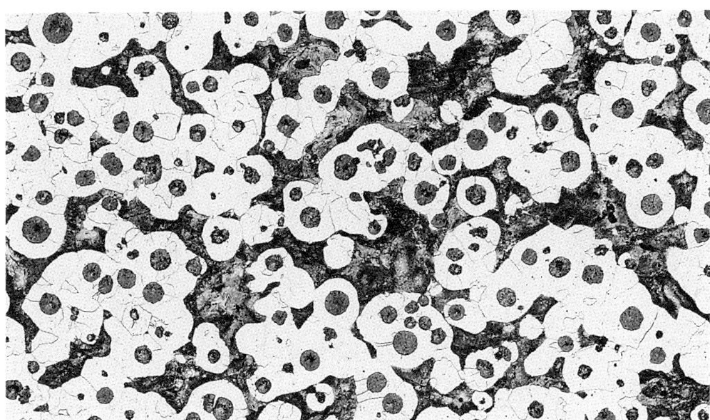

Fig. 2-7d. Grade 65-45-12 (approx. 30% pearlite), as cast 30/70: T.S.= 69,800 psi; Y.S. = 47,200; elongation = 16%; BHN = 179; 2% nital etched, ×100.

This grade is used in gears, cams, bearings, dies, pistons, crankshafts, sheaves, sprockets, wear and strength applications for automotive, aeronautical, diesel, agricultural, heavy machinery, mining, paper, textile and other related industries.

Typical Mechanical Properties

- Tensile Strength: 80,000 psi
        proportional limit: 30,100 psi
        0.1% yield strength: 48,000 psi
        0.2% yield strength: 55,000 psi
        0.5% yield strength: 57,000 psi
        elongation: 6% (minimum) up to 12%

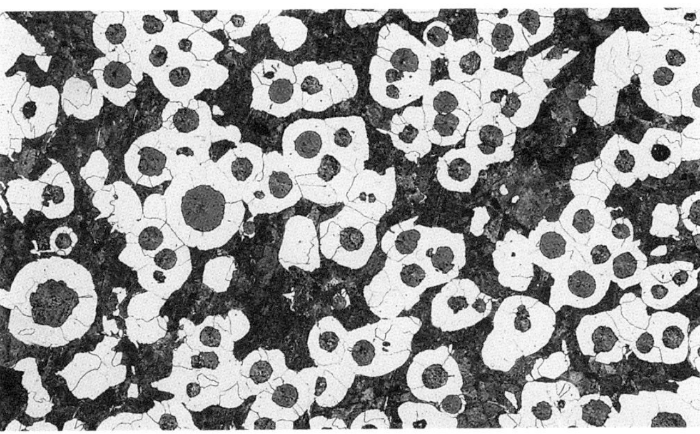

Fig. 2-7e. Grade 80-55-06 (approx. 50% pearlite): T.S. = 82,500 psi; Y.S. = 56,700 psi; elongation = 10%; BHN = 202; 2% nital etched, ×100.

- Compressive Strength:
    proportional limit: 40,000 psi
- Shear Strength (approximately 0.9 × tensile strength): 72,000 psi.
- Torsional Strength (approximately 0.9 × tensile strength): 72,000 psi
    proportional limit (0.75 × 0.1% yield strength): 36,000 psi
    0.1% yield strength: 39,000 psi
    0.2% yield strength: 41,000 psi
    0.5% yield strength: 43,000 psi
- Modulus of Elasticity
    tension: 25,500,000 psi
    compression: 25,000,000 psi
- Modulus of Rigidity (0.39 × modulus of elasticity in tension): 9,750,000 psi
- Poisson's Ratio: 0.280
- Fatigue Limit, unnotched: 32,000–37,000 psi
- Hardness: 170–250 Brinell
- Specific Gravity: 7.15
- Density: 0.250–0.260 lbs/in^3

IMPACT PROPERTIES (CHARPY)

- Ductile to Brittle Transition Temperatures
    notched: +14F to +86F as tensile increases
    unnotched: –76F to +14F as tensile increases
- Charpy Impact Values at 72F
    minimum, notched: 5 ft-lb

COEFFICIENT OF THERMAL EXPANSION: GRADES 80-55-06 AND 100-70-03

Temperature Range	Mean Coefficient of Expansion
68–212F	5.9×10^6 in./°F/in.
68–392F	6.5×10^6 in./°F/in.
68–572F	6.9×10^6 in./°F/in.
68–752F	7.3×10^6 in./°F/in.
68–932F	7.4×10^6 in./°F/in.
68–1112F	7.5×10^6 in./°F/in.
68–1400F	8.2×10^6 in./°F/in.
68–1600F	8.5×10^6 in./°F/in.

THERMAL CONDUCTIVITY, GRADES 80-55-06 AND 100-70-03 (CAL/SEC/CM^2/°C/CM)

    20C (68F): 0.060–0.070

    100C (212F): 0.070–0.075
    400C (752F): 0.065-0.070

MAGNETIC AND ELECTRIC PROPERTIES

- Maximum Magnetic Permeability (at 5700 gauss): 800–1300 oersteds
- Remanent Magnetism (induction at 10,000 gauss): 5800 gauss
- Coercive Force: 5.8 oersteds
- Hysteresis Loss per Cycle for B (induction at 10,000 gauss): 17–24 ergs/cm^3
- Specific Electrical Resistivity: 50×10^6 ohm cm

DESIGN STRESS

- Direct Tension: 22,500 psi
- Direct Compression: 29,800 psi
- Alternating Fatigue: 10,900 psi (unnotched)

■■■ DUCTILE IRON GRADE 100-70-03

This grade of ductile will have a pearlitic matrix. It is usually normalized and tempered, although it may be alloyed to produce the properties shown without heat treatment. It offers an excellent combination of strength, toughness and wear resistance. It is readily flame or induction hardened.

Gears, crankshafts, camshafts, dies, pistons, agricultural implement parts, bolsters, bolster forks, ratchets, governor weights, truck shoes, tractor brake drums and mining machinery are all applications for this grade.

TYPICAL MECHANICAL PROPERTIES

- Tensile Strength: 100,000 psi
    proportional limit: 42,100 psi
    0.1% yield strength: 60,000 psi
    0.2% yield strength: 70,000 psi
    0.5% yield strength: 78,000 psi
    elongation (3% minimum) up to 6%
- Compressive Strength
    proportional limit: 48,000 psi
- Shear Strength (approximately 0.9 tensile strength): 90,000 psi
- Torsional Strength (approximately 0.9 × tensile strength): 90,000 psi
    proportional limit: 40,000 psi
    0.1% yield strength: 41,000 psi
    0.2% yield strength: 44,000 psi

0.5% yield strength: 48,000 psi

- Modulus of Elasticity
    tension: 25,500,000 psi
    compression: 25,00,000 psi
- Modulus of Rigidity (0.39 × modulus of elasticity in tension): 9,900,000 psi
- Poisson's Ratio: 0.280
- Fatigue Limit (unnotched): 40,000-44,000 psi
- Hardness: 241-300 Brinell
- Specific Gravity: 7.2
- Density: 0.250–0.260 lbs/in^3

IMPACT PROPERTIES (CHARPY)

- Ductile to Brittle Transition Temperatures
    unnotched: 212–250F, as tensile increases
- Charpy Impact values at 72F
    unnotched ductile fracture: 5-30 ft-lb

COEFFICIENT OF THERMAL EXPANSION

—approximately the same as 80-55-06

THERMAL CONDUCTIVITY

—approximately the same as 80-55-06

MAGNETIC AND ELECTRIC PROPERTIES

- Maximum Magnetic Permeability (at 5700 gauss): 290–500 oersteds
- Remanent Magnetism (induction at 10,000 gauss): 6200 gauss
- Coercive Force: 11.0 oersteds
- Hysteresis Loss per Cycle for B (at 10,000 gauss): 24–27 ergs/cm^3
- Specific Electrical Resistivity (room temperature): 54 × 10^6 ohm cm

DESIGN STRESS

- Direct Tension: 25,000 psi
- Direct Compression: 34,500 psi
- Alternating Fatigue: 13,500 psi (unnotched)

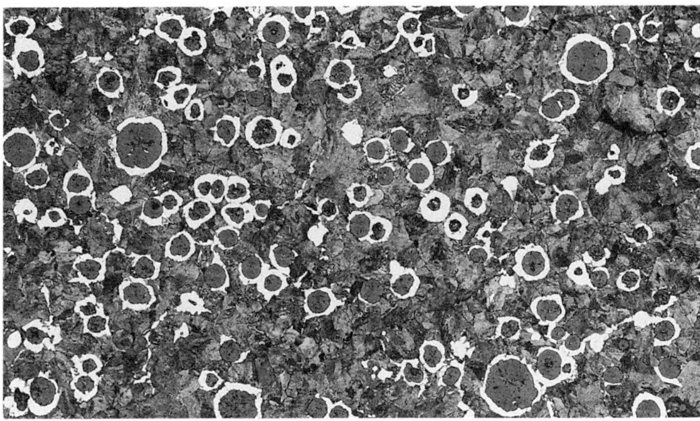

Fig. 2-7f. Grade 100-70-03, (approx. 95% pearlite), as-cast pearlitic microstructure: T.S. = 104,100 psi; Y.S. = 71,000 psi; elongation = 6.1%; BHN = 255; 2% nital etched, ×100.

DUCTILE IRON GRADE 120-90-02

This heat treatable grade is usually oil quenched and tempered, although sections up to 1.5 in. thick may be normalized and tempered. For this reason mechanical properties and physical constants cover a very wide range, and accurate data cannot be given. The reader should refer back to comments made for specific properties.

Alloying elements such as nickel and molybdenum should be used to obtain adequate hardenability in heavy sections. It is readily flame or induction hardened to 58 Rc (618 Bhn).

Pinions, gears, crankshafts, cams, dies, machine guides, track rollers, idlers, tractor steering gear arms, drill press columns, pump liners and clutch drums have all been made from this metal.

HEAT RESISTANT TYPE (2.80–6.00% SI)

This type offers maximum resistance to oxidation and growth. Sections less than 1 in. thick are usually annealed for maximum dimensional stability. Higher silicon contents increase hardness and strength, but reduce ductility and toughness at room temperature, although normal toughness is achieved at elevated temperatures. This grade is not recommended for applications involving severe thermal shock.

Primary applications are in furnace doors and frames, grate bars, blast furnace parts, sinter pots, reduction pots, lead pots, aluminum pouring troughs, glass molds and plungers, annealing pots, slag runners, pallets, manifolds, turbocharger housings, etc.

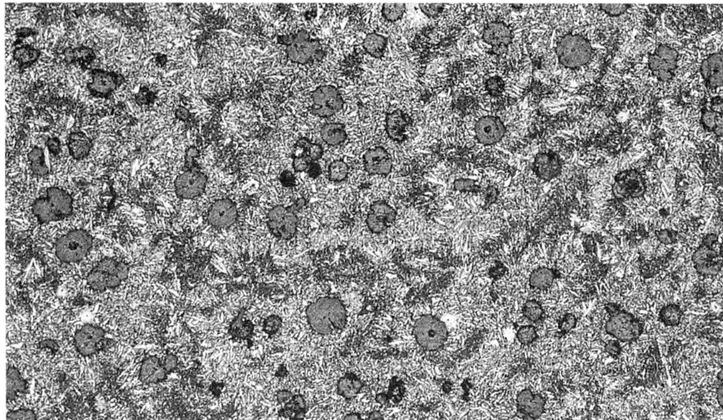

Fig. 2-7g. Grade 120-90-02 showing a martensitic matrix: T.S. = 131,800 psi; Y.S. = 92,000 psi; elongation = 3.4%; BHN = 352 (38 RC); 2% nital etched, ×100.

■ AUSTEMPERED DUCTILE IRONS

ADI is used in gears, crankshafts, connecting rods, agricultural and construction ground engaging parts such as plow points, digger teeth and track shoes; shafts; cams; mining and grinding equipment such as balls and hammers; mill liners and rollers; spring hanger brackets and railroad wear applications.

The different grades are produced primarily by varying the austempering temperature and duration of the heat treatment cycle. The highest strength is produced with the lowest austempering temperatures (approximately 450–500F). The highest ductility and toughness values are obtained at austempering temperatures of 650–750F. These grades are used where combinations of high fatigue strength, good toughness, and excellent wear resistance properties are needed. It is a replacement for steel castings, forgings, and fabrications.

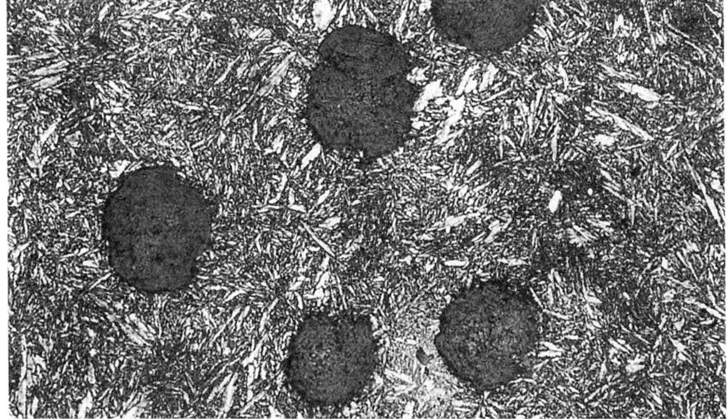

Fig. 2-7h. Grade 120-90-02, tempered martensite, T.C. = 3.60%; Si = 2.70%; Mn = 0.38%; Ni = 0.48%; Cu = 0.15%; Mg = 0.045%; 2% nital etched, ×400.

The matrix of ADI irons will be a mixture of acicular ferrite and high-carbon austenite with a minimum amount of pearlite. It is used where high strength and ductility are required, or where very high strength and good wear resistance is necessary, depending on the grade desired. (Complete data is not available as of publication.)

TYPICAL MECHANICAL PROPERTIES

- Tensile Strength Range: 125,000 to over 230,000 psi
- Yield Strength Range: 80,000–185,000 psi
- Elongation Range: 16%–0%
- Compressive Strength Range
    —accurate data is not available
- Torsional Strength Range: 0.8–0.9 (approx.) × tensile strength
- Shear Strength Range: 0.8–0.9 (approx.) × tensile strength
- Torsional Strength Range: use the same values as for shear strength
- Modulus of Elasticity: 24,000,000–25,500,000
- Modulus of Rigidity Range:
    —accurate data is not available
- Poisson's Ratio: 0.280
- Fatigue Limit Range: Unnotched 50,000–65,000 psi
- Hardness Range: 270–550 Brinell
- Specific Gravity: 7.2
- Density: 0.25–0.26 lbs/in^3

IMPACT PROPERTIES (CHARPY)

- Ductile to Brittle Transition Temperatures
    —accurate data is not available
- Impact Values at 72F
    unnotched: –120 to 0 ft-lb (Highest values are for the lowest strength grades.)

COEFFICIENT OF THERMAL EXPANSION

    —accurate data is not available

THERMAL CONDUCTIVITY

    —accurate data is not available

DESIGN STRESS

    —accurate data is not available

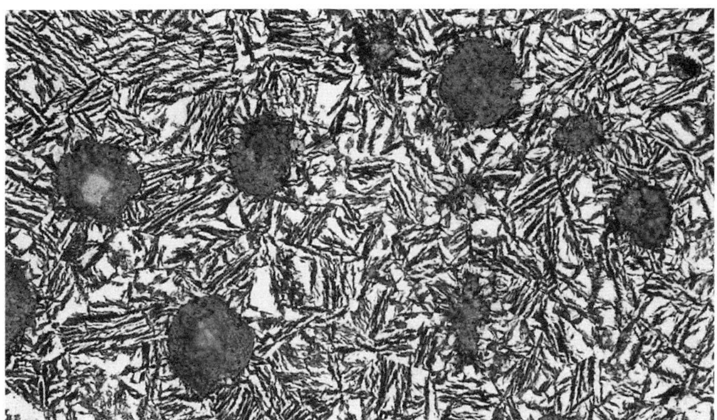

Fig. 2-7i. Grade 125-80-10 austempered ductile iron: T.S. = 145,800 psi; Y.S. = 110,000 psi; elongation = 11%; BHN = 302; 2% nital etched, ×400.

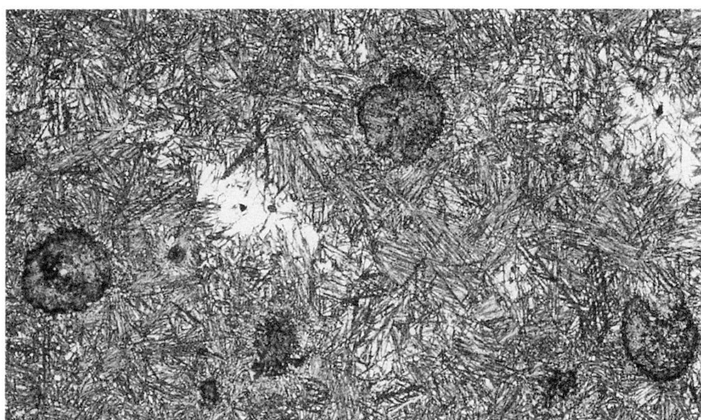

Fig. 2-7j. Grade 230-185-_ austempered ductile iron (tempered martensite): T.S. = 240,000 psi; Y.S. = 200,000 psi; elongation = 1.0%; BHN = 480; 2% nital etched, ×400.

SECTION III: NI-RESIST AUSTENITIC DUCTILE CAST IRON

■ INTRODUCTION

Austenitic ductile irons are a series of nickel-bearing cast irons that contain from 18–36% nickel, and have been treated with magnesium to bring about the formation of nodular graphite. These irons have tensile strengths ranging from 55,000 psi to 80,000 psi and elongations from 4-40%. Castings made from this group of alloys have the following special qualities:

- high resistance to erosion and corrosion;
- high resistance to heat and oxidation;
- low thermal sensitivity;
- nonmagnetic response;
- good machinability;
- good castability.

In addition, high strength and ductility are available over a wide temperature range.

These high-nickel alloyed ductile irons are made in a number of different compositions to produce the desired properties. While conventional foundry practices are used in the production of Ni-resist ductile iron castings, special precautions not normally used must be taken. Treating practices, pouring temperatures and gating practice must be modified considerably from those used in conventional ductile iron production. For this reason, design engineers and Ni-resist ductile iron producers should review proposed casting designs if minimum cost and maximum product reliability are to be obtained.

■ MECHANICAL AND PHYSICAL PROPERTIES

Ni-resist ductile irons are the high-alloy counterpart of ductile iron, which contains sufficient magnesium to change the graphite from conventional flake graphite to the spheroidal form. This change in graphite form minimizes the notch effect of graphite and thus permits strength and ductility of the austenitic matrix to prevail. The microstructure of this iron is shown in Figure 2-8.

Alloy irons that contain appreciable amounts of chromium and molybdenum contain carbides as shown in Figure 2-9. These carbides tend to increase tensile properties at the expense of some ductility. Those Ni-resist irons that are chromium free tend to be more ductile. Table 2-10 shows composition and mechanical property requirements of various types of Ni-resist ductile irons in the as-cast condition from ASTM A439. Table 2-11 shows composition and mechanical property requirements for castings used to contain pressure suitable for low temperature service, from ASTM A571.

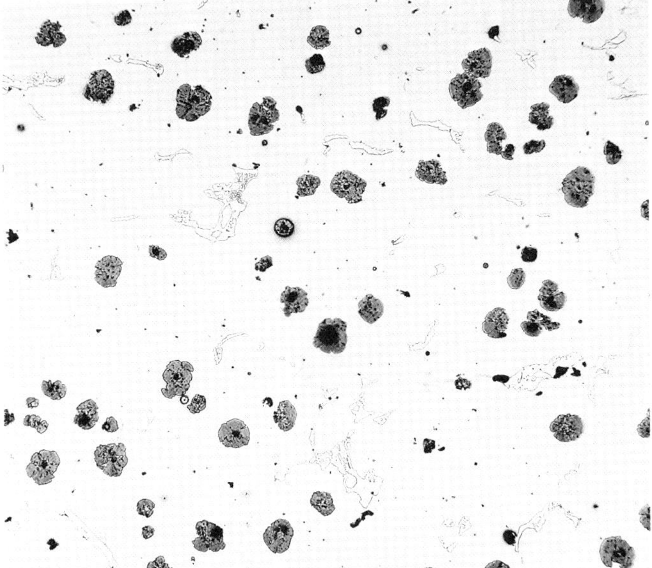

Fig. 2-8. Ni-resist D-2C containing 0.28% chromium (×250; etched).

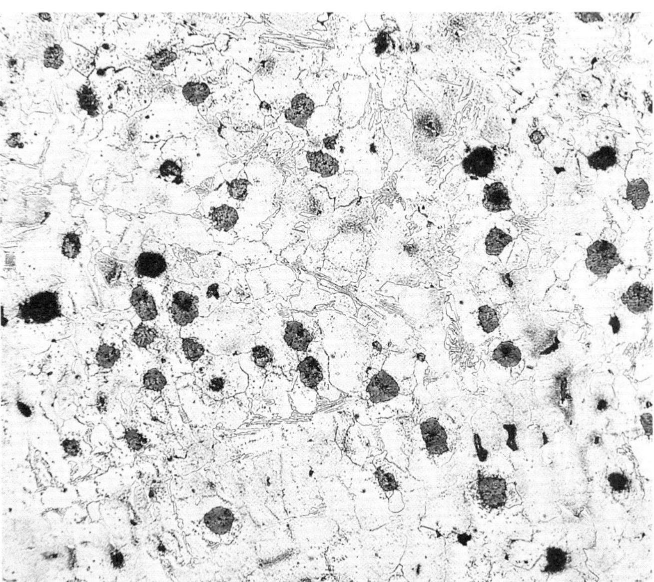

Fig. 2-9. Ni-resist D-2 containing 2.2% chromium (×250; etched).

For special service requirements the castings may be defined in more detail than in one of the accepted specifications. Such cases should be handled as separate and special agreements between the buyer and producer.

TENSILE STRENGTH

The tensile strength is essentially the same for all types of sand-cast Ni-resist ductile irons. The austenitic matrix, which is common to all the types, determines the level of strength. It is possible, however, to improve the tensile strength by dispersing the carbides with a high-temperature heat treatment. Table 2-12 shows the effect of such carbide dispersal. The addition of molybdenum will raise the room temperature tensile values slightly and will have a significant effect above 800F (425C).

With a quench from 1700–1850F (927–1010C) in oil or water, the tensile strength of Ni-resist ductile iron castings can be raised 10,000 to 15,000 psi.

YIELD STRENGTH

The yield strength is also about the same for all types of Ni-resist ductile iron. The 0.2% offset values of Table 2-11 tend to be slightly higher as the chromium increases. The yield strength may also be raised with additions of molybdenum. (See Table 2-13.) Yield strengths can be raised slightly by using the carbide dispersion heat treatments or by quenching from 1700–1850F (927–1010C). (See Table 2-12.)

ELASTICITY

A stress-stain curve of the Ni-resist ductile alloys shows a relatively straight line up to the proportional limit similar to that of steels. Practical values of proportional limit and modulus of elasticity can therefore be measured on these alloys.

The proportional limit of the Ni-resist ductile irons varies appreciably in the as-cast condition. Values from 10,000–19,000 psi are obtainable on sand castings. The highest values are from the higher chromium and molybdenum-containing alloys, and from those with the most rapid cooling rates. By quenching from 1700–1850F (927–1010C) in oil or water, the proportional limit can be increased to about 25,000–30,000 psi. (See Table 2-14.)

The moduli of elasticity of Ni-resist ductile irons do not show any appreciable increase (Table 2-14) over that of the flake graphite types of Ni-resist irons, due to the dominant influence of the austenitic matrix. This is in contrast to the ferritic or pearlitic ductile irons, which show an appreciable increase in the moduli values over those of the flake graphite gray irons.

	D-2	D-2B	D-2C	D-3	D-3A	D-4	D-5	D-5B	D-5S
C, MAX	3.00	3.00	2.90	2.60	2.60	2.60	2.40	2.40	2.30
Si,	1.50 – 3.00	1.50 – 3.00	1.00 – 3.00	1.00 – 2.80	5.0 – 6.0.0	1.00 – 2.80	1.00 – 2.80	1.00 – 2.80	4.90 – 5.50
Mn	0.70 – 1.25	0.70 – 1.25	1.80 – 2.40	1.00 MAX	1.00 MAX	1.00 MAX	1.00 MAX	1.00 MAX	1.00 MAX
P	0.08 MAX	0.08 MAX	0.08 MAX	0.08 MAX	0.08 MAX	0.08 MAX	0.08 MAX	0.08 MAX	0.08 MAX
Ni	18.0 – 22.0	18.0 – 22.0	21.0 – 24.0	28.0 – 32.0	28.0 – 32.0	28.0 – 32.0	34.0 – 36.0	34.0 – 36.0	34.0 – 37.0
Cr	1.75 – 2.75	2.75 – 4.00	0.50 MAX	2.50 – 3.50	1.00 – 1.50	4.50 – 5.50	0.10 MAX	2.00 – 3.00	1.75 – 2.25
TENSILE STRENGTH PSI MIN (MPa)	58,000 (400)	58,000 (400)	58,000 (400)	55,000 (379)	55,000 (379)	60,000 (414)	55,000 (379)	55,000 (379)	65,000 (449)
*YIELD STRENGTH PSI. MIN (MPa)	30,000 (207)	30,000 (207)	28,000 (193)	30,000 (207)	30,000 (207)	–	30,000 (207)	30,000 (207)	30,000 (207)
**ELONGATION % MIN.	8.0	7.0	20.0	6.0	10.0		20.0	6.0	10.0
BRINELL HARDNESS (3000 KG)	139–202	148–211	121–171	139–202	131–193	202–273	131–185	139–193	131–193

* YIELD STRENGTH – 0.2 PERCENT OFFSET METHOD   ** ELONGATION IN 2 INCHES (50 MM) GAGE LENTH

TABLE 2-10. STANDARD SPECIFICATION FOR AUSTENITIC DUCTILE IRONS

(From ASTM A439-83, reprinted with permission, from the *Annual Book of ASTM Standards,* copyright, American Society for Testing and Materials.)

CHEM	TYPE D-2M	PROPERTY (MIN) (HEAT TREATED)	CLASS 1	CLASS 2
C*	2.2 – 2.70	TENSILE (KSI)	65	60
Si	1.5 – 2.50			
Mn	3.75 – 4.5	*YIELD	30	25
P	0.008 MAX			
Ni	21.0 – 24.0	ELONGATION (%)	30	25
CR*	0.20 MAX			
		BHN HARDNESS	121 – 171	111 – 171
*C MAX BE ADJUSTED TO 2.90% FOR CASTINGS WITH SECTIONS UNDER 1/4 INCH		CHARPY V-NOTCH (FT-LBS.		
		– MIN. AVG. 3 TESTS	15	20
**NOT INTENTIONALLY ADDED		– MIN. AVG. INDIVIDUAL TEST**	12	15

* YIELD STRENGTH – 0.2 PERCENT OFFSET METHOD

** NOT MORE THAN ONE TEST IN A SET OF THREE MAY BE BELOW THE MIN. AVG. REQUIRED FOR THE SET OF THREE.

TABLE 2-11. STANDARD SPECIFICATIONS FOR AUSTENITIC DUCTILE IRON CASTINGS FOR PRESSURE CONTAINING PARTS SUITABLE FOR LOW-TEMPERATURE SERVICE

(From ASTM A571-84, reprinted with permission, from the *Annual Book of ASTM Standards,* copyright, American Society for Testing and Materials.)

PROPERTIES	SAND CAST	CAST IN METAL MOLD REHEATED TO 1850°F AND FURNACE COOLED TO 1800°F WATER QUENCH	CAST IN METAL MOLD REHEATED TO 1850°F AND COOLED IN AIR	CAST IN SHELL MOLD REHEATED TO 1850°F AND FURNACE COOLED TO 1800°F WATER QUENCH
TENSILE STRENGTH, PSI	55,000–69,000	76,500	75,900–83,500	75,000
YIELD STRENGTH, PSI 0.2% OFFSET	32,000–36,000	43,500	42,500–46,200	39,000
PROPORTIONAL LIMIT, PSI	16,500–18,500	23,000	27,500–31,000	24,000
ELONGATION, % IN 2 IN	8–20	9.5	13.5–30	13
BRINELL HARDNESS	140–180	179	165–190	170

TABLE 2-12. EFFECT OF FINE CARBIDES AND HEAT TREATMENT ON STRENGTH AND TOUGHNESS OF TYPE D-2 NI-RESIST

ELONGATION

The ductility of Ni-resist iron castings can be varied considerably depending upon the chromium, molybdenum and/or silicon content. In the chromium-free types, D-2C and D-5, the elongation, as cast, is 25–40%. This affords the high level of weldability, impact, etc., usually associated with elongations of this magnitude.

Types D-2, D-2B, D-3, D-5B and D-5S, containing nominally 1.75–3.5% chromium, have as-cast elongation values in the range of 6–20%.

Annealing generally has only a slight effect on increasing the elongation. When the carbides are finely divided by annealing, the elongation values are generally increased by an additional 5–10%.

IMPACT

The various Ni-resist ductile alloys show impact values as measured by conventional Charpy tests to be intermediate between that of cast iron and steel. The impact values at room temperature for the chromium-free types are about double those for the chromium containing types. Also, because of the austenitic matrix, there is no sharp embrittlement at sub-zero temperatures. As the temperature falls below room temperature to –320F (–196C), the impact of the various types changes at different rates. (See Table 2-15.)

WEAR AND GALLING RESISTANCE

The presence of dispersed graphite, as well as the work hardening character of Ni-resist ductile castings affords a high level of resistance to frictional wear and galling. Types D-2, D-3C, D-3A and D-4 offer good wear properties, not only at room temperature and with a wide variety of other

TABLE 2-13. ELEVATED TEMPERATURE PROPERTIES OF NI-RESIST DUCTILE IRONS

	TYPE D-2	TYPE D-2 WITH MO	TYPE D-2C	TYPE D-3	TYPE D-3 WITH MO	TYPE D-4	TYPE D-4 WITH MO	TYPE D-5B	TYPE D-5B WITH MO	TYPE D-5S
TENSILE STRENGTH, PSI										
ROOM TEMP	59,400	61,500	62,200	58,300	61,000	64,000	60,500	60,800	61,200	70,400
800°F (427°C)	54,000	–	52,400	–	–	–	–	–	–	60,600
1000°F (538°C)	47,700	37,000	42,000	48,000	46,800	60,700	54,500	47,200	48,800	56,300
1200°F (649°C)	35,600	38,500	28,000	41,700	44,000	48,000	45,500	40,600	46,400	38,700
1400°F (706°C)	22,100	24,900	17,200	26,500	28,800	21,800	22,300	24,900	31,100	17,300
YIELD STRENGTH, PSI										
ROOM TEMP	35,000	37,500	34,100	39,300	40,000	44,300	43,000	41,000	41,000	34,100
800°F (427°C)	28,000	–	26,200	–	–	–	–	–	–	26,800
1000°F (538°C)	28,000	29,200	23,100	28,400	28,500	41,400	38,400	25,800	28,600	26,900
1200°F (649°C)	25,600	25,200	24,200	27,400	29,000	34,000	35,500	24,200	29,900	25,300
1400°F (706°C)	17,000	17,100	16,600	15,200	21,800	18,500	18,600	18,600	24,300	9,700
IMPACT STRENGTH (UNNOTCHED CHARPY FT-LBS)										
ROOM TEMP	26	23*	–	–	–	–	–	–	–	–
1000°F (538°C)	35	21*	–	–	–	–	–	–	–	–
1200°F (649°C)	32	24*	–	–	–	–	–	–	–	–
1500°F (816°C)	35	27*	–	–	–	–	–	–	–	–
STRESS RUPTURE (1000 HRS), PSI										
1000°F (538°C)	28,000	–	–	–	–	–	–	–	–	--
1100°F (593°C)	(18,000)	26,000	(13,500)	23,500	30,500	17,000	19,000	25,000	32,000	--
1200°F (649°C)	12,000	(15,500)	9,000	(15,000)	(18,000)	(9,500)	(11,000)	(15,000)	(18,500)	--
1300°F (704°C)	(8,500)	11,000	(6,000)	9,700	11,000	6,200	7,500	10,000	12,000	--
1400°F (706°C)	(5,500)	(6,000)	(4,000)	(6,000)	(7,000)	(3,000)	(4,000)	(5,500)	(6,500)	3,600
CREEP DATA (0.0001%/HR), PSI										
1000°F (538°C)	23,000	(27,000)	13,000	–	(29,000)	–	–	(27,000)	(28,000)	–
1100°F (593°C)	(13,100)	16,000	(9,000)	–	18,000	–	–	(16,000)	17,000	–
1200°F (649°C)	8,000	(9,600	5,700	–	(12,000)	–	–	(9,600)	(10,500)	–
1300°F (649°C)	(4,800)	5,600	(3,400)	–	7,000	–	–	8,000	(6,000)	–
(0.00001%/HR)										
1000°F (538°C)	9,000	(18,000)	–	–	(21,000)	–	–	–	(18,000)	–
1100°F (593°C)	(5,600)	11,000	–	–	(8,000)	–	–	10,000	11,000	–
1200°F (649°C)	3,400	(6,000)	–	–	(8,000)	–	–	–	(6,700)	–
1300°F (649°C)	(2,100)	3,600	–	–	5,000	–	–	5,600	4,000	–
ELONGATION (%) STRESS RUPTURE										
1000°F (538°C)	6	–	14	–	–	–	–	–	–	–
1100°F (593°C)	–	5.5	–	7	5	10.5	10	6.5	5.5	–
1200°F (649°C)	13	–	13	–	–	–	–	–	–	–
1300°F (704°C)	–	11.5	–	12.5	16	25	21	13.5	11.5	–
SHORT TIME TENSILE										
ROOM TEMP	10.5	8.5	35	7.5	7	3.5	2.5	7	7.5	–
800°F (427°C)	12	–	23	–	–	–	–	–	–	–
1000°F (538°C)	10.5	1.5	19	7.5	7	4	3.5	9	7.5	–
1200°F (649°C)	10.5	3	10	7	4	11	8	6.5	6.5	–
1400°F (760°C)	15	14.5	13	18	13	30	24.5	24.5	12.5	–

*Type D-2B has no Mo. (Values in parentheses are extrapolated or interpolated.)

TABLE 2-14. PHYSICAL AND MECHANICAL PROPERTIES OF NI-RESIST DUCTILE IRON

	TYPE D-2	TYPE D-2B	TYPE D-2C	TYPE D-3	TYPE D-3A	TYPE D-4	TYPE D-5	TYPE D-5B	TYPE D-5S
SPECIFIC GRAVITY	7.41	7.45	7.41	7.45	7.45	7.45	7.68	7.72	6.96
DENSITY, LBS. PER CU. IN	.268	.27	.268	.27	.27	.27	.278	.279	.252
MELTING POINT °F	2250	2300	2250	2250	2250	2200	2250	2250	2160
THERMAL EXPANSION 70–400°F, MILLIONTHS PER °F	10.4	10.4	10.2	7.0	–	8.0	3.9	–	6.84
THERMAL CONDUCTIVITY CAL/CC/SEC/°C	.032	–	–	–	–	–	–	–	–
ELECTRICAL RESISTIVITY MICROHMS PER CC	102	–	–	–	–	–	–	–	–
MAGNETIC RESPONSE (ROOM TEMPERATURE)	NON-MAGNETIC	SLIGHTLY MAGNETIC	NON-MAGNETIC	DEFINITELY MAGNETIC	DEFINITELY MAGNETIC	SLIGHTLY MAGNETIC	DEFINITELY MAGNETIC	DEFINITELY MAGNETIC	DEFINITELY MAGNETIC
MAGNETIC PERMEABILITY H-300 AT ROOM TEMP	1.02–1.04	1.04–1.08	1.93	–	–	1.0	–	–	–
PROPORTIONAL LIMIT, PSI	16,500 18,500	16,000 19,000	12,000 16,000	16,000 19,000	15,000 19,000	12,000 16,000	9,500 11,000	10,500 13,000	15,000 20,000
MODULUS OF ELASTICITY, PSI X 10^6	16.5 18.5	16.5 19.0	15.0 –	13.5 14.5	16.0 18.5	13.0 –	16.0 20.0	16.0 17.5	14.6 –

TABLE 2-15. TYPICAL CHARPY V-NOTCH IMPACT PROPERTIES (FT-LB)

TEMPERATURE	TYPE D-2	TYPE D-2B	TYPE D-2C	TYPE D-3	TYPE D-3A	TYPE D-4	TYPE D-5	TYPE D-5B	TYPE D-5S
68°F (20°C)	12.0	10.0	28.0	7.0	14.0	—	17.0	6.0	—
0°F (– 18°C)	11.5	—	15.0	—	14.0	—	15.0	5.5	—
-100°F (– 73°C)	10.0	—	6.5	6.5	13.0	—	14.0	4.5	—
-320°F (–196°C)	4.5	—	4.0	3.5	7.5	—	11.0	4.0	—

metals, but also at subzero and elevated temperatures to 1500F (816C). Types D-2B and D-3 are not recommended for maximum frictional wear resistance because the structure contains fairly massive carbides that are likely to abrade the mating metal.

Johnson, et. al.,[1] report that both Types D-2 and D-2C showed the lowest rate of wear from room temperature up to 1000F (538C), compared with bronze, ductile iron and Inconel™ nickel-chromium alloy. They attributed the resistance to wear at low temperatures to the graphite-spheroid distribution, and at high temperatures to the nickel oxide film.

When wearing-in these alloys, it is recommended that a lubricating film such as molybdenum sulfide be used. With this procedure, a "glazed" surface results, which has been work hardened and is ideal for metal-to-metal service against itself, or other corrosion-resisting metals such as stainless steels, Monel™, nickel-copper alloy, etc. Ni-resist ductile castings are useful where the service demands resistance to corrosion and rubbing wear.

™Trade Mark Reg. International Nickel Co., Inc.

■ EROSION RESISTANCE

GENERAL

In service requiring resistance to erosion and corrosion, Ni-resist ductile iron castings offer excellent service possibilities, particularly in those types containing 3.0–5% Cr (Types D-2B, D-3, and D-4). In handling wet steam, salt slurries and relatively high-velocity corrosive liquids, all of the types containing 2% or more chromium have shown superior service.

Erosion and corrosion testing of turbine materials in wet oxygenated steam shows that a cobalt-base alloy and Type D-3 Ni-resist ductile irons were the most resistant of all the metals tested in wet steams at 300–340F (149–171C) at a pH of 7, with both low and high oxygen contents.

CAVITATION EROSION

In resisting cavitation erosion, the strength level of these alloys makes them particularly useful for pump impellers or propellers for small harbor and coastal boats. The higher chromium types (D-2B, D-3 and D-4) are recommended for severe cavitation erosion resistance. Service results reported from pump impellers and harbor tugboat propellers show Type D-2 to be superior in resisting cavitation compared to the straight chromium stainless steels or bronzes previously used in these applications.

CORROSION RESISTANCE

GENERAL

A number of corrosion tests have been completed comparing Ni-resist ductile austenitic irons with conventional flake graphite types of Ni-resist irons. Additional tests are continually being conducted. From the data already in, it appears that the corrosion resistance of the Ni-resist ductile austenitic irons is of the same magnitude as that of corresponding types of conventional Ni-resist irons. Table 2-16 shows the data from several environments such as acids, alkalies and salt waters. It should be remembered that in most corrosive media it is desirable to alloy with 2% or more chromium. Therefore, Types D-2C, D-3A and D-5 are not usually recommended for applications where a high level of corrosion resistance is desired.

Until more complete data is available, we may assume that the corrosion rates of these ductile iron types will exhibit about the same magnitude as the corresponding types of flake graphite Ni-resist irons. The reader is urged, therefore, to refer to the International Nickel Company's bulletin "Engineering Properties and Application of Ni-resist" where corrosion data for over 400 environments are listed.

GRAPHITE CORROSION

In environments where non- or low-alloyed cast irons corrode leaving a graphite layer on the surface as a black coating, graphitic corrosion occurs, which further increases the corrosion rate of the base metal. The Ni-resist ductile alloys do not develop this graphite layer to any appreciable extent in most media because of their inherent resistance to corrosion and the reduced amount of free carbon as graphite. Thus, the use of Ni-resist ductile irons offers marked improvement in environments where graphite corrosion may be a problem.

GALVANIC CORROSION

The relatively high voltage potential between cast iron and bronze (e.g., a cast iron pump or valve body with bronze parts in touch with the body) can cause excessive corrosion of the cast iron. Because the Ni-resist alloys have only about half this voltage potential when coupled with cast iron, the galvanic corrosion rate on the cast iron is reduced to as little as one-tenth that of the cast-bronze combination. In pumps, Ni-resist valve stem bushings have been used in place of bronze to prevent excessive corrosion of the cast iron or steel yoke.

HIGH TEMPERATURE PROPERTIES

HEAT RESISTANCE AND OXIDATION

The chromium-containing types of Ni-resist ductile irons (Types D-2, D-2B, D-3, D-4, D-5B and D-5S) resist oxidation and maintain satisfactory mechanical properties to about 1400F (760C). These characteristics have led to applications such as furnace parts, turbocharger and gas turbine parts, engine exhaust lines, and valve guides. If service temperatures in excess of 1300F (704C) are contemplated, Types D-2B, D-3, D-4 and D-5S should be considered. The higher chromium content of these types improves the resistance to oxidation effects.

Table 2-17 shows some oxidation data of some types of these alloys compared with conventional Ni-resist irons and unalloyed ductile iron. One of the useful features of the high nickel irons at high temperatures is the adherence of the oxide scale to the casting. For those engine parts in contact with combustion and exhaust gases—turbochargers in particular—the Ni-resist ductile irons minimize abrasion from any free scale particles.

Type D-4 has a high order of oxidation resistance, and if high toughness is not required, it will be found economical, at least to 1500F (816C). In service where appreciable sulfur is present (e.g., fuels containing 1% or more sulfur) the maximum service temperature is about 1000F (538C). Type D-4, with higher silicon and chromium, will withstand the presence of sulfur and higher temperatures quite well.

A more recent development, D-5S (also called In-854) has provided a material with exceptional dimensional stability and oxidation resistance. When these two characteristics are desired at temperatures to 1600F (871C), Type D-5S is recommended. The improved properties are achieved by maintaining the silicon content at about 5.5%.

MECHANICAL PROPERTIES

The high-temperature mechanical properties of the various Ni-resist ductile alloys are listed in Table 2-13.

Table 2-16. Corrosion Resistance of Ni-Resist Austenitic Ductile Irons
(In Inches Penetration per Year [IPY])

Reactive Agent	Constituents, Exposure and Duration	Ductile Type D-2C	Ductile Type D-2	Flake Graphite Type 2 (Ni–Resist)
Ammonium chloride sol.	10% NH_4Cl, pH 5.15, 13 days at 86F(30C), 6.25 ft/min	0.0280	0.0168	—
Amonium sulfate sol.	10% $(NH_4)_2SO_4$, pH 5.7, 15 days at 86F (30C), 6.25Ft/min	0.0128	0.0111	0.0095
Ethylene vapors and splash	38% ethylene gylcol, 50% diethylene glycol 4.5% H_2O, 4%Na_2SO_4, 2.7% NaCl, 0.8% Na_2CO_3 + trace NaOH, pH 8 to 9, 85 days at 275–300F (135–149C)	0.0023	0.0019*	0.0013
Fertilizer	commercial "5–10–5", damp, 290 days at atmospheric temperature	—	0.0012	0.0025
Nickel chloride solution	15% $NiCl_2$, pH 5.3, 7 days at 86F(30C) 6.25 ft/min	0.0062	0.0040	0.0040
Phosphoric acid	85% aerated at 30C (86F), velocity 16 ft/min,12 days	0.213	0.235	0.087
Raw sodium chloride brine	300 gpl of chlorides, 2.7 gpl CaO 0.06 gpl NaOH, traces of NH_3 + H_2S, pH 6–6.5, 61 days at 50F, 1–2 fps	0.0023	0.0020*	0.0020
Sea water	at 26.6C (86F), velocity 27 ft/sec, 60 day test	0.039	0.018	0.016
Soda and brine	15.5% NaCl, 9.0% NaOH, 1.0% $NaSO_4$, 32 days at 180F (82C)	0.0028	0.0015	0.0025
Sodium bisulfate sol.	10% $NaHSO_4$, pH 1.3, 13 days at 86F(30C), 6.25 ft/min	0.0431	0.0444	0.0612
Sodium chloride sol.	5% NaCl, pH 5.6, 7 days at 86F(30C), 6.25 ft/min	0.0028	0.0019	0.0021
Sodium hydroxide	50% NaOH + heavy conc. of suspended NaCl, 173 days at 55C (131F), 40 gal/min	0.0002	0.0002	0.00018
	50% NaOH, 10 days at 260F, 4 days at 70F (21C)	0.0048	0.0049	0.0046
	30% NaOH + heavy conc. of suspended NaCl, 82 days at 185F(85C)	0.0004	0.0005	0.0004
	74% NaOH, 19–3/4 days at 260F(127C)	0.005	0.0056	0.006
Sodium sulfate sol.	10% $NaSO_4$, pH 4.0, 7 days at 86F(30C), 6.25 ft/min	0.0136	0.0130	0.0132
Sulfuric acid	5% at 30C (86F), aerated, velocity 14 ft/min, 4 days	0.120	0.104	0.112
Synthesis of sodium + sodium bicarbonate by Solvay process	44% solid $NaHCO_3$, slurry plus 200 gpl NH_4CL, 100 gpl NH_4HCO_3, 80 gpl NaCl, 8 gpl $NaHCO_3$, 40 gpl CO_2, 64 days at 86F(30C)	0.0009	0.0003	0.0006
Tap water	aerated at 30F (–1C), velocity of 16 ft/min, for 28 days	0.0015	0.0023	0.0045
Vapor above ammonia liquor	40% NH_3, 9% CO_2, 51% H_2O, 109 days at 185F(85C), low velocity	0.011	0.025	0.017
Zinc chloride sol.	20% $ZnCl$, pH 5.25, 13 days at 86F(30C), 6.25 ft/min	0.0125	0.0064	—

*Contains 1% Cr.

Included are 1000 hour, stress-rupture values with resulting elongations, short-time tensile, yield, elongation values, creep stress data, and elevated temperature impact strengths. In addition, Figure 2-10 illustrates, graphically, the short-time, high temperature tensile properties of Type D-2.

The stress rupture cures for Types D-2, D-3 and D-5B, with and without molybdenum, are show in Figures 2-11, 2-12 and 2-13, along with the stress rupture values of cast HF grade stainless steel (19 Cr-9Ni). Some limited creep data are graphically shown in Figure 2-14, with a comparison to cast CF-4 grade stainless steel (18 Cr-8 Ni). The Rockwell A hot hardness values of Types D-2, D-3, D-4, and D-5B, with and without molybdenums, are illustrated in Figure 2-15. The standard conversions to Brinell Hardness Numbers are listed in parentheses.

It is important to note that the addition of approximately 1% molybdenum to Ni-resist ductile iron raises the evaluated temperature properties with only slight accompanying reductions in elongation, in almost all instances. The stress rupture and creep properties benefit the most from the molybdenum additions (see Table 2-13 and Figures 2-11 to 2-14) such that the resultant alloy is equal to or better than the cast stainless steel grades HF and CF-4. The addition of molybdenum also raises the cast hardness, except in Type 4, and maintains it at elevated temperatures. (See Figure 2-15.)

RESISTANCE TO CRACKING AND WARPING

In cycle heating and cooling service at 1250F (677C) or above, ferritic irons and steels encounter a "critical range" that frequently results in warping or cracking in the casting. This is because of the sudden volume changes that occur while passing through this critical range. The Ni-resist alloys, being austenitic, do not have a critical range; therefore no sudden volume change occurs as the metal is heated or cooled through these temperature ranges. This fact, together with their high order of mechanical properties at 1100–1400F (593–760C), and moderate moduli of elasticity, explain why the Ni-resist ductile irons have excellent resistance to warping or cracking.

Where ductile Ni-resist is to be used for either static or cyclic elevated temperature service at 900F (482C) or above, a stabilizing heat treatment should be used to prevent any growth and warpage during the service. This heat treatment consists of heating to 1600F (871C) and holding for a minimum of 2 hours, followed by furnace cooling to 1000F (538C) and then air cooling. This treatment should be done prior to final machining.

STEAM SERVICE

As previously mentioned, for steam service requiring

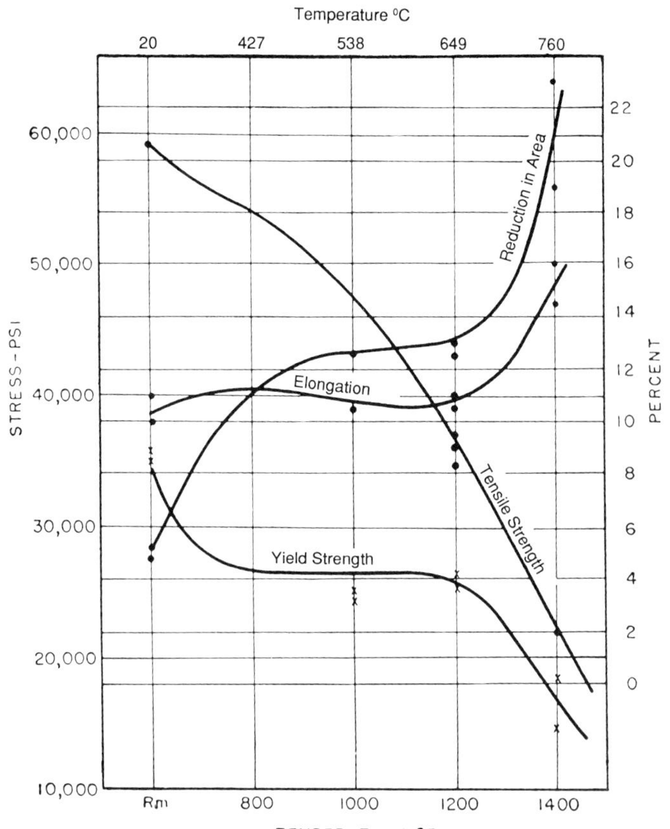

Fig. 2-10. Properties of Type D-2 at elevated temperatures.

TABLE 2-17. OXIDATION RESISTANCE

	INCHES PENETRATION PER YEAR	
	TEST 1	TEST 2
DUCTILE IRON (2.5 Si)	.042	.50
DUCTILE IRON (5.5 Si)	.004	.051
NI RESIST DUCTILE IRON TYPE D-2	.042	.175
NI RESIST DUCTILE IRON TYPE D-2C	.07	–
NI RESIST DUCTILE IRON TYPE D-4	.004	0.0
CONVENTIONAL NI RESIST TYPE 2	.098	.30
TYPE 309 STAINLESS STEEL	0.0	0.0

TEST 1 FURNACE ATMOSPHERE - AIR, 4000 HRS AT 1300°F (704°C)

TEST 2 FURNACE ATMOSPHERE - AIR, 600 HRS. AT 1600-1700°F (871-927°C), 600 HOURS BETWEEN 1600-1700°F, AND 800-900°F, 600 HOURS AT 800-900°F (427-483°C)

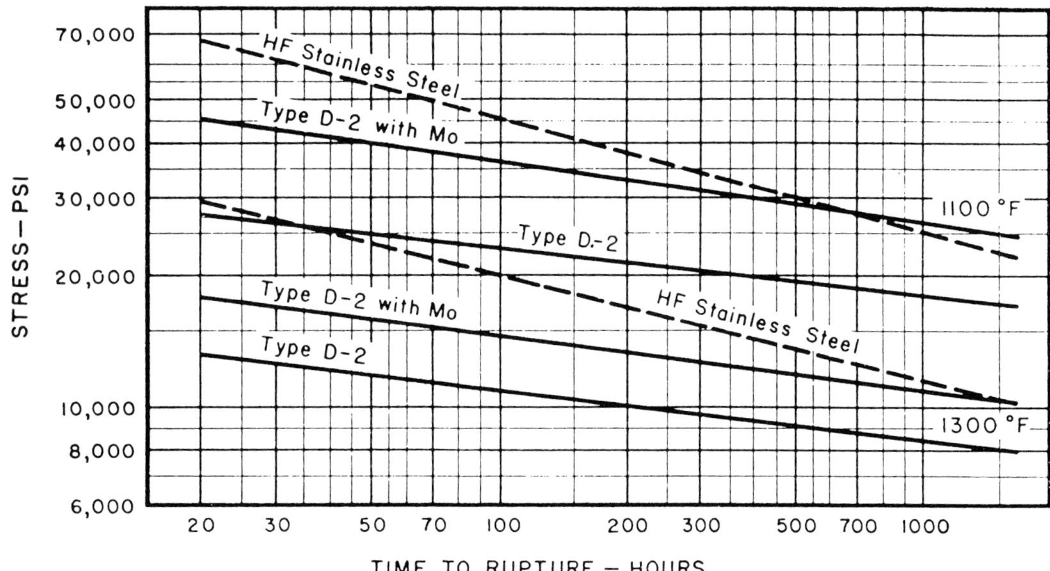

Fig. 2-11. Stress-rupture data for Type D-2.

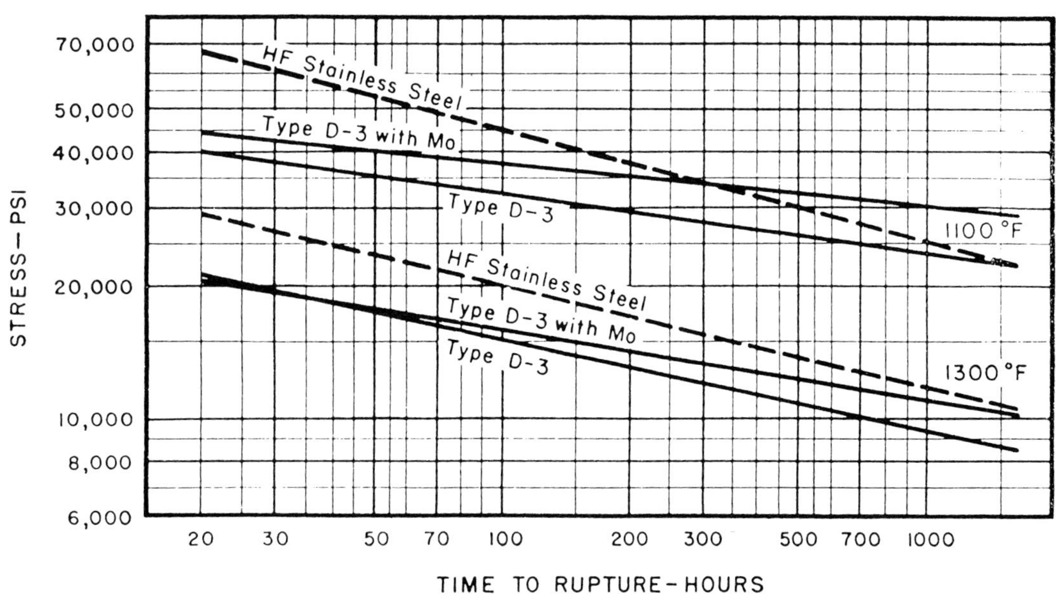

Fig. 2-12. Stress-rupture data for Type D-3.

resistance to wet steam erosion at lower temperatures, a cobalt-base alloy and Type D-3 were the most resistant of the metals tested. At the higher steam temperatures, Types D-2 and D-3 offer excellent resistance to growth and scaling. Applications include steam turbine components such as diaphragms, shaft seals and control valves. Table 2-13 and Figure 2-14 illustrate the creep properties, and Table 2-18 lists data on the growth of Ni-resist in steam as compared to gray iron.

THERMAL EXPANSION

General Ranges

The data of Table 2-14 and Figures 2-16 and 2-17, show that the different types of Ni-resist ductile irons have different mean thermal expansion coefficients varying between about 2.5–9.6 millionths/°F between 50 and 200F (10–93C).

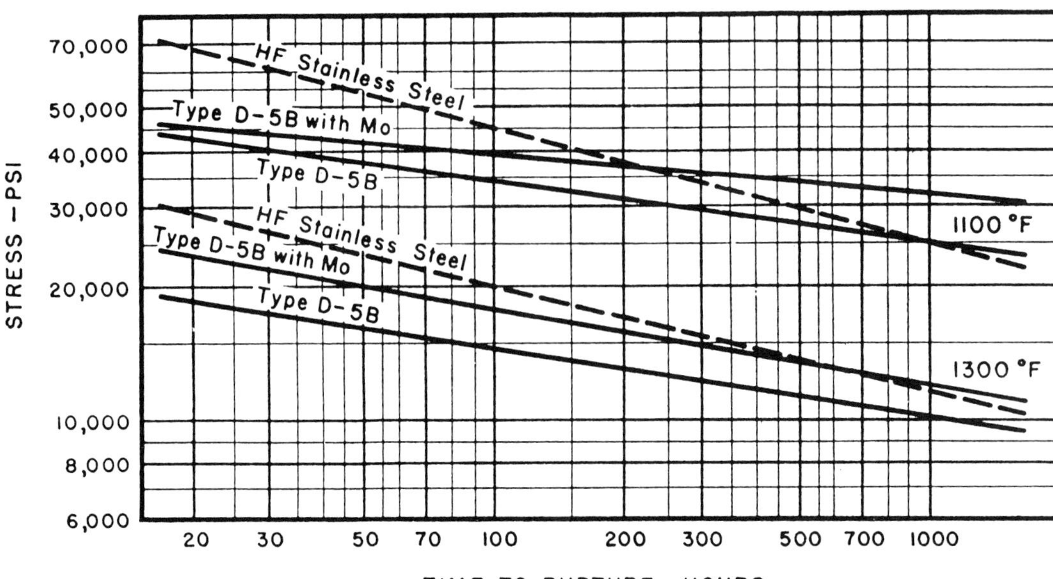

Fig. 2-13. Stress-rupture data for Type D-5B.

By selecting a suitable nickel content or a particular type of Ni-resist ductile iron, a variety of thermal expansivities are available. This wide range of thermal expansivities has been utilized in several important ways as outlined below.

HIGH EXPANSION

Type D-2 is often selected for high expansivity to be used in conjunction with metals such as aluminum, copper, bronze and austenitic stainless steels, which generally have expansivities in the range of 9–12 millionths/°F. As seen in Figure 2-16, the thermal expansion of Type D-2 lies in this range.

Type D-2 Ni-resist ring bands and piston caps in aluminum pistons are examples of this practice of matching the expansivities of dissimilar metals to prevent warping and/or failure at the interface between the two metals. Another application involves engine turbocharger diaphragms, wherein the vanes of 18%Cr-8%Ni stainless steel are

mounted in a Type D-2 Ni-resist retainer. Type D-4 has an expansivity in this range at the higher temperatures (Figure 2-16), so that it, too, has found use in the aforementioned applications. It is especially recommended where greater resistance to oxidation and scaling are needed.

CONTROLLED EXPANSION

Type D-2 Ni-resist alloy is quite often used to meet a variety of intermediate expansivities. This is illustrated by Figure 2-17, which indicates the effect of varying nickel content on the resultant expansivity. Thus, the thermal expansion can be controlled to match steels and cast irons as well as straight chromium stainless steels, nickel and Monel™ alloys.

Examples of applications include the matching of Type D-2 with 12% chromium stainless steel in throttling valves on a steam turbine and in bearing separators for high-speed unlubricated jet engine bearings.

TABLE. 2-18. GROWTH IN STEAM COMPARISON*

	AFTER 500 HRS.	AFTER 1000 HRS.	AFTER 2500 HRS.
GRAY IRON	.0023	.0052	.014
CONVENTIONAL NI RESIST TYPE 2	.0005	.0010	.0015
CONVENTIONAL NI RESIST TYPE 3	.0003	.00045	.00048
NI RESIST DUCTILE IRON TYPE D-2	.0003	.0005	.0005
NI RESIST DUCTILE IRON TYPE D-3	.0003	.0000	.0000

*Growth at 900F (482C) in inches per inch.

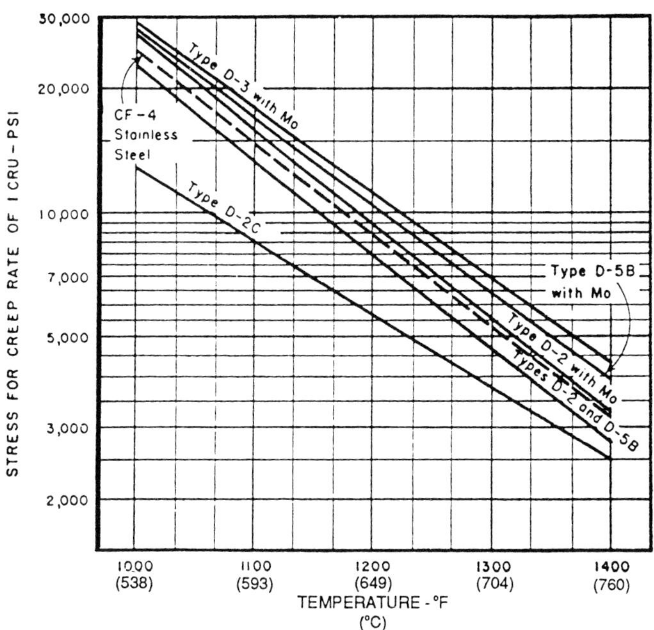

Fig. 2-14. Comparison of creep strengths of several Ni-resist ductile irons.

Minimum Expansion

Where minimum or low thermal expansion is desired for dimensional stability, Types D-5, D-5B or D-5S, are recommended. Applications include machine tool parts, glass molds and gas turbine housings. Type D-5B Ni-resist gas turbine housings are also an example of the application of these properties.

Glass molds made of Type D-5B maintain clean, smooth mold surfaces, and their dimensional stability gives true dimensions to the finished product. This, together with resistance to cracking, wear and freedom from sticking of the glass to the molds, makes the material especially suitable for applications where the volume of the glass container is important, such as milk bottle molds. The chromium-containing type, D-5B, in addition to having low thermal expansion, has good heat resistance, and is therefore well adapted for low expansion and minimum distortion at elevated temperatures.

Low thermal expansion is usually accompanied by the need for extreme dimensional stability. In this type of application, the following heat treatment is recommended: heat to 1600F (871C) and hold for at least 2 hours minimum plus 1 hour per inch of section, followed by furnace cooling at a maximum rate of 100F (38C)/hr down to 1000F (538C); hold at 1000F (538C) for 1 hr per inch of section, and then cool uniformly. After rough machining, reheat to 850–900F (470C) and hold for 1 hour per inch of section.

Effect of Thermal Expansion on Corrosion Resistance

Improved corrosion resistance in cycle temperature service can be obtained by the use of low expansivity alloys. When metals are subjected to temperature cycle service, the corrosion-protective film on the casting is often flaked off by the expansion and contraction movements. This exposes new metal to the corrosive until a new film is produced. In this way, the corrosion may proceed. However, by using the low-expansion alloys, the expansion and contraction movements are minimized, so that the protective film adheres to the metal, thus maintaining maximum protection and giving improved service life.

THERMAL SHOCK SERVICE

The thermal shock resistance of the Ni-resist ductile alloys is superior to the corresponding flake graphite types because the strength and toughness level of ductile iron is considerably higher. In many industrial processes, sudden changes in temperatures are frequently encountered, such as when cold solutions are pumped into hot equipment, or if a pump is utilized to mix hot and cold liquids; or in equipment handling for example, when caustic soda at 300F (149C) is suddenly flushed out with tap water. Under these conditions, the sudden change in temperature can cause high thermal stresses that may result in premature failure of the metal by warping and/or cracking during repeated cycling. It is recommended, therefore, that the low-expansion, Type D-3, should be used, since the stresses can be held to a safe level in most industrial services involving temperature changes up to 400F (204C).

When large equipment, such as pumps or valves, is to be subjected to sudden temperature changes up to 250F (121C) or higher, it is strongly recommended that Type D-3 be specified. In the above instances, and any others where the thermal shock service is known to be unusually severe, it is recommended that the minimally expansive Types D-5 or D-5B Ni-resist iron be considered.

As previously noted, in service involving cycle heating and cooling at 800F or above, the high mechanical properties of these high-nickel ductile irons and their moderate moduli of elasticity offer an excellent combination of properties to prevent warping and/or cracking.

THERMAL CONDUCTIVITY

The thermal conductivities of Type D-2 Ni-resist ductile iron, gray iron and some steels are listed in Table 2-19. The structure of the Ni-resist ductile alloys causes the thermal conductivity to be reduced, so that it is of the order of one-third the value of flake graphite Ni-resist and one-fourth that of gray cast iron, or slightly less than the values for the austenitic stainless steel.

LOW TEMPERATURE PROPERTIES

The austenitic metal matrix structure of Ni-resist ductile irons allows them to retain their room-temperature ductility or toughness to fairly low temperatures. Types D-2, D-3 and D-5 gradually lose impact toughness as the temperature falls. At –100F (–73C) Type D-2C has lost much of its room temperature impact toughness, and at –320F (–196C) it shows the same order of toughness as the chromium-containing types. Type D-3A maintains its room temperature impact strength to –100F (–73C) and still has half of this amount at –320F (–196C). The chromium-free Type D-5, Minovar', shows good levels of impact resistance down to –320F (–196C). The impact value of Type D-5 at –3200F (–196C), for instance, is only 6 ft-lb less than its room temperature value. (See Table 2-15.)

However, D-2M has superior cryogenic properties to any other Ni-resist ductile iron. The impact properties are 23 ft-lb at –320F (–196C) for the D-2M alloy. This is more than twice as high as any other Ni-resist ductile iron. In

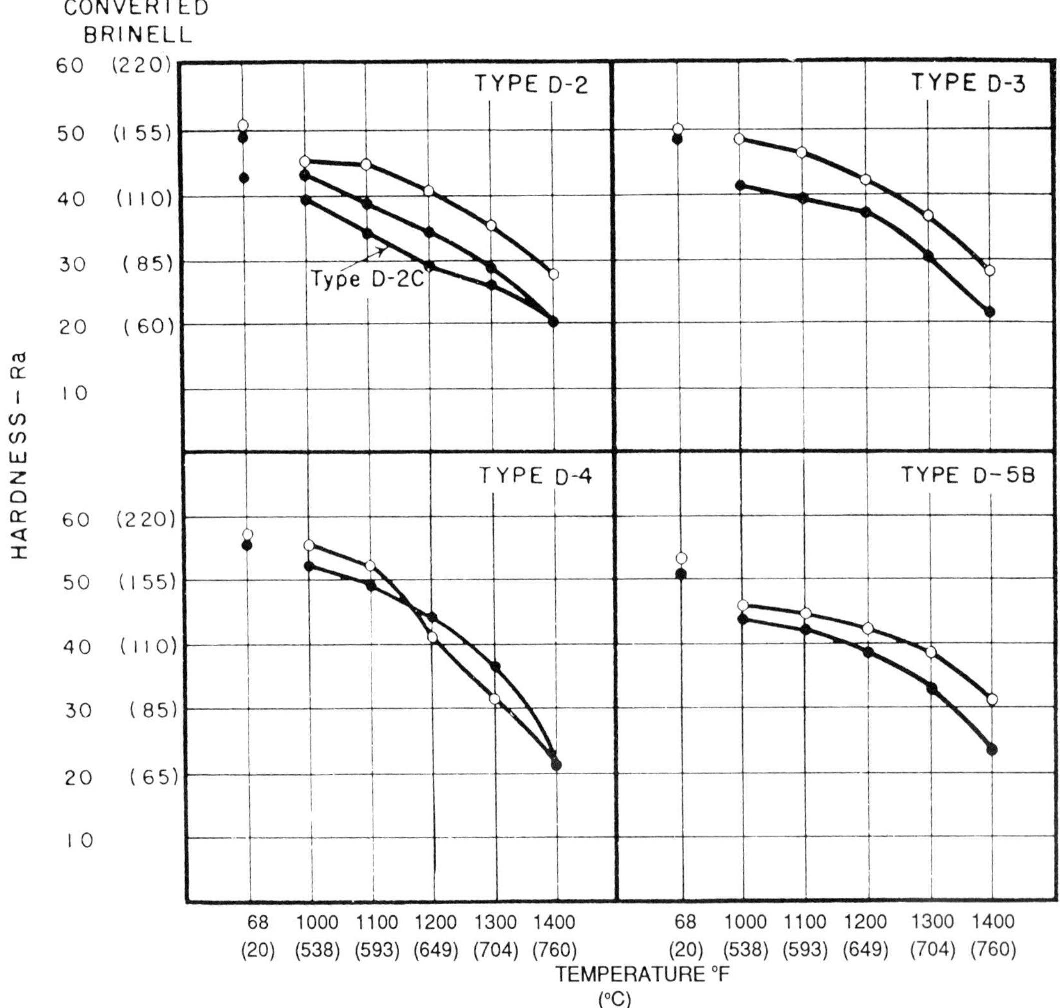

Fig. 2-15. Hot hardness of Ni-resist ductile irons. (Solid symbols represent the standard material. Open symbols represent the corresponding material containing 0.7–1% Mo.)

addition, mechanical properties increase at –320F (–196C) as follows:

Tensile	110,000 psi
Yield	68,000 psi
Elongation	25%.

ELECTRICAL AND MAGNETIC PROPERTIES

The electrical resistivity of Type D-2 (Table 2-20) is less than that of the corresponding flake graphite Ni-resist alloy, and of the order of about twice that of straight chromium stainless steels. Table 2-16 lists a few comparisons in resistivity of Type D-2 with gray iron and various steels.

The nonmagnetic character of Types D-2 and D-2C has been applied in several industrial applications where magnetic permeability must be kept at a minimum in order to prevent excessive heat generation and power losses from eddy currents. Table 2-21 indicates that the magnetic permeability of Type D-2 and D-2C are of the same order as the various stainless steels. As the chromium is increased, there is a slight increase in the magnetic response. Types D-2B and D-4, therefore, would show a slightly higher magnetic permeability, as indicated in the table, than Types D-2 and D-2C at room temperature.

It should be mentioned that, if the alloy content of nickel, chromium and manganese for Types D-2 and D-2C are permitted to go below the specified alloy minima, the magnetic response will increase. This is because below the specified alloy content the austenite stability at subzero temperature is reduced, and ferromagnetic martensite is formed. This would cause an attendant increase in hardness.

Table 2-21 lists the magnetic permeability for all Ni-resist ductile nonmagnetic alloys. Values for Types D-3, D-3A, D-5 and D-5B are not indicated, as they are ferromagnetic.

The following descriptions of the general characteristics and suggested applications for each grade of Ni-resist ductile iron can be used by foundrymen to assist customers in choosing suitable grades of Ni-resist for castings.

CHARACTERISTICS OF DUCTILE IRON NI-RESIST IRONS

D-2 (20%Ni-2%Cr) (Mil-I-18397) AMS 5394

Excellent resistance to corrosion, erosion, frictional wear. Re-

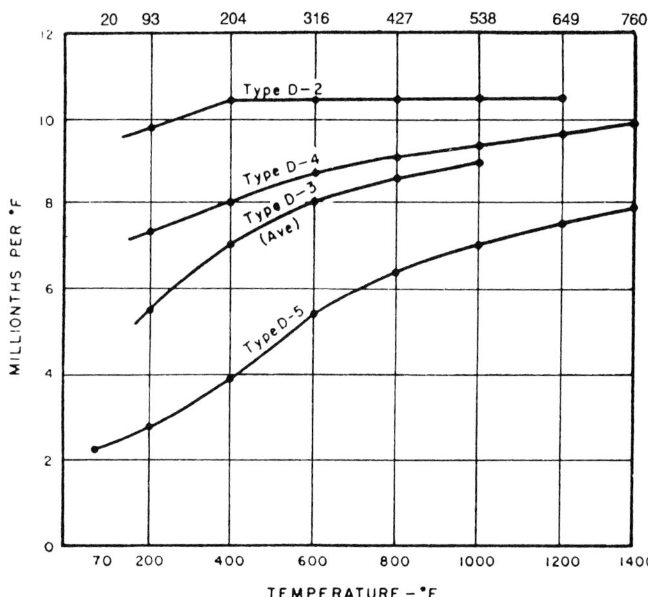

Fig. 2-16. Mean coefficient of thermal expansion of various types of Ni-resist ductile irons.

sists heat up to 1400F (760C). This type also has a high thermal expansivity and good machinability. Addition of 0.7-1.0% Mo increases mechanical properties above 800F (427C).

Recommended for valve parts, exhaust manifolds, pump bodies, and other pressure castings. Other applications are paper pulp grinding machinery, piston rings, diesel engine cylinder liners, nonmagnetic components, pump impellers for refinery service and ship castings.

D-2B (20%Ni-3%Cr)

Offers superior resistance to neutral and reducing salts, and used where higher resistance to erosion and oxidation than that afforded by Type D-2 is required. This material is slightly magnetic and is resistant to cavitation.

Recommended for pump impellers, propellers for small boats, turbocharger housings, diesel engine manifolds and high-pressure brine pump parts.

D-2C (22%N-0.50%Cr(max)) (Mil-I-18397 (ships)) AMS 5395

Used where resistance to heat and corrosion is less severe and where high ductility is required. This material is nonmagnetic.

Primarily for parts operated at temperatures up to 800F (427C) requiring an austenitic material with good castability and corrosion resistance, and capable of being welded.

D-2M (23%Ni-4.25Mn-0.50%Cr(max))

This grade is recommended for low-temperature applications down to – 320F (–196C). This modified version of austenitic ductile iron combines excellent low-temperature metallurgical and mechanical properties with superior castability.

The alloy is suitable for many cryogenic applications, possessing excellent casting properties for making excellent parts. Some applications are pumps, valves, compressors, and pipes and fittings for handling liquefied gases.

D-3 (30%Ni-3%Cr)

Recommended for thermal shock service where thermal expansivity should match that of ferritic stainless steel. This type, in addition to having excellent elevated temperature properties, also offers high resistance to erosion service of wet steam and corrosive slurries.

Superior cavitation resistance for applications such as pump impellers and tug boat propellers. Recommended for wet steam and corrosive slurry applications. Also used for turbocharger castings.

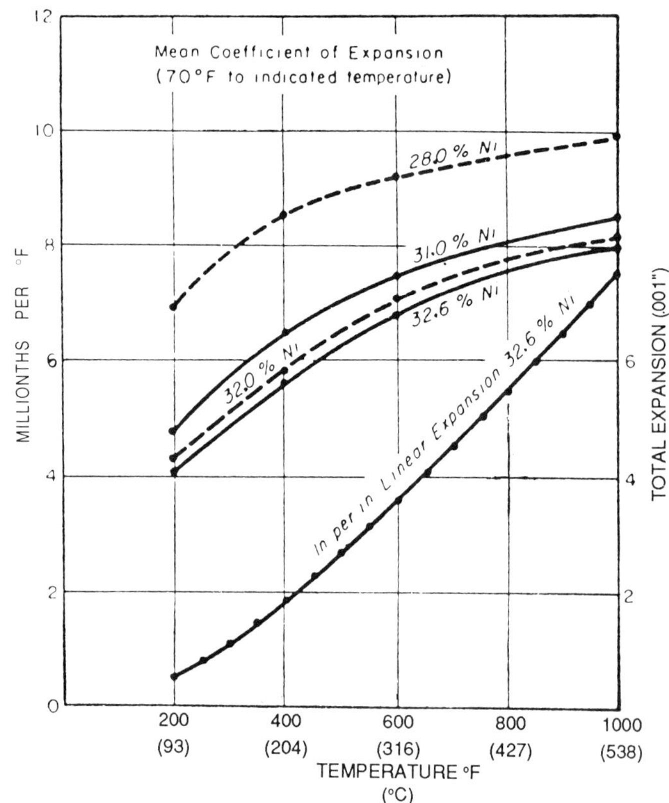

Fig. 2-17. Effect of nickel content of Type D-3 on thermal expansivity.

D-3A (32%Ni-1.5%Cr)

Recommended for use where a high degree of wear and galling resistance are required along with intermediate thermal expansion.

Recommended for gas turbine housings and other gas turbine components requiring expansion characteristics which can be controlled to match steel, ductile and gray iron.

TABLE 2-19. THERMAL CONDUCTIVITY COMPARISON

MATERIAL	Cal/cm^3/sec/°C (0 TO 100°C)	Btu/Hr/Ft^2/in/°F (50–200°F)
NI RESIST DUCTILE IRON TYPE D-2	0.032	93
FLAKE GRAPHITE NI RESIST	0.090–0.085	260–275
GRAY CAST IRON	0.108–0.131	313–380
STEEL	0.15–017	435–495
12% Cr STEEL	0.045	130
18% Cr-8% Ni STEEL	0.039	113

D-4 (30%Ni-5%Cr-5%Si)

Recommended for a high order of corrosion and oxidation resistance, superior to D-2 and D-3. Service applications for engine parts exposed to combustion and exhaust gases. Used on turbocharger and gas turbine applications up to 1500F (815C), where a scale-free material is required. Can withstand exposure to 1% sulfur fuel at temperatures up to 1000F (538C).

D-5 (34–36%Ni-0.10%Cr(max))

Can be used wherever minimum thermal expansion is desired. It may be preferred over other types of Ni-resist to reduce thermal stresses.

Recommended for use wherever castings with low expansivity are required, such as machine tool parts, glass molds and gas turbine housings. D-5B (34–36% Ni-3% Cr) is recommended for applications requiring a very low order of thermal stress.

■■■ REFERENCES

1. Johnson, Swikert and Bisson; "The Effects of Sliding Velocity and Temperature on Wear and Friction of Several Materials," *Lubricating Engineering*, p. 164 (May-June 1955).

Table 2-20. Electrical Resistance

MATERIAL	ELECTRIC RESISTANCE (MICROHMS/CC)
NI RESIST DUCTILE IRON TYPE D-2	102
FLAKE GRAPHITE NI RESIST	130-170
GRAY CAST IRON	75-100
PLAIN MEDIUM C STEEL	18
12% Cr STEEL	57
18% Cr-8% Ni STEEL	70

Table 2-21. Magnetic Permeability

MATERIAL	PERMEABILITY
NI RESIST DUCTILE IRON TYPE D-2	1.01-1.04
NI RESIST DUCTILE IRON TYPE D-2B	1.04-1.08
NI RESIST DUCTILE IRON TYPE D-2C	1.03
NI RESIST DUCTILE IRON TYPE D-4	1.10
FLAKE GRAPHITE NI RESIST	1.03
GRAY CAST IRON*	125
MILD STEEL*	150
12% Cr STEEL	FERROMAGNETIC
10-14% Mn STEEL	1.03-1.10
18% Cr-8% Ni STEEL	< 1.001
BRONZES	< 1.001

*Initial permeability.

Production of Ductile Ni-Resist

Warren M. Spear

Consultant
Basking Ridge, New Jersey

■ INTRODUCTION

Ni-resists are a family of cast irons containing sufficient nickel to produce an austenitic matrix structure similar to that of austenitic stainless steel. The austenitic structure provides a base for improved heat- and corrosion-resistance properties in comparison to unalloyed and low-alloy gray and ductile cast irons. Castability of the alloys is comparable to that of gray and ductile cast irons.

Nickel content of the Ni-resists varies from 15–36%. Most grades contain chromium to improve strength and corrosion resistance. Types 1 and 1b contain copper as a lower cost alternative to nickel to provide corrosion resistance. There is no ductile grade of the copper-containing Type 1 Ni-resist.

A stable austenite structure in the ductile Ni-resist types D-2, D-2B, D-4 and D-5S depends mainly upon the balance between the nickel and the silicon content. Unstable austenite, due to the combination of a low nickel content and a high silicon content, will result in the presence of pearlite and/or martensite in the matrix; inferior machinability, corrosion and oxidation resistance, and inferior properties at elevated temperatures.

■ SELECTING PROPER TYPE OF NI-RESIST

The selection of the proper type of Ni-resist depends upon the service in which the casting will be used. Suitability of the different grades of Ni-resist for various applications is described in the "Engineering Properties and Applications of Ni-resist" brochure.[1] Additional information is also available in Chapter 2 of this handbook.

■ MELTING FURNACES

Ductile Ni-resist has been melted in channel and coreless induction furnaces, direct and indirect arc furnaces, cupolas, and gas-fired crucible furnaces. Present production, however, is predominantly from coreless induction furnaces. Channel furnaces are used in some high-production foundries and in foundries producing very large castings, but rapid turnover of metal is important in order to avoid deterioration in melt quality due to excessive holding time.

---

[1]Inco Alloys International, Inc., Huntington, West Virginia, USA

TABLE 3-1. CHEMICAL REQUIREMENTS FOR DUCTILE NI-RESISTS, ASTM A 439

(Reprinted with permission from the *Annual Book of ASTM Standards*, copyright, American Society for Testing and Materials.)

Element	Composition								
	D-2	D-2B	D-2C	D-3	D-3A	D-4	D-5	D-5B	D-5S
Total carbon, max.	3.00	3.00	2.90	2.60	2.60	2.60	2.40	2.40	2.30
Silicon	1.50–3.00	1.50–3.00	1.00–3.00	1.00–2.80	1.00–2.80	5.00–6.00	1.00–2.80	1.00–2.80	4.90–5.50
Manganese	0.70–1.25	0.70–1.25	1.80–2.40	1.00 max.	1.00 max.	1.00 max.	1.00 max.	1.00 max.	1.00 max.
Nickel	18.0–22.0	18.0–22.0	21.0–24.0	28.0–32.0	28.0–32.0	28.0–32.0	34.0–36.0	34.0–36.0	34.0–37.0
Chromium	1.75–2.75	2.75–4.00	0.50 max.	2.50–3.50	1.00–1.50	4.50–5.50	0.10 max	2.00–3.00	1.75–2.25
Phosphorus, max.	0.08	0.08	0.08	0.08	0.08	0.08	0.08	0.08	0.08

TABLE 3-2. MECHANICAL REQUIREMENTS FOR DUCTILE NI-RESISTS, ASTM A 439

(Reprinted with permission from the *Annual Book of ASTM Standards*, copyright, American Society for Testing and Materials.)

Property	Type								
	D-2	D-2B	D-2C	D-3	D-3A	D-4	D-5	D-5B	D-5S
Tensile strength, min. ksi (MPa)	58 (400)	58 (400)	58 (400)	55 (379)	60 (414)	55 (379)	55 (379)	55 (379)	65 (449)
Yield strength (0.2% off.), min. ksi (MPa)	30 (207)	30 (207)	28 (193)	30 (207)	30 (207)	—	30 (207)	30 (207)	30(207)
Elongation in 2 in., or 50 mm, min. %	8.0	7.0	20.0	6.0	10.0	—	20.0	6.0	10.0
Brinell Hardness (3000 kg)	139–202	148–211	121–171	139–202	131–193	202–273	131–185	139–193	131–193

■ CHARGE MATERIALS

The base charge for ductile Ni-resists should be made up preferably from primary materials—low sulfur pig iron, nickel, high-carbon ferrochromium, foundry-grade ferrosilicon and high-carbon ferromanganese—plus carbon steel scrap, alloy steel scrap of known analysis, identified ductile Ni-resist scrap and ductile Ni-resist returns. Cerium-free nodulizing alloys should be used for the magnesium addition.

Lead, aluminum, and wet, oily and rusty scrap are the main contaminants causing difficulty in producing ductile Ni-resist. Lead, in very small quantities (<0.003%), produces Widmenstatten graphite that substantially reduces mechanical properties. Aluminum has the same deleterious pinholing effect on ductile Ni-resist as in gray and ductile iron.

Wet, oily and rusty charge material should never be charged in induction melting when there is a molten heel in the furnace. Even when wet, oily and rusty materials are charged on top of unmelted material, convection will draw some of the evaporating gas to the melt surface. Red rust is particularly detrimental because it is made up of iron hydroxide, which resists breakdown into water vapor and iron oxide until a temperature of about 600F (316C) is reached.

GENERAL MELTING PRACTICE

The Ni-resists are similar to gray and ductile irons in response to melting practice, except that the results of poor practice are more pronounced in Ni-resist production. The effects of excessive superheating, excessive holding time, excessive recycling of returns, poor inoculation practice, and too high a pouring temperature are much more obvious in Ni-resist production. The most obvious effects of deteriorating castability are increased shrinkage, increased carbide content and decreased machinability.

There is an obviously greater tendency for gassing during melting of ductile Ni-resist than in the melting of ductile iron, particularly when using ductile Ni-resist returns in induction melting. Because of the much slower oxidation of residual magnesium from the recycled returns, which are retained in the Ni-resist melt, the sulfur content of the bath is tied up as a refractory magnesium-sulfide compound. As a result, there is a very low effective sulfur content in the bath, which opens up the bath to rapid hydrogen pickup from water vapor in the atmosphere. This is generally not encountered when melting ductile iron returns, because of the much more immediate loss of residual magnesium during melting. Avoiding this effect is discussed later in this chapter under individual melting practices. High residual magnesium in the returns (above 0.06%), or high percentages of ductile Ni-resist returns in the charge, accentuate the gassing problem.

Solution of graphite from returns and pig iron during melting is slower for the ductile Ni-resists than for ductile iron. Graphite-containing charge materials, such as pig iron and returns, should, therefore, be completely melted prior to superheating the induction furnace bath. If a late addition of pig iron or Ni-resist returns has to be made to the furnace, an extra 10-min. hold at the superheat temperature will assure complete solution of the graphite.

MAGNESIUM TREATMENT

Making the same percentage magnesium addition to Ni-resist as to ductile iron at the same metal temperature will result in a 20–30% greater recovery of magnesium. In addition, residual magnesium fade in the ladle during transportation, holding, and pouring, is slower for Ni-resist than for ductile iron.

Ductile Ni-resist has the same minimum residual magnesium threshold requirement of 0.035% minimum for a fully spheroidal structure as in unalloyed ductile iron.

Similarly, when the residual magnesium content exceeds about 0.06%, ductile Ni-resist exhibits an increased occurrence of oxide scum defects.

Higher magnesium residuals have generally been recommended for the ductile Ni-resists in comparison to ductile iron because of the somewhat improved graphite shape. Spheroid shape and ductility of ductile Ni-resist are somewhat improved with increasing magnesium residuals, up to about 0.10%. Nodule count in ductile Ni-resist is usually substantially lower than that of ductile iron. The significantly higher pouring temperatures normally used in producing ductile Ni-resist are largely responsible for the difference in nodule count and spheroid quality. A balanced combination of lower residual magnesium (0.04–0.05%) and lower pouring temperature, together with good ferrosilicon inoculation practice, results in a similar nodule count and spheroid quality as that of ductile iron at the same minimum residual magnesium level.

INOCULATION

Inoculation practice is much more critical for the chromium-containing Ni-resists than for ductile iron. Those grades of Ni-resist containing substantial amounts of chromium present problems in variable shrinkage and machinability. Chromium is added to various grades for a number of reasons: room temperature tensile strength; elevated temperature rupture strength and creep resistance; and oxidation and corrosion resistance. Chromium carbides are present because the amount of chromium in the alloy exceeds the solubility limit for chromium carbide in the matrix. Melt practice and inoculation are aimed at minimizing the amount of carbides in the austenite matrix while distributing the carbides present the form of fine, evenly scattered particles.

Maximum superheat in the induction furnace should be limited to the maximum temperature required for solution of graphite, and for handling and pouring. For maximum inoculation effect, all of the more oxidizable makeup elements (Cr, Mn and Si) should be added as late as possible to the furnace; 0.5% Si added as foundry grade 75% ferrosilicon in the ladle; and a final 0.02% silicon added as powdered* foundry grade 75% ferrosilicon at the bottom of the sprue, or to the stream, during pouring of molds. The tap temperature of the metal should be adjusted so that the metal is at the proper pouring temperature in the ladle when the metal reaches the mold. Fade time for the inoculation effect appears to be more rapid for Ni-resists than for ductile iron.

***Editors' Note: It is recommended that this material be purchased with a top and bottom size limitation, e.g., 20M × 70M or 30M × 100M.**

▬ GATING AND RISERING

Solidification shrinkage for ductile Ni-resists that do not contain chromium is similar to that of ductile iron. The high-chromium ductile Ni-resists have solidification shrinkage volumes approaching those of cast steel.

Nickel content reduces the solid solubility of carbon during solidification so that, in spite of the lower total carbon content of Ni-resists, a similar total amount of graphite forms during solidification of Ni-resists as during solidification of ductile iron. However, the chromium in the chromium-containing grades combines with carbon in the late stages of solidification to form cell boundary carbides, resulting in a late spurt of solidification shrinkage. In addition, the amount of late solidification shrinkage will vary with the amount of returns in the charge, superheat temperature, holding time in the furnace, and timing and effectiveness of late silicon inoculation.

The percent chromium targeted should be at the bottom of the specified chromium range unless there is some reason for a higher analysis. Chromium increases solidification shrinkage, particularly with repeated recycling of returns and with increasing percentages of returns being used in the charge.

Ductile iron rigging practices can be used for Ni-resist grades D-2C, D-5 and D-3A, since they exhibit solidification shrinkage similar to that of ductile iron. Those with the highest chromium ranges—the high-chromium grades, D-2B, D-3 and D-4—exhibit solidification shrinkage approaching that of steel. Riser feeding distances for these grades, however, will be six or seven times the riser diameter. As a result, while riser volume required for feeding approaches that for steel, fewer but larger and more widely spaced risers can be used. The wide spacing prevents cross feeding between risers.

Poor Ni-resist melting practice will result in an increase in solidification shrinkage, loss of fluidity, persistent oxide scum, increased carbide in the microstructure, and loss of machinability. This effect is much more pronounced for the Ni-resists than for unalloyed gray and ductile irons.

Gray and ductile Ni-resists containing chromium are subject to abrupt increases in solidification shrinkage when the eutectic carbon equivalent is exceeded. This effect is most evident in thin-section castings, decreasing progressively with increasing section size. The increased solidification shrinkage due to hypereutectic composition has a tendency to appear as severe centerline or extensive sponge shrinkage. Hypereutectic solidification apparently interferes with normal feeding characteristics.

This hypereutectic increase in solidification shrinkage is caused by the shift from graphite precipitation and growth to the formation of eutectic iron-chromium carbides at a late stage in the solidification. The small amount of remaining molten metal solidifies with steel-like shrinkage. The more rapid solidification of small castings results in an increased percentage of carbide formation and, hence, greater solidification shrinkage.

Carbon equivalent calculations for the Ni-resists, which allow for the effect of nickel in reducing the eutectic carbon equivalent, are made according to the formula:

$$\text{C.E.} = \text{C}\% + 0.33 \times \text{Si}\% + 0.047 \times \text{Ni}\% - 0.0055 \times \text{Ni}\% \times \text{Si}\%$$

Another effect that can produce a marked difference in solidification behavior for Ni-resist in comparison to ductile iron is its much higher short-time hot strength. The high-strength shell of Ni-resist, which solidifies against the mold and cores, resists inward movement so that solidification shrinkage can appear to be greater than that for ductile iron. At other times, the strong shell, preventing inward movement of the casting walls, facilitates feeding of remote heavy sections.

Risering should be done with side risers (at the parting line) and top risers. Since the horizontal feeding distance for risers is at least six times the riser diameter, widely spaced, large risers are employed to avoid cross feeding. Generally, heavy sections that are low in the mold will be fed over the long vertical distances by the top and side risers because of the tendency of liquid metal pressure to hold wall sections open. Very heavy sections, low in the mold, should be chilled using graphite or gray iron chills. Because the Ni-resists have excellent short-time high-temperature strength (in comparison to gray or ductile iron), the tendency for inward wall movement during solidification due to atmospheric pressure is greatly reduced.

All risers should be atmospheric. Insulated sleeves should be used on the upper half of risers on large castings. After pouring, risers should be covered with insulating compound to prevent freezing over. A riser height-to-diameter ratio of 1.5:1 is recommended. If practicable, the height of the riser above the top of the casting should be 50% of the casting height.

▬ POURING PRACTICE

While carbon equivalents for the Ni-resists, after allowing for the nickel effect, are equivalent to those for gray and ductile irons, nickel raises the liquidus and, consequently, the pouring temperature by 30–50°F (17–28°C).

The level of residual magnesium, the pouring temperature, and the pouring rate are interrelated. Higher magnesium residuals require either higher pouring temperature, higher pouring rate, or a combination of both. A low pouring temperature with high-residual magnesium will result in an excessive amount of oxide scum defects. Increasing pouring temperature to avoid oxide scum defects results in loss of directional solidification effects and poor nodule morphology.

The nickel content of the Ni-resists reduces the magnesium vapor pressure and rate of magnesium fade in the ladle during handling and pouring, in comparison to ductile iron. This improves control over residual magnesium content so that, in large-scale production of castings, the residual magnesium can be held in an optimum narrow range. Ductile Ni-resist, therefore, is more amenable to the use of automatic pouring devices than is ductile iron.

■ REFINING DURING INDUCTION MELTING

Ni-resists are more sensitive to melting practice than are gray or ductile irons. Increasing percentages of returns, slow melting, excessive superheat, and long holding times increase the tendency for gas defects, solidification shrinkage, and tendency for cracking, and result in poorer machinability and inferior corrosion resistance. The deterioration in metal quality appears to be related to the ability of austenitic cast irons to absorb considerably more oxygen than the gray and ductile irons together, as well as a lower solid solubility limit for hydrogen.

Rapid melting, minimum superheat, late addition of the more oxidizable elements (Cr, Mn, Si), immediate tap at the minimum practical superheat temperature, and effective inoculation—both in the ladle and in the pouring stream—produce consistent, high-quality Ni-resist. The effect of long holding time in the furnace can be minimized by holding at a low silicon level, so that a steady, mild boil occurs at the holding temperature. The slow boil will maintain oxygen equilibrium at a low level and reduce the tendency to pick up hydrogen.

As previously noted, recycling ductile Ni-resist returns when air induction melting leads to increased susceptibility of hydrogen pickup from the moisture in the air. The problem is particularly evident in semi-continuous melting where ductile returns are backcharged onto a molten heel of low sulfur-base Ni-resist. During the brief time required to melt the ductile Ni-resist returns, magnesium, retained in the bath from the returns, compounds with the bath sulfur, opening the bath up to rapid gas absorption from the air.

The possible cures for this condition are: (1) an argon protective cover while the backcharge is melting and superheating, or (2) argon injection into the bath to remove the hydrogen after the returns have melted in. Argon injection appears to be the more practical solution.

■ SUGGESTED COMPOSITION AND FOUNDRY PRACTICE FOR TYPE D-2 NI-RESIST

Carbon equivalent aim should always be below the eutectic carbon equivalent, even for thin-section castings. A typical microstructure for D-2 ductile Ni-resist is shown in Figure 3-1. Table 3-3 compares the specified A 439 analysis for Type D-2 Ni-resist to the suggested foundry analysis.

A characteristic of chromium-containing Ni-resists is the abrupt increase in solidification shrinkage when the eutectic carbon equivalent is exceeded. Nickel acts in a manner similar to silicon (substituting for carbon) when calculating carbon equivalent. Carbon equivalent for Ni-resists, based on the 4.3 eutectic for unalloyed iron, can be calculated according to the formula:

$$C + 0.33Si + 0.047Ni - 0.0055NiSi = C.E.$$

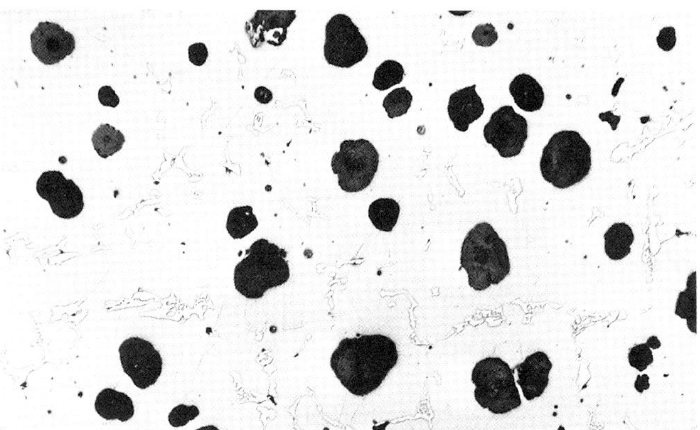

Fig. 3-1. Typical microstructure for D-2 ductile Ni-resist; 2% nital etched, ×150.

TABLE 3-3. SPECIFIED AND SUGGESTED FOUNDRY TARGET ANALYSIS FOR TYPE D-2 NI-RESIST

	Specified ASTM A 439	Suggested Analysis
Total Carbon	3.0 max.	2.7
Silicon	1.5–3.0	2.0
Manganese	0.7–1.25	0.8
Nickel	18.0–22.0	20.0
Chromium	1.75–2.75	1.85
Phosphorus	0.08 max.	LAP
Iron	bal.	bal.
Magnesium	—	0.05

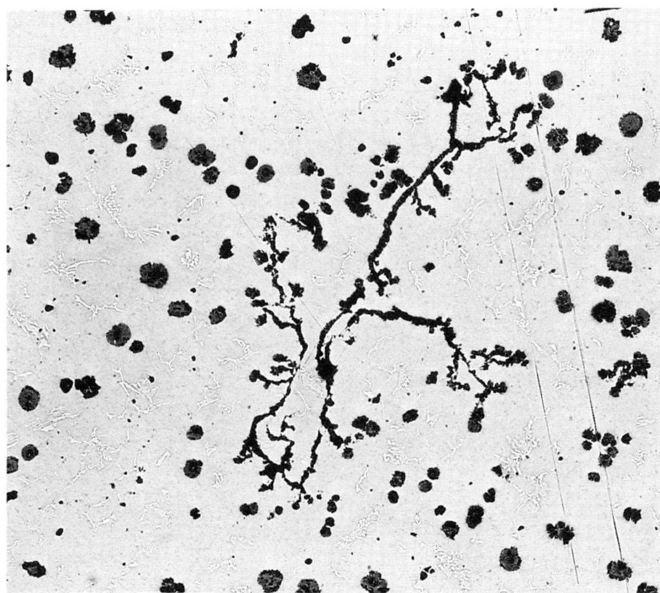

Fig. 3-2. Slag stringers in D-2 Ni-resist with 0.09% Mg; unetched, ×100.

Using the previous D-2 Ni-resist aim analysis results in the calculated carbon equivalent as shown:

$$(2.7) + 0.33(2.0) + 0.047(20.0) - 0.0055(20)(2.0) = 4.08 \ (C.E.)$$

Carbon equivalent does not have to be adjusted for varying casting section sizes of ductile Ni-resist, except for very thin and very heavy sections. For thin-section castings (<1/2 in.), a composition just below the eutectic is advantageous for castability and minimizing the volume of carbide phase. For heavy-section castings where shape of the graphite spheroids improves with reduced carbon equivalent, a carbon equivalent analysis of at least 0.5% below the eutectic carbon equivalent is suggested. A high carbon content with a relatively low silicon content has a favorable effect in maintaining melt quality when continually recycling returns.

Manganese content just above the 0.7% minimum specified is adequate to prevent secondary graphitization, particularly where the Ni-resist is to be use at elevated temperature. Manganese analysis should be maintained below 1.0% in order to minimize gassing problems when high percentages of returns are recycled.

Phosphorus content should be maintained below 0.08%, maximum, in order to minimize the tendency for leakers. When Ni-resists with phosphorus content exceeding 0.08% are subjected to certain corrosives, phosphide segregation to cell boundaries can result in an attack similar to intergranular corrosion in stainless steel. There is no appreciable advantage for adding phosphorus within the specified 0.08% maximum, so none should be added.

Chromium should be maintained near the bottom of the specified range unless the application dictates otherwise. Chromium content at the lower end of the specified range reduces solidification shrinkage, improves machinability and improves response to melt recycling. Higher chromium contents progressively improve resistance to cavitation erosion and to elevated temperature oxidation.

Magnesium content should be maintained within the minimum residual range that will produce a satisfactory spheroidal graphite structure throughout the casting. Excessive magnesium results in oxide scum stringers with precipitated graphite as shown in Figure 3-2.

INDUCTION FURNACE MELT PRACTICE FOR D-2 NI-RESIST

1. Charge, in order:
   a. Graphite
   b. Returns
   c. Pig iron
   d. Nickel
   e. Steel (clean and dry)
2. Melt down—superheat to minimum temperature required to insure complete solution of graphite and for handling and pouring.
3. Add to furnace, in order, with low power for stirring:
   a. High-carbon FeCr to aim analysis.
   b. FeMn to target analysis.
   c. Foundry grade 75% ferrosilicon to 0.50% below final targeted analysis.
   d. Mg as nickel-magnesium (4.5Mg-95Ni). At 2700F (1482C) addition temperature, 85–90% recovery should be expected. If magnesium is added in the ladle, use of 15Mg-balNi additive is suggested with 65–75% recovery expected at 2700F (1482C).
4. Tap immediately into hot ladle adding 0.50%Si as 75% ferrosilicon in bottom of ladle or to tap stream.
5. Pour at 2600–2550F (1427–1399C) for small and thin-section castings; 2550–2500F (1399–1371C) for larger castings. Add 0.02% 75% ferrosilicon powder* to the sprue or pouring stream for mold inoculation.

***[Ed. N.: It is recommended that this material be purchased with a top and bottom size limitation, e.g., 20×70 mesh or 30×100 mesh.]**

There should be no difficulty in recycling up to at least 50% of the returns, indefinitely, without detrimental effect on castability or properties. Recycling higher percentages of returns, if necessary, should be limited to short runs. The returns are charged first during induction melting in order to oxidize out residual magnesium as soon as possible. This is necessary in order to avoid tossing the metal. If gassing problems are encountered, argon purging through a graphite lance, fitted with a porous plug, is recommended. The lance should be inserted about 10 inches into the bath, with the melt temperature held at 2450–2500F (1343–1371C) during purging.

SUGGESTED COMPOSITION AND FOUNDRY PRACTICE FOR D-5S NI-RESIST

MECHANICAL PROPERTY REQUIREMENTS FOR D-5S

A comparison of the ASTM (A 439) specifications to suggested foundry target analysis for D-5S Ni-resist is given in Table 3-4; the mechanical property requirements are given in Table 3-5.

Carbon provides castability, but the optimum analysis for carbon will vary with section size and must be reduced for heavier sections. For a 1-in. section, the suggested 2.00% carbon analysis is satisfactory; for a 4-in. section, a 1.50% analysis is suggested in order to maintain a good spheroidal graphite structure.

Silicon is the most potent element responsible for the outstanding oxidation resistance of D-5S. Ability to retain a high-integrity oxide skin, under all heating and cooling conditions, requires a minimum silicon content of 4.9%, supplemented by critical chromium and manganese contents. Silicon, in excess of its solubility limit in the austenitic matrix, appears in the microstructure as silicides.

A minimum manganese content of 0.4% is needed in order to avoid secondary graphitization in cell boundary areas during cooling in the sand mold and in elevated temperature service. Figure 3-3 shows the deleterious secondary graphitization that occurred during mold cooling of D-5S Ni-resist in a 1-in. double keel block containing insufficient manganese (0.15%Mn).

Nickel provides the stable austenitic matrix that is the alloying base for high-temperature strength and stability, and outstanding oxidation resistance. The nickel content must be sufficient to assure a stable austenitic matrix in service.

TABLE 3-4. SPECIFIED AND SUGGESTED FOUNDRY TARGET ANALYSIS FOR TYPE D-5S NI-RESIST

	Specified ASTM A 439	Suggested Target Analysis
Total Carbon	2.30 max.	2.00
Silicon	4.90-5.50	5.00
Manganese	1.00 max.	0.60
Nickel	34.0–37.0	35.0
Chromium	1.75–2.25	2.00
Phosphorus	0.08 max.	LAP
Magnesium	—	0.05

TABLE 3-5. MECHANICAL PROPERTY REQUIREMENTS FOR D-5S NI-RESIST

	A 439 (Min.)	Typical
Yield strength, psi	30,000	30–35,000
Tensile strength, psi	65,000	65–70,000
Elongation, %	10	20–35

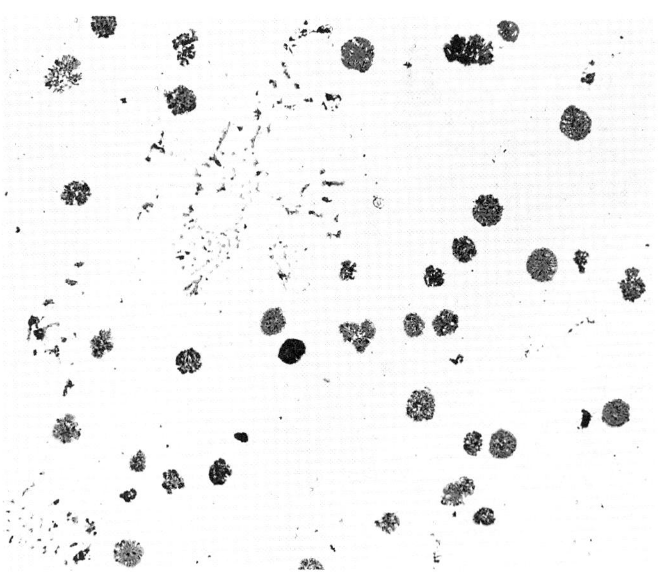

Fig. 3-3. Effect of insufficient manganese (0.15% Mn) on the formation of secondary graphite in as-cast D-5S ductile Ni-resist; unetched, ×75.

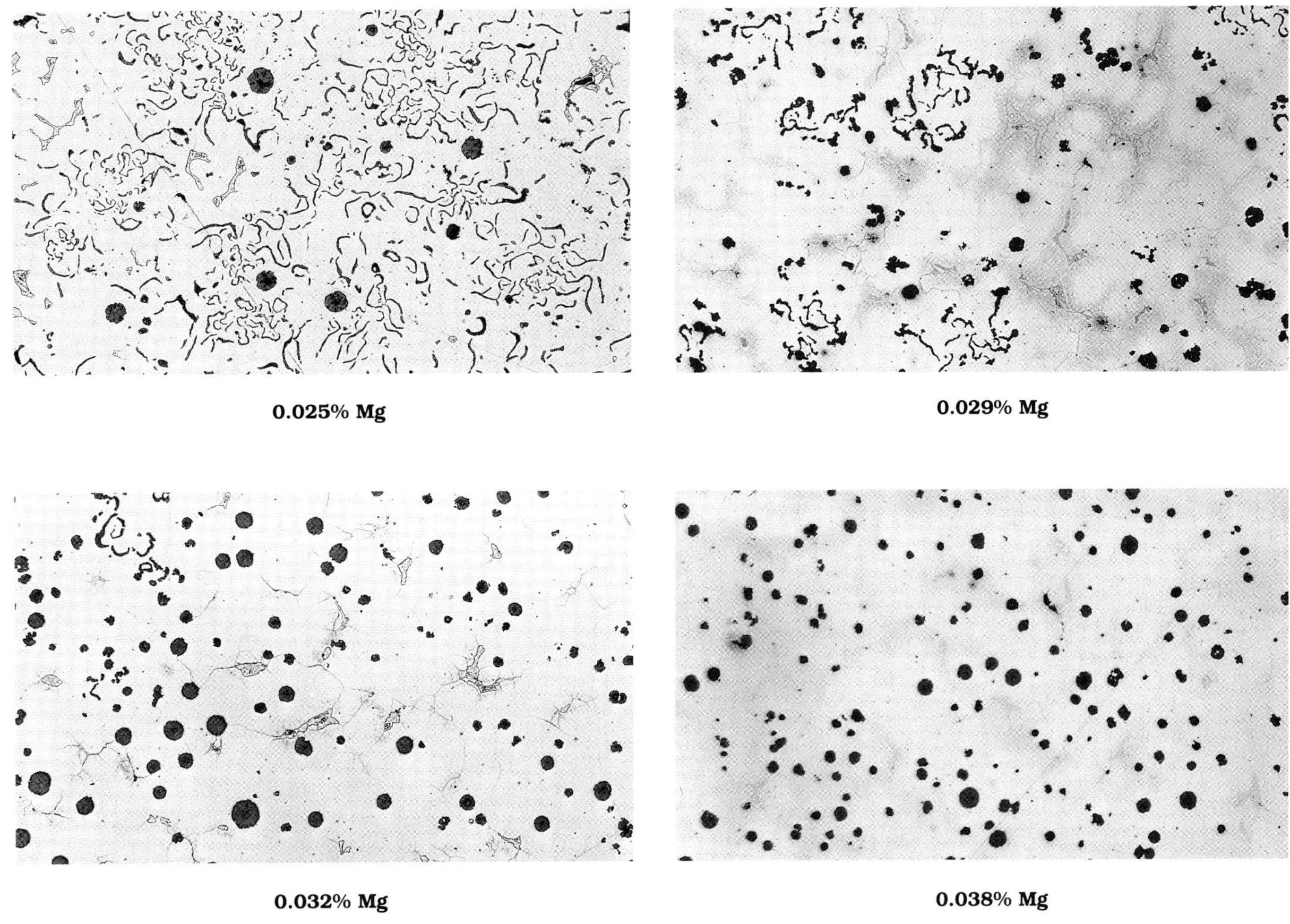

Fig. 3-4. Microstructures showing that a minimum of 0.035% magnesium is required to produce a fully spheroidal graphite structure; ×100 (all).

Chromium supplements silicon in providing resistance to high temperature oxidation and for the development of a retentive scale. The limit for chromium solubility in the matrix at 1500F (816C) is about 2%. Above 2% chromium, chromium carbides will appear in the matrix.

Phosphorus, below the 0.08% maximum specified, has no appreciable effect on the properties and castability of D-5S.

Minimum magnesium residual requirements for fully spheroidal graphite in D-5S are the same as for ductile iron (i.e., 0.035% min.). A residual magnesium analysis range of 0.040–0.050% should insure proper treatment for a 100% spheroidal structure and minimize presence of ox-ide stringers. The graphite in the microstructure of D-5S for magnesium levels of 0.025, 0.029, 0.033 and 0.038% is shown in Figure 3-4. The photomicrographs show the progression from mostly compacted graphite at 0.025% residual magnesium to a completely spheroidal graphite structure at 0.038% residual magnesium.

INDUCTION MELT PRACTICE FOR D-5S NI-RESIST

1. Charge, in order:
   a. Graphite or pig iron for carbon makeup
   b. Returns

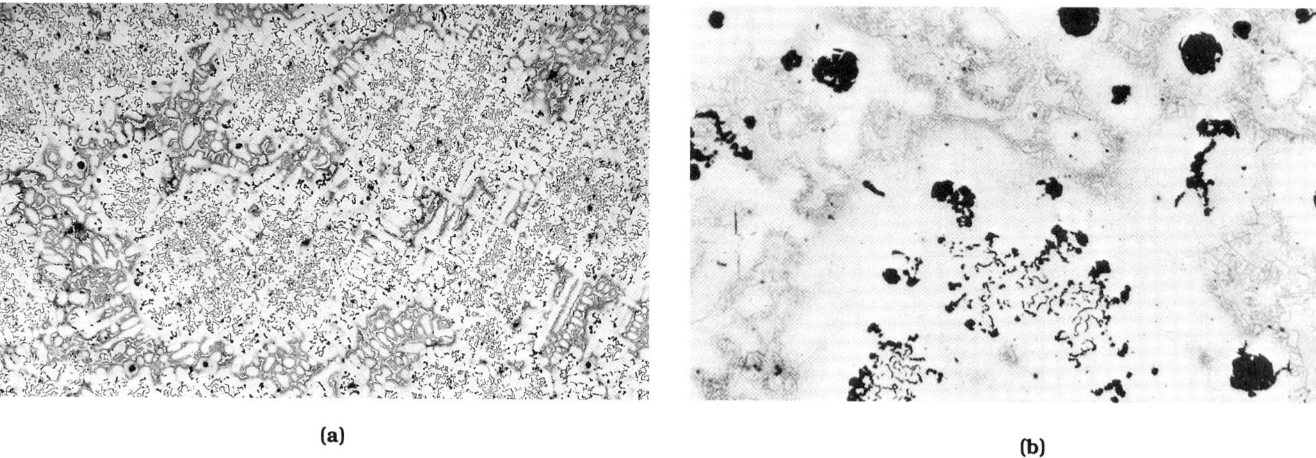

(a)   (b)

Fig. 3-5. Cell boundary silicides and dendritic flake graphite resulting from poor melting practice; 2% nital etched; (a)×20, (b)×100.

   c. Steel (clean and dry)

   d. Nickel

2. Melt down—superheat to minimum temperature required to insure complete solution or graphite and for handling and pouring.

3. Add to furnace, in order, with low power for stirring:

   a. High-carbon FeCr to aim analysis.

   b. High-carbon FeMn to aim analysis.

   c. Foundry grade 75% ferrosilicon to 0.50% below final aim.

   d. Mg as nickel-magnesium (no Ce).

4. Tap immediately into a hot, clean ladle, adding 0.5% silicon as foundry grade 75% ferrosilicon to the bottom of the ladle or to the tap stream.

5. Pour at 2600–2550F (1427–1399C) for small and thin section castings; lower temperatures for heavier section castings. Mold or stream inoculate with 0.02% 75% ferrosilicon powder.*

***[Ed. N.: It is recommended that this material be purchased with a top and bottom size limitation, e.g., 20×70 mesh or 30×100 mesh.]**

At least 50% of the returns can be recycled indefinitely, without deterioration in castability and quality, provided that the previously suggested air induction melting practice is followed and that superheat and time in the furnace is kept to a minimum. Up to 80% of the returns have been intermittently recycled by backcharging on a molten heel and by using minimum time and temperature in the furnace. If gassing problems are encountered, argon purging is suggested, as described in the D-2 Ni-resist practice.

D-5S is more sensitive to poor melt practice than are other grades of ductile Ni-resist because of its high silicon content. Long holding times in the induction furnace result in compacted graphite in the primary austenite dendrites, and in heavy segregation of silicide and carbide to the cell boundaries. This structure is shown in Figure 3-5.

In spite of the low carbon content in comparison to other cast irons, D-5S Ni-resist has solidification shrinkage similar to that of normal, high-carbon-equivalent ductile iron. The high silicon and nickel contents reduce the solid solubility of carbon to the point where almost all of the carbon is precipitated as graphite during solidification. The alloy has excellent fluidity because of the high silicon content, and has excellent feeding characteristics because of the high carbon equivalent.

HEAVY-SECTION PROPERTIES OF D-5S

With care in selecting raw materials, and good melting and pouring practice, there is only a minor reduction in mechanical properties with increasing section size. Results for D-5S heavy-section tests are shown in Table 3-6.

With good foundry practice, yield and tensile strengths are relatively unaffected by the change in section size. Elongation is reduced somewhat in the heavier section, but

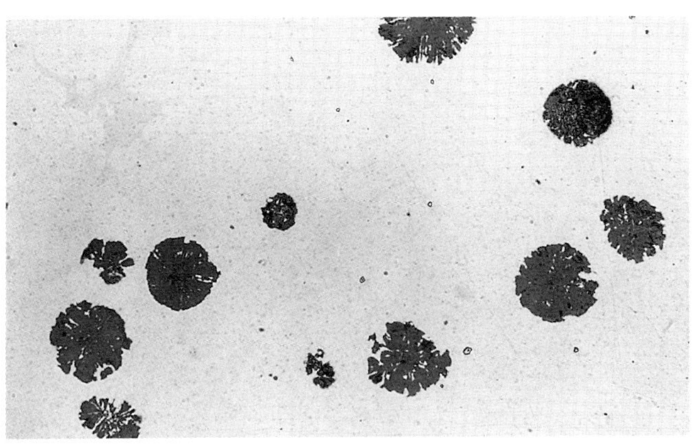

Fig. 3-6. Microstructure of D-5S in a 1-in. double-keel block. Heat treated at 1800F for 2 hr; as-cast, 2% nital etched, ×250.

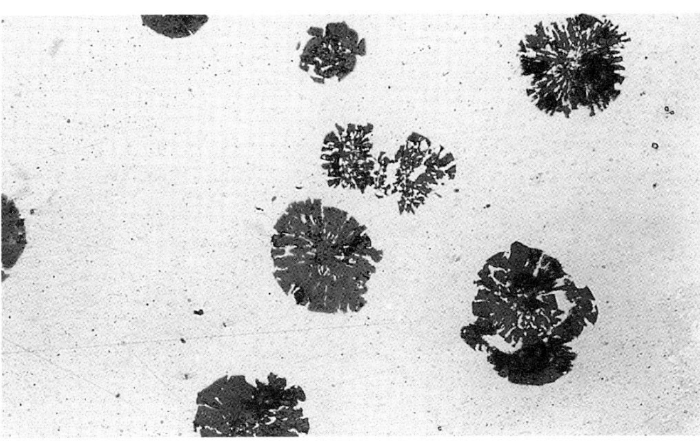

Fig. 3-7. Microstructure of D-5S in a 4-in. single-keel block, as cast, 2% nital etched; ×250.

Composition Analysis					
Heat	TC	Si	Mn	Ni	Cr	Mg
F	1.40	5.1	0.67	35.6	2.2	0.52
G	1.61	5.1	0.63	33.9	2.2	0.63

Mechanical Properties					
Heat	Section	Condition	YS (0.2% off)	UTS (psi)	Elongation
F	1 in. keel	As Cast (AC)	35,300	68,100	24.0
F	1 in. keel	1800 F/2 hr (AC)	35,000	74,100	30.0
F	4 in. keel	AC	35,200	64,000	19.5
F	4 in. keel	1800 F/2 hr (AC)	35,200	70,100	26.0
G	1 in. keel	AC	32,500	65,900	29.5
G	1 in. keel	1800 F/2 hr (AC)	34,000	68,600	29.0
G	4 in. keel	AC	32,200	57,000	19.0
G	4 in. keel	1800 F/2 hr (AC)	33,600	64,500	23.0

SUGGESTED COMPOSITION AND FOUNDRY PRACTICE FOR D-2C NI-RESIST

This grade of ductile Ni-resist is commonly used in steam turbines. A comparison of ASTM specifications and suggested foundry analysis is presented in Table 3-7. Freedom from carbides and excellent spheroidal graphite is required. Keeping the residual chromium content to a minimum is required in order to avoid the presence of carbides. A high manganese content is specified in order to prevent secondary graphitization in the cell boundaries.

The melt practice suggested for D-2 Ni-resist should be followed. Effective inoculation is essential in order to meet the microstructural requirements. Effect of insufficient superheat for complete solution of graphite in the charge material is shown in Figure 3-8.

SUGGESTED COMPOSITION AND FOUNDRY PRACTICE FOR D-2M NI-RESIST

Chromium and phosphorus contents should be as low as possible for best elongation and impact results. A comparison of ASTM specifications and suggested foundry analysis for D-2M is presented in Table 3-8, and for D-3, D-4 and D-5 in Tables 3-10, 3-11, and 3-12, respectively. Mechanical property requirements for D-2M are shown in Table 3-9. The magnesium residual aim should be controlled to the minimum required for a fully spheroidal graphite structure.

values remain well above the 10% minimum specified in ASTM A439. Microstructures of the 1-in. double, and 4-in. single keel blocks are shown in Figures 3-6 and 3-7.

It is necessary to lower the carbon equivalent of ductile Ni-resists in order to maintain good tensile properties in heavier casting sections. Since silicon must be maintained above the specified 4.9% minimum level in D-5S, the carbon equivalent is lowered by reducing the carbon content to the 1.4–1.6% range.

TABLE 3-7. SPECIFIED AND SUGGESTED FOUNDRY TARGET ANALYSIS FOR TYPE D-2C NI-RESIST

	Specified ASTM A 439	Suggested Analysis
Total Carbon	2.9 max.	2.6
Silicon	1.0–3.0	1.8
Manganese	1.8–2.4	2.0
Nickel	21.0–24.0	22.5
Chromium	0.50 max.	LAP
Phosphorus	0.08 max.	LAP
Magnesium	—	0.35–0.5

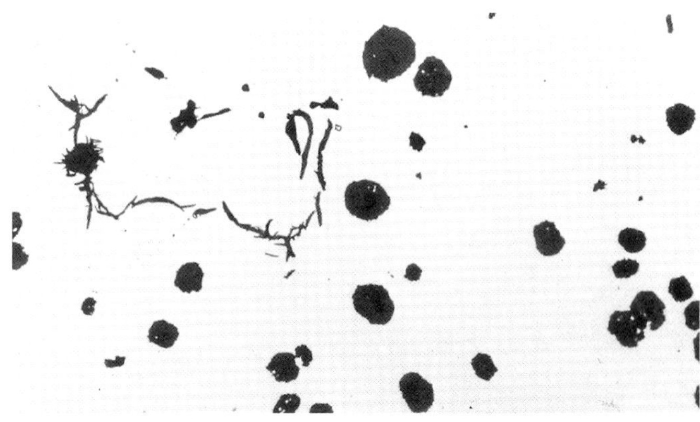

Fig. 3-8. Undissolved flake graphite in cell boundaries due to insufficient superheat of melt; ×100.

TABLE 3-8. SPECIFIED AND SUGGESTED FOUNDRY TARGET ANALYSIS FOR TYPE D-2M NI-RESIST

	Specified ASTM A 571	Suggested Analysis
Total Carbon	2.20–2.70	2.50
Silicon	1.50–2.50	2.00
Manganese	3.75–4.50	4.00
Nickel	21.0–24.0	22.5
Chromium	0.20 max.	LAP
Phosphorus	0.08 max.	LAP
Magnesium	—	0.05

D-2M castings are applied in cryogenic applications. The nickel and high manganese contents insure austenite stability at temperatures down to liquid hydrogen (–486F/ –252C). The CVN impact test requirement is applicable at temperatures down to –383F (–195C).

INDUCTION FURNACE MELT PRACTICE FOR NI-RESIST D-2M

Melting practice is critical in meeting the impact and elongation requirements. For best results, induction furnace heats should be made up of primary materials. The melted alloy is extremely sensitive to hydrogen pickup from moisture in the atmosphere because of the high manganese content. Consequently, returns should be applied as a charge material elsewhere; melting should be done under an argon cover, or the heat should be thoroughly purged with argon.

1. Charge in order:
    a. Carbon raiser or pig iron
    b. Steel scrap
    c. Nickel
2. Melt down, superheat to 2700–2750F (1482–1510C).
3. With power on low for mixing, add low hydrogen manganese metal.
4. Add foundry grade 75% ferrosilicon to 0.50% below final aim.
5. Add 4.5%Mg-balNi additive to furnace.
6. Tap into hot ladle adding 0.50% silicon in bottom of ladle or to tap stream.
7. Pour castings at lowest practical temperature. Mold or stream inoculate with 0.02% 75% ferrosilicon powder.*

***[Ed. N.: It is recommended that this material be purchased with a top and bottom size limitation, e.g., 20×70 mesh or 30×100 mesh.]**

◼ CLEANING ROOM PRACTICE

Similarity to gray and ductile iron rigging practice results in similar cleaning room practice for the Ni-resists. Salvage by welding deserves additional consideration. The better weldability and the greater value of Ni-resist castings, in comparison to ductile iron, justifies a separate evaluation to the extent that labor-intensive upgrading by welding should be applied. The technology of weld repair is complicated by the fact that end use as corrosion- and heat-resisting castings is a critical factor in deciding whether to weld or not, and in the choice of filler metal.

■■ INSPECTION

MICROSTRUCTURE

The Ni-resists should be fully austenitic. In the nonmagnetic grades (D-2 and D-2B), magnetic response to a hand magnet indicates that pearlite or martensite is present in the microstructure. This can be the result of underalloying with nickel and chromium, or of overalloying with silicon.

Examination of the microstructure will reveal whether the magnetic response is due to the presence of pearlite or martensite. For the magnetic alloys (D-3, D-4 and D-5), the microstructural examination is the only way to insure the proper, stable, austenitic microstructure.

AUSTENITE STABILITY

In applications where Ni-resist must perform over a range of temperatures, thermal tests may be required to insure stable austenite. The most common test is a low-temperature stability test, in which a casting from each batch is subjected to a temperature below the lowest expected operating temperature, then tested for transformation by magnetic response. This test is applicable only to the nonmagnetic grades.

The low-temperature test can also be used to test for instability during machining, where work-hardening by machining can cause a martensite transformation of the matrix with consequent difficulty in machining.

TABLE 3-9. MECHANICAL PROPERTY REQUIREMENTS FOR D-2M NI-RESIST

	Class 1	Class 2
Yield Strength, min, Mpa	205	170
Tensile Strength, min, Mpa	450	415
Elongation, min., %	30	25
Charpy V-notch, ft-lbs		
Min., average 3 tests	20	27
Min., individual test	16	20

TABLE 3-10. SPECIFIED AND SUGGESTED FOUNDRY TARGET ANALYSIS FOR TYPE D-3 NI-RESIST

	Specified ASTM A 439	Suggested Analysis
Total Carbon	2.60 max	2.30
Silicon	1.00–2.80	2.00
Manganese	1.00 max	0.60
Nickel	28.0–32.0	30.0
Chromium	2.50–3.50	3.00
Phosphorus	0.08 max	LAP
Magnesium	—	0.04–0.06

TABLE 3-11. SPECIFIED AND SUGGESTED FOUNDRY TARGET ANALYSIS FOR TYPE D-4 NI-RESIST

	Specified ASTM A 439	Suggested Analysis
Total carbon	2.60 max	2.40
Silicon	5.00–6.00	5.50
Manganese	1.00 max	0.60
Nickel	28.0–32.0	30.0
Chromium	4.50–5.50	5.00
Phosphorus	0.08 max	LAP
Magnesium	—	0.04–0.06

TABLE 3-12. SPECIFIED AND SUGGESTED FOUNDRY TARGET ANALYSIS FOR TYPE D-5 NI-RESIST

	Specified ASTM A 439	Suggested Analysis
Total Carbon	2.40 max	2.20
Silicon	1.00–2.80	1.50
Manganese	1.00 max	0.60
Nickel	34.0–36.0	35.0
Chromium	0.10 max	LAP
Phosphorus	0.08 max	LAP
Magnesium	—	0.04–0.05

HEAT TREATMENT OF DUCTILE NI-RESIST

Most castings can be used in the as-cast condition. However, castings may be heat treated to reduce residual stresses, reduce susceptibility to stress corrosion cracking, improve machinability, improve elevated temperature stability and improve dimensional stability

Stress Relief Treatment

Ductile Ni-resist can be stress relieved at 1150–1250F (621–677C) to remove residual stresses. Large, relatively thin-section castings must be heated and cooled at rates that will not result in appreciable temperature differentials. In air-circulating furnaces, the heating and cooling rate should not exceed 400F (204C) per hour at temperatures above 600F (316C). With noncirculating furnaces, the heating and cooling rates should not exceed about 200°F (111°C) per hour above 600F (316C).

Large Ni-resist castings that are exposed to seawater or to brackish water treated with hypochlorite should always be stress relieved. For other services, stress relief by mold cooling to 600F (316C) is adequate.

Annealing for Machinability

High-temperature annealing is used to improve machinability. Since the chromium grades of Ni-resist will have chromium carbides in the microstructure, a high temperature heat treatment at 1800–1900F (982–1038C) is required for solution of a significant part of the carbide content. Annealing temperature and duration should be varied according to section size, with a half-hour for sections less than 1/2 inch; two hours for 1-in. sections and four hours for heavier sections. The annealing heat treatment will reduce hardness and increase elongation and toughness with minor reduction in tensile and yield strength.

Heat Treatment for Elevated Temperature Stability

A stabilizing heat treatment is recommended for castings subjected to service at elevated temperatures where retention of machined dimensions is required. Heat treatment depends upon the service application temperature. For castings in use at temperatures below 1000F (538C), a four-hour anneal at 1400F (760C) followed by furnace cooling to 600F (316C) is suggested. For castings to be used in the 1000–1500F (538–816C) temperature range, castings should be held for a minimum of two hours at 1650F (899C), then furnace cooled to 600F (316C).

For application temperatures below 1000F (538C), the 1400F (760C) treatment will produce the required stability by precipitating excess carbon from the austenitic matrix. Carbon solution and rate of diffusion is negligible below 1000F (538C).

For elevated temperature applications (above 1000F/538C), the 1650F (899C) heat treatment dissolves the less stable carbides from the cellular carbides, then precipitates the carbon taken into solution on graphite spheroids, or as fine chromium carbides in the matrix during furnace cooling. Without the high-temperature stabilization, the gradual dissolution of unstable carbides and graphite precipitation at elevated temperatures can result in casting growth in service. The treatment of two hours at 1650F (899C) plus furnace cool should be repeated for maximum high-temperature dimensional stability. Dilatometer studies show that hysteresis and dimensional instability is negligible after the second heat treatment.

Heat Treatment for Ambient Temperature Dimensional Stability

Ni-resist types D-5 and D-5B are occasionally specified for applications such as machine tool parts and mounting plates requiring a high degree of dimensional precision. For machining to precise dimensions or optical flatness requirements, the castings must be totally free of residual stresses.

For maximum dimensional stability at ambient temperatures, castings should be soaked for a minimum of two hours at 1550F (843C), cooled to 600F (316C) at not more than 100°F (56°C)/hr, then cooled to room temperature in still air. After rough machining, castings should be stress relieved for one hour at 700–900F (371–482C), then furnace cooled to room temperature.

MACHINING

The chromium-free ductile Ni-resists (D-2C and D-5) are readily machinable, with machinability equivalent to that of pearlitic ductile iron. Types D-5B and D-5S also can have excellent machinability because, with good foundry practice, the low chromium content will result in the presence of little or no carbide in the matrix. The rest of the ductile Ni-resists will contain carbides, and machinability will be affected by shape and quantity of carbide. The high chromium Ni-resist D-4 has the poorest machinability, because of the large amount of chromium carbide in the matrix.

WELDING

Ni-resists can be gas welded, arc welded and Heliarc welded. Ni-resists Type 1, 2 and D-2, when applied for corrosion resistance, can be either welded with matching composition filler or with the 55Ni/Fe/C rod or wire. The 55Ni/Fe/C type of welding deposit is more corrosion resistant than Types 1, 1b, 2, 2b, D-2 and D-2B Ni-resist base metals. A 600F (316C) general preheat is suggested in order to avoid stress cracking or heat-affected zone cracking. As soon as the welding is completed the casting should be transferred, while still hot, to a furnace for a two-hour stress relief treatment at 1250F (677C), furnace cooled at 200°F (111°C)/hr to 600F (316C), then air cooled.

Type 55Ni/Fe/C rod or wire is generally suitable for cosmetic repairs in noncritical areas for all the ductile Ni-resists. However, for critical weld repair of castings subjected to high service temperatures, or where it is critical to match corrosion resistance, or physical or mechanical properties, gas welding with a matching filler may be considered.

Ductile Ni-resists are generally more weldable than ductile iron by both arc and gas welding. Choice of filler depends upon application of the part. Type D-2 Ni-resist can be repaired by arc welding with the 55Ni/Fe/C filler, when the application involves corrosion resistance. For high-temperature applications, gas welding with a matching filler is suggested. Since there is no matching filler commercially available, a foundry would have to cast its own matching rod for gas welding. A somewhat higher magnesium content (0.080–0.10%) is suggested for the gas weld filler in order to insure that the residual magnesium retained in the weld deposit will exceed the 0.035% minimum needed to insure a fully spheroidal graphite structure. The silicon content of the rod should be increased to about 2.50% in order to minimize carbides in the weld deposit.

Weldability of the D-2 Ni-resist with the 55Ni-balFe rod or wire can be improved by keeping the phosphorus content below 0.040%, the residual magnesium below 0.05%, chromium to the low end of its specified range, and silicon in the 1.70–2.20% range. Alternatively, the addition of 0.12–0.17% columbium to D-2 Ni-resist will further improve weldability so that arc welding can be accomplished without preheat. All defects must be removed to expose completely sound metal, and the weld metal deposited at a low rate in comparison to deposition rates allowable with 600F (316C) or higher preheat. Castings must be stress relieved after welding. Welds made with the 55% nickel rod or wire are magnetic.

Ductile Ni-resists for heat resisting applications can be gas welded using cast rod of matching composition. Since there are no commercially available matching rods, foundries intending to make repair welds must cast their own welding rods and develop qualification procedures for weld repairs.

Stainless steel rod or wires (E310 or E308) with matching coefficient of expansion can also be used to repair heat-resisting ductile Ni-resist castings. A 600–900F (316–482C) preheat is suggested for welding with stainless steel.

Since large or complex castings frequently have defects from the casting process, a prohibition on all repair practices places a difficult burden on the procurement of castings. Qualification for weld repair of a Ni-resist casting depends upon the criticality of the area to be welded and the casting application, and should be performed by agreement with the customer.

CONCLUSION

The information in this chapter is intended to present the experience and illustrate the procedures that have been developed over many years in the production of ductile Ni-resist. The information is presented mostly in the form of a comparison to ductile iron production experience, since most of the Ni-resist producing foundries have extensive experience in the production of ductile iron.

4

Composition of Ductile Irons

Richard B. Gundlach

Climax Research Services
Farmington Hills, Michigan

Carl R. Loper, Jr.

Dept. of Materials Science and Engineering
University of Wisconsin—Madison
Madison, Wisconsin

Bernardo Morgensteren

Casting Division
Ford Motor Company
Dearborn, Michigan

■ INTRODUCTION

Many factors control the microstructure and properties of ductile iron. The major factors affecting these properties are: (1) molten metal processing, (2) chemical composition and (3) solidification rate and cooling rate of the solid. Chemical composition is the primary factor affecting graphite shape control, and it has a major influence on the microstructure of the metallic matrix. Each element has a unique effect on either the solidification structure (graphite morphology) or the metallic matrix microstructure. The many elements present in ductile irons can be classified by their influence on microstructure. They include:

- primary elements—C, Si, Mn, S and P;
- spheroidizing elements—Mg, rare earths, Ca, etc.;
- alloying elements—Cu, Ni and Mo;

- residual and special purpose elements—Al, Bi, Pb, Sb, etc.;
- pearlite and carbide-promoting elements—As, B, Cr, Sn and V;
- gases—H, N, and O.

It should be emphasized that composition control is fundamental to achieving a spheroidal graphite structure with the desired metallic matrix microstructure.

DUCTILE IRON AND THE FE-C PHASE DIAGRAM

The solidification process in ductile iron consists of the formation of two solid phases from the liquid iron: graphite and austenite (the solid metallic phase). The desired solidification process for ductile iron consists of a sequence of events. Initially, precipitation of solid graphite nuclei from the liquid iron should occur. This is followed immediately by the "eutectic" reaction, which is the simultaneous formation of two solid phases from the liquid iron, i.e., the formation of graphite and austenite (the metallic iron phase). For this solidification sequence to occur, the iron must be of a hypereutectic composition.

The iron-carbon phase diagram is very useful for describing the influence of composition on the solidification process. The Fe-C diagram, as shown in Figure 4-1, defines the phases present in iron-carbon alloys (steel or cast iron) at any given temperature and carbon content under equilibrium conditions. Table 4-1 contains a list of definitions of the microconstituents and features described in Figure 4-1.

The eutectic composition, which is 4.26%C for a pure Fe-C alloy, is the unique composition that, when solidifying, will transform from liquid to solid by forming two solid phases—graphite + austenite—simultaneously at a single temperature (the eutectic temperature).

To satisfy the need to promote precipitation of graphite at the beginning of solidification, most ductile irons are produced from alloys that have hypereutectic compositions. Hypereutectic iron compositions are those that have carbon contents somewhat greater than the eutectic composition, i.e., carbon contents to the right of the eutectic composition (see Figure 4-1).

For *hyper*eutectic iron compositions, the liquid cools through the two-phase field—liquid + graphite—before reaching the eutectic temperature. Consequently, graphite is expected to precipitate from the melt prior to eutectic solidification, when both graphite and austenite form from the melt at the same time.

For *hypo*eutectic iron compositions, iron chemistries to the left of the eutectic composition, the liquid cools through the two-phase, liquid + austenite field before reaching the eutectic temperature. Under these circumstances, primary austenite dendrites precipitate from the melt prior to eutectic solidification.

CARBON EQUIVALENT—THE C-SI RELATIONSHIP

Many elements found in commercial cast irons influence the eutectic composition. They either lower or raise the carbon content of the eutectic composition, primarily through their influence on the solubility of carbon in the liquid iron. The most important element having an influence on the eutectic carbon content is silicon. Silicon lowers the solubility of carbon and, thus, lowers the carbon content of the eutectic composition, as illustrated in the Fe-2%Si-C phase diagram in Figure 4-2.

The effect of silicon is very predictable and is often described as a substitutional element for carbon in a "carbon equivalent" relationship. While several different carbon equivalent (CE) relationships have been derived to describe various aspects of the silicon effect, the most commonly used relationship is:

Carbon equivalent (CE) = %C + (%Si/3)

Using this equation, the foundryman can go to the Fe-C phase diagram to determine whether the composition of his iron is hypoeutectic or hypereutectic, i.e., below or above 4.26% CE.

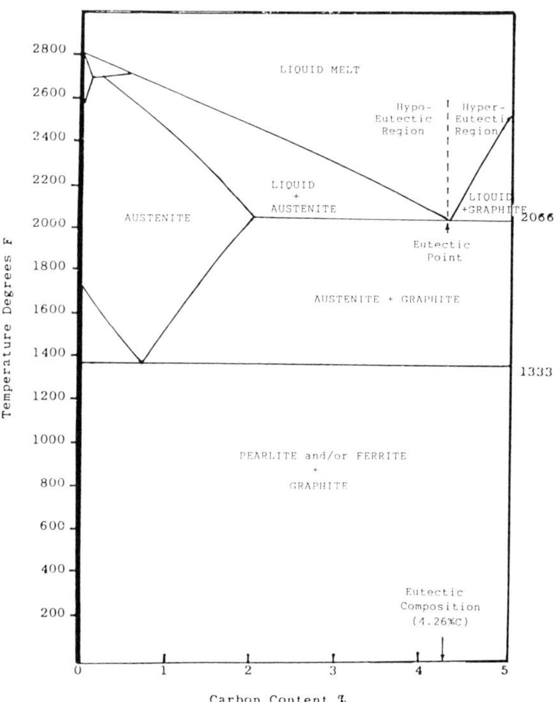

Fig. 4-1. Simplified iron-carbon equilibrium diagram at 0.0% silicon.

TABLE 4-1. DEFINITION OF TERMS IN FIGURES 4-1 AND 4-2

TERM	DEFINITION
Liquidus temperature	The temperature at which solidification of an alloy begins, i.e., temperature when primary austenite or graphite first precipitates from the melt.
Solidus temperature	The temperature at which solidification of an alloy is completed. (Temperature at which eutectic solidification is complete.)
Eutectic	The solidification reaction when the liquid iron solidifies to form both graphite spheroids and austenite, simultaneously. The eutectic composition is a specific chemistry or carbon equivalent (4.26%CE). The eutectic reaction occurs at a fairly constant temperature.
Graphite	A crystalline form of carbon having a hexagonal structure. It is one of the two solid phases freezing out of liquid iron during eutectic solidification. It is the spheroidal, gray constituent displayed in Figure 4-3.
Austenite	Austenite is the solid metallic phase that forms when cast iron solidifies. It can dissolve up to 2%C at the solidification temperature, and is normally stable only at elevated temperatures With high alloy additions it can be stabilized at room temperature, see microstructure in Figure 4-4.
Ferrite	Ferrite is the matrix microconstituent stable at room temperature. It is a single phase having essentially no solubility for carbon. It is a transformation product of austenite and the principal constituent (white) of the matrix illustrated in the microstructure of Figure 4-3.
Pearlite	Pearlite is a transformation product of austenite consisting of two phases, ferrite and cementite (iron carbide). It is stable at room temperature. It has a fine, lamellar structure consisting of alternate platelets of ferrite and carbide, and is dark-etching. It is the dark matrix microconstituent displayed in Figure 4-5.

The extent to which the composition is raised above the eutectic composition in ductile iron is dependent on the difficulty associated with precipitating graphite in the liquid iron prior to eutectic solidification. The thermodynamic driving force for graphite precipitation is low. Therefore, for graphite precipitation to occur, there is a need to (1) promote nucleation of the graphite, and (2) produce a high chemical driving force. The former is produced by inoculation of the molten metal, and the latter produced by increasing the carbon content above the eutectic composition. As the solidification rate increases or section size decreases, the CE must be raised to higher values above the eutectic composition to assure that adequate graphite precipitation will occur and carbides will be avoided.

It will be necessary to increase the carbon equivalent with decreasing section size. For example, the recommended CE for sections of 2 inches is 4.4%, while for 1/4-in. sections it is 4.9%. Likewise, if the iron composition is excessively hypereutectic for the section size and degree of inoculation, premature graphite precipitation can occur, and graphite flotation and agglomeration in the cope surface develops. Consequently, the carbon equivalent must be controlled within a specific range for each casting and molten metal handling practice.

AUSTENITE TRANSFORMATION

The matrix microstructure in a ductile iron casting is determined by the mode of transformation that occurs when the austenite matrix cools through the critical temperature range. Depending on the ductile iron grade, the desired microstructure can vary from a fully ferritic structure, to mixed ferrite and pearlite, or a fully pearlitic microstructure. Tempered martensitic microstructures are attained upon heat treatment. Whatever the desired microstructure may be it is achieved by controlling the transformation of the austenite phase, which is controlled largely by chemical composition.

As the solid casting cools from the freezing point to room temperature, the austenitic metallic matrix will transform to other phases that are more stable at low temperatures. This transformation will not take place until the austenite has cooled below the upper critical temperature, i.e., the upper boundary of the "austenite + ferrite + graphite" field shown in Figure 4-2. The alloy content of the iron, the number and distribution of graphite nodules, and cooling rate are the dominant factors determining the transformation products of austenite on cooling to room temperature.

Given adequate time to cool through this transformation region, the microstructure will be fully ferritic. With more rapid cooling rates, a mixture of pearlite and ferrite will occur. A largely ferritic microstructure can be achieved in the as-cast condition, when the composition is properly controlled. A fully pearlitic matrix microstructure will develop, providing sufficient pearlite-promoting elements are added to the iron.

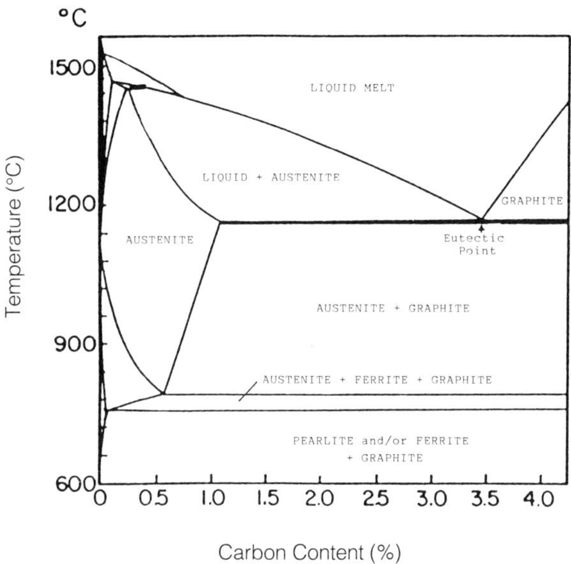

Fig. 4-2. The Fe-2.4%Si-C phase diagram.

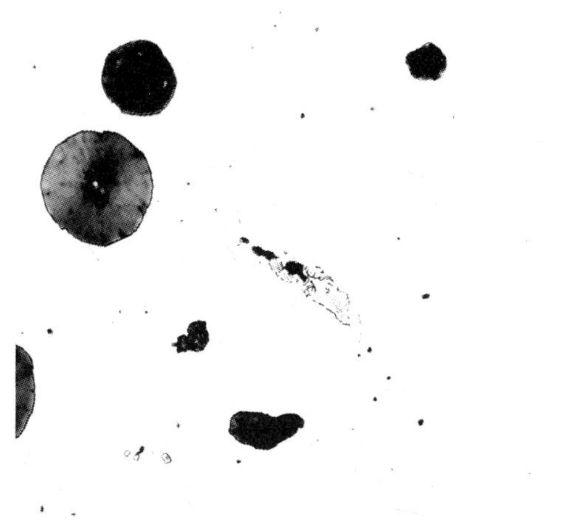

Fig. 4-4. Ni-Resist D-2C containing 0.28% chromium; etched, ×250.

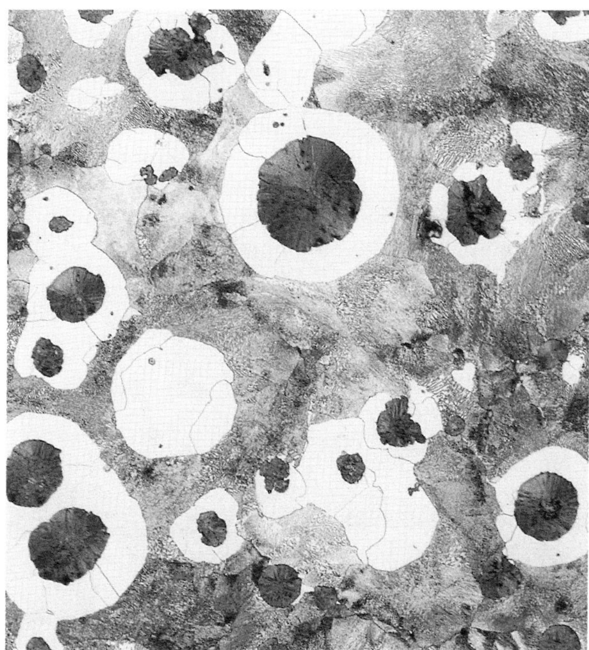

Fig. 4-3. Graphite and ferrite formations in ductile iron; 2% nital etched, ×200.

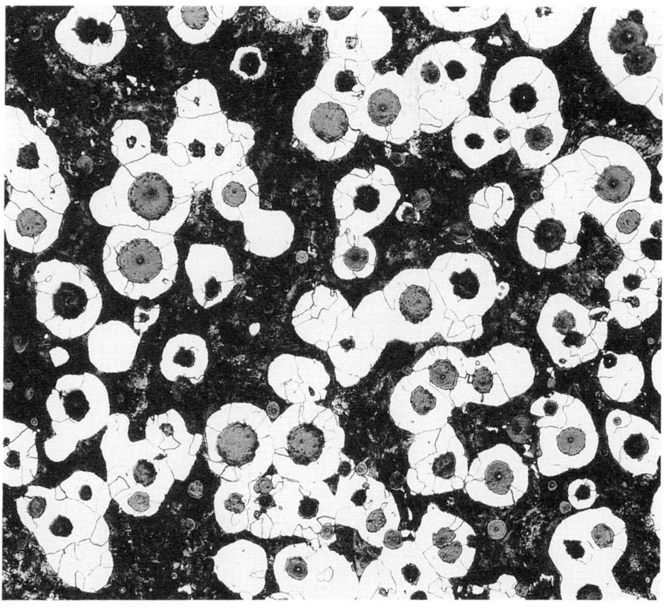

Fig. 4-5. 50% pearlite–50% ferrite; etched, ×100.

69

When high-strength grades are desired, heat treatment is usually required in order to develop martensitic microstructures. In these circumstances, alloying is required to increase hardenability and, thereby, inhibiting the transformation to ferrite and pearlite. With sufficient alloying, acicular microstructures free of ferrite and pearlite can be produced in the as-cast condition.

Austenitic ductile irons, with a stable austenitic matrix, can also be produced with sufficient alloying.

■ DUCTILE IRON CLASSIFICATION

ASTM issued standards for ductile iron graphite classification based on type, size and distribution. ASTM A 247-67 (reapproved 1984) "Standard Method for Evaluating the Microstructure of Graphite in Iron Castings," consists of a method in which characteristic features of graphite particles observed in a microstructure are compared with the idealized microstructures published in charts or plates from which a rating based on similarity to the ideal is reported. Examples of graphite classifications are shown in Table 4-2 and Figures 4-6 and 4-7.

TABLE 4-2. CLASSIFICATION OF GRAPHITE BASED ON SIZE[1]

Size Class	Maximum Dimension (mm) × 100
I	128
2	64
3	32
4	16
5	8
6	4
7	2
8	1

■ GRAPHITE SPHEROIDS— MICROSTRUCTURES

GRAPHITE SIZE (NODULE SIZES)

As shown in Table 4-2, the size class is designated by a numeral, 1 through 8. The numbers represent a maximum dimension (in mm) of Type I and II nodules measured in a micrograph at 100 magnifications (×100). Often, more than one size will be present in the same micrograph, in which case the sizes may be reported as percentages of the total graphite area shown, or more commonly, indicating the sizes more predominant with the presence of few nodules of the other sizes. (Example: Predominantly sizes 5 and 6 with some 4 and 7.)

GRAPHITE FORM (NODULE TYPES)

The graphite form types are designated by Roman numerals, as shown in Figure 4-6. Reporting of nodule types is usually done by scanning the sample micrograph at ×100, and noting the graphite shapes that more closely match the shapes in Figure 4-6. The percentages of each type are either estimated visually, or counted individually. The sum of the percentages add up to 100.

GRAPHITE POPULATION (NODULE COUNT)

Nodule counts are expressed in number of nodules per square millimeter. Figure 4-7 shows typical micrographs at ×100, with nodule counts in increments of 25 and 50 per mm². Manual estimation consists of counting particles in the microscope-superimposed grid and visually comparing this to the selected standard. More accurate results are obtained with image-analyzing microscopes.

ANOMALIES IN GRAPHITE FORMATIONS

Aligned Graphite: Distinct rows of graphite nodules formed on long dendrites that can cause reduction of mechanical properties. Poor nucleation and high pouring temperatures can cause these formations.

Agglomerated Graphite: Groupings or nodule clusters, generally comprised of small-size nodules in over-inoculated fields. Sometimes associated with shiny/hard spots of undissolved inoculant seen on machined surfaces.

Carbon Flotation: Free graphite that forms prior to solidification in hypereutectic iron and floats to the top of the

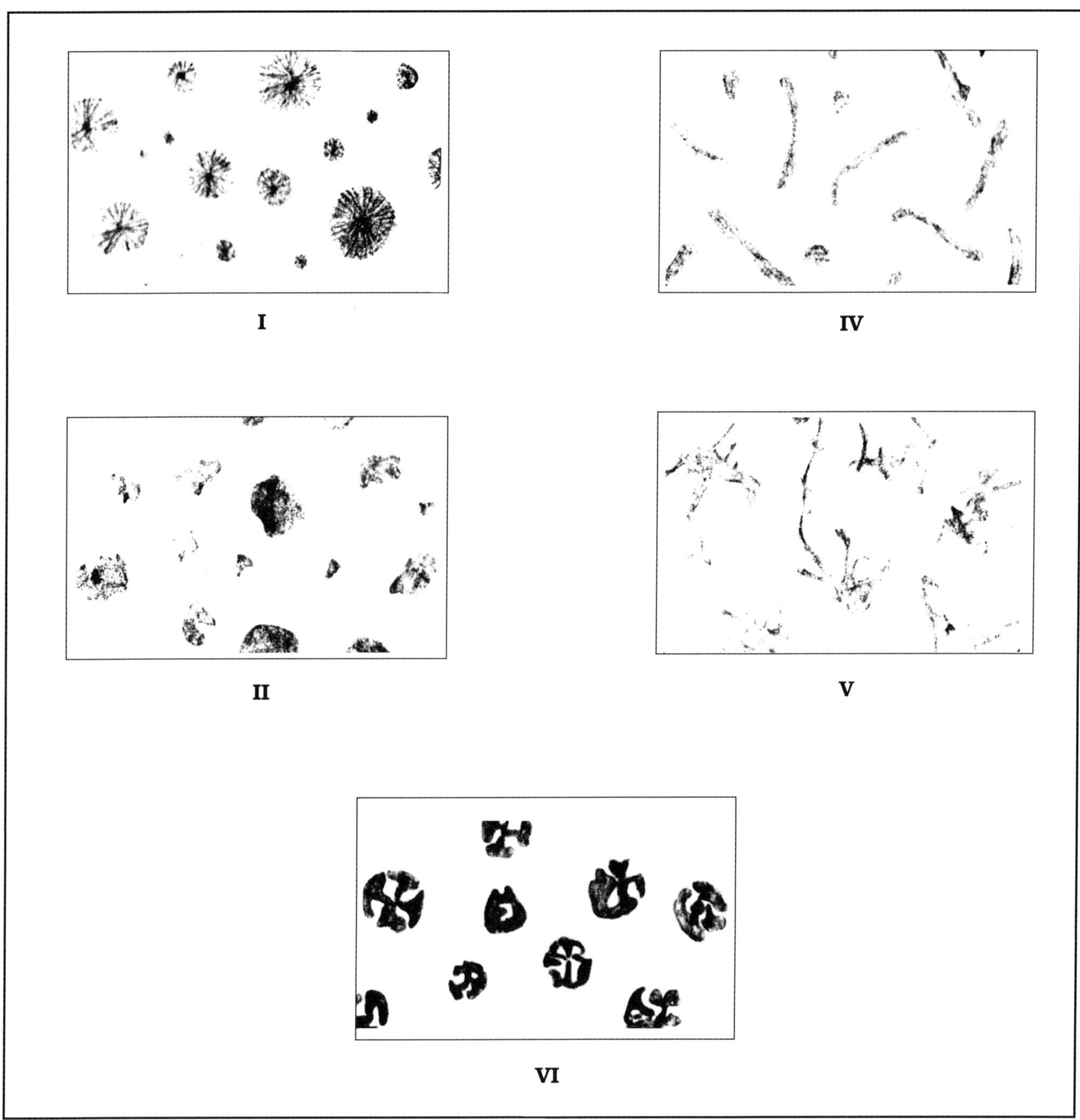

Fig. 4-6. Classes of ductile iron graphite forms, ×100.

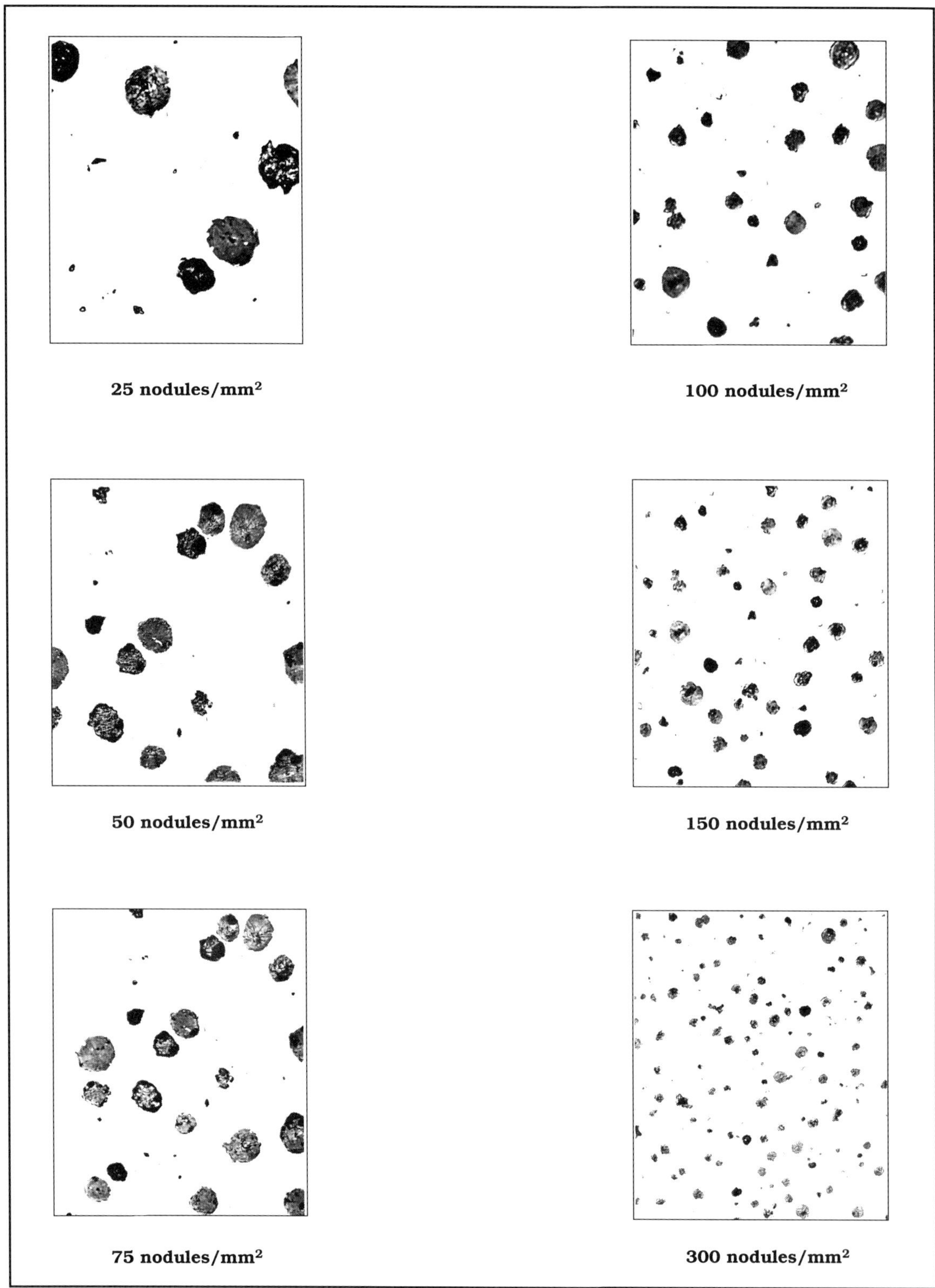

Fig. 4-7. Ductile iron graphite classification based on nodules/mm².

casting (or to the inside surface of a centrifugal casting) as the casting solidifies. Figure 4-8.

Compacted Graphite: An intermediate form of graphite characterized by short, thick interconnected flakes with rounded ends. Vermicular is similar in appearance. Figures 4-9, 4-10, 4-11, 4-12 and 4-13.

Crab Graphite: A degenerate form of nodular graphite (caused by magnesium over-treating and/or subversive elements) characterized by distorted nodules with extensions or thick flakes, resembling crab legs. Figure 4-14.

Exploded Graphite: A form of graphite usually found near the cope surface of castings, in heavy-section hypereutectic irons with high magnesium residuals. Also caused by heavy/late inoculation. Aggravated by excessive amounts of rare earths. Figures 4-15, 4-16.

Flake Graphite in Nodular Iron: Could be present as a surface layer or as "islands of gray" immersed in the casting. Generally attributed to mold-metal interface sulfur reactions, sulfur reversions of dross, or untreated metal splashes in the mold cavity. Figure 4-17.

Irregular Graphite: Distorted shape nodules mostly associated with lack of nodulizing agent or late postinoculation. Excessive cerium and over treatment can also cause nodule deformities. Mechanical work (hot or cold) can change nodule shapes. Figures 4-11, 4-12, 4-18, 4-19 and 4-20.

Nodular Graphite: Rounded, spheroidal form of graphite created by treatment of low-sulfur-base iron with nodulizing agents. Figure 4-21.

Secondary Graphite: Graphite particles that form after solidification by precipitation of previously combined carbon. It can be seen in ferritizing treatments or in sufficiently slow-cooled castings. Secondary graphite may form a concentric sphere surrounding primary (or original) nodules.

Spiky Graphite: Nodules with one or more sharp protrusions, similar to crab graphite.

Vermicular Graphite: Degenerated form of graphite consisting of thick, rounded-wormlike flakes, usually resulting from low magnesium residuals. Similar in appearance to compacted graphite.

■ INFLUENCE OF THE PRIMARY ELEMENTS IN DUCTILE IRON

The primary elements in ductile iron are generally considered to consist of carbon, silicon, sulfur, phosphorus and manganese, since they are present to some extent in all alloys produced. Their influence will be discussed in terms of the individual elements. However, the acceptable level of one element often affects the tolerable level of another element. A typical ductile iron composition is: 3.65% C, 2.40% Si, 0.01% S, 0.02% P, and 0.40% Mn. It is apparent that the carbon equivalent (CE) is much higher and the sulfur content is much lower than that encountered in gray cast irons. The interaction of carbon and silicon contents makes it necessary to discuss these elements

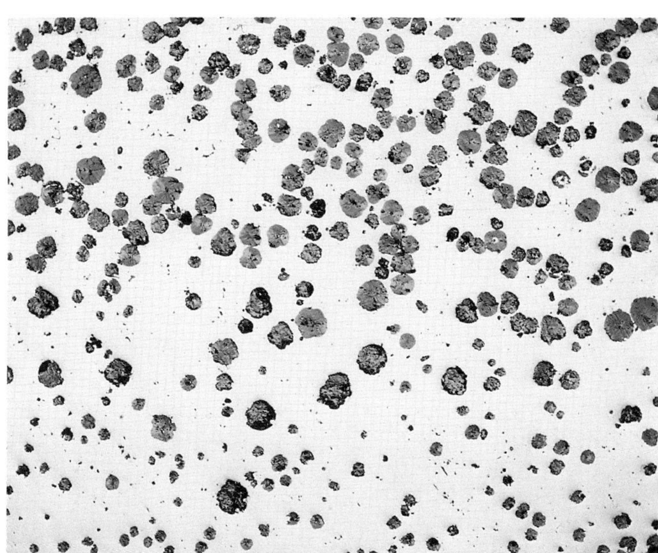

Fig. 4-8. Carbon flotation shown in the cope area of a casting section; unetched, ×50.

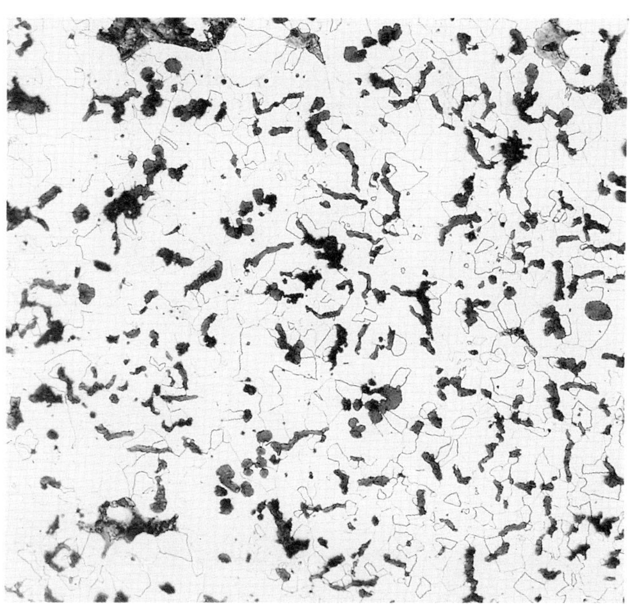

Fig. 4-9. A ductile iron with a 50% nodularity rating.

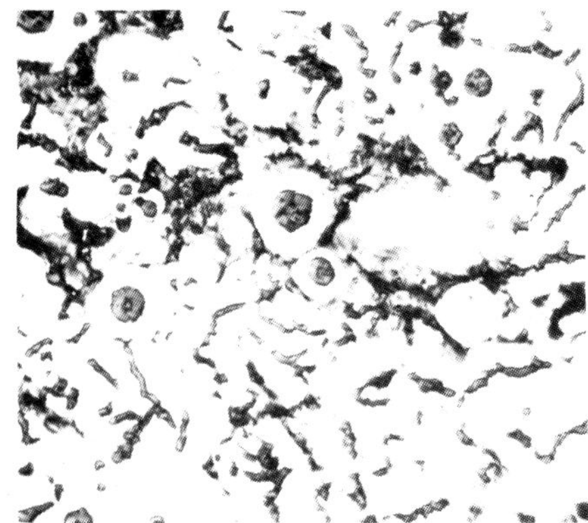

Fig. 4-10. Vermicular graphite (C.G. iron) with scattered nodules in a primarily ferritic matrix, ×100, 2% nital etched.

Fig. 4-11. Vermicular graphite showing a few irregular nodules in a pearlitic-ferrtic matrix, ×250, nital etched.

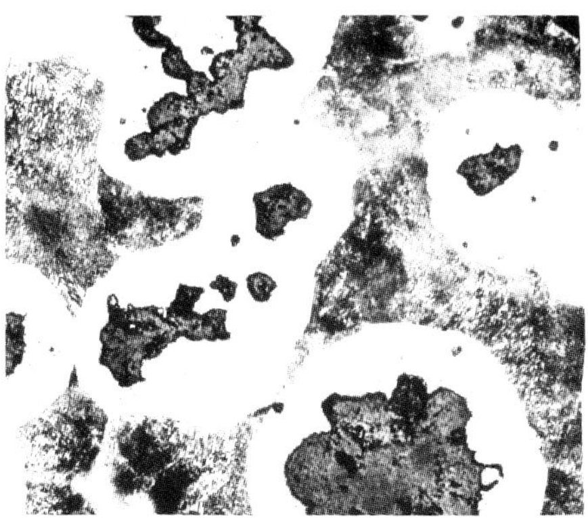

Fig. 4-12. Extremely coarse, irregular nodules showing a vermicular formation in a pearlitic-ferrtic matrix, ×250, 2% nital etched.

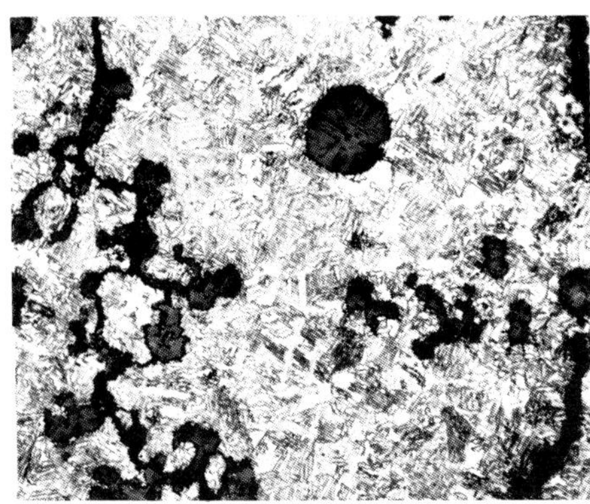

Fig. 4-13. A crack following a vermicular graphite pattern in the matrix, ×100, 2% nital etched.

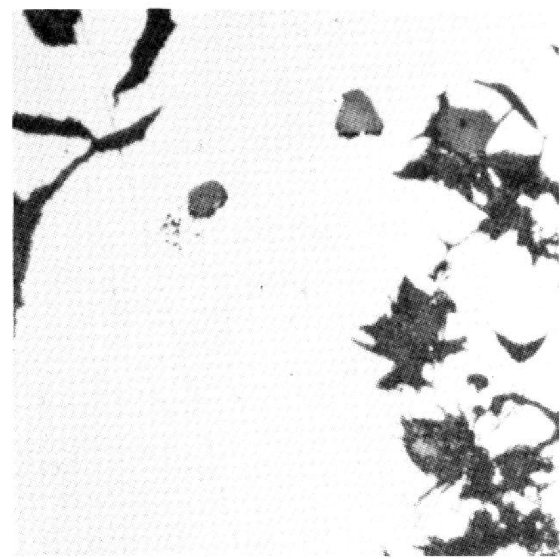

Fig. 4-14. Crab graphite is shown at ×1000, unetched.

Fig. 4-15. Exploded graphite spheroids in a heavy-section hypereutectic ductile iron casting, ×100, 2% nital etched.

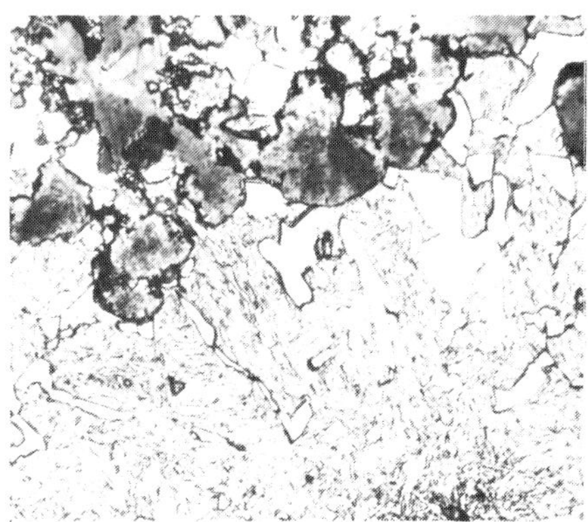

Fig. 4-16. Exploded graphite at ×250, nital etched.

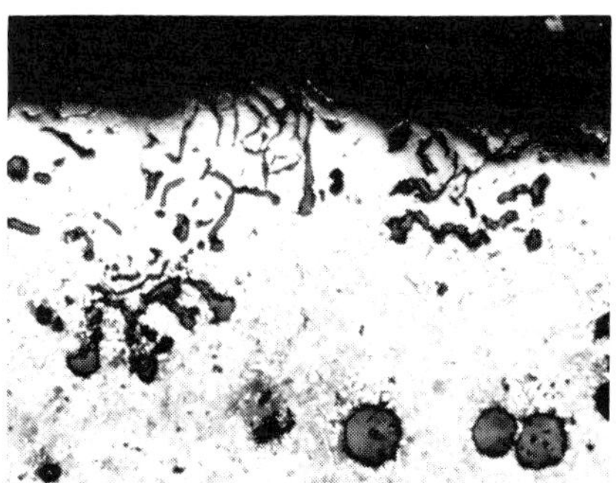

Fig. 4-17. Surface-sand reaction (flake skin) is shown. It commonly occurs to a depth of about 0.02 in. at the cast surfaces of heavy ductile iron castings, or at much greater depths in castings poured in molds made with certain organic chemical binders, ×100, 2% nital etched.

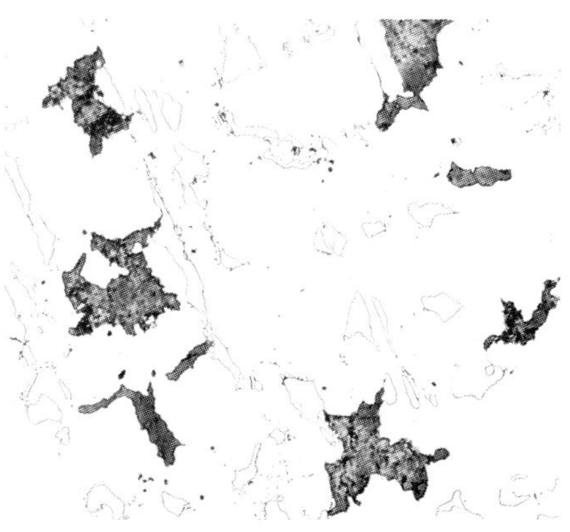

Fig. 4-18. Irregular graphite in a carbidic matrix caused by over treatment with cerium (0.20% Ce), ×250, 2% nital etched.

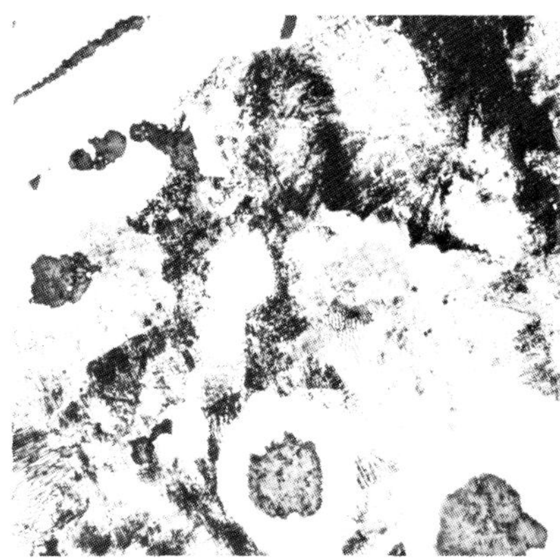

Fig. 4-19. Irregular graphite nodules and chunk graphite, ×250, 2% nital etched.

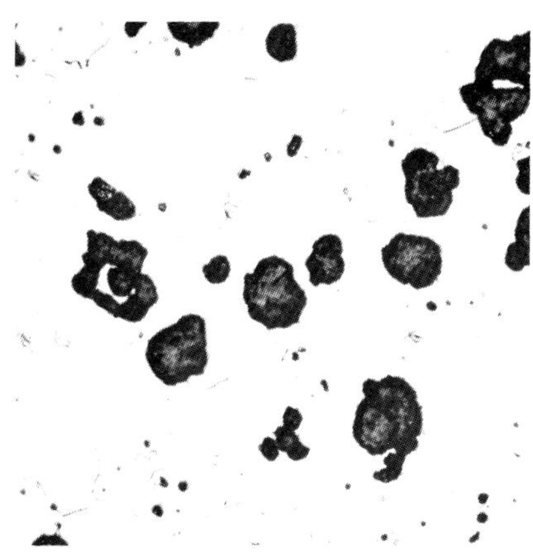

Fig. 4-20. Irregular graphite in a ferritic matrix, ×250, 2% nital etched.

Fig. 4-21. Graphite nodule surrounded by ferrite and pearlite, ×1500.

together, even though there are significant individual effects that must also be considered. A "Summary of the Significant Effects of Elements on Ductile Iron" is provided at the end of this chapter (Table 4-19).

CARBON AND SILICON

The combined effects of carbon and silicon may be described using the Henderson diagram, Figure 4-22. While the carbon content present in commercial ductile iron ranges from below 3.0% to over 4.0%, it is more typical to encounter carbon content between 3.5 and 3.9% with the higher values typical of thin-section castings (e.g., below 3/8 in.) and the lower values found in heavy-section castings (e.g., over 1.5 in.). The combination of high carbon content (more accurately, carbon equivalent) and low solidification cooling rates will result in graphite flotation and the presence of degenerate graphite. For thin-section (less than 1/2 in.) castings a 4.55% CE is recommended; for moderate sections (between 1/2 and 1-1/2 in.) a 4.35 to 4.45% CE may be used; and for heavy sections (over 2 in.) the CE may be limited to a maximum of 4.30%. In centrifugal ductile iron castings, the carbon content may be limited to a maximum of 3.5% to minimize the formation of surface wrinkles.

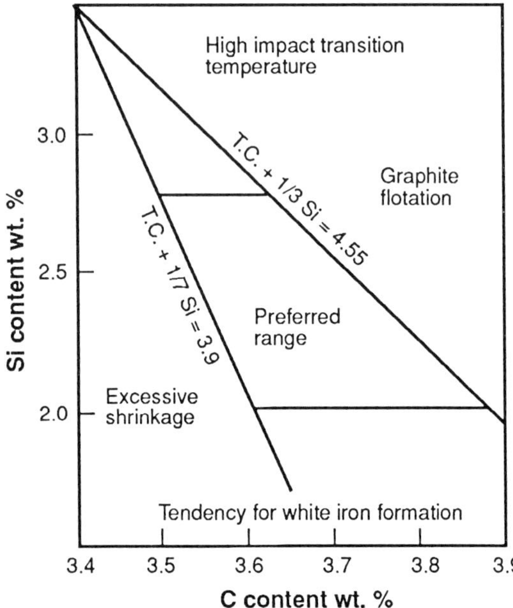

Fig. 4-22. Typical carbon and silicon ranges for ductile irons as limited by processing factors (After H. Henderson).

During the solidification of ductile irons, the formation of graphite is accompanied by a volumetric expansion, so that, as the carbon (and silicon) content is reduced, the feed metal requirements of a casting may be increased and feeding within a casting may be impaired. The Henderson diagram suggests a lower limit of 3.9 (%C + 1/7 %Si) to avoid excessive shrinkage, but this parameter must be considered only a guideline, since effective feeding of a casting depends upon a number of additional factors other than carbon and/or silicon content.

Typical silicon content of ductile irons range from 1.80–2.80%, although lower and higher levels (e.g., oxidation-resistant alloys containing up to 6.0% Si) are encountered. Silicon is a potent graphitizer, so that, at low silicon content, there is an increased tendency for carbide formation encountered as chilled edges, intercellular carbides and centerline carbides. Silicon increases the nodule count, decreases the eutectic cell size and results in decreased carbide formation—although these effects are time dependent (fading) and are realized through effective inoculation procedures.

Note that, as shown in the Henderson diagram, an excessive silicon content results in a high-impact transition temperature. It should be noted that this is not an effect realized at a given silicon content, but that the impact properties are gradually reduced as the silicon content is increased. Furthermore, this effect is of significance primarily in ferritic ductile irons, as illustrated in Figures 4-23 and 4-24. The impact properties of pearlitic irons are relatively low. The difference between the upper- and lower-shelf energies is relatively small, and the transition temperature is increased with increased pearlite content, so that the effect of silicon is substantially muted. The highest impact properties (and lowest transition temperatures) are obtained in fully ferritic irons having low silicon and phosphorus content.

The influence of carbon and silicon on the mechanical properties of fully spheroidal ductile iron must be considered in terms of the following variables:

- the effect of these elements on the nodule count of ductile iron;
- the effect of these elements on the pearlitic hardenability of ductile iron (the formation of ferrite and/or pearlite in the as-cast, annealed or normalized condition);
- the effect of these elements on the properties of ferrite present in the matrix structure;
- the effect of these elements on the hardenability of ductile iron (the formation of martensite and/or bainite during quench and temper heat treatments).

The nodule count effect has previously been described. Increased nodule count generally accompanies higher carbon and silicon content, pretreatment of the base iron with graphite and/or ferrosilicon, and inoculation of treated iron with ferrosilicon.

During relatively slow cooling (e.g., as-cast, normalized and annealed) of ductile iron, through the critical temperature range, austenite will transform to ferrite around the graphite spheroids. Therefore, the amount of ferrite formed depends upon the cooling rate (slower cooling rates yield higher ferrite content), the composition of the austenite (all silicon is present in the austenite and higher silicon contents yield higher ferrite content) and the cooling rate (slower cooling yields higher ferrite content).

Below the critical temperature, all of the silicon is dissolved within the ferrite, thereby affecting the properties of the ferrite. Silicon hardens and strengthens the ferrite, and this effect is directly responsible for the embrittlement previously discussed. Silicon also moderately increases the hardenability of ductile iron.

As a general rule, carbon has little effect on the mechanical properties of ductile irons. The tensile and yield strengths are reduced about 350 psi and 375 psi, respectively, for each 0.10% increase in carbon content, and the hardness is decreased about 5 HB for each 0.15% increase in carbon content. Increased carbon content also increases elongation values and improves impact properties. These characteristics are all related to the ferrite content of the ductile iron. In addition, the modulus of elasticity is modestly affected in proportion to the volume of graphite present.

Silicon raises the tensile and yield strengths about 1000 psi and 1600 psi, respectively, and decreases the elongation (about 0.3%) and hardness (about 3 HB) for each 0.10% increase in ferritic irons. The impact properties' effects are shown in Figures 4-23 and 4-24.

Examples of the effect of silicon on tensile properties are shown in Table 4-3, and the effect of silicon on microstructures is illustrated in Figures 4-25 and 4-26. Figures 4-25 (heat WJA43, 2.01% Si) and 4-26 (heat WJA37, 2.89% Si) illustrate the effect of silicon at the same carbon equivalent and at 0.38–0.39% Mn. Note that the pearlite-forming effect of manganese and phosphorus has been overcome by the high silicon content. The change in weight of a series of ductile irons of different carbon and silicon content is shown in Figure 4-27.

It is important that the effect of silicon on properties be related to the manganese content of the ductile iron (Figures 4-28a and 4-28b). At low manganese levels, silicon increases the tensile and yield strength of as-cast ferritic irons, while at 0.40% Mn, increasing the silicon content decreases the tensile strength as the pearlite content of the matrix is increased. When annealed, the influence of silicon on the strength and hardness is the range of manganese levels. The combined effects of silicon and manganese on matrix structures of different section thicknesses of as-cast and annealed ductile irons is illustrated in Figures 4-29 and 4-30, respectively.

SULFUR

Desulfurization of the melt is an essential segment of the process of producing ductile iron because magnesium reacts with sulfur to form magnesium sulfide. Economical treatment of the base iron, therefore, dictates the use of low-sulfur raw materials and melting processes, which do not raise the base-iron sulfur content. The base-iron sulfur content significantly affects the amount of magnesium required to promote and maintain graphite spheroidization.

Magnesium sulfide will form in the iron as a dross, which will tend to float to the melt surface, but could be entrapped within the casting or adhere to ladle or mold walls. If exposed to oxygen the magnesium sulfide (MgS) will oxidize to magnesia (MgO) and may combine with silicon to form magnesium silicate ($MgSiO_3$) (both of which are also dross formers), while the sulfur from the magnesium sulfide may also oxidize to form SO_2 or revert to the melt, where it will consume additional magnesium. This reversion or fading can be minimized by maintaining a low base-iron sulfur level, or by using elements (e.g., rare earths) that form stable sulfides.

Chemical analyses are not effective indicators of the effective magnesium content of a treated iron, because techniques such as spectrochemical analysis do not differentiate between magnesium in solution or in the form of compounds (e.g., MgS, MgO, $MgSiO_2$, etc.). Typically,

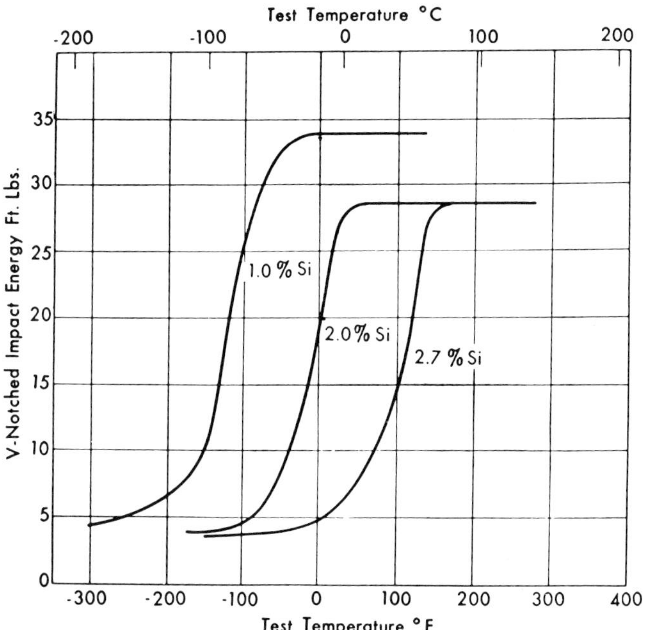

Fig. 4-23. Influence of silicon content on the impact properties (NCIRA Charpy V-notch test bars) and transition temperature of annealed, fully ferritic ductile irons. (GIRI)

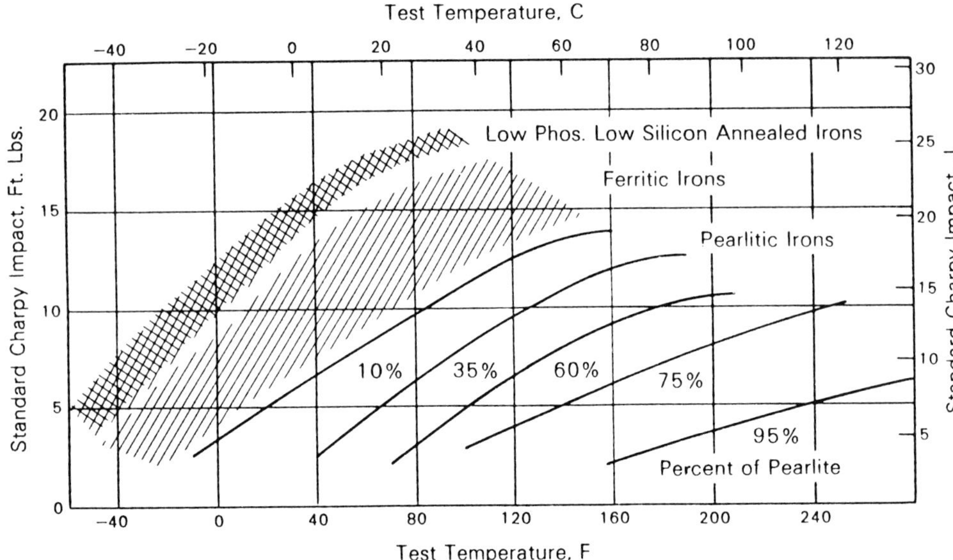

Fig. 4-24. Influence of temperature on the Charpy impact fracture energy of various types of ductile irons.

TABLE 4-3. EFFECT OF SILICON ON MECHANICAL PROPERTIES

Melt	WJA 37	WJA 43
Total Carbon	3.42%	3.66%
Silicon	2.89	2.01
Manganese	0.39	0.38
Nickel	0.06	0.01
Chromium	0.025	0.025
Magnesium	0.048	0.045
Copper	0.01	0.01
Phosphorus	0.033	0.030
Sulfur	0.015	0.012
Carbon Equivalent	4.29	4.26
As Cast		
% Ferrite	80	40
%Pearlite	20	60
Tensile Strength, psi	76,200	79,100
Yield Strength, psi	59,600	53,700
Elongation, %	18.5	11.8
BHN	179	179
Annealed		
Tensile Strength, psi	67,800	61,600
Yield Strength, psi	51,600	42,900
Elongation, %	22.1	26.7
BHN	149	143

high-quality ductile irons are produced at a sulfur content of less than 0.02%. With a base iron of 0.008–0.010% S, excellent nodularity can be achieved with a residual of only 0.018% Mg. In those cases where the base iron contains 0.025–0.035% S, about 0.04% Mg may be required. And, with 0.085% S base irons, a residual Mg level of 0.055% will be required.

Experience gained in the magnesium treatment of base irons with very low sulfur content has demonstrated that the base iron should not be desulfurized to a level below about 0.006% S prior to magnesium treatment. Base irons with lower sulfur contents are less responsive to spheroidization, exhibit lower nodule counts and have a greater tendency for carbide formation.

PHOSPHORUS

In gray cast irons, phosphorus segregates to the eutectic cell boundaries. A similar segregation of phosphorus occurs during the solidification of ductile iron, except that cell boundaries essentially surround each graphite spheroid. This segregation of phosphorus forms a brittle phosphide (steadite) network, which adversely affects ductility and toughness. Phosphorus, however, stabilizes and refines pearlite, increasing the hardness and the yield-to-tensile strength ratio. At high levels (0.08% P) the transition temperature is raised.

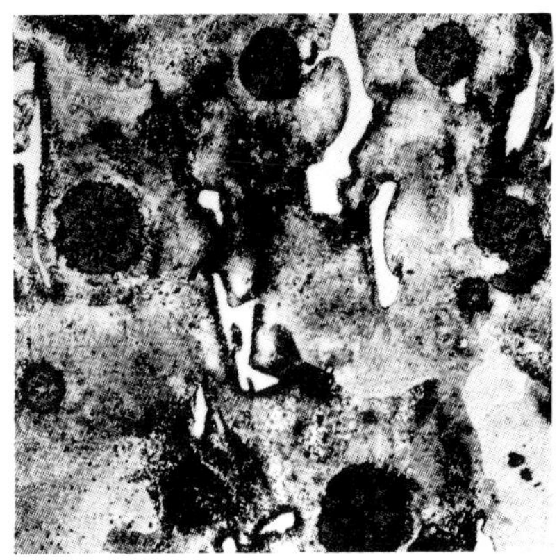

Fig. 4-25. As-cast ductile iron produced with 2.01% Si (Alloy WJA43, Table 4-3) showing carbide formation, nital etched, ×250.

Fig. 4-26. As-cast ductile iron produced with 2.89% Si (Alloy WJA37, Table 4-3) at the same carbon equivalent as Fig. 4-25; nital etched, ×250.

The effect of phosphorus on mechanical properties for as-cast and annealed irons is shown in Figures 4-31 and 4-32, respectively, and in Table 4-4.

MANGANESE

Ductile irons can be produced over a wide range of manganese contents, however, 0.50–0.70% Mn is typical for most as-cast pearlitic grades (e.g., 80-55-06), and a 0.20% Mn will usually result in an as-cast 60-40-18 grade. (These results are very sensitive, though, to casting section size.) Since sulfur is essentially not present in ductile irons, manganese functions as an alloying element, increasing the hardness and strength of the ferrite, stabilizing and refining the pearlite, etc.

This influence of manganese on the as-cast matrix structure is evident, even at low manganese contents. When these irons are annealed to a fully ferritic structure, this level of manganese has very little effect on mechanical properties. The effect of manganese at this level can be offset by silicon, as illustrated in Figures 4-28 through 4-30. The influence of silicon on the ferrite content of ductile iron, in various as-cast sections, is illustrated in Figure 4-33 for low (0.08–0.09%) manganese and in Figure 4-34 for higher (0.49–0.57%) manganese content. Suggested maximum manganese content at various silicon contents have been proposed as a function of casting thickness (Figure 4-35). Increasing the manganese content results in significant increases in as-cast tensile and yield strengths, and in modest increases in these properties in the fully annealed condition. Manganese is about five times more effective than nickel in forming pearlite. Increasing the

manganese content at various silicon contents results in the structural and property changes noted in Table 4-5.

Manganese is also an inexpensive method of increasing the pearlitic hardenability in normalized ductile irons, and martensitic hardenability in quenched and tempered irons. For example, alloying with 1.0% Mn and 1.0% Ni permits a seven-inch-thick casting to be normalized to 300 HB. However, this combination of alloying elements results in a fine intercellular carbide network, Figure 4-36, reducing elongation values. When the manganese content exceeds about 0.70% (depending upon the nodule count and the presence of other alloying elements), this carbide network structure may generally be anticipated.

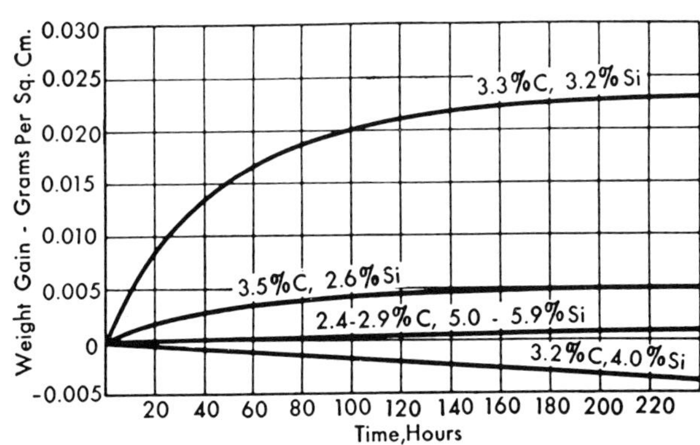

Fig. 4-27. Change in weight in several as-cast nodular irons when held in air at 1600F. (GIRI)

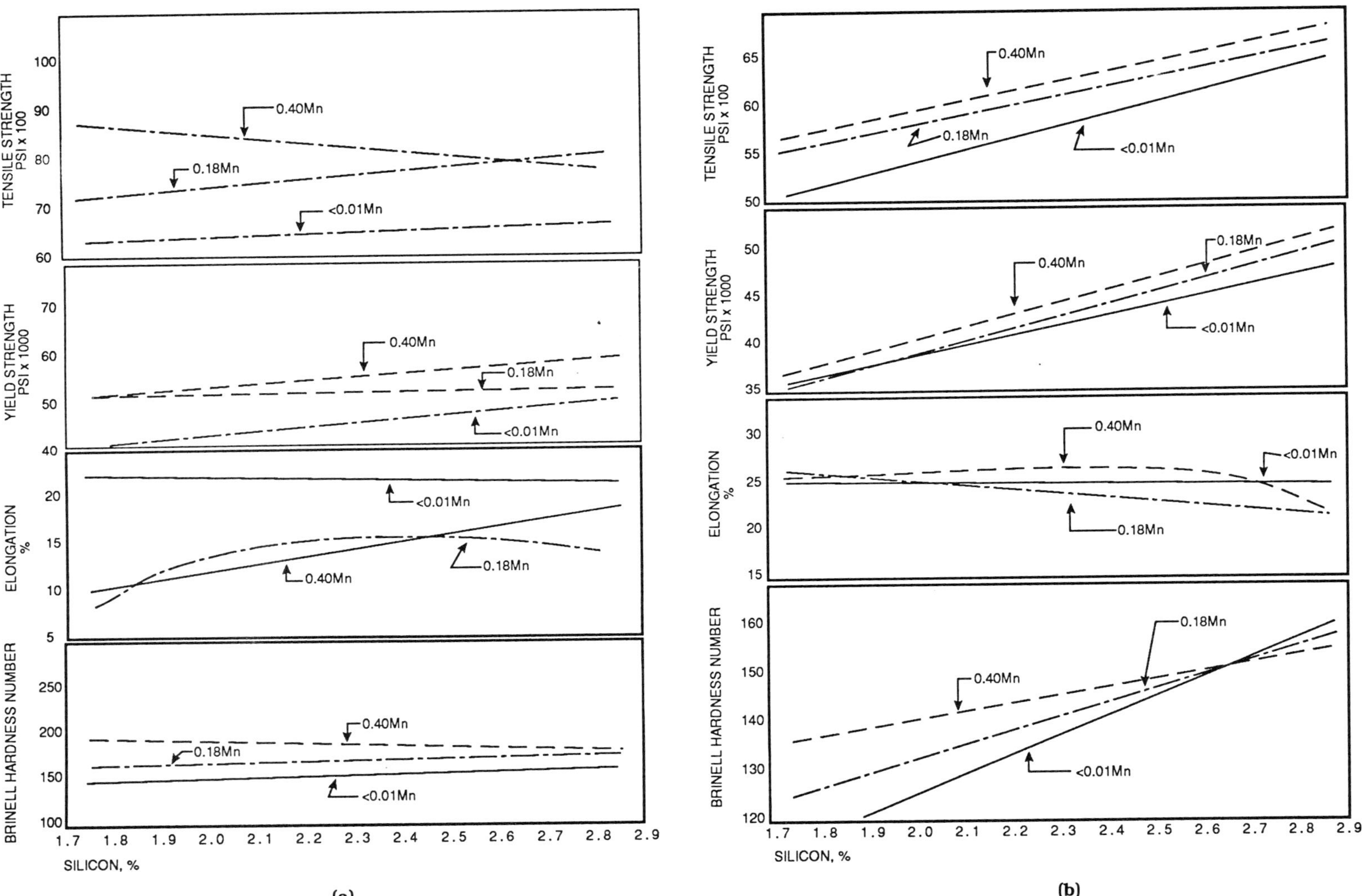

Fig. 4-28. Effect of silicon and manganese contents on the mechanical properties of ductile irons: (a) as cast; (b) annealed.

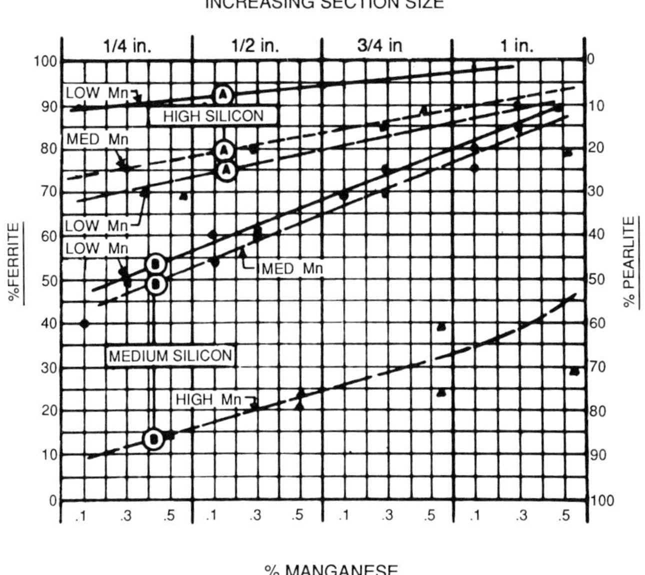

Fig. 4-29. Ferrite/pearlite content of ductile iron in different section sizes as affected by silicon and manganese contents.

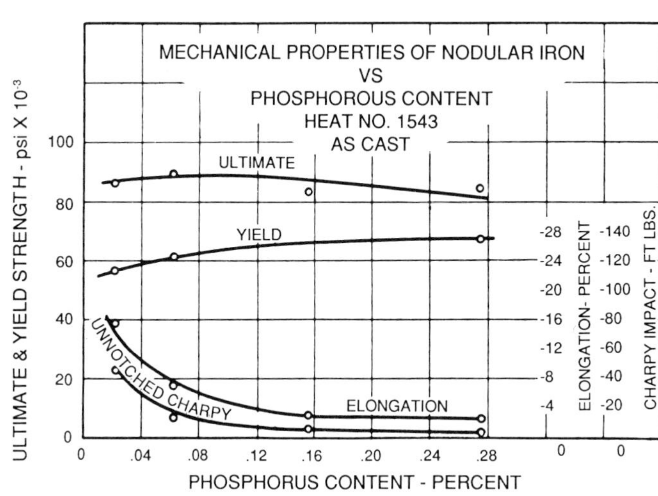

Fig. 4-31. Influence of phosphorus on the mechanical properties of as-cast ductile iron.

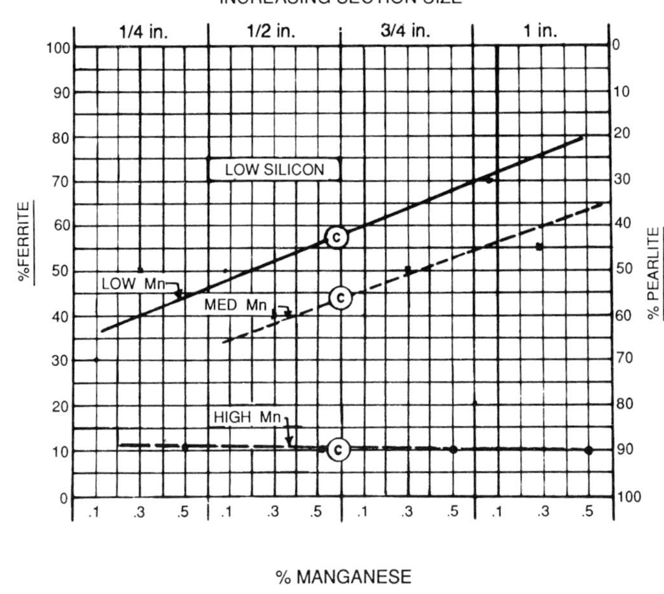

Fig. 4-30. Influence of silicon and manganese contents on the ferrite-pearlite matrix structure of ductile iron castings of various section thickness.

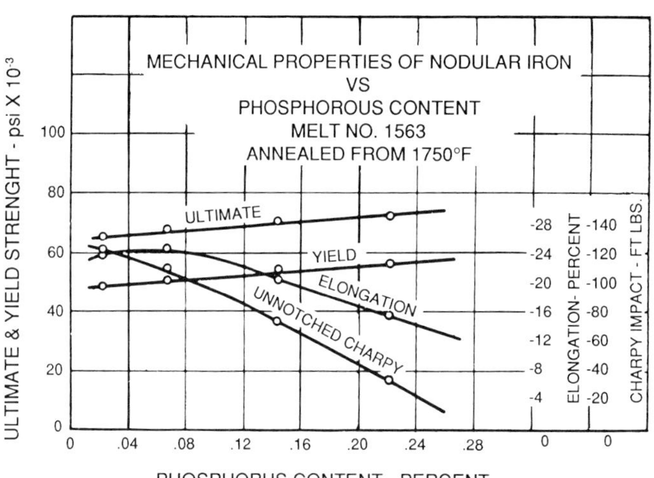

Fig. 4-32. Influence of phosphorus on the mechanical properties of annealed ductile iron.

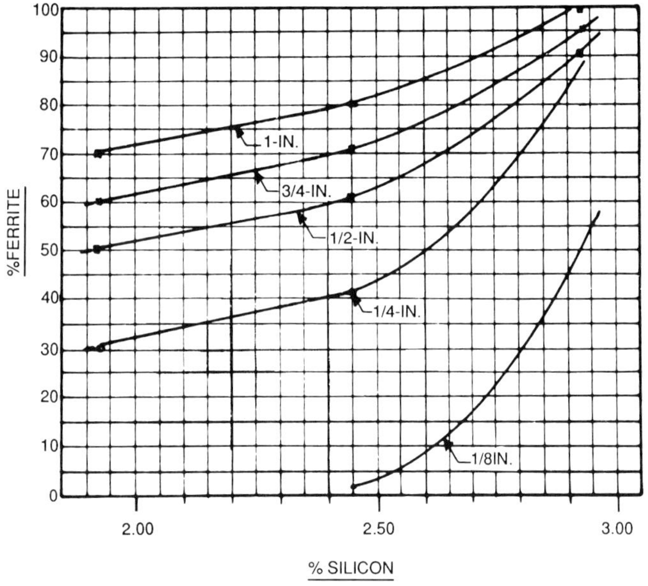

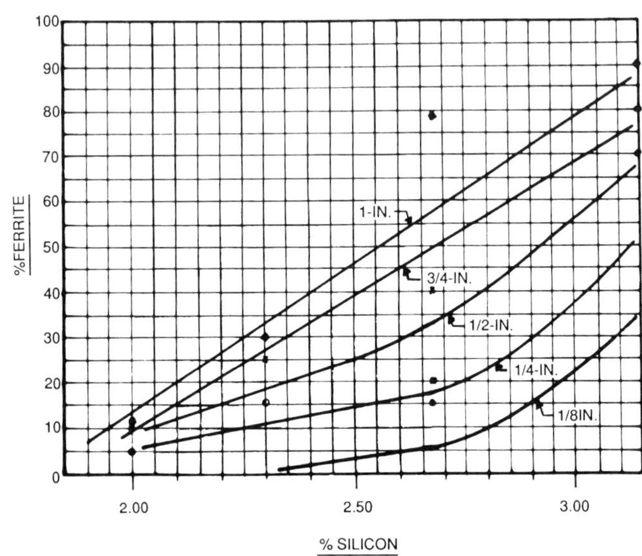

Fig. 4-33. Effect of silicon content and section thickness on the ferrite content of a low manganese content (0.08–0.09%) ductile iron.

Fig. 4-34. Effect of silicon content and section thickness on the ferrite content of a higher (0.49–0.57%) manganese content ductile iron.

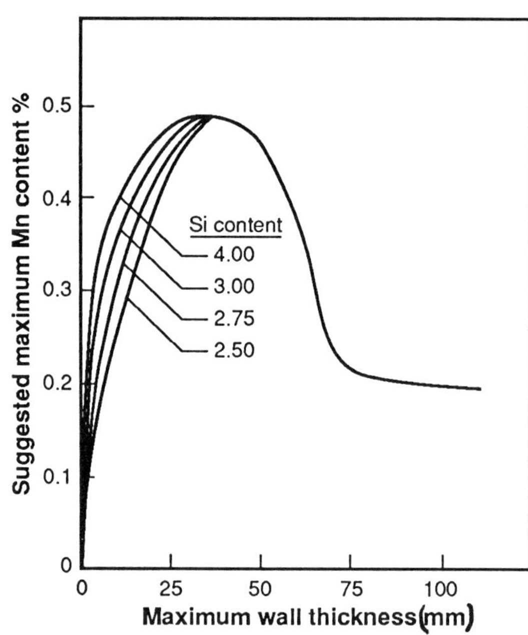

Fig. 4-35. Suggested maximum manganese content for various silicon levels based on maximum wall thickness. (From: *Cast Iron Technology*, R. Elliott, Butterworth—Heinemann Limited [1988]. Reprinted with the publisher's permission.)

Fig. 4-36. Example of the formation of intercellular carbides in a 1.02%Mn-1.0%Ni as-cast ductile iron cast into a 7-in.-thick section, ×2000, etched.

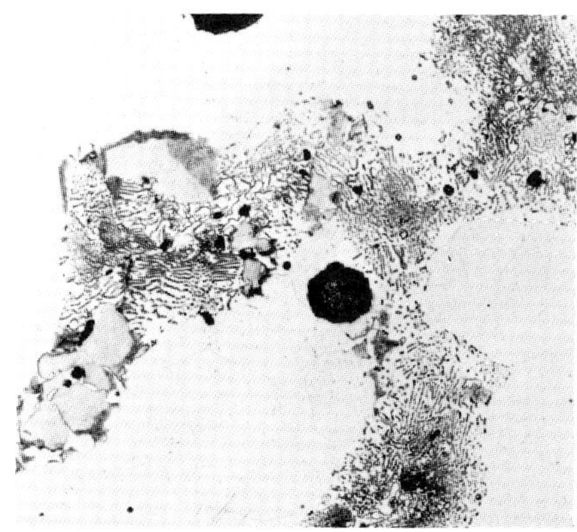

Fig. 4-37. Example illustrating the stability of the intercellular structure in a 1.18% Mn-content ductile iron annealed at 1700F for 4 hr, 2% nital etched, ×250.

Melt	ICP-1	ICP-2	ICP-3
Total Carbon (approx.)	3.60%	3.60%	3.60%
Silicon	2.22	2.29	2.32
Manganese	0.08	0.09	0.11
Phosphorus	0.03	0.06	0.225
Sulfur (approx.)	0.01	0.01	0.01
Nickel	0.03	0.07	0.035
Magnesium	0.023	0.051	0.058
Tensile Strength, psi	71,200	71,900	78,000
Yield Strength, psi	47,400	47,500	61,600
Elongation, %	20.2	11.5	3.0
BHN	159	167	229

The time required for ferritizing ductile iron, as well as the annealing temperature, is increased by the manganese content. This is particularly evident when the aforementioned intercellular carbide network has formed. For example, refer to Figures 4-37 and 4-38 illustrating the tenacity of the intercellular network during annealing.

desulfurize the melt, to alter the graphite shape, and to promote the formation of a carbide eutectic. As a general rule, the treatment functions in that order, so that, if the base iron and the treatment are not properly controlled, excessive oxygen and/or sulfur could consume the magnesium, leaving an insufficient amount to spheroidize the

INFLUENCE OF THE SPHEROIDIZING ELEMENTS

While a wide range of elements has been demonstrated to result in the formation of spheroidal graphite when they are added to a low sulfur content hypereutectic base iron, the addition of magnesium is the commercially acceptable procedure for producing ductile iron. All of these elements exhibit limited solubility in the melt, and are effective deoxidizers and desulfurizers. Examples include Group IA elements (lithium, sodium, potassium), Group IIA elements (beryllium, magnesium, calcium, strontium, barium), and Group IIIB elements (yttrium, rare earths). In all cases, only very small amounts of these elements are required to modify the graphite shape.

Magnesium

Ductile iron magnesium contents that will produce completely spheroidal graphite typically range from 0.02–0.08%, depending upon the sulfur content. When added to molten iron, magnesium acts to deoxidize and

Fig. 4-38. Same structure as shown in Fig. 4-37, 2% nital etched, ×1000.

graphite. Likewise, excessive magnesium could result in a highly carbidic casting.

Magnesium is an effective deoxidizer of the base iron, and a magnesium-treated melt will combine with oxygen, preferentially, whenever oxygen is made available to the melt (holding, turbulence, etc.). A base iron oxygen content of 0.0135% will be reduced to 0.003%, when treated with magnesium. Accordingly, a highly oxidized base iron (excessive melt temperature, high steel content charge, moisture in the charge, or melting environment, etc.) will consume magnesium during treatment and correspondingly affect the magnesium recovery obtained.

The result of this deoxidation is the production of magnesium oxide (MgO), which is very stable and has a high melting temperature (remains solid), low density (floats in molten iron), low solubility in iron, and is white in color. This MgO readily separates from the melt floating to the surface, where it can be collected. Formation of excessive MgO dross formation, however, will result in its accumulation on the cope surface of the casting, entrainment within the melt, etc. Because of the reactivity of other elements with oxygen, the composition of the dross typically contains Fe, Si, etc., in addition to Mg.

Magnesium is also an effective desulfurizer of the base iron—about 1 lb of magnesium will remove about 1.5 lb of sulfur by forming magnesium sulfide (MgS). Magnesium sulfide has low solubility in the melt and is of low density, so that it floats to the melt surface where it enters the dross. Magnesium sulfide is not very stable and will react with oxygen to release the sulfur to the atmosphere or to return sulfur to the melt. Maintaining the MgS-containing dross on the melt will result in reversion of the sulfur to the melt and a corresponding fading of nodularity as it recombines with magnesium.

Magnesium is an effective modifier of the graphite shape during solidification of the treated melt: as little as 0.018% Mg has been shown to result in fully spheroidal graphite structures, when the base iron is low in oxygen and sulfur. As the amount of magnesium added to the base iron is increased, the graphite structure changes from a conventional Type A flake, to a highly undercooled flake shape, then to a vermicular or compacted shape, and finally to spheroidal graphite. Similarly, as the magnesium content of the iron is reduced, due to vaporization of this very volatile element or reaction with oxygen and/or sulfur, the graphite will revert back toward a Type A morphology (Fig. 4-39).

Magnesium is an effective and very potent carbide former. As the amount of magnesium is increased, the eutectic is undercooled, and the melt eventually solidifies with the formation of carbides. If the nodule count of the iron is low, then portions of the melt will solidify with carbides (intercellular carbides, centerline carbide, etc.). But if a high nodule count is maintained, solidification will be completed before the carbide formation temperature is reached. Carbide formation in ductile irons is, therefore, a function of the magnesium content, the nodule count and the solidification cooling rate. Typically, these carbides can be decomposed during heat treatment, but the presence of other elements may increase carbide stability.

Magnesium is a highly volatile element, boiling at 2025F (1107C), well below the treatment temperature. The magnesium content then decreases with increasing melt temperature and with increasing holding time. The rate of magnesium loss will depend upon the ability of magnesium to escape to the atmosphere. For example, the loss rate will be greater in smaller ladles than in larger ladles, due to the greater surface area-to-volume ratio, and will be greater in uncovered ladles than in covered ladles, etc. Sufficient magnesium must be maintained to assure adequate speroidization of the last metal poured.

The effects of magnesium are also subject to the presence of other elements with which it may form compounds, or that affect its ability to spheroidize graphite. These detrimental, deleterious or subversive elements reduce the effective amount of magnesium in the melt. They will be considered in a later section of this chapter.

CERIUM AND RARE EARTH ELEMENTS

Cerium was the element used by Henton Morrogh of the British Cast Iron Research Association to form spheroidal graphite in hypereutectic cast irons. This process was never found to be as satisfactory as the magnesium treatment process developed by Millis and Gagnebin at the International Nickel Company. As a result, cerium and the rare earths are used primarily to supplement magnesium treatment by increasing the graphite nodule count, or to neutralize the effects of deleterious elements, such as Pb, Bi, Ti, etc.

Typically, rare earths are added to the melt in the form of mischmetal, rare earth silicide or proprietary Mg-bearing FeSi alloys. Two types of rare earth materials are usually used in alloys produced in the United States. The first, based upon the use of mischmetal, contain approximate ratios of 50% cerium, 33% lanthanum, 12% neodymium, 4% praseodymium, plus other rare earths and yttrium. The second, often referred to as high-cerium rare earths (because other rare earths have largely been removed) contain approximate ratios of 92% cerium, 5% lanthanum, 2% neodymium, 1% praseodymium, plus yttrium and other rare earths. Because of the differences in the ratio of rare earths, somewhat different results might be anticipated using these materials, particularly since it is typical to identify the amount of rare earths added to the iron either as the total amount of rare earths or as the

TABLE 4-5. EFFECT OF MANGANESE ON MECHANICAL PROPERTIES

Melt	WJA 3A	WJA 52	WJA 35	ABF 1	ABF 2	ABF 3
Total Carbon	3.44%	3.65%	3.48%	3.50%	3.50%	3.50%
Silicon	1.90	1.87	1.98	2.50	2.50	2.50
Manganese	<u>0.01</u>	<u>0.18</u>	<u>0.38</u>	<u>0.27</u>	<u>0.76</u>	<u>1.28</u>
Nickel	0.01	0.01	0.04	1.03	1.03	1.03
Chromium	0.025	0.022	0.021	—	—	—
Magnesium	0.059	0.033	0.044	0.053	0.053	0.053
Copper	0.01	0.01	0.01	—	—	—
Phosphorus	0.028	0.030	0.033	0.034	0.034	0.034
Sulfur	0.017	0.015	0.011	0.012	0.012	0.012
Cerium	—	—	—	—	—	—
Carbon Equivalent	4.01	4.21	4.07	4.25	4.25	4.25
As-Cast 1-in. "Y" Block						
% Ferrite	95	60	20	—	—	—
% Pearlite	5	40	80	—	—	—
Tensile Strength, psi	64,200	74,400	90,000	94,860	106,300	107,800
Yield Strength, psi	41,400	50,400	53,000	58,900	66,680	77,200
Elongation, %	22.6	12.1	13.3	10.5	5.0	2.0
BHN	152	163	197	196	241	286
Annealed As-Cast 4-in. "Y"						
Tensile Strength, psi	53,250	55,500	58,900	—	78,800	88,360
Yield Strengtth, psi	37,800	36,300	40,900	—	57,920	68,900
Elongation, %	25.9	25.4	26.1	—	3.0	2.0
BHN	121	128	138	—	269	269
Melt		HM- 32*	HM- 33*	HM- 34*	HM- 35*	HM- 36*
Total Carbon		3.34%	3.34%	3.34%	3.34%	3.48%
Silicon		2.13	2.07	2.11	2.15	2.18
Manganese		0.02	0.12	0.25	0.44	0.72
Sulfur		0.012	0.011	0.010	0.011	0.010
Phosphorus		0.029	0.029	0.029	0.029	0.021
Nickel		0.70	0.60	0.65	0.66	0.84
Magnesium		0.048	0.049	0.045	0.044	0.052
Carbon Equivalent		3.98	3.96	3.97	3.99	4.13
As-Cast 1-in. "Y" Block						
Tensile Strength, psi		62,700	65,200	72,600	74,500	84,900
Elongation, %		18.0	15.0	13.0	5.0	2.0
BHN		164	166	173	199	235
Microstructure		Nodular	Nodular	Nodular	Nodular	Nodular
% Ferrite		90	90	85	40	25
% Pearlite		10	10	15	60	75

*H. Morrogh, *AFS Transactions*, vol 60, pp 439-52 (1952).

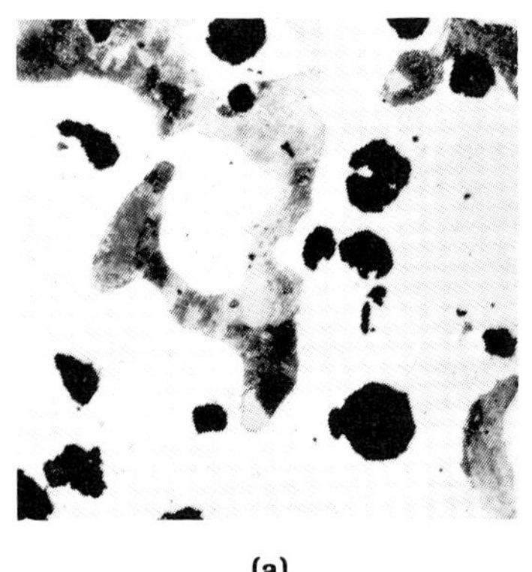 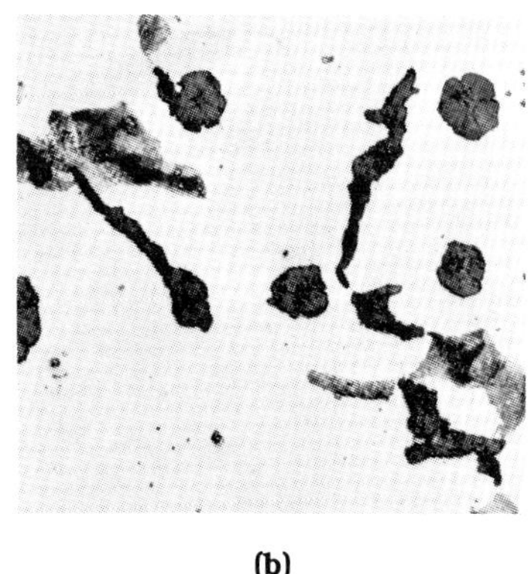 

(a) (b) (c)

Fig. 4-39. Influence of magnesium treatment on graphite structure: (a) base iron (0.016% S) treated with 0.03% Mg; (b) same iron as in (a) faded to 0.025% Mg; (c) same iron as in (a) faded to a flake graphite structure.

cerium content. The problem is more significant in alloys produced outside of the United States, since a variety of materials having different ratios of rare earths and yttrium may be encountered.

Cerium

Cerium, like magnesium, is a strong deoxidizer and desulfurizer. However, cerium is not highly volatile (boiling point, 4362F/2406C), and its addition involves no pyrotechnics; the oxides and sulfides that form are stable and generally not subject to reversion.

About 0.035% cerium will produce spheroidal graphite in a hypereutectic iron. Because cerium is such a potent carbide stabilizer, the base iron must be high in carbon (above 3.80%), and the treated iron must be heavily inoculated with ferrosilicon. The structures obtained are quite section-size sensitive, and it is difficult to obtain a fully spheroidal graphite- and carbide-free microstructure.

Cerium can be used to partially replace magnesium in the production of quality ductile iron. In an iron treated with 0.015–0.020% Mg, an addition of 0.02% Ce results in a completely nodular graphite structure. These combined additions are less subject to fading, and because of the lower magnesium level, dross formation is significantly reduced.

Perhaps of greater interest is the fact that cerium additions result in a substantial increase in nodule count (Fig. 4-40). Results are different for the low- and high-cerium rare earth alloys, but when the total amount of rare earths are considered, peak nodule counts are obtained at 0.016–0.019% total rare earths. The lower rare earth alloys were

less subject to carbide formation and produced higher nodule counts.*

> ***[Editors' Note: The literature contains additional information contrary to this, showing improved results with high-Ce rare earths. Foundries should test both types of rare earths in their operations.]**

Finally, cerium (as well as other rare earths) is an important agent in controlling the influence of subversive elements that may be present. About 0.005% cerium will usually overcome the deleterious effect of subversive elements (e.g., titanium, lead, antimony, arsenic, etc.) and make it possible to produce a fully spheroidal graphite structure. On the other hand, excessive cerium (or rare earths), particularly in heavier-section castings, result in increased graphite flotation, exploded graphite formation, and the development of chunky graphite in the thermal center of large castings. In these cases, it has been proposed that the excess cerium be balanced by the addition of a controlled amount of antimony.

Lanthanum

While some spheroidal graphite reportedly has been produced using lanthanum, a structure consisting of some spheroids within vermicular graphite is more typical. Even with 0.05% lanthanum, a high percentage of nodules was not obtained. Lanthanum is not as potent a carbide former as cerium, and when lanthanum is added in combination with MgFeSi, a high nodule count (>800 N/mm^2) is obtained, and a fully ferritic matrix can be achieved.

Neodymium

Nodular graphite was not obtained with up to 0.1% neodymium added to a low-sulfur content base iron, although vermicular graphite and carbides were promoted. It has been reported that neodymium is an extremely strong carbide former.

Praseodymium

Vermicular graphite, with a few nodules, was obtained with 0.030–0.050% praseodymium additions. These irons were carbidic, and did not respond to a 0.60% silicon inoculation.

Yttrium

Additions of 0.12–0.20% yttrium have been used to produce spheroidal graphite in a low-sulfur content base iron. The reaction is not violent, but the cost of yttrium is such that the use of this element is not practical for ductile iron casting production.

CALCIUM

Calcium is an important element to be considered in the production of ductile irons. While spheroidal graphite has been produced in irons that have been treated only with calcium, this element is of greater interest in reducing the volatility of the magnesium reaction and increasing the effectiveness of inoculation.

Calcium, like other spheroidizing agents, reacts readily with oxygen and forms calcium oxide (CaO), which floats to the melt surface. High-calcium-content alloys will yield greater amounts of calcium oxide, so that these alloys are considered to be dirty. Calcium is not used as the significant nodulizing element in iron, except in some foundries in the Orient.

In general, as the calcium content of the MgFeSi alloys is increased, the volatility of its reaction is decreased considerably. Typically in the United States, 0.80–1.50% calcium is present in most MgFeSi alloys, in order to suppress the violence of Mg treatment and improve magnesium recovery. When the level of calcium added exceeds about 0.03%, the carbide-forming tendency of magnesium appears to be increased. But, below this level, calcium is effective in increasing nodule count and reducing nodule count fading. Calcium is also utilized in ferrosilicon in order to improve the effectiveness of these alloys in inoculation.

Fading, in ductile irons, is considered to be the loss of nodularity with time. Fading may occur due to the loss of magnesium from the melt as a result of its volatility, or due to reaction of magnesium with oxygen and/or sulfur. This is *magnesium fading.* Fading may also occur due to a reduction in nodule count attributable to treatment and not to a change in magnesium content. This is *nodule count fading.*

BARIUM

The principle use of barium in ductile iron production is in ferrosilicon alloys, where 2.00–2.50% barium is reportedly used to promote high nodule counts during inoculation.

Some MgFeSi alloys, containing up to 6.0% barium, have also been developed. In these alloys, the use of barium is intended to increase nodule count and may reduce the need to inoculate.

OTHER SPHEROIDIZING ELEMENTS

As noted, many more elements have been found to form completely or partially spheroidal graphite structures in cast irons. With the possible exception of sodium, none of these elements has received sustained industrial use. The treatment of a low-sulfur content iron with sodium was the basis of the Allis Chalmers process used in the 1950s and early 1960s. While fully spheroidal graphite structures were obtained, and the iron was used to produce a wide range of castings (including sections up to 15-in. thick), the process economics and environmental characteristics rendered it uncompetitive to magnesium treatment of the iron.

■ INFLUENCE OF ALLOYING ELEMENTS

NICKEL

Nickel is added to ductile irons at levels ranging from 0.5–36%. It is added for a variety of purposes including the following:

- for hardenability—to avoid transformation to pearlite (used at levels of 0.5–4%);
- for austenite stability—to promote stable austenitic matrix microstructures (used at levels of 18–36%);
- for low-temperature applications—to develop reasonably high strength ferritic irons having low silicon content (used at levels of 1–2%).

Nickel is commonly added as metallic nickel pellets (>90% Ni) or as nickel-magnesium or nickel-magnesium-silicon alloys. Because of the high solubility of Mg in nickel, nickel-mag alloys dissolve in molten iron with a minimum of flare and turbulence.

Nickel is soluble in cast iron at all levels. It is a graphitizing element; it reduces the solubility of carbon in liquid

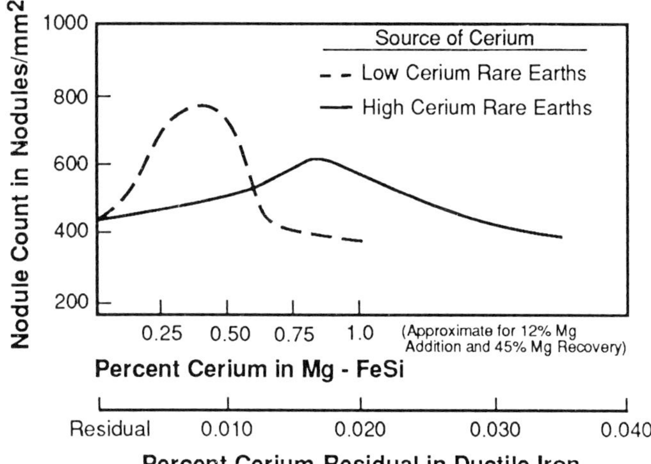

Fig. 4-40. Composite diagram illustrating the effects of low- and high-cerium, rare earth additions on the graphite nodule count.

iron and solid iron, and it lowers the eutectic carbon content by 0.06% for each 1% nickel. Nickel raises the stable austenite-graphite eutectic temperature and lowers the metastable austenite-iron carbide eutectic temperature and thus reduces chill and the propensity for carbides. It is said to have one-third the graphitizing ability of silicon.

Nickel also is an austenite stabilizer; it lowers the gamma-alpha transformation temperature. Austenitic ductile irons are stabilized with 18–36% nickel. The use of nickel in austenitic ductile irons is reviewed separately in Chapter 3.

When used at lower levels, nickel alloying produces solid-solution hardening of the ferrite phase. Brinell hardness of ferrite increases about 15 HB units for each 1% nickel, and yield strength increases about 5000–6000 psi (40 MPa) per 1% nickel addition.

Nickel is a weak pearlite promoter and is usually avoided where an as-cast ferritic structure is desired. With a 3% nickel content, pearlitic structures can be achieved having tensile strengths up to 120 ksi with better than 3% elongation. Mechanical properties for various nickel-alloyed ductile irons are given in Table 4-6.

Nickel is effective in delaying transformation to pearlite and is used to produce high-strength acicular irons as-cast, and to produce martensitic structures on heat treatment. The contribution of nickel to hardenability is significantly enhanced by additions of molybdenum, and as-cast acicular irons are commonly alloyed with combinations of nickel and molybdenum. The graphs in Figure 4-41 illustrate the Ni requirements to achieve acicular microstructures in as-cast cylinders up to four inches in

diameter. The hardness data in Figures 4-42 and 4-43 illustrate the influence of alloy content on the properties of as-cast and normalized ductile irons in sections up to six inches.

Nickel is also used effectively in concentrations from 0.5–2.5% to avoid pearlite in austempering. The influence of nickel on hardenability in continuous cooling transformation is illustrated in Figure 4-44.

Nickel is used in ductile irons for low-temperature applications, where low ductile-to-brittle transition temperatures are required. Ferritic irons having reduced silicon contents (1.5%) are alloyed with 1–2% nickel. Nickel is added specifically to raise the strength of these irons, which can have transition temperatures well below zero.

MOLYBDENUM

Molybdenum is added to ductile iron for one of two purposes:

- *for hardenability*—to achieve pearlite-free microstructures, either as cast or after heat treatment (used at levels up to 1%);
- *for elevated temperature strength*—to increase elevated temperature tensile strength, stress-rupture and creep strength, and thermal fatigue resistance (used at levels up to 2%).

Molybdenum is a mild carbide former* and has a very modest effect on chill. At levels of 0.5% or more, some grain boundary carbides can be expected. This carbide-stabilizing effect will be accentuated by segregation in slowly solidifying thick-section castings and by the presence of other carbide-stabilizing elements, such as Cr, Mn, and V.

Molybdenum is a powerful hardenability agent and is very effective in delaying transformation to pearlite, as illustrated in Figures 4-42, 4-43 and 4-45 through 4-47. In the presence of Ni or Cu, molybdenum is even more effective, as shown in the continuous cooling transformation diagrams of Figures 4-44 and 4-48. Molybdenum is essential to producing as-cast acicular irons (see Fig. 4-49); composition limits to achieve acicular structures are illustrated in Figure 4-41.

Molybdenum has been reported to mildly promote ferrite or pearlite, depending on base chemistry and section size. Hardness and yield strength in ferritic irons increase with molybdenum content by solid-solution strengthening. Strength increases approximately 6000 psi and hardness about 15 HB units for each 1% Mo. At the same time, tensile elongation is decreased by about 8%.

[*Ed. N.: There is currently disagreement among metallurgists concerning the actual formation of carbides due to the presence of Mo.]

Table 4-6. Effect of Nickel on Mechanical Properties

Melt	IE-2	IE-4	ABI-1	ABI-2	ABF-5	ABF-6	ABF-7
Total C	3.55%	3.62%	3.70%	3.68%	3.37%	3.37%	3.37%
Si	2.46	2.01	2.45	2.37	2.40	2.40	2.40
Mn	0.01	0.01	0.22	0.44	0.24	0.24	0.24
Ni	0.08	0.63	0.64	1.70	1.00	2.34	3.25
Cr	0.04	0.04	0.05	0.05	0.05	0.05	0.05
Mg	0.062	0.062	0.06	0.055	0.096	0.096	0.096
Cu	0.08	0.03	T	0.05	0.08	0.08	0.08
P	0.020	0.018	0.029	0.034	0.038	0.038	0.038
S	0.006	0.006	0.006	0.007	0.011	0.011	0.011
Ce	0.003	0.003	—	—	—	—	—
Carbon	4.37	4.39	4.52	4.47	4.17	4.17	4.17
Equivalent	4.29	4.22	4.30	4.39	4.09	4.09	4.09
As-Cast Keel block 1-in. "Y"							
%Ferrite	98	95	90	70	80	—	—
%Pearlite	2	5	10	30	20	—	—
Tensile Strength, psi	63,700	61,000	68,400	100,800	93,600	96,000	118,800
Yield Strength, psi	41,700	37,000	46,300	63,700	59,400	65,200	87,400
Elongation, %	27.5	27.0	20.0	10.0	10.0	8.5	6.5
BHN	141	140	170	241	207	228	269
Annealed			**As-Cast 4-in. "Y"**				
Tensile Strength, psi	62,400	59,250	—	—	—	83,500	101,400
Yield Strength, psi	41,800	37,700	—	—	—	61,400	86,900
Elongation, %	29.0	28.5	—	—	—	4.0	1.0
BHN	143	132	—	—	—	228	286

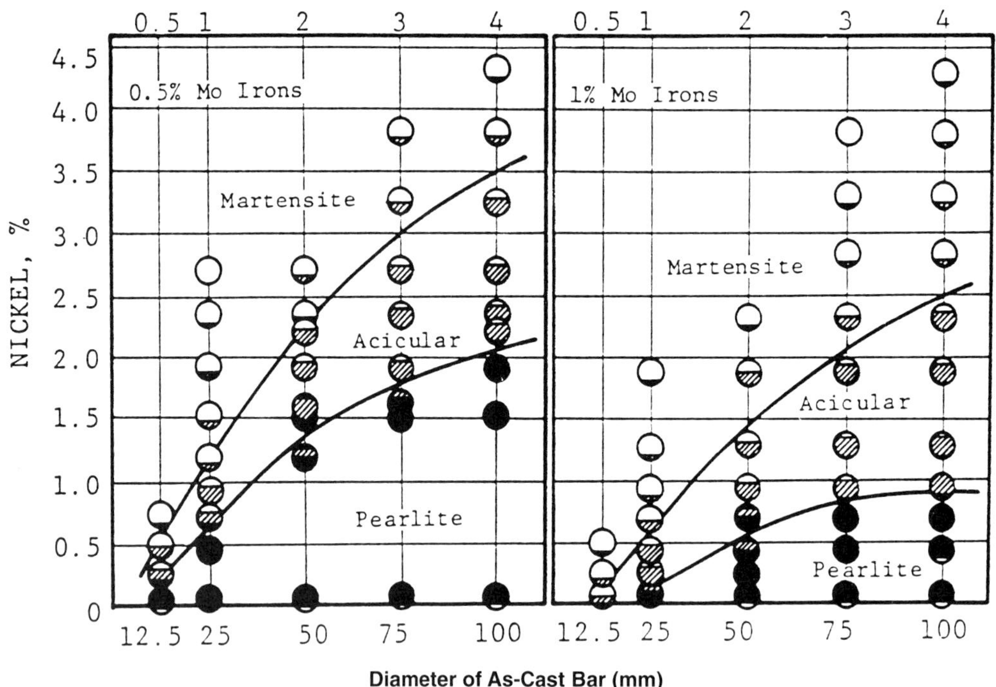

Fig. 4-41. Composition limits to obtain acicular structures in cast cylinders of indicated diameters. Ductile iron of base analysis 3.3% C, 1.8% Si, 0.85% Mn; containing 0.5% or 1.0% Mo with 0–4.3% Ni (Tokunaga).

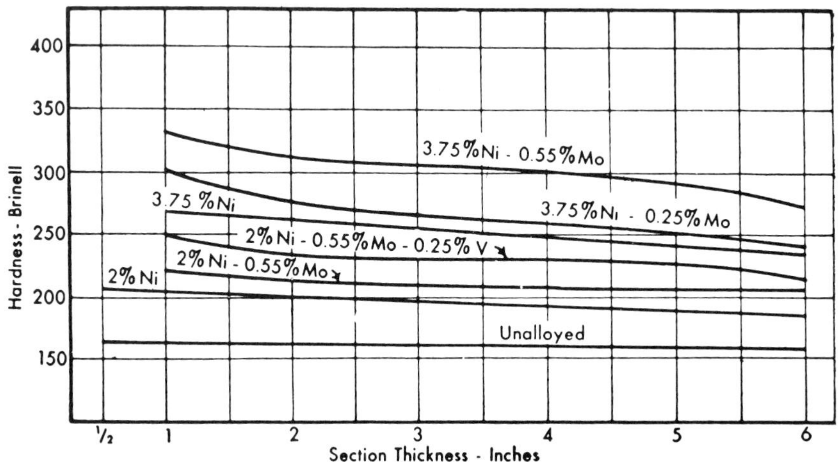

Fig. 4-42. The effect of alloy content and section thickness on the as-cast hardness of ductile iron.

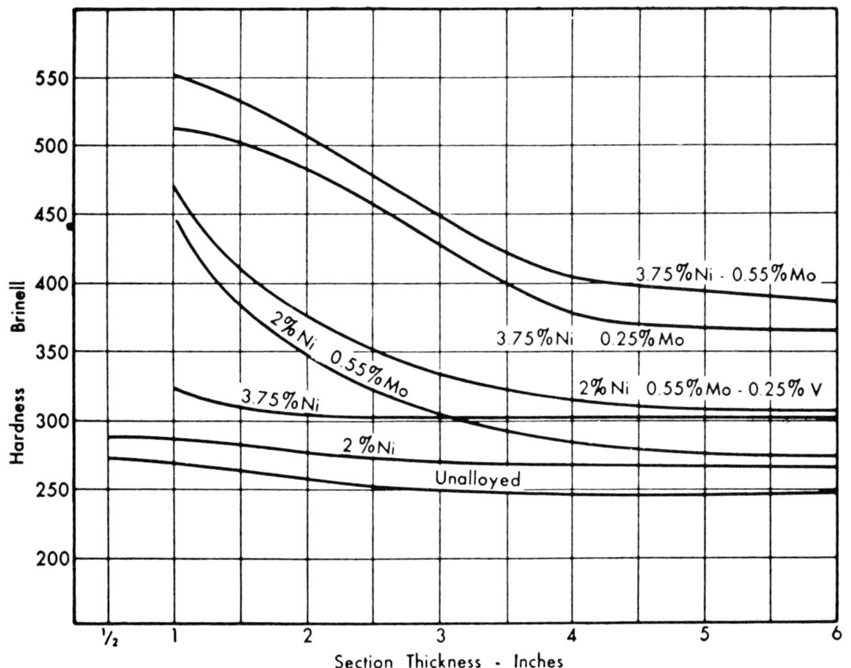

Fig. 4-43. The effect of alloy content on the hardness of normalized ductile iron castings up to six inches in thickness.

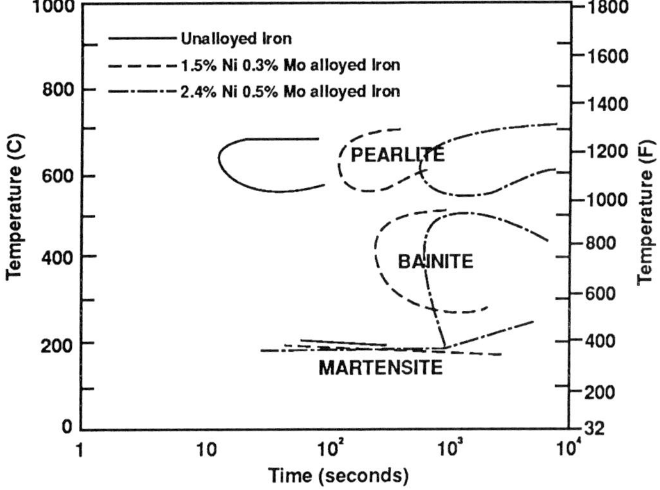

Fig. 4-44. CCT diagram comparing the increase in hardenability of 1.5%Ni-0.3%Mo and 2.4%Ni-0.5%Mo alloys to an unalloyed ductile iron.

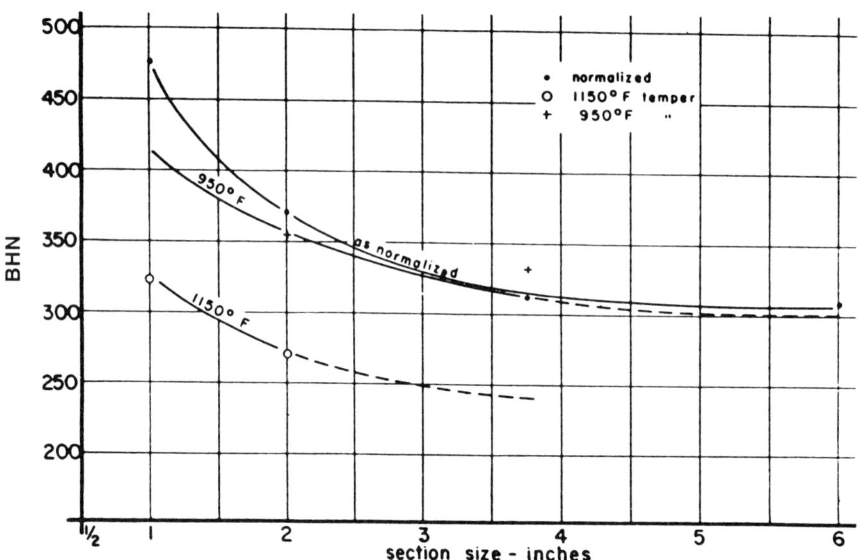

Fig. 4-45. Effect of 1-hr temper on normalized 2%Ni-0.55%Mo-0.25%V ductile irons in various section sizes.

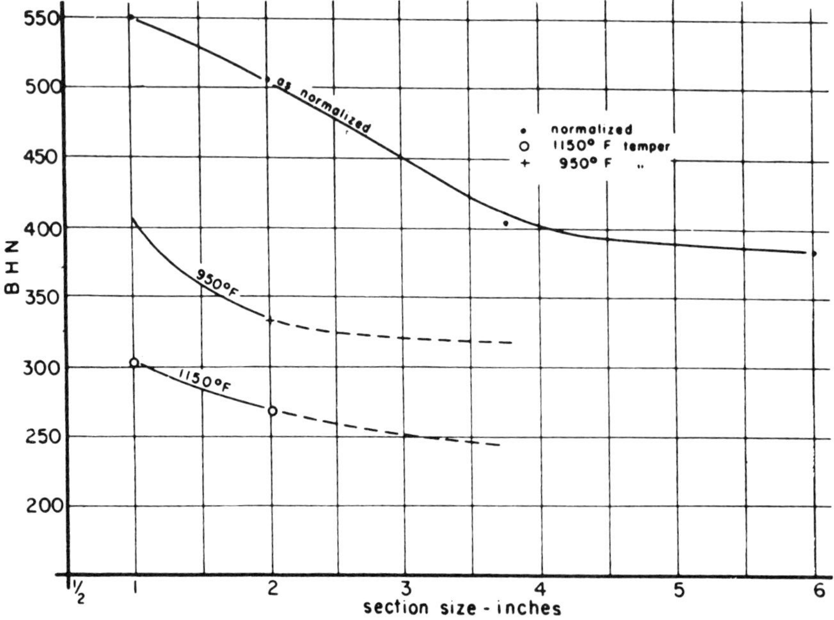

Fig. 4-46. Effect of 1-hr temper on normalized 3.75%Ni-0.55%Mo ductile irons in various section sizes.

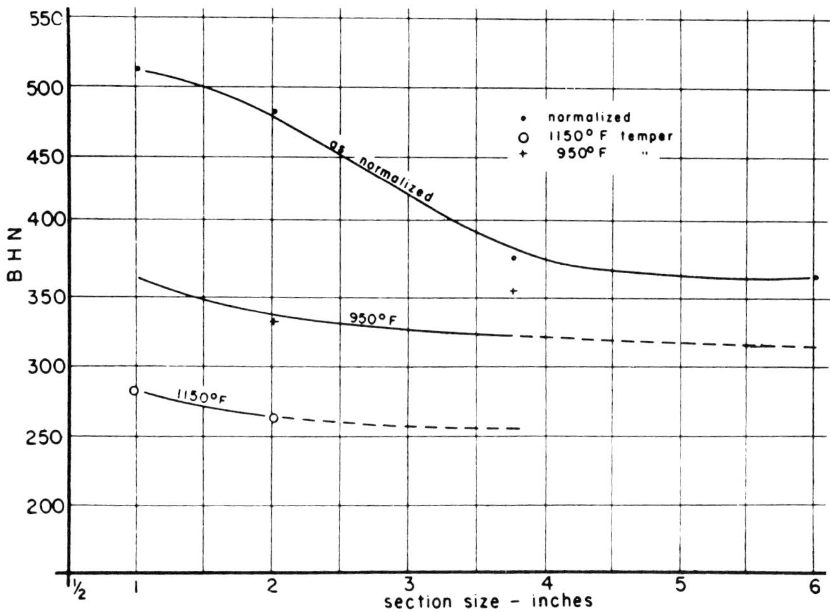

Fig. 4-47. Effect of 1-hr temper on normalized 3.75%Ni-0.25%Mo ductile irons in various section sizes.

Molybdenum increases the elevated temperature strength of both ferritic and austenitic irons. The influence of molybdenum on elevated temperature tensile properties in annealed ferritic ductile irons of two silicon levels is shown in Table 4-7. Molybdenum is very effective in increasing the creep-rupture strength and thermal fatigue resistance, as shown in Figures 4-50 and 4-51.

Molybdenum is usually added as the alloy ferromolybdenum (62% Mo), a relatively low-melting near-eutectic alloy. The less expensive moly-oxide briquettes, which are self-reducing, can also be used in some foundries.

COPPER

Copper is a graphitizer and a pearlite promoter. It is most commonly used to develop pearlitic microstructures. Because of its strong pearlite-forming tendencies, it is normally restricted to levels of 0.03% maximum in ferritic grades. As an individual alloy addition, copper imparts some hardenability to ductile iron; but, in combination with molybdenum copper is significantly more effective. Combinations of molybdenum and copper are commonly used in the production of acicular iron castings. They are also used to enhance the hardenability of irons for heat treatment, where austempering or martensitic structures are desired. The influence of copper and molybdenum on the transformation characteristics of ductile iron are illustrated in Figure 4-48.

Copper, like nickel, is an austenite stabilizer and is used to partially replace nickel in austenitic ductile irons. It has limited solubility in cast irons and can be

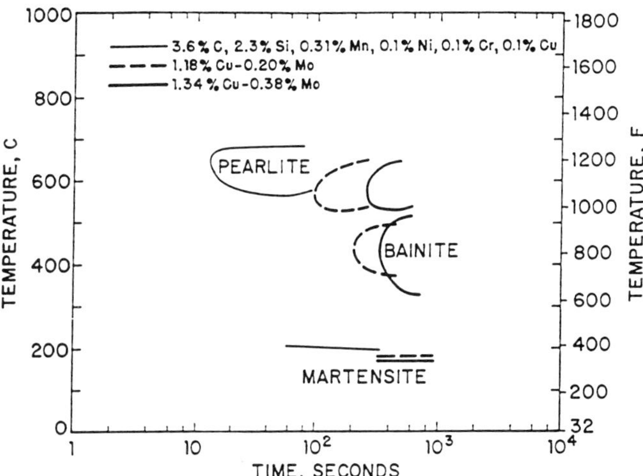

Fig. 4-48. Continuous-cooling transformation diagram showing the increases in hardenability obtained with a combined addition of copper and molybdenum to an unalloyed ductile iron.

TABLE 4-7. TENSILE PROPERTIES AT ROOM TEMPERATURE FOR TWO DUCTILE IRONS WITH DIFFERENT PERCENTAGES OF SILICON

2.5% Si Ductile Iron Annealed at 730C (1350F)			
Nominal Composition	**0.2% Yield Strength MPa (ksi)**	**Ultimate Tensile Strength MPa (ksi)**	**Tensile Elongation, %**
Room Temperature			
2.5%Si-0%Mo	288 (41.7)	432 (62.6)	27.1
2.5%Si-0.4%Mo	290 (42.0)	436 (63.3)	23.9
2.5%Si-0.6%Mo	312 (45.3)	461 (66.8)	22.0
650C (1200F)			
2.5%Si-0%Mo	72 (10.5)	85 (12.3)	61.0
2.5%Si-0.4%Mo	97 (14.1)	119 (17.3)	47.1
2.5%Si-0.6%Mo	98 (14.2)	118 (17.2)	38.5

4%Si Ductile Iron Annealed at 790C (1450F)			
Nominal Composition	**0.2% Yield Strength MPa (ksi)**	**Ultimate Tensile Strength MPa (ksi)**	**Tensile Elongation, %**
Room Temperature			
4%Si-0%Mo	443 (64.3)	565 (81.9)	19.5
4%Si-0.5%Mo	470 (68.2)	596 (86.4)	17.0
4%Si-1%Mo	474 (68.8)	609 (88.4)	14.5
650C (1200F)			
4%Si-0%Mo	67 (9.7)	83 (12.1)	59.0
4%Si-0.5%Mo	112 (16.3)	130 (18.8)	71.5
4%Si-1%Mo	111 (16.1)	130 (18.8)	53.0

dissolved to levels of about 2.5%. Its solubility is increased by the presence of nickel; copper solubility increases approximately 0.4% Cu for each 1% Ni. Solubility in the ferrite phase is significantly lower, and precipitation hardening is possible in irons containing upwards of 1% copper.

The influence of copper on mechanical properties is given in Table 4-8, and illustrated graphically in Figures 4-52 and 4-53. Representative microstructures of copper-alloyed irons are presented in Figures 4-54, 4-55 and 4-56.

Copper decreases the ferrite content in favor of pearlite and is about twice as potent a pearlite stabilizer as manganese. Copper retards annealing of pearlite, particularly at lower silicon content. Copper decreases impact resistance and raises the ductile-to-brittle transition temperature. Copper above about 2% becomes insoluble and reportedly precipitates at the grain boundaries, however, it does not appear to contribute to the formation of primary carbides.

Frequently observed is the increased sensitivity of copper-bearing irons to the influence of subversive elements on graphite degeneracy. Copper can also be the source of deleterious elements such as lead (see Fig. 4-57). Treatment with rare earths usually restores nodularity.

High-purity copper must be employed to best take advantage of the beneficial contributions of this element. Secondary shot must be of high purity and free from the typical copper contaminants: lead, arsenic, tellurium, tin, and hydrogen. Even electrolytic copper of very high purity can contain hydrogen.

Copper substantially increases as-cast strength and hardness through increased pearlite formation and pearlite refinement (see Fig. 4-52). Even the hardness and strength of annealed ferritic irons increases with copper additions, by solid-solution hardening (see Fig. 4-53).

Copper-bearing ductile irons can be precipitation-hardened when from 1–1.5% copper is present. Irons containing 1.25–1.75% copper can be precipitation-hardened by heating to a temperature between 900 and 1050F (482–566C). Normalized castings that have been machined can be hardened to improve wear resistance and tensile strength without serious distortion.

THE INFLUENCE OF DELETERIOUS ELEMENTS

The formation of spheroidal graphite can be assured if the iron is treated with an "effective" amount of magnesium. However, there are a number of elements that may interfere with the development of acceptable spheroidal graphite structures. For example, consider the role of magnesium, which is essential to the effective treatment of ductile irons. Over treatment to residual levels of about 0.1% will be the cause of intercellular films or flakes and of spiky graphite in heavier-section iron castings. As was noted previously, cerium is used to increase the number of graphite nodules, but when present in excessive amounts, particularly in heavy sections, it causes the formation of chunky graphite. When the carbon and/or silicon content is excessive, graphite flotation with exploded graphite formation is realized.

Those elements that, when present in minor amounts sufficient to interfere with the formation of spheroidal graphite, are referred to using terms such as deleterious, subversive, residual, trace, etc. These elements, for the most part, are highly surface active and tend to concentrate at the graphite-melt and graphite-solid interface, where they have a significant effect on the graphite morphology. However, some of these elements, while *desirable* if present at one concentration, may be the cause of *undesirable* graphite if present at other concentrations.

These elements may be classified into three groups as shown in Table 4-9. Elements in the first column promote chunky graphite as a result of their segregation toward the thermal center of the casting section. Those in the second column segregate in an intercellular manner and reach concentrations sufficient to form flake graphite. The elements in the third column appear to have a direct influence on the growth of graphite by interfering with magnesium.

The undesirable effects of the flake graphite-promoting elements in the second column can be neutralized by the addition of an element from the first column that promotes chunky graphite, e.g., Ce. And, the formation of chunky graphite can be eliminated by the addition of elements that would otherwise promote intercellular flake graphite. These effects are particularly significant in heavy-section castings, where the long solidification times permit this segregation to develop.

The effect of the elements in the third column is to either react chemically with magnesium to reduce the "effective" amount necessary to insure spheroidal graphite growth, or to interfere directly with magnesium in the growth of spheroidal graphite, thereby producing a degenerate form of graphite. The influence of these elements is counteracted by either maintaining them at exceptionally low concentrations in the base iron, or by increasing the amount of nodulizer (magnesium) to overcome their effect.

The influence of these subversive elements is such that the presence of one of them decreases the tolerance for others. For example, it has been noted that the presence of lead limits the tolerance for titanium, antimony and bismuth, and that the adverse effects of titanium are observed more strongly in copper-alloyed irons.

While ASTM A247 has classified graphite shapes into seven types, that classification scheme is not adequate to describe the effects encountered due to the presence of

TABLE 4-8. EFFECT OF COPPER AND NICKEL ON MECHANICAL PROPERTIES

Melt	ABF-8	ABF-9	ABF-10	ABF-11	
Total Carbon	3.56%	3.56%	3.56%	3.56%	
Silicon	2.55%	2.55%	2.55%	2.55%	
Phosphorus	0.38	0.38	0.38	0.38	
Sulfur	0.012	0.012	0.012	0.012	
Manganese	0.34	0.34	0.34	0.34	
Nickel	1.00	1.00	1.00	1.00	
Magnesium	0.071	0.071	0.071	0.071	
Copper	0.08	0.33	0.88	1.33	
Molybdenum	—	—	—	—	
Carbon Equivalent	4.33	4.33	4.33	4.33	
As-Cast 1-in. "Y"					
Tensile Strength, psi	81,200	118,800	—	—	
Yield Strength, psi	55,600	77,200	—	—	
Elongation, %	14.5	4.0	—	—	
BHN	196	286	—	—	
As-Cast 4-in. "Y"					
Tensile Strength, psi	—	95,400	103,850	105,750	
Yield Strength, psi	—	70,900	78,400	80,200	
Elongation, %	—	3.0	2.0	2.0	
BHN	—	255	286	302	
Melt	**HM-1***	**HM-2***	**HM-3***	**AB1-3**	**AB1-4**
Total Carbon	3.22%	3.22%	3.22%	3.70%	3.70%
Silicon	2.12	2.01	1.95	2.35	2.35
Phosphorus	0.025	0.025	0.025	0.031	0.031
Sulfur	0.012	0.015	0.016	0.008	0.008
Manganese	0.02	0.02	0.02	0.32	0.32
Nickel	0.80	0.79	0.76	1.30	1.30
Magnesium	0.10	0.065	0.062	0.060	0.060
Copper	T	0.27	0.60	0.38	0.74
Molybdenum	—	—	—	—	—
As-Cast 1-in. "Y"					
% Ferrite	95	50	0	5	5
% Pearlite	5	50	100	95	95
Tensile Strength, psi	63,800	82,900	110,500	107,100	112,350
Yield Strength, psi	—	—	—	66,000	70,850
Elongation, %	—	—	—	4.0	3.0
BHN	152	213	262	255	269

*H. Morrogh, *AFS Transactions*, vol 60, pp 439-52 (1952).

minor amounts of damaging elements. It should also be noted that the impact of these graphite forms on the mechanical properties depends upon the specific graphite shape involved. For example, increasing the amount of vermicular/compacted graphite leads to a gradual deterioration of properties. On the other hand, only an extremely small amount (at times difficult to detect, metallographically) of intercellular flake graphite or of spiky graphite is necessary to drastically reduce ductility and dynamic properties. Acceptance, therefore, of 80% nodularity, without regard to the morphology of the remaining 20% graphite, can be problematic. With these points in mind, the influence of selected elements will be considered.

TELLURIUM

While tellurium is utilized in many foundries as a method of controlling pinhole formation (about 1–2 gm/100 lb), when used in excessive amounts it has a deleterious effect on the graphite structure and, at even higher levels, promotes the formation of carbide. The effect of tellurium is similar to that of sulfur and selenium, in that magnesium will react chemically to form a telluride (sulfide, selenide), lowering the effective magnesium content. This effect can be offset by adding additional magnesium or, preferably, by adding a stronger desulfurizer such as cerium.

When cerium is present in the treatment alloy, tellurium increases the nodule count, but may result in some graphite spheroid deterioration. Table 4-10 illustrates this effect. These irons are essentially identical except for the addition of one ounce of tellurium per 50 lb of iron in melt ABF25, resulting in a slight reduction in hardness (increased ferrite), and a reduction in tensile properties (graphite structure).

LEAD

The effect of extremely small amounts of lead on the graphite structure of spheroidal graphite is to promote the formation of a highly deteriorated graphite, as illustrated in Figure 4-57 and Table 4-11. As little as 0.002% lead in the base iron will counteract the spheroidizing effect of magnesium, producing a flake form of graphite. Lead may enter the base iron through the melt stock, fluorspar, or refractories and, since such a small amount of lead is necessary to cause this deleterious graphite formation, it is difficult to monitor these effects by chemical analysis. The effect of lead is much more damaging in heavier-section castings, but the use of cerium, or cerium-containing MgFeSi, will neutralize the effect of lead (believed to occur as a result of a Ce-Pb compound formed).

TITANIUM

Titanium is present in minor amounts in almost all ductile irons, entering through the melt stock most typically in the pig iron or in certain structural steel. The effect of titanium is section-size sensitive. As much as 0.07% can be tolerated in thin-section castings, but 0.02% has had a damaging effect in heavy-section castings. A maximum of about 0.035% titanium is preferred for general use. The effect of titanium is to counteract the spheroidal graphite growth control of magnesium and to cause vermicular graphite formation. In fact, the balanced usage of magnesium and titanium is a preferred method of producing compacted/vermicular graphite cast irons. Table 4-12 illustrates the effect of titanium in ductile irons.

ALUMINUM

Aluminum is present in all ductile irons. It enters the iron through the charge materials and through ferroalloys used in treatment and inoculation. Typically, 0.6–1.5% aluminum is present in FeSi and MgFeSi alloys, and aluminum is present in most steels as a deoxidizer. Up to about 0.05% aluminum can be tolerated in ductile irons, but, like titanium, this effect is section-size sensitive and more pronounced in heavier-section castings. While not as deleterious as titanium, aluminum promotes the formation of vermicular graphite. This effect can be neutralized by the addition of cerium and as little as 0.01% Ce (0.014% rare earths) will neutralize the effect of 0.50% aluminum. Another reason for limiting the amount of aluminum is that of pinhole formation, aggravated by the presence of

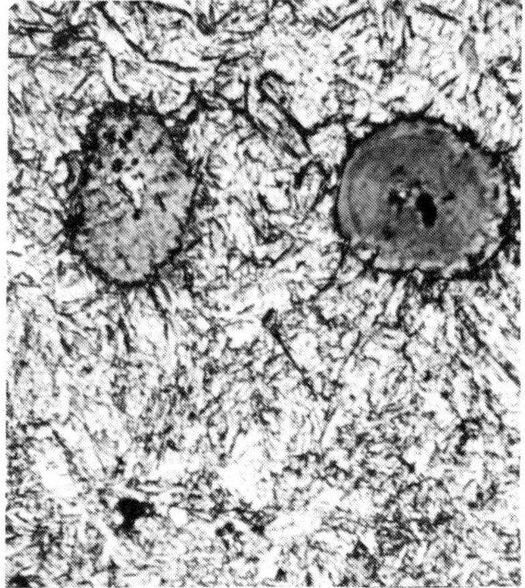

Fig. 4-49. Martensitic structure of ductile iron is shown, oil quenched, 1650F, etched, ×250.

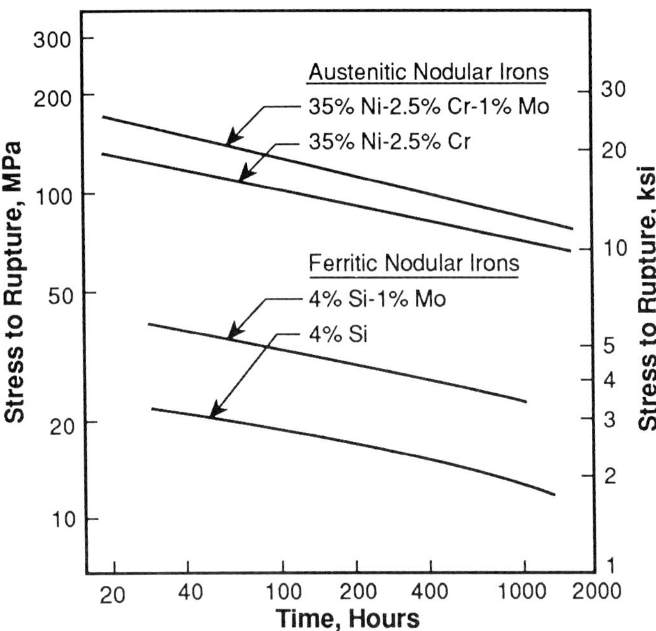

Fig. 4-50. Effect of molybdenum addition on the stress-rupture properties at 705C (1300F) of a 4% Si ferritic nodular iron and a 35%Ni-2.5%Cr austenitic nodular iron.

aluminum. The effect of aluminum on mechanical properties is illustrated by the data in Table 4-13.

ANTIMONY

It has been stated that over 0.004% antimony cannot be tolerated in ductile irons without a severe reduction in the quality of graphite present. The effect of antimony is also section-size sensitive, tending to segregate to the intercellular regions where it promotes the formation of a mesh-type flake graphite. The deleterious effects of antimony can also be offset by the addition of cerium. Even smaller levels of antimony (0.002%) have been noted to increase the pearlite content of the iron. The combined results of deteriorated graphite, increased pearlite, and the neutralizing effect of cerium are illustrated in Table 4-14.

BISMUTH

As with many of these elements, bismuth exhibits both favorable and damaging characteristics in ductile irons. The presence of as little as 0.003% bismuth results in the formation of intercellular flake or mesh graphite, which then results in a significant deterioration of mechanical properties. Only 0.005% bismuth is required to almost eliminate

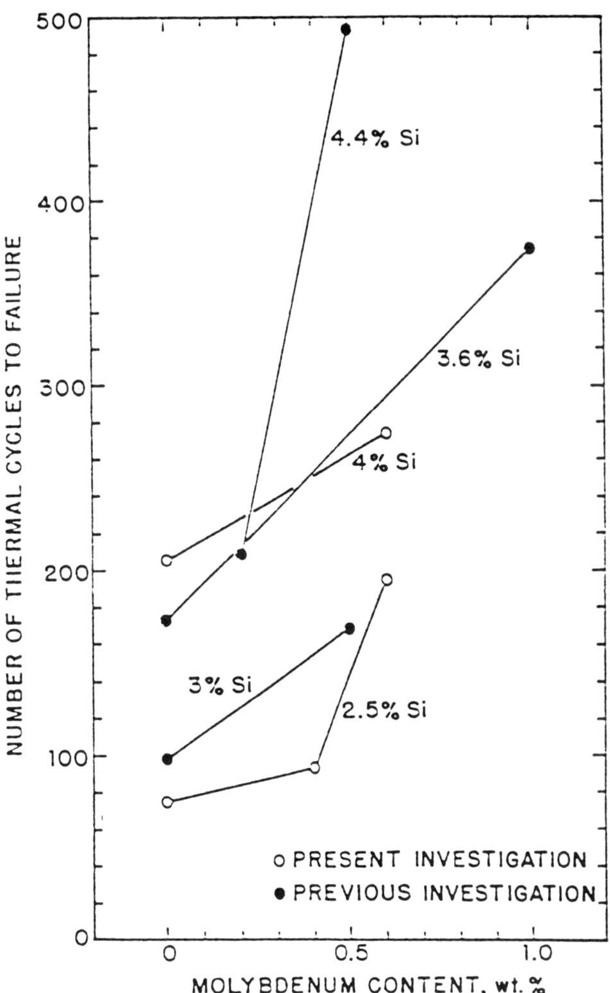

Fig. 4-51. Correlation of thermal fatigue life with molybdenum and silicon contents in ferritic ductile irons.

spheroidal graphite formation. The effect of bismuth, however, can be offset by cerium additions. This combined introduction of bismuth and cerium has been used to obtain appreciable increases in graphite nodule count, particularly in smaller castings where carbide prevention is important. In larger castings, the Bi-Ce combination can be used to promote a larger number of smaller nodules, which segregate more slowly toward the cope surface during solidification (per Stokes' Law), thus reducing the tendency to carbon flotation. The effect of bismuth is illustrated in Table 4-15.

OTHER DELETERIOUS ELEMENTS

As seen in Table 4-9, a number of other elements can have

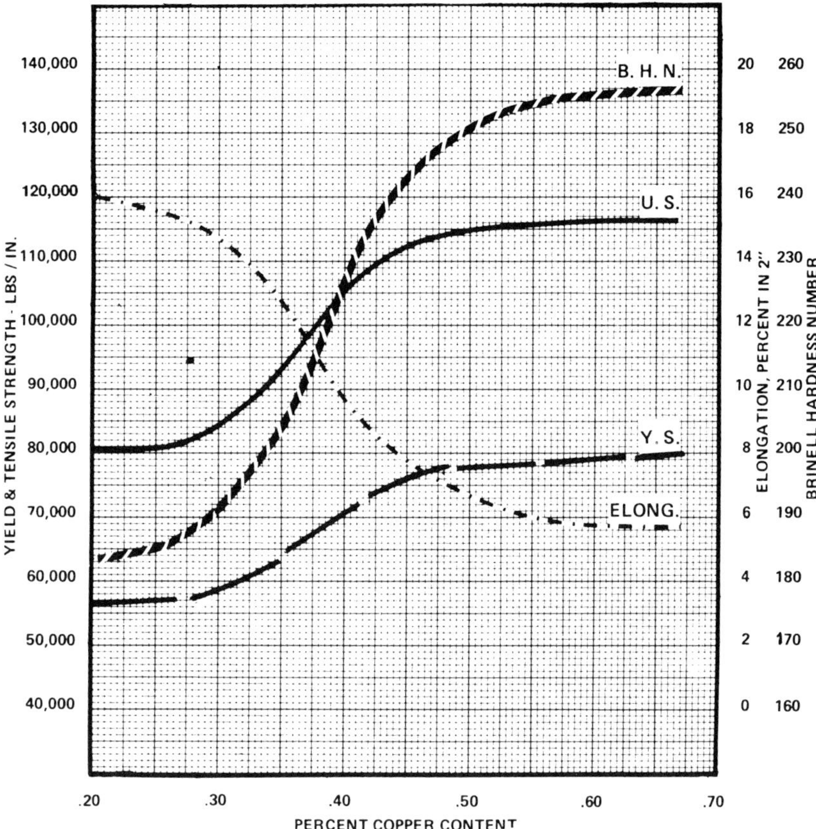

Fig. 4-52. The influence of copper additions on the mechanical properties of as-cast nodular iron are shown.

damaging effects on the structure and properties of ductile irons. Their individual effect falls into the classifications cited in this table. Elements within the same column act in a similar manner and would have similar effects on the properties. The amounts that can be tolerated, however, will be a function not only of the specific element and the presence of other elements, but also one of section size, with the impact being greater in those castings having a longer solidification time.

THE INFLUENCE OF PEARLITE- AND CARBIDE-PROMOTING ELEMENTS

These elements are utilized (intentionally or unintentionally) in ductile iron production to promote the formation of carbides during solidification of the iron, or to promote formation of pearlite during cooling of the casting through the critical temperature range. These elements may be directly involved in the formation of carbides or pearlite (e.g., vanadium, chromium, niobium, etc.) or they may be indirectly involved (e.g., tin, copper, etc.). In addition, these elements may influence the hardenability (martensite, bainite) in a manner different from their effect on either carbide formation or pearlitic hardenability (pearlite formation).

CHROMIUM

Chromium is a particularly potent carbide former and pearlite promoter in ductile irons. Chromium is introduced into the iron through the metallic charge materials, particularly from steel where it is a tramp element. In those cases where chromium in excess of the residual level is desired, chromium is usually added as ferrochrome.

The tolerance level for chromium depends upon the type of ductile iron produced and the matrix structure desired. For example:

99

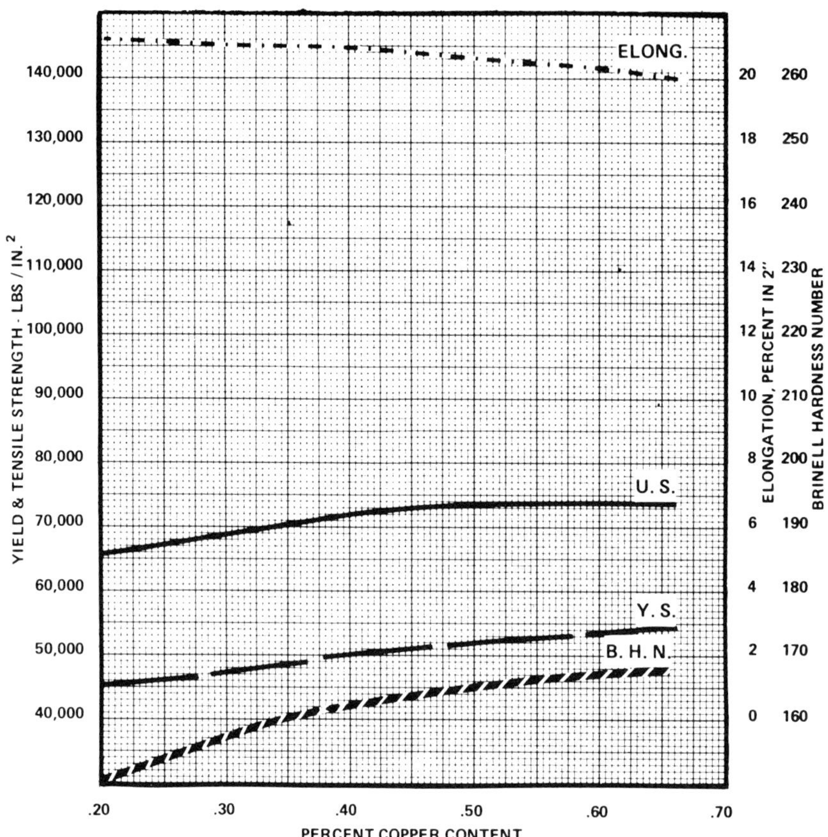

Fig. 4-53. The influence of copper additions on the mechanical properties of annealed ductile iron are shown.

TABLE 4-9. PROMOTION OF CHUNKY, INTERCELLULAR FLAKE AND DEGENERATE GRAPHITE IN SPHEROIDAL GRAPHITE STRUCTURES AS INFLUENCED BY MINOR ELEMENTS

Chunky	Intercellular Flake		Deleterious
Ce, Ca	Bi,	Cu, Al, Pb	Zr, Zn, Se
Si, Ni	Sb,	Sn, As, Cd	Ti, N, S, 0

Ferritic as-cast ductile iron. Chromium should be as low as possible. Up to 0.04% can be tolerated, but this depends upon the nodule count (higher chromium with higher nodule counts) and section thickness (lower chromium with longer solidification times). Chromium will result in pearlite formation in the intercellular regions of the structure.

Pearlitic ductile iron. Where a fully pearlitic matrix is desired, 0.10% chromium can be tolerated. However, this also depends upon the nodule count and solidification time, since chromium will segregate during solidification to the intercellular regions where it can cause intercellular carbide network formation.

Wear- and abrasion-resistant ductile irons. Up to 0.30% chromium can be added in these applications, within the framework of the segregation effects noted above.

Austenitic ductile irons. The chromium level in these irons is specified by the grade. Chromium increases the oxidation and corrosion resistance, and higher silicon contents may be employed to minimize carbides and to improve mechanical properties.

The presence of chromium is strongly observed when annealing ductile irons. Chromium retards annealing, both when carbide removal is the aim and when ferritizing is desired. Irons with 0.10% chromium content cannot be effectively annealed to a fully ferrite structure.

TABLE 4-10. EFFECT OF TELLURIUM ON MECHANICAL PROPERTIES

Melt	ABF-24	ABF-25
Total Carbon	3.62	3.62
Silicon	2.48	2.48
Phosphorus	0.03	0.03
Sulfur	0.011	0.011
Manganese	0.23	0.23
Nickel	0.65	0.65
Magnesium	0.056	0.056
Chromium	0.05	0.05
Copper	0.08	0.08
Tellurium	ND	0.052
Carbon Equivalent	4.36	4.36
As-Cast 1-in. "Y"		
Tensile Strength, psi	75,200	70,000
Yield Strength, psi	55,400	48,300
Elongation, %	17.0	13.0
BHN	179	170

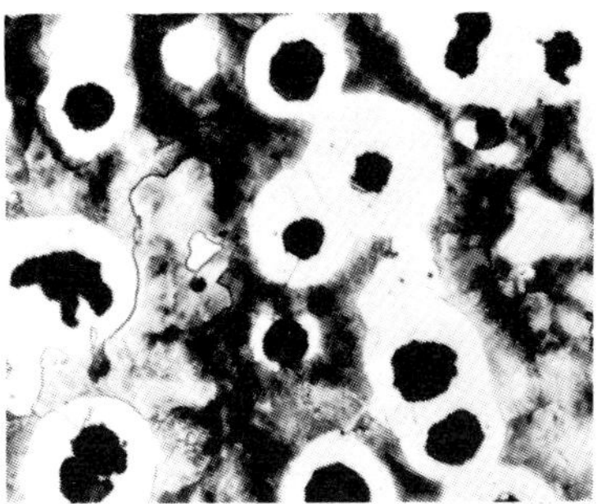

Fig. 4-54. Ductile iron microstructure containing 0.11% Cu, 3.51% T.C. and 2.48% Si; etched, ×150.

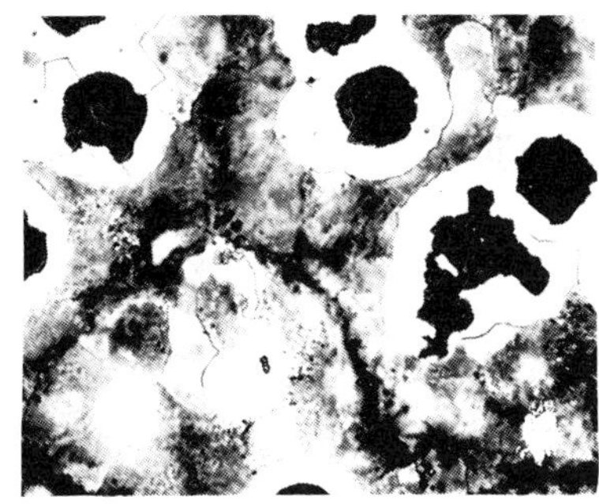

Fig. 4-55. Microstructure of ductile iron containing 0.48% Cu, 3.51% T.C. and 2.42% Si; etched, ×150.

VANADIUM

Vanadium exhibits an effect in ductile iron similar to that of chromium and may be used in amounts up to 0.02%, along with nickel and molybdenum, to increase the strength and hardness of casting sections over three inches. Like chromium, the effect is section-size sensitive and massive carbides will form at levels over 0.20%.

BORON

As in other ferrous alloys, the influence of boron is encountered at very small levels of this element. Boron segregates very strongly in ductile irons, forming boron carbides in the grain boundaries—and an actual network-like structure at higher levels of boron. Segregation will also result in inverse chill formation in hypereutectic compositions. Boron is one of the most powerful carbide-forming elements, and as little as 0.002% boron can result in intercellular carbides and a deterioration in mechanical properties (Table 4-16). Boron carbides are extremely stable and will not be removed by annealing.

Boron is usually introduced into the iron inadvertently from metallic charge materials (boron alloyed steel, vitreous enamel, malleable iron scrap, etc.), refractory linings, calcined anthracite coal, etc.

TIN

Tin is a powerful pearlite promoter in ductile irons, but does not promote the formation of carbides even at levels as high as 0.15%. Tin is effective as a pearlite promoter because it accumulates preferentially at the surface of the graphite spheroid and, thus, interferes with carbon diffusion, thereby preventing the formation of ferrite. Probably,

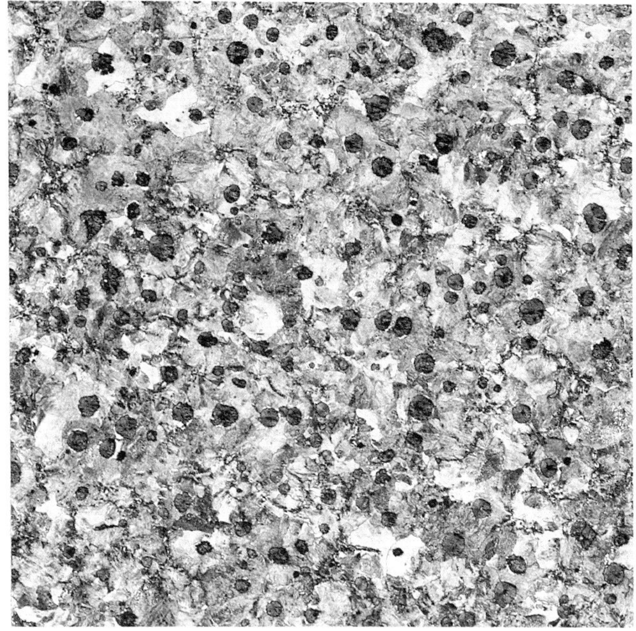

Specifications for the copper-alloyed ductile iron shown.

3.63% C	0.07% Ni
2.09% Si	0.020% P
0.38% Mn	0.005% S
0.74% Cu	0.054% Mg
Tensile Strength	119,4000 psi (823 MPa)
0.2% Yield Strength	68,000 psi (469 MPa)
Elongation	4.1%
Hardness	277 HB3000

Fig. 4-56. Microstructure of a copper-alloyed ductile iron; 2% nital etched, ×50.

because of this diffusion barrier, the nodule shape obtained when tin is added is improved slightly. The pearlite-promoting influence of tin is about ten times that of copper and about six times that of chromium.

Limitations on the amount of tin depend upon the matrix structure desired:

Ferritic ductile irons. Only about 0.01% tin can be required, but this is a function of the nodule count.

Pearlitic ductile irons. About 0.03–0.08% tin is used to produce a fully pearlitic matrix, the amount varying with casting section thickness. The tolerable level of tin is also a function of the presence of other elements, and excess tin segregates to the intercellular regions, where fine, difficult-to-observe carbide films will form, having a detrimental effect on properties. It is essential to carefully monitor residual tin levels to avoid this effect.

Table 4-17 illustrates the effect of tin on the mechanical properties of ductile iron. In addition, a 0.20% manganese ductile iron, which would be ferritic as cast, will be completely pearlitic with a 0.05% tin addition, yielding 101 ksi tensile strength, 4.6% elongation and 260 HB.

ARSENIC

Arsenic is generally not encountered in ferrous alloys produced in the United States, but it is a factor in certain alloys produced elsewhere, and in the ferrous charge materials obtained from those locations. Arsenic was included as an intercellular flake former in discussing undesirable elements in ductile irons, but the degradation of spheroidal graphite that occurs is rather minor and is easily neutralized by the use of cerium in the treatment alloy. While the maximum amount of arsenic tolerated in ductile irons is about 0.02%, and while about 0.045% arsenic will result in about 95% pearlite, the effect on nodularity is sufficient to limit the use of arsenic to 0.02%. The influence of arsenic on mechanical properties is shown in Table 4-18.

■ INFLUENCE OF GASES

The gases of concern in ductile iron production are oxygen, hydrogen and nitrogen, since they all have some solubility in molten iron and considerably reduced solubility in the solidified iron. Little information concerning the influence of these gases on the structure and properties of ductile irons is available due to the lack of reliable data on gas content in iron. In addition, the dissolved gas content at the solidification temperature is usually of primary interest in relating these values to solidification phenomena, but there is only limited data of this type available. Ductile cast irons are unusual, compared to other cast ferrous alloys, in that treatment of the melt with magnesium decreases the gas content of the melt as the magnesium vapor is released.

OXYGEN

Since magnesium is so effective as a deoxidizer, the oxygen content of the base iron is reduced to low levels during magnesium treatment. However, the oxygen content that was present in the base iron appears to influence, to some degree, the solidification process.

The oxygen content of the base iron varies with the melting process. For example, a basic cupola yields a base iron of about 0.009% (90 ppm) oxygen, while an acid cupola produced an iron of 0.0135% (135 ppm) oxygen. When this melt was magnesium-treated, the oxygen content was reduced to 0.0015% and 0.0033%, respectively.

TABLE 4-11. EFFECT OF LEAD ON MECHANICAL PROPERTIES

Melt	HM-22*	HM-23*	HM-24*	HM-25*
Total Carbon	3.44%	3.44%	3.44%	3.33%
Silicon	1.96	2.04	1.98	2.09
Manganese	0.02	0.02	0.02	0.02
Sulfur	0.018	0.014	0.013	0.014
Phosphorus	0.026	0.026	0.026	0.026
Nickel	0.74	0.72	0.81	0.81
Magnesium	0.084	0.059	0.054	0.061
Cerium	—	—	—	0.021
Lead	0.004	0.009	0.013	0.014
Carbon Equivalent	3.98	4.03	4.03	3.96
As-Cast 1-in. "Y"				
Tensile Strength, psi	67,200	61,620	47,450	64,720
Yield Strength	—	—	—	—
Elongation, %	19.0%	6.0%	None	20.0%
BHN	169	172	203	152
Comment	Nodular	Flake present with spheroids	Flake graphite; scattered spheroids	Flake present with spheroids

*H. Morrogh, *AFS Transactions*, vol 60, pp 439-52 (1952).

The oxygen content in induction furnace-melted irons may be considerably lower, or higher, than that of cupola-melted irons depending upon the melting practice. The use of charges that have a high steel content (particularly where the steel is thin and has been preheated to a high temperature), high furnace temperatures, etc., has been demonstrated to result in high base-iron oxygen levels. It should be noted that the oxygen content of the base iron affects the liquidus arrest temperature/carbon equivalent relationship, requiring corrections to be made.

Oxygen was classified as a deleterious element in ductile irons, tending to promote a vermicular/compacted graphite structure. The control of oxygen in ductile iron requires monitoring of melting stock, melting practice, melt temperature and melt handling.

HYDROGEN

Hydrogen is recognized to be a potent carbide promoter as a result of the undercooling of the iron encountered with increased hydrogen content. Very low levels must be main-tained to achieve reproducible structural control. However, it is difficult to correlate specific hydrogen content to structural features, due to the complexity of obtaining reliable hydrogen analyses. Centerline carbides, or inverse chill, are promoted by increased hydrogen levels, and by the segregation of hydrogen into the last portions of the casting to solidify (Figure 4-58). Where the hydrogen content is quite high, the presence of pinholes is often observed in association with this centerline carbide.

NITROGEN

Nitrogen is soluble in molten iron, yet molecular nitrogen can be used in injection and degassing techniques with no deleterious effects. This is attributed to the fact that only atomic nitrogen is soluble in molten iron. Dry nitrogen gas is commonly used in injection lancing, the porous plug, and other melt processing procedures. If molecular nitrogen is dissociated by electrical or chemical reaction in or at the melt surface, the nitrogen content of the melt may be increased. This may be encountered when using nitrogen

TABLE 4-12. EFFECT OF TITANIUM ON MECHANICAL PROPERTIES

Melt	HM-8*	HM-9*	ABF-14	ABF-15	ABF-16
Total Carbon	3.60%	3.60%	3.68%	3.68%	3.68%
Silicon	2.06	2.12	2.43	2.43	2.43
Manganese	0.02	0.02	0.21	0.21	0.21
Sulfur	0.012	0.013	0.012	0.012	0.012
Phosphorus	0.024	0.024	0.035	0.035	0.035
Nickel	0.76	0.76	0.63	0.63	0.63
Magnesium	0.056	0.068	0.042	0.042	0.042
Titanium	—	0.03	T	0.037	0.047
Carbon Equivalent	4.22	4.24	4.41	4.41	4.41
As-Cast 1-in. "Y"					
Tensile Strength, psi	68,800	70,750	79,900	71,700	71,900
Yield Strength, psi	—	—	51,900	45,000	39,500
Elongation, %	19.0%	15.0%	14.0%	13.5%	14.5%
BHN	170	178	179	179	179
Comment	Essentially nodular	Some flake evidenced	—	—	—

*H. Morrogh, *AFS Transactions*, vol 60, pp 439-52 (1952).

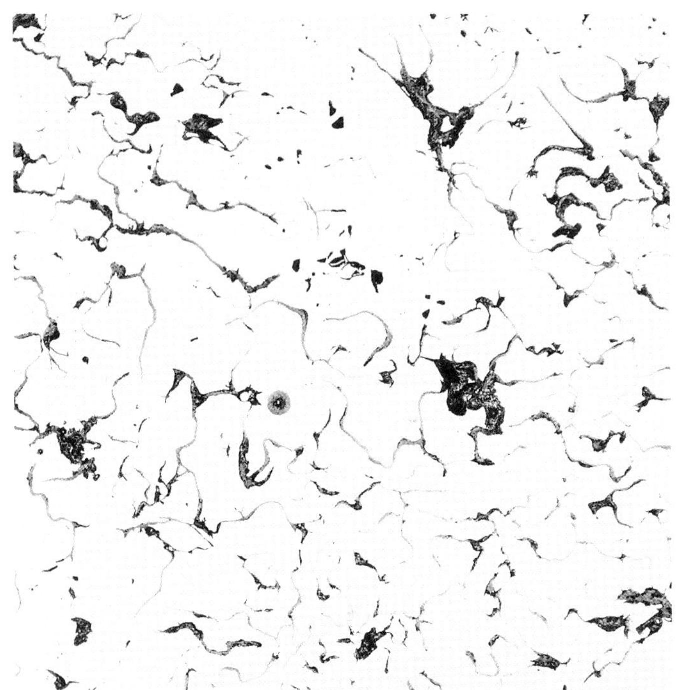

Fig. 4-57. Flake films typical of magnesium-treated iron contaminated with lead (0.0086%) and antimony (0.0057%); unetched, ×100.

gas to stir the melt during carbide desulfurization, or during electric arc melting (particularly during holding or superheating) of the base iron. As a result, it would appear that some nitrogen is dissolved in all melting processes.

The solubility limit of nitrogen in the solid at solidification is about 0.008–0.009%, and little effect is observed on the microstructure at levels below this amount. Nitrogen is a pearlite former, and it has been added to ductile irons in the form of calcium cyanimide or nitrogen-bearing (0.03-0.05%) ferromanganese. With ductile irons of lower silicon content, an increase in tensile strength (about 1200 psi per 0.001% nitrogen) is obtained.

Controlling nitrogen levels is critical to reduce pinhole formation. In addition to the charge materials and melt atmosphere, sources for nitrogen include carbon additives and core and sand binders. If the nitrogen content of the melt exceeds the solubility limit of 0.008–0.009%, or where segregation has raised the concentration of nitrogen to that limit near the thermal center of the casting, nitrogen porosity can be anticipated.

TABLE 4-13. EFFECT OF ALUMINUM ON MECHANICAL PROPERTIES

Melt	ABF-12	ABF-13	HM-4*	HM-5*	HM-6*	HM-7*
Total Carbon	3.46%	3.46%	3.48%	3.48%	3.36%	3.36%
Silicon	2.48	2.48	2.07	2.10	2.05	2.01
Manganese	0.20	0.20	0.02	0.02	0.02	0.02
Sulfur	0.008	0.008	0.008	0.011	0.010	0.013
Phosphorus	0.035	0.035	0.019	0.019	0.029	0.029
Nickel	0.74	0.74	0.74	0.73	0.86	0.90
Magnesium	0.110	0.110	0.068	0.070	0.076	0.078
Chromium	0.05	0.05	—	—	—	—
Copper	0.08	0.08	—	—	—	—
Aluminum	0.02	0.55	T	0.04	0.08	0.13
Carbon Equivalent	4.20	4.20	4.08	4.11	3.98	3.96
As-Cast 1-in. "Y"						
Tensile Strength, psi	74,700	83,450	66,800	68,550	67,400	62,300
Yield Strength, psi	49,600	55, 800	—	—	—	—
Elongation, %	16.5	12.0	20.0	18.0	20.0	6.0
BHN	179	186	162	174	158	164

*H. Morrogh, *AFS Transactions*, vol 60, pp 439-52 (1952).

■ REFERENCES

1. L.L. Wyman, G.G. Moore; "Quantitative Metallographic Evaluations of Graphite Microstructures," *Modern Casting*, vol 43, no. 1, p 7 (Jan 1963).

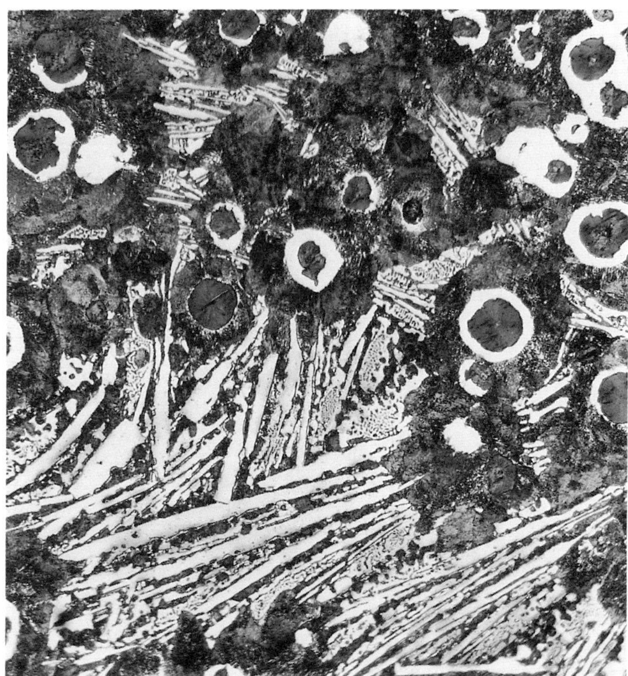

Fig. 4-58. Micrograph showing a heavy concentration of carbides located in a central cross section of a casting. Carbides are not attacked by ammonium persulfate; all other constituents are darkened. Note that carbides assume both acicular and cellular form; picral etched, × 250.

TABLE 4-14. EFFECT OF ANTIMONY ON MECHANICAL PROPERTIES

Melt	HM-18*	HM-19*	HM-20*	HM-21*
Total Carbon	3.42%	3.42%	3.42%	3.45%
Silicon	2.06	2.20	2.16	2.27
Manganese	0.02	0.02	0.02	0.02
Sulfur	0.014	0.013	0.013	0.011
Phosphorus	0.024	0.024	0.024	0.023
Nickel	0.72	0.79	0.76	0.79
Magnesium	0.046	0.055	0.047	0.081
Cerium	—	—	—	0.015
Antimony	—	0.004	0.012	0.022
Carbon Equivalent	4.04	4.08	4.07	4.13
As-Cast 1-in. "Y"				
Tensile Strength, psi	70,550	66,340	81,720	76,200
Yield Strength, psi	—	—	—	—
Elongation, %	17.0	2.0	None	12.0
BHN	178	212	232	207
Comment	Good nodularity—essentially ferrite	Some flake with increasing pearlite	Mostly pearlite; some flake	Good nodularity with more pearlite than HM-19

*H. Morrogh, *AFS Transactions*, vol 60, pp 439-52 (1952).

Melt	HM-26*	HM-27*	HM-28*	HM-29*
Total Carbon	3.44%	3.44%	3.44%	3.39%
Silicon	2.07	2.16	2.27	2.26
Manganese	0.02	0.02	0.02	0.02
Sulfur	0.023	0.012	0.011	0.014
Phosphorus	0.024	0.024	0.024	0.024
Nickel	0.65	0.72	0.70	0.78
Magnesium	0.049	0.072	0.054	0.096
Cerium	—	—	—	0.021
Bismuth	<u>0.005</u>	<u>0.003</u>	<u>0.005</u>	<u>0.006</u>
Carbon Equivalent	4.06	4.09	4.12	4.07
As-Cast 1-in. "Y"				
Tensile Strength, psi	69, 850	63,600	43,720	73,000
Yield Strength, psi	—	—	—	—
Elongation, %	17.0	10.0	None	16.0
BHN	174	155	187	173
Comment	Good nodules; essentially ferrite	Flake graphite present with pearlite	Flake with occasional spheroids	Good nodularity

TABLE 4-15. EFFECT OF BISMUTH ON MECHANICAL PROPERTIES

*H. Morrogh, *AFS Transactions*, vol 60, pp 439-52 (1952).

Melt	ABF-17	ABF-18	ABF-19	ABF-20
Total Carbon	3.68%	3.68%	3.68%	3.68%
Silicon	2.43	2.43	2.43	2.43
Manganese	0.21	0.21	0.21	0.21
Sulfur	0.012	0.012	0.012	0.012
Phosphorus	0.035	0.035	0.035	0.035
Nickel	0.63	0.63	0.63	0.63
Magnesium	0.042	0.042	0.042	0.042
Chromium	0.05	0.05	0.05	0.05
Copper	0.08	0.08	0.08	0.08
Boron	<u>T</u>	<u>0.018</u>	<u>0.041</u>	<u>0.105</u>
Carbon Equivalent	4.41	4.41	4.41	4.41
As-Cast 1-in. "Y"				
Tensile Strength, psi	79,900	82,400	76,700	66,450
Yield Strength, psi	51,900	53,450	43,450	44,450
Elongation, %	14.0	10.5	5.0	1.0
BHN	179	196	207	228

TABLE 4-16. EFFECT OF BORON ON MECHANICAL PROPERTIES

Melt	HM-13*	HM-14*	HM-16*	HM-17*
Total Carbon	3.49%	3.49%	3.49%	3.34%
Silicon	2.08	2.05	2.02	2.11
Manganese	0.02	0.02	0.02	0.02
Sulfur	0.010	0.010	0.011	0.011
Phosphorus	0.023	0.023	0.022	0.022
Nickel	0.66	0.69	0.66	0.66
Magnesium	0.047	0.049	0.044	0.044
Tin	—	0.014	0.016	0.022
Carbon Equivalent	4.11	4.11	4.09	3.99
As-Cast 1-in. "Y"				
Tensile Strength, psi	62,500	86,500	91,800	99,500
Yield Strength, psi	—	—	—	—
Elongation, %	13.0	10.0	4.0	4.0
BHN	156	208	239	249
Microstructure	Nodular	Nodular	Nodular	Nodular
% Ferrite	90	30	5	2
% Pearlite	10	70	95	98

TABLE 4-17. EFFECT OF TIN ON MECHANICAL PROPERTIES

Melt	HM-10*	HM-11*	HM-12*
Total Carbon	3.43%	3.51%	3.51%
Silicon	2.10	2.01	2.08
Manganese	0.02	0.02	0.02
Sulfur	0.011	0.012	0.013
Phosphorus	0.023	0.023	0.023
Nickel	0.74	0.62	0.60
Magnesium	0.047	0.045	0.045
Arsenic	—	0.025	0.046
Carbon Equivalent	4.06	4.11	4.13
As-Cast 1-in. "Y"			
Tensile Strength, psi	68,100	79,950	93,370
Elongation, %	13.0	12.0	3.0
BHN	177	185	266
Microstructure	Nodular	Nodular	Nodular
% Ferrite	90	50	5
% Pearlite	10	50	95

TABLE 4-18. EFFECT OF ARSENIC ON MECHANICAL PROPERTIES

*H. Morrogh, *AFS Transactions*, vol 60, pp 439-52 (1952).

TABLE 4-19. SUMMARY OF THE SIGNIFICANT EFFECTS OF ELEMENTS IN DUCTILE CAST IRON

ELEMENTS	TYPICAL AMOUNT	MAXIMUM for MATRIX		POSITIVE EFFECTS	DELETERIOUS EFFECTS
		Ferrite	Pearlite		
Spheroidizing Elements					
Mg	0.02–0.08%	Sufficient to assure spheroidal graphite		Lowers sulfur and oxygen contents; causes graphite to form spheroids	Excess promotes carbides
Rare Earths (RE)	0–0.30%	About 0.035%	About 0.035%	Promotes nodule count and quality in combination with magnesium; neutralizes subversive elements	Excess promotes carbides in thin sections and chunky graphite in heavy sections
Ca	Not detected	Essentially insoluble	Essentially insoluble	Increases nodule count and improves nodule quality; optimizes inoculation	Excess promotes carbides
Ba	Not detected	Essentially insoluble	Essentially insoluble	Increases nodule count; optimizes inoculation	—
Primary Elements					
C	3.00–4.00%	3.00–4.00%	3.00–4.00%	Present as spheroids or carbides	Excess results in graphite formation
Si	1.80–3.00%	1.80–3.00%	1.80–2.75%	Promotes graphitization during solidification and matrix formation	Hardens and strengthens ferrite; increases nil-ductility temperature
P	About 0.02%	0.035% max	0.05% max	Kept as low as possible	Forms intercellurlar carbide network; promotes pearlite
S	0.01–0.02%	0.02% max	0.02% max	Combines with magnesium and rare earths	Limits efficiency of magnesium treatment process
Mn	0.00–1.20%	0.20%	0.80% max	Promotes pearlite in as-cast and normalized iron	Intercellular carbides when over 0.70%
Alloying Elements					
Ni	0.01–2.00%	As low as possible for as-cast	To specification	Employed for hardenability (e.g. pearlitic)	—
Mo	0.01–0.75%	0.03% max	To specification	Promotes hardenability	Excess promotes intercellular carbides
Cu	0.01–0.90%	0.03% max	To specification	Promotes pearlitic hardenability	No significant effect on nodule count or quality
Deleterious Elements					
Te	<0.005%	0.02% max	0.02% max	Used to control pinholes	Promotes spheroid degeneration in absence of rare earths
Pb	—	0.002% max	0.002% max	Kept as low as possible	Promotes intercellular flake graphite
Ti	<0.07%	0.03% max	0.07% max	Kept as low as possible	Promotes vermicular graphite
Al	0.003–0.06%	0.05% max	0.05% max	Used in ferroalloys to suppress chill	Promotes vermicular graphite effect—greater in heavy sections; promotes pinholes
Sb	<0.005%	0.001% max	0.001% max	Strong pearlite former; counteracts chunky graphite in heavy sections	Nodule degeneration at high levels when rare earths are not present
Bi	<0.01%	0.002% max	0.002% max	Increases nodule count and quality when rare earths are present	Promotes vermicular graphite in absence of RE
Zr	<0.01%	0.10% max	0.10 max	Kept as low as possible	Promotes vermicular graphite
Carbide and Pearlite Forming Elements					
Cr	0.02–0.15%	0.04% max	0.10 max	Very powerful carbide former	Carbides resistant to annealing
B	<0.0005%	0.002% max	0.002% max	Kept as low as possible	Forms intercellular carbides that resist annealing
Sn	<0.10%	0.01% max	0.08% max	Very potent pearlite former	At <0.10%, forms intercellular structure with flake graphite
As	0.01% max	0.02% max	0.05% max	About 0.08% required for pearlitic matrix	—
V	<0.04%	0.04% max	0.05% max	Forms very stable carbides	Retards annealing
Gaseous Elements					
O	<0.005%	About 0.003%	About 0.003%	Kept as low as possible	Combines with magnesium
H	0.0002–0.0015%	About 0.0003%	About 0.0003%	Kept as low as possible	Promotes centerline carbides and inverse chill; promotes pinholes
N				Kept as low as possible	Mild carbide forming tendency; may contribute to porosity

5

Melting Practices for Ductile Iron Production

C. Ronald Kern

American Cast Iron Pipe, Co.
Birmingham, Alabama

Richard Kryzanek

John Deere Foundry
East Moline, Illinois

Timothy Zeh

Intermet Corp.–Columbus
Foundry
Columbus, Georgia

William L. Powell

Waupaca Foundry
Waupaca, Wisconsin

■ INTRODUCTION

All iron melting units capable of melting casting iron, be it gray or malleable irons, may be used to melt iron that will be converted to ductile iron. Melting units of every conceivable size, fuel source and operating practice have produced acceptable ductile iron castings. The prime consideration in selecting a melting unit should be that it can meet ductile iron's special and unique needs. These are:

- low sulfur content—0.02% or less to reduce magnesium treatment usage;
- relatively low silicon content—1.00–1.50%—if conventional magnesium ferrosilicon alloys are to be utilized (2.0–2.5% for pure magnesium treatment methods);
- high carbon content—usually 3.60–4.00%;
- high tapping temperature—2650–2850F (1454–1566C), depending on the treatment method;
- low residual elements—low levels of chromium, copper, molybdenum, vanadium, etc.—within the charge materials;
- consistency of operation—good ductile iron production is based on rigid controls of chemistry, tap temperature, holding times, etc.

Melting units that usually meet the above requirements are cupolas and induction melting furnaces. Arc furnaces, reverberating furnaces, air furnaces and gas melting units are used to a lesser degree. The final choice of which melting unit to use is a complex decision, based on factors such as available capital, plant layout, available charge materials, knowledgeable labor and operating experience.

CUPOLA MELTING

Cupola melting can generally be divided into two basic methods: acid or basic slag practice. Each can be further divided into lined or unlined, water cooled or dry shell, coke/gas-fired or electric (plasma), hot or cold blast and so forth. The varieties of cupola systems are nearly endless.

Acid practice with external desulfurization currently produces the major tonnage of cupola-melted ductile iron. The majority of practice, here in the United States, is the hotblast, unlined cupola. Basic slag cupolas still produce some ductile iron and are favored by some melters.

The typical cupola is a vertical shaft supported by a bedplate and four legs. It varies in diameter from about 24 to 150 inches and in height from 8 to 35 feet. The shaft is subdivided into a number of regions, i.e., well, melt zone, body, charge zone, etc. Air is fed to the cupola from blowers, through blast ducts, the wind box, downcomers and, finally, tuyeres. Iron and slag exit the cupola from a tap hole into the front spout and separate by a slag skimmer and iron dam.

Bottom doors and bottom sand support the bed of incandescent coke, alternating layers of metallics and running coke. Hot, dirty gases are cleaned in an adjoining emission system, rendering the entire system environmentally acceptable. These cupola members can be seen in Figures 5-1 and 5-2. Production rates of a cupola can be calculated based on diameter, stock height, blast air volume and blast temperature. Widely varying tonnages can be obtained from similar cupolas, due to variations in charge materials, shell heat loss, tap chemistry and temperature, etc. Cupola manufacturers can accurately assess the effects of these and many other variables on melt rate, chemistry and tap temperature. By defining various parameters, a suitable cupola may be designed or modified to fit a given production situation.

ADVANTAGES OF CUPOLA MELTING

The advantages of the cupola for the production of ductile base iron are numerous, accounting for its wide-spread popularity. Some advantages are:
- good carbon and silicon control;
- ability of a single melting unit to melt gray or ductile iron;
- low-cost operation;
- continuous supply of iron available;
- wide variety of charge materials usable;
- high tapping temperatures achievable.

One disadvantage of the acid-lined cupola is the high sulfur content of the iron. Sulfur contained in the molten iron will typically be from 0.07–0.17%, the exact amount of sulfur being a function of the amount of coke used and its sulfur content. This high sulfur content must be reduced prior to treatment with magnesium, or excessive magnesium amounts will be required. (Techniques of desulfurization will be discussed in a later chapter.)

CUPOLA OPERATION

The cupola is readied for operation by first returning it to its original condition. All slag and spent refractories from the previous campaign are removed. Appropriate refractories are applied to the body, well and spout sections. Heat

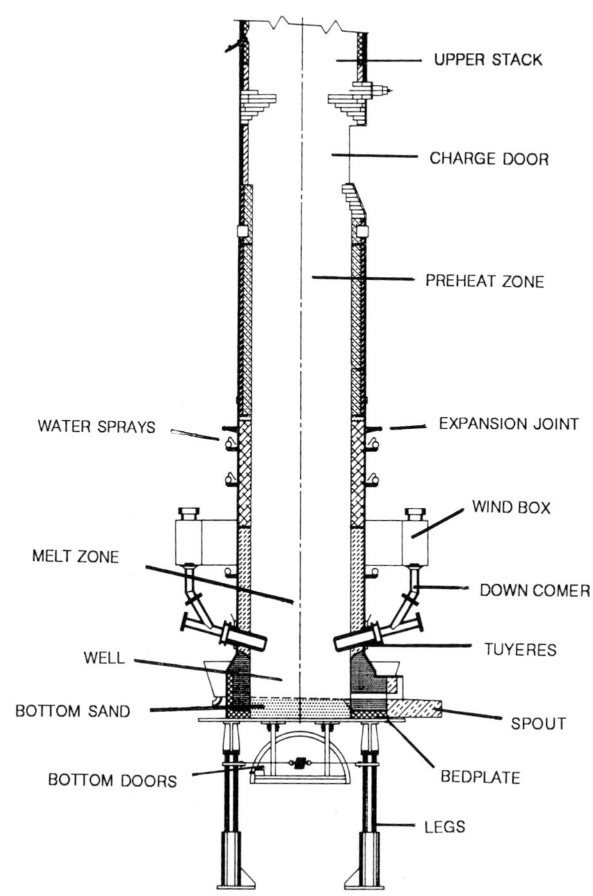

Fig. 5-1. Major components of a typical cupola are shown. (Drawing courtesy of Modern Equipment Co.)

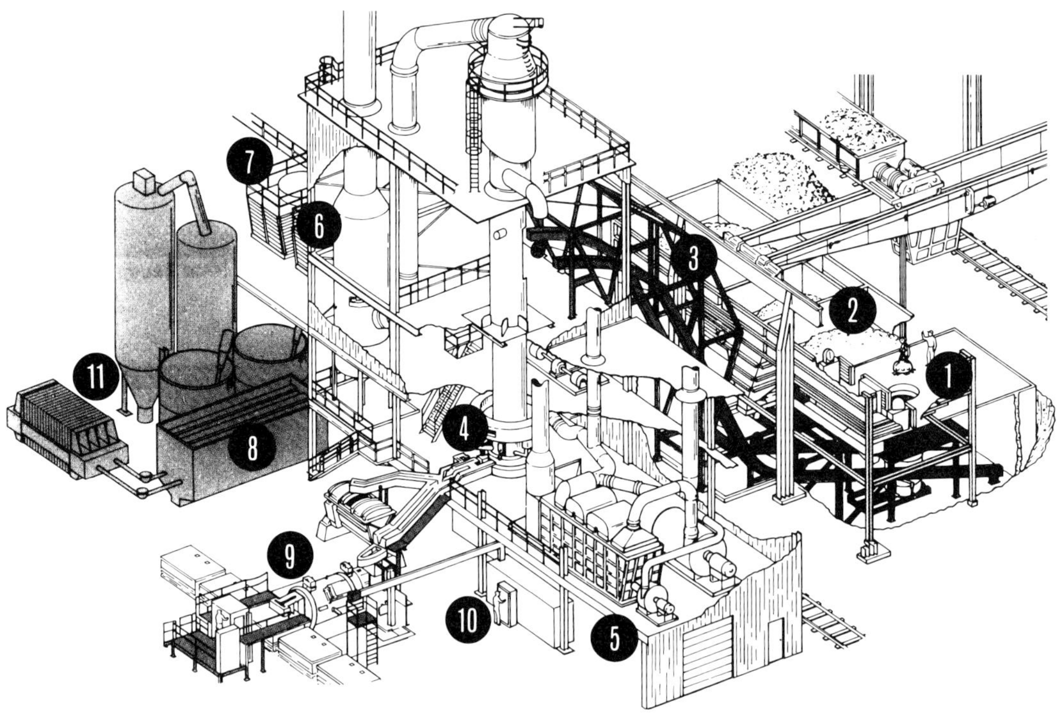

1. *Metal Batching*—Weight hopper and scale system for batching of metallics from rail car, truck or bin.

2. *Nonmetallics Batching*—Storage bins with automatic batching and weighing of coke, stone, flux and alloys.

3. *Cupola Charging*—Charge materials conveyed by skip charger for bucket or feeder loading of cupola.

4. *Cupola*

5. *Cupola Blowing System*—Recuperative hot blast system supplying tuyere air complete with blower controls and ductwork.

6. *Air Pollution Control*—Cupola gases cleaned to code requirements by wet or dry collectors with below charge or top gas takeoffs.

7. *Water Cooling*—Recirculating noncontact cooling water system, minimizing water consumption and treatment requirements.

8. *Water Treatment*—Neutralization, flocculation and clarification recycle systems for air pollution control equipment.

9. *Hot Metal Handling*—Manual and automatic delivery and pouring systems.

10. *Systems Control*—Control and computer interface with equipment and auxiliaries interlocked and sequenced.

11. *Sludge Handling*—Wastewater treatment sludge handling systems, providing high solids content and stabilization for disposal.

Fig. 5-2. A drawing of a typical, fully integrated melting system. (Drawing courtesy of Modern Equipment Co./ecologix inc.)

is applied to assure proper and thorough drying of all refractories. The bottom doors are closed and a sand bottom rammed in place with pneumatic rammers. A bed of coke is placed in the melting zone and burned to a bright red color. Fresh coke is added to a predetermined height above the tuyeres. Measurements of this bed height are critical to provide an adequate amount of coke for sufficient temperature and proper chemistry at tap-out. The cupola is charged with alternate layers of metallic charge and coke. Blast air is directed into the cupola and melting begins. After a short time interval (10–20 minutes) the tap hole is opened and a flow of metal and slag is established. (Exact tap-out methods should be established for each particular cupola operation.)

As blast air is directed into the cupola, iron will continue to melt. Charges consisting of scrap iron, pig iron, returns, coke, flux and alloys will continuously be charged at the cupola top. Accurate weighing of each item is essential to good cupola control. The use of electronic scales and a computer-compensating program greatly enhances proper weight control.

Cupola melting is a combustion process. Operational changes will be reflected within the combustion process and should always be viewed in that light. Consistency of blast air control is vital. Avoiding major changes in blast rate, frequent on-off blast periods, or wide variation in oxygen usage will be manifested in change of melt rate, temperature, carbon pickup, sulfur pickup, silicon loss and other items.

The melting campaign is terminated by melt out and bottom drop. While techniques will vary, safety must be the primary consideration. All traces of water must be removed from the drop zone. Combustible materials should be removed from both the drop area and tapper's platform. All the iron is melted from the cupola. Provision to drain all the slag and iron from within the cupola well must be considered and a suitable drop area must be provided. Bottom doors are opened, and the bottom sand, along with the remaining coke, ash and other materials, are allowed to drop.

ACID PRACTICE

The acid cupola is a versatile operation most usually used in gray iron melting. For ductile iron production, external desulfurization must be employed. Even with this disadvantage it is the most popular ductile iron production method. A typical acid slag contains 40–50% SiO_2, 10–20% Al_2O_3, 25–40% CaO and MgO, 1–5% MnO_2, 1–8% FeO and 0.05–0.08% S.

Using the relationship:

$$\text{Basicity} = (MgO + CaO) / (SiO_2 + Al_2O_3)$$

acid practices will be below 1 and usually in the 0.6–0.9 range. (Some melters do not consider the Al_2O_3 factor.) The flux employed is limestone, with typical composition as shown in Table 5-1. While usually a local material, consideration should be given to optimizing its composition.

Stone creates the fluxing action to clean the coke ash. Typical stone usage is 20–35% of the coke weight or 2–4% of the metallic charge weight, depending on how the operator chooses to control the stone use. Excess stone use will cause refractory wear, especially in the tap hole and well areas. Low stone usage will cause excess coke use and low metal temperatures.

Refractories used in the cupola lining can be either silica- or alumina-based. Typically, they are gunned on the walls in the melt zone and either gunned or bricked higher up in the stack. The lining contributes little to the slag chemistry beyond the first four to six hours. At this time, refractory in the melt zone has eroded to an equilibrium thickness and will be maintained in relation to heat loss.

Bare-shell cupolas can also be operated in an acid mode. Here, stone use is adjusted to give a slightly more neutral ratio. Refractory practice in the well and spout areas usually favor a carbon-based material.

TABLE 5-1. COMPOSITION OF LIMESTONE FOR CUPOLA MELTING

	Ideal Composition	Stone #1	Stone #2
$CaCO_3$	+45%	52.26	49.10
$MgCO_3$	+30%	44.17	37.70
SiO_2	<2%	2.95	10.78
FeO_3	<2%	0.40	0.83
Al_2O_3	Low	0.17	1.52

BASIC PRACTICE

Basic cupola practice and basic slag control are more complex and require a greater degree of supervision than acid practices. The objective of basic practice is the melting of a base iron containing a low sulfur content not requiring external desulfurization. While once the dominant melting technique for ductile base iron, it now has relatively few users. The slags contain a predominance of CaO and MgO over Al_2O_3 and SiO_2. Sulfur in the slags can reach levels of 0.5–2.5%. Typical slag chemistries are shown in Table 5-2. Iron sulfur can be as low as 0.02%.

Basic slag practices offer the user several advantages:

- low-sulfur base iron without external desulfurizing practices;
- very low oxygen content irons;
- higher carbon contents due to better carbon pickup.

Some disadvantages are:

- basic refractory costs are higher than acid or neutral;
- high silicon losses;
- slower melting rates due to high coke usage;
- more difficult operations due to a larger volume of high-melting-point slag.

TABLE 5-2. TYPICAL SLAG ANALYSIS, VERY BASIC TO ACID

Analyses		SiO_2%	Al_2O%	CaO%	MgO%	FeO%	MnO%	CaO+MgO	SiO_2+Al_2O_3	Basicity
Very basic (0.008% S in melt)(very reducing) (disintegrates leaving white powder)		16.1	3.45	62.4	8.1	0.28	Trace	70.5	19.6	3.6
Very basic (very reducing) (0.008% S in melt) (disintegrates to a gray slag)		19.7	3.2	61.45	7.1	0.27	0.06	68.6	22.9	3.0
Basic (oxidizing) (0.021% S in melt) (sandy powder)		26.0	3.9	65.4	2.2	0.33	0.30	67.6	29.8	2.3
Basic (oxidizing) (0.03% S in melt)	I=Dark	30.5	3.9	46.9	14.3	1.44	0.36	61.2	34.4	1.8
	II=Light	28.7	6.9	47.9	12.7	1.73	0.35	60.6	35.6	1.7
Basic (oxidizing) (0.04% S in melt (dull sandy color)		29.3	9.5	55.4	2.2	0.32	1.38	57.6	38.8	1.5
Neutral (0.05–0.07% S in melt)		33.6	13.1	34.6	12.4	1.5	2.7	47.0	46.7	1.0
Acid (typical) (0.09–0.10% S in melt) (glassy)		46.2	11.0	37.2	1.4	1.1	1.4	38.6	57.2	0.7
Acid (oxidizing) (0.09–0.12% S in melt) (viscous and black)		47.1	12.1	22.0	1.6	6.9	4.6	23.6	59.2	0.4

Basic cupola practice is usually done in either a bare shell (unlined) cupola or one with a melt zone of neutral or basic refractories. Magnesite brick and carbon blocks are the preferred lining materials. Silica refractories are to be avoided, because their erosion will reduce the basicity value of the slag. A basicity ratio of greater than 1.0 indicates a basic cupola operation.

Ratios of 1.4–1.8 are usually used, with some going as high as 3.5 for very basic operations. Silicon losses can approach 40% under some melting conditions. Metal:coke ratios for cupolas can vary with shell diameter, but 5:1 to 7:1 are typical. Control of melt temperature becomes a greater challenge as basicity climbs. Consistent operation of the cupola is mandatory for good control.

CHARGE MATERIALS

Charge materials for cupola melting of ductile iron are much like charge materials from any other cupola-melted iron. The important characteristics are size, thickness, chemistry, consistency and degree of contamination. Depending on the grade of ductile iron to be produced, alloying elements like copper, manganese, nickel, or molybdenum may be tolerated. Elements like chromium, lead, aluminum and titanium should be avoided or held to low values.

Typical charge percentages of steel scrap, pig iron, returns, etc. are very difficult to cite. A ductile pipe producer may use 80–90% scrap steel, while a captive small-parts producer will use as little as 15–20%. Pig iron, used to control chemistry and achieve the desired carbon content, can be used in amounts varying between zero and 40%. Returns, usually used as available, are limited to the highest allowable silicon content of the base iron. This return usage is usually in the 40–60% range for non-pipe producers.

Steel scrap is typically recent, industrially produced and of a known composition. Older scrap may be used only if its chemical composition is known and is relatively stable. Knowledge of the scrap source and a close working relationship with the scrap preparer is essential. Steel scrap should be cut to about one-third of the cupola diameter or smaller. The scrap size and shape should not bridge or cause air flow channeling. Steel scrap is the most likely source of metallurgical contamination due to lead, manganese, arsenic, chromium, tellurium, etc.

Pig iron is purchased, based on chemical analysis, to provide an essentially pure iron source. Principally an iron, carbon, silicon material, it can be purchased with varying degrees of manganese, phosphorous and sulfur. One of its functions is to dilute other charge materials and detrimental elements. Its ability to pretreat or condition the iron is often a topic of debate. Pig iron melts easily and contributes to an increased melt rate.

Foundry returns are a product of the foundry. Foundries producing several grades of ductile iron, by base iron chemistry or alloy composition, must segregate by grade. In a basic operation, the adhering sand can cause a reduction in basicity ratio.

Foundry coke for cupola melting is the same for both gray and ductile irons. All U.S.-produced foundry cokes are essentially of the same analysis. Typical coke composition is seen in Table 5-3. This is the principal source of the sulfur in the base iron, especially in acid practice. Foundry coke should be consistent in size and carefully handled throughout the charging practice. This will preserve the as-shipped coke size and reduce losses due to breakage. Handling damage can be expected to be 1–3% by truck or rail delivery, 3–5% in yard storage and up to 10–15% in rough handling conditions.

Silicon sources are diverse. They can be in the form of lump 50% FeSi, silicon carbide briquets or blocks, silvery pig irons, or FeSi briquets or blocks. Consideration of residual element levels of chrome, manganese, copper, etc. must be made. Significant buildup levels of these elements can result if not properly controlled. A wide range of carbon additives are available as briquets, which can increase the carbon content of the alloy from 1 to 35% C. This can impact coke ratios in some instances. Silicon additives come in many forms to meet any charging method requirements, be it manual or automated, with addition by piece count or weight.

NONTRADITIONAL CUPOLAS

Over the past few years, several variations of the cupola have emerged. The two most prominent are the plasma cupola and the gas-fired cupola. Both have had very limited use within the United States, but the gas-fired version has a significant number of foreign users. Both use alternate fuels, i.e., electricity and natural gas, to replace some or all of foundry coke as a fuel source. In both cases, the basic cupola design is similar to a conventional cupola. The significant change is the tuyere area where burners are introduced.

Both of these developments offer distinct advantages to the ductile iron melter. First, very low sulfur is present in the base iron, due to elimination or low usage of foundry coke. Second, very low emissions (requiring smaller air emission control devices) are possible due to coke reduction or elimination. These two items make ductile iron more environmentally acceptable to produce.

The plasma cupola utilizes highly sophisticated plasma torch technology. The plasma torch is an electrical device used to produce a high-temperature, ionized, conductive gas stream, as shown in Figure 5-3. This stream, mixed with a suitable process gas, produces a torch capable of temperatures varying from 1800F (982C) to over 9000F (4968C).

The advantages of the plasma fired cupola are:

- greatly reduced coke usage;
- lower silicon losses;
- use of loose cast iron chips;
- increased control of temperature;
- reduced energy consumption.

The two major disadvantages are high capital cost and very limited experience within the melting community. Presently, only one operating cupola is known to exist. The experiences gained from it will set some of the design and operational patterns of the future.

The gas-fired cupola enjoys a wider use in Europe and the Middle East countries, where natural gas is plentiful and good foundry coke is expensive or unavailable. Gas burners replace the tuyeres and refractory balls or scrap bricks replace the coke bed. Water-cooled pipes support the refractory and charge materials above the burners, providing a large void for iron collection and firing. Figures 5-4 and 5-5 show the two typical types of gas-fired or "cokeless" cupolas.

The burners are conventional, with computer control of the air-fuel ratio and a short flame length. This high degree of control provides maximum flame temperature. However, even at best, iron temperatures are colder than usable unless a heated forehearth is employed. Carbon in the form of low-sulfur graphite is added to the metal stream as it enters the forehearth. Good normal mixing techniques (i.e., turbulence, appropriate material sizing, clean bath, etc.) must be used to obtain adequate recovery of carbon. In the case of low melting temperatures, a heated forehearth of unusually high electrical power input to metal holding capacity must be employed.

TABLE 5-3. COMPOSITION OF TYPICAL FOUNDRY COKE

Fixed Carbon	91.5–94.5%*
Volatile	0.1–1.0%
Ash	5.0–7.5%
Sulfur	0.4–0.8%
Moisture	0.1–4.0%
Shatter (% retained in 2 in. screen)	92.5–99.5%

*statistical analysis representing a 3-sigma range

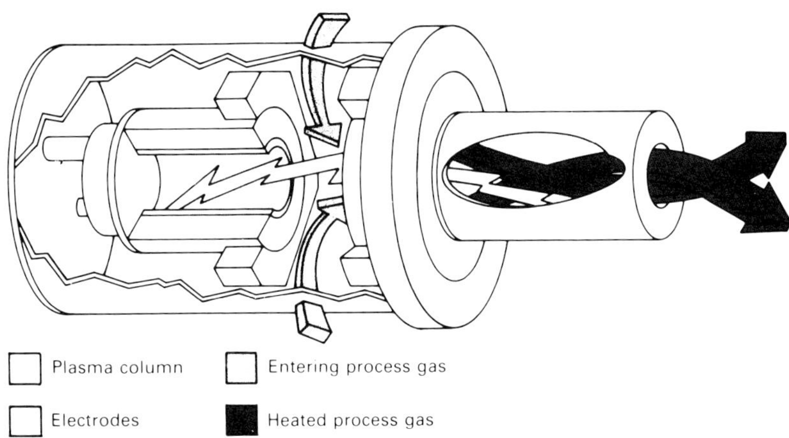

Fig. 5-3. A plasma torch as utilized in the plasma cupola is shown. (Westinghouse "Descriptive Bulletin" 27-501, p 4, October 1984.)

■ VERTICAL CHANNEL MELTING FURNACES—DUPLEXING

The channel furnace is most often employed as a holding furnace for duplexing. Duplexing is the process of melting ductile base iron in one furnace and then transferring it to another furnace for further treatment. This further treatment includes such operations as chemical analysis change, superheating, homogenizing the chemical analysis, holding the iron for use at a later time, or performing nodulization treatment. Duplexing is normally employed by high-production shops, because of their demand for high productivity and high efficiency. Duplexing is one method of attaining a consistent supply of iron that is of the desired chemistry and temperature.

Duplexing is used by many producers of ductile iron. In most installations the primary melter is a cupola or a direct arc electric furnace. The melt is then duplexed into a coreless induction furnace or a channel induction furnace. The cupola offers good thermal efficiency for melting, but not for superheating. Induction furnaces offer good thermal efficiency for superheating liquid iron. Induction furnaces are excellent for making alloy changes to the iron, whereas cupolas are not. Induction furnaces can also provide a continuous supply of iron when the primary melter is the intermittent tapping electric arc furnace.

Properly engineered and suitably powered, the vertical channel furnace can be an excellent ductile iron melter. Typically sized in 20- to 50-ton holding capacity units, melting rates vary from 1 to 6 tons per hour. The large holding capacity and slow melt rates provide extreme consistency.

The advantages of the vertical channel furnace are:

- a large volume of iron available;
- excellent metal control because of a high holding-capacity-to-melt rate ratio;
- long term campaigns (up to one year and longer);
- off-peak melting for reduced electrical cost;
- low-sulfur iron without desulfurizing.

A disadvantage is that iron must be left in the furnace at all times, adding to holding costs. The working capacity of a vertical channel melter is approximately two-thirds of the total holding capacity.

The vertical channel furnace is composed of two basic sections: the upper body section and the inductor. Typical furnace construction is seen in Figure 5-6. The lid or roof area swings back for charging, slagging and refractory repair. If placed above floor level, operator comfort and safety is enhanced during all phases of operation.

The body is a steel shell, refractory lined with brick or dry vibratable materials. Unlike the coreless induction furnace, high energy stirring does not take place. Thus, refractory wear is more even and very slow. Typically, both pour-down and charging proceed slowly, so thermal shock to the lining is also minimized. Specific refractories have been developed to provide the necessary insulating and chemical requirements of the different melting zones of the vertical channel furnace.

Refractory usually consists of insulating brick at the shell. This is followed by a layer of backup brick consisting of alumina brick between 70% and 90% Al_2O_3 brick or a suitable dry vibratable mix. Floor and throat areas may be brick and dry vibratable refractory or all dry vibratable material.

The inductor is the working portion of the vertical channel furnace. The inductor is usually either a twin-loop or single-loop design. While both offer claims of improved operation, in fact, both melt ductile base iron very well. Melters usually are powered from 1500–3000 kVA with melt rates varying from 2–6 tons per hour. Power requirements vary with the needs of the user. For estimating inductor size, the following rules of thumb apply: (1) one ton of molten iron can be raised 100°F (55.5°C) with 12–15 kW; (2) one ton of scrap will require 500–600 kW to melt it and bring it to pouring temperature.

The inductor is composed of a steel case to contain the refractory, a loop for molten iron to pass through and electrical components, such as the core and coil, to transmit the energy to the iron. Each component is balanced with the others to provide a precise electrical system. Maintenance of the design parameters of each component is mandatory for optimum operation of the furnace. Core, bushing, and case maintenance should receive particular attention. Rigidity and alignment should be checked with each rebuild of the inductor, as well as the electrical insulation of each item. Movement of the case or coil during operation can cause refractory malfunction and furnace damage.

Electrical components include a transformer, power cabinet and control system. The transformer provides the voltage step-down from incoming line voltage and increased currents. It is located near the power cabinet. The power cabinet is located adjacent to the inductor to allow for short power cable runs and minimal line losses. Its function is to provide the switching functions and variable kilowatt input to the furnace. Power input will vary from very low start-up levels to full rated capacity. A variety of control techniques are available to the ductile iron melter.

The control cabinet is located in an area accessible to the operator. It is the input device to drive the power cabinet. It usually contains the required metering systems required to monitor the furnace condition and operation.

Charge material requirements for the vertical channel furnace are similar to the coreless furnace. They should be clean, free of nonferrous materials, dry and with no closed compartments. Composition of materials should be well controlled and free of contaminating elements in order to achieve the desired chemistry. Steel scrap, pig iron and foundry returns are the usual charge components. Ferrosilicon and carbon additions are required to adjust the final chemistry to predetermined ranges. Melt losses or individual elements will vary somewhat, but nearly complete recovery of most elements can be expected. The exception is carbon. Due to the lack of stirring in the vertical channel melting furnace, recoveries will vary from 60–80%, depending on carbon type and source.

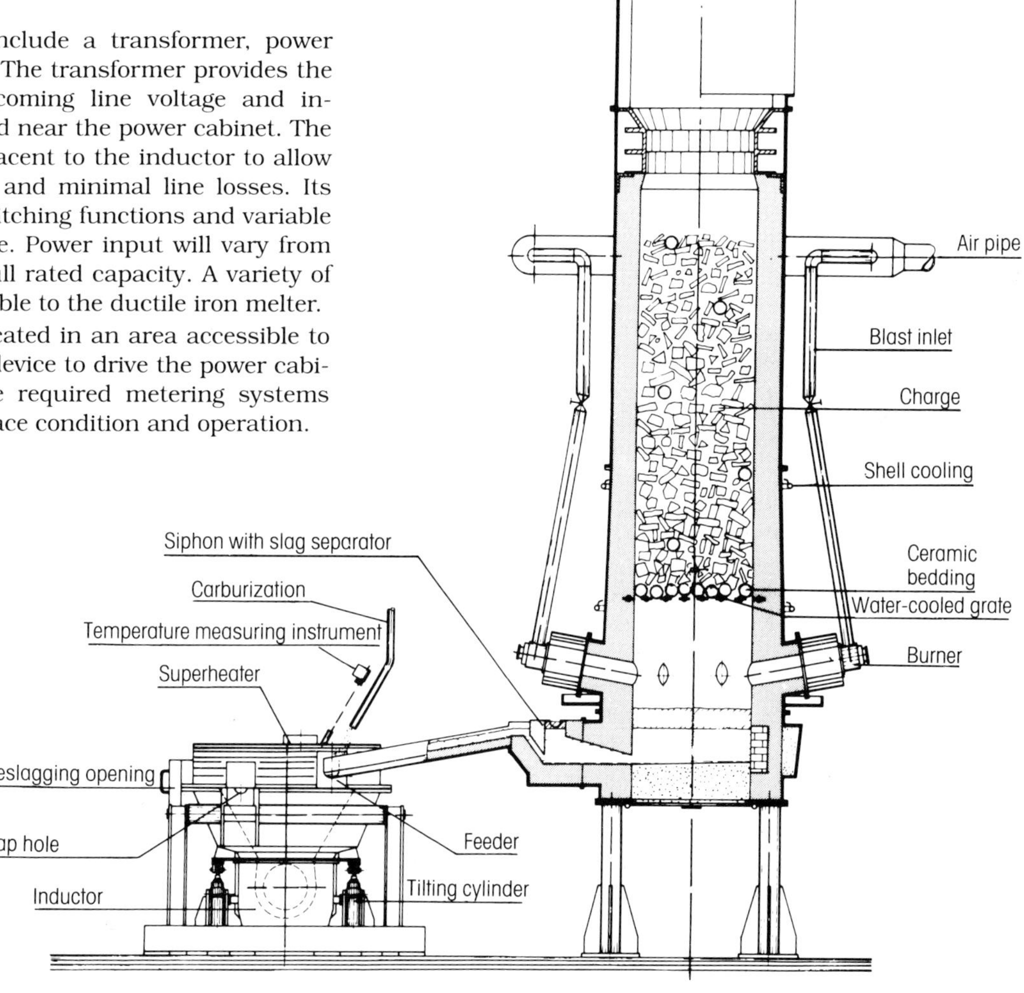

Fig. 5-4. A gas-fired cokeless cupola currently in use in Europe. (Drawing courtesy of Cokeless Cupola, Ltd. and Dücker.)

The vertical channel furnace can meet the ductile iron melter's requirements if it is properly designed and attention is given to operating practices. Its large storage capacity, low turbulence and long campaigns provide a consistent source of metal.

CORELESS INDUCTION FURNACE MELTING

Coreless induction melting has proven to be an economic melting system for many ductile iron producers. Induction furnaces provide rapid and efficient melting, and are capable of very accurate chemistry and temperature control. Problems associated with fuel (coke) in a cupola, such as sulfur pickup, poor carbon recovery and expensive dust collection, are minimized with coreless induction melting.[1-2]

OPERATING PRINCIPLES

Induction furnaces operate on the principle that, when alternating current is passed through a coil, a magnetic field is created, which induces eddy currents in the metal placed within that field. These currents react with the resistance in the metal, which causes the temperature to rise. The degree of heating achieved is dependent on the rate of variation (or frequency) of the magnetic field and on its intensity. Frequency is fixed for most furnaces at either line frequency (60 Hz), medium frequency (180–1000 Hz), or high frequency (above 1000 Hz).

Future furnace designs will probably incorporate variable frequency options to maximize power input during melting at 120–1000 Hz levels, and then maximize stirring of late alloy additions at the end of the heat using 60 Hz. As frequency increases, the depth of current penetration in individual charge pieces is reduced, and higher induced voltages are generated. Coupling between the components is, therefore, much improved, making possible a greater power input than could be achieved with low frequencies.

The intensity of the magnetic field is dependent on the power density of the furnace, or the amount of power that can be applied to a given amount of iron. The newer, medium-frequency furnaces can apply more power and, therefore, melt faster.[11]

These concepts are incorporated in several different designs of induction furnaces. For iron melting applications the most versatile and practical design is the coreless or crucible furnace, in which the induction coil induces current into the whole charge. Figures 5-7 and 5-8 show diagrams of typical coreless induction furnace installations.[1,8]

The induction process works only with metals; no heat is induced in the refractory crucible or lining, since it is not an electrical conductor. Temperature limits of the furnace are determined solely by the softening point of the refractory. For example, a vibrated silica lining can withstand continuous operation at 2900F (1593C) and intermittent service at higher temperature. Other lining materials, such as alumina, zircon, or magnesia (magnesite), will withstand considerably higher temperatures, but are usually not cost effective in iron production.

A special characteristic of the coreless induction furnace is the agitation that takes place in the iron bath. This is illustrated in Figure 5-9. The stirring action is produced by the electromagnetic field and its intensity is controlled by the frequency, power density (kW/ton) and relationship of the iron bath level to the top of the coil of the furnace. The frequency of the furnace affects the stirring action of the molten iron, which varies indirectly with the square root of frequency; low frequencies, therefore, stir more, whereas higher frequencies stir less.

Adequate stirring is very important to the production of ductile iron to properly distribute additives. Low-frequency furnaces are power delimited (250–350 kW/ton of furnace capacity), since splashing and excessive lining wear can result from violent stirring. Medium-frequency furnaces can have higher power densities (600–1000 kW/ton), because of the lower stirring action associated with increased frequencies. It is important that medium-frequency furnaces have sufficient power densities to produce the stirring action necessary to mix alloys and melt efficiently. The homogeneous mixing in induction furnaces lends itself to the production of special irons requiring large alloy additions and to the preparation of ductile irons from charges consisting of returns, steel scrap, silicon carbide and carbon additives. Sufficient stirring is necessary to efficiently dissolve the carbon additions and silicon carbide used in the production of ductile base iron.

MELTING EFFICIENCIES

The efficiency of any induction furnace can be measured as theoretical versus actual power required to melt one ton of iron. The theoretical heat content of molten iron is approximately 350 kWh/ton at 2750F (1510C). Since current coil designs convey about 80% of the induced field into the bath, the actual electric power required is approximately 435–440 kWh/ton at 2750F (1510C). The theoretical power is reduced 8 kWh for every 100°F (55.5°C) of preheat applied to the charge. Therefore, a 1000F (538C) preheat will lower the energy required to reach 2750F (1510C) into the 340 kWh/ton range. Actual power required in practice is much higher—due to conductive losses to the refractory lining and the water-cooled coil and radiant losses at the

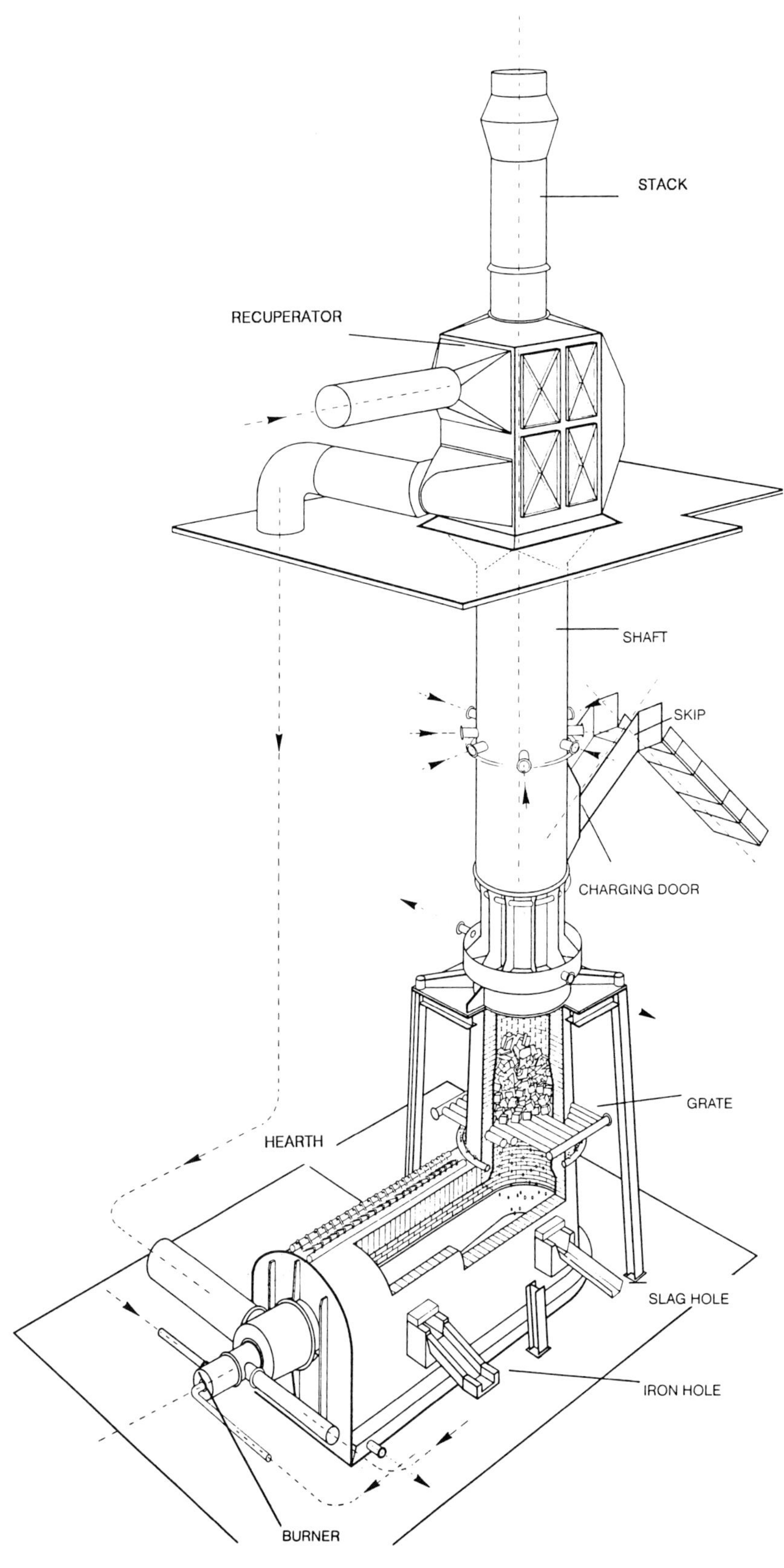

Fig. 5-5. A cokeless cupola utilizing a single burner and gas-geated forehearth. (Drawing courtesy of KGT Giessereitechnik GmbH.)

bath surface—increasing the power required back to 380–400 kWh/ton. Melting efficiency is also reduced by operational losses, which include: sampling, slagging, tapping and other off-line operations. In a batch-melting operation, approximately 90% of the induction field is transmitted to the charge while it is still magnetic. Above the Curie point, the input drops back to the 80% level. Therefore, in a batch-melt, cold-charged, medium-frequency melter, the average power input from the coil will be about 85%. Some of this 5% improvement will be lost due to additional lining losses when the lining is exposed after tapping. The high power densities available in medium-frequency melters have also reduced the need for larger furnaces with their associated thermal losses.

Electrical efficiencies have increased over the years with new solid-state electronics. New polypropylene capacitors with dielectric fluids have increased capacitor efficiencies to 90%. Overall, new technology has increased electrical efficiencies in melting furnaces to 97% versus 75% for systems available during the early 1970s.

The output and power consumption of coreless melters is dependent on the furnace design, but for any one furnace it is largely dependent upon the foundry operating practices. One of the simplest operating practices effective in increasing melting efficiencies of any furnace is keeping the molten iron bath covered to reduce thermal losses. Records should be kept on at least a weekly basis to monitor the kWh/ton of iron used for each furnace. This information can be useful in evaluating operating techniques and process improvements.[3,4,10]

CHARGING MATERIALS

The fact that ductile base iron with carbon content up to 4.30% and sulfur below 0.025% can be easily produced makes coreless induction melters particularly suitable for the production of ductile iron. Tables 5-4 and 5-5 are examples of typical charges used to make as-cast pearlitic and ferritic iron, respectively. Carbon additives must be carefully chosen to prevent excessive sulfur and, therefore, subsequent desulfurization. Close control of charge materials is necessary, because induction furnaces cannot refine out unwanted elements. Unlike the cupola, a number of undesirable elements in the melt are not oxidized and removed in the slag. The induction melter normally recovers most of the metallic components of the charge, in addition to a large portion of the nonmetallics. Nickel, chromium, molybdenum and lead are completely recovered from the charge, some of which can be detrimental to lining life and the quality of ductile iron. Some elements, such as aluminum and zinc, may oxidize and be removed in the slag, while others, like lead, normally remain in the bath and castings. Of particular concern is the fact that zinc and lead, found on some plated or painted steel scrap,

may pose a significant health risk unless proper ventilation is used.

The physical size and configuration of charge materials is also important. *Never charge wet materials or closed containers, such as crimped pipe, shock absorbers, or hydraulic cylinders.* Size is important in achieving proper charge density in order to reduce the time necessary to charge the furnace. To ensure a safe and efficient operation, only desirable charge materials should be used in electric melting.[2]

LINE-FREQUENCY MELTERS

Line-frequency (low-frequency) induction furnaces operate most economically when the charge is added to a molten bath. The volume of liquid metal retained in the furnace after tapping should be approximately 65–70% of the holding capacity. Operating below this level will result in reduced electrical efficiency and may cause damage to the furnace power components. For these reasons, line-frequency furnaces are operated as heel melters and are emptied only when relining is required.

When heel melting, charge materials should be added to a relatively slag-free bath in the following sequence:

1. Carbon additives and silicon carbide are charged into the furnace on top of molten metal using sufficient electrical power to quickly stir them into the bath.
2. Metallic materials are then charged, with power on or off based on shop needs and furnace manufacturer's recommendations.
3. Heavier materials should go on top, when possible, to push the lighter materials into the molten iron and prevent bridging.

When this procedure is used, recoveries in the area of 90–95% can be obtained from properly sized carbon additives. Under these conditions, it is possible to recover the major portion of the silicon and carbon in the silicon carbide—making silicon carbide (under these conditions) one of the most economical silicon sources available.[9]

After refractory installation, a line-frequency induction furnace can be started with a cold charge, provided that temperature and power inputs are closely monitored. Failure to follow recommended control parameters can result in damage to equipment or linings. Typical start-up methods consist of loading the furnace with starter blocks, pig iron, or sprue to obtain the greatest possible charge density. Starter blocks are solid iron discs about three to four inches less than the furnace diameter. The starter blocks are placed in the bottom of the furnace and low power is applied. Whenever sintering a new lining, a thermocouple should be used to monitor the temperature. The furnace power is regulated, so that the furnace temperature rises at a specified rate. Specific start-up procedures should be

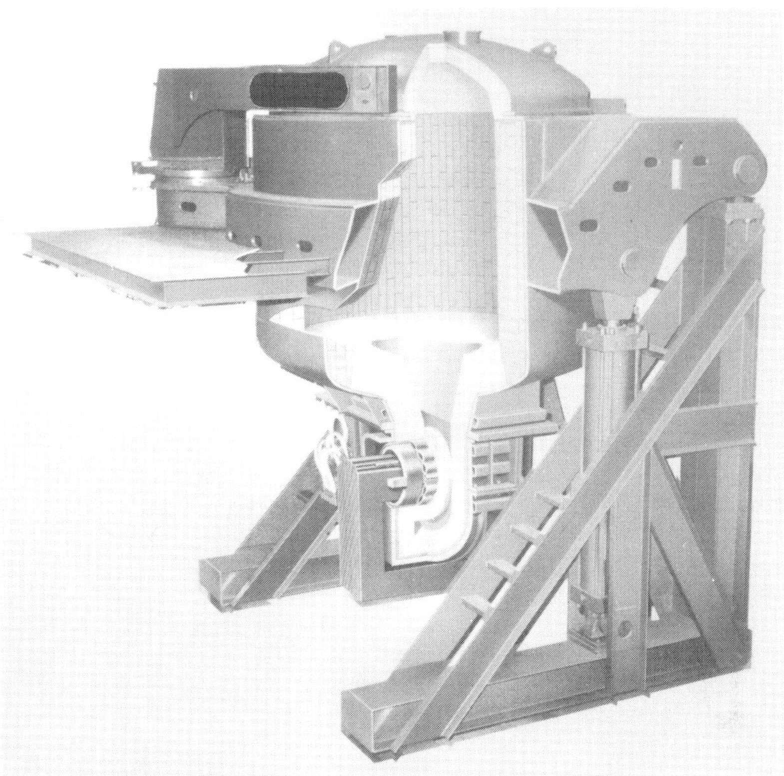

Fig. 5-6. A typical vertical channel furnace with single loop inductor. (Drawing courtesy of Inductotherm.)

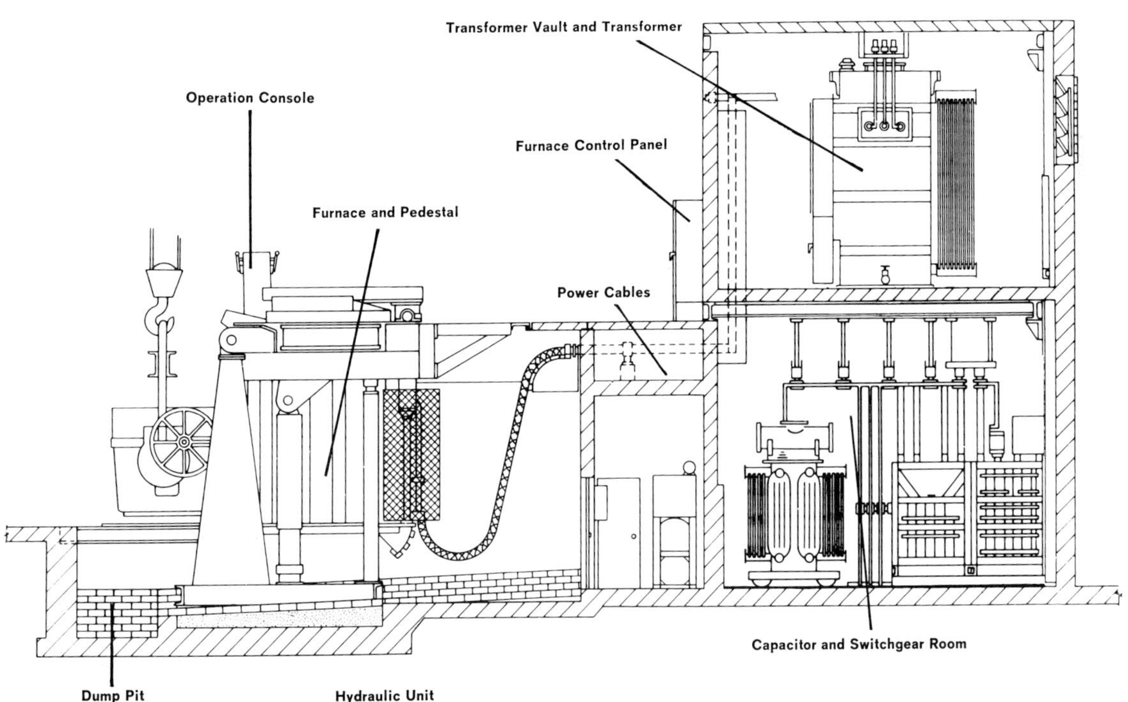

Fig. 5-7. Layout of a coreless induction furnace installation.

established for both new and previously sintered linings. These procedures should be strictly followed. Contact the furnace manufacturer and the refractory supplier for recommended procedures.

MEDIUM-FREQUENCY MELTERS

Medium-frequency melters can work more efficiently as batch melters, where the entire bath is emptied and cold-charged. This method substantially increases overall efficiencies and production at lower operating costs, due to lower thermal losses associated with smaller furnaces. The charge materials do not have to be preheated, because the materials are charged, using a heated vibratory feeder, onto solid materials.

Power densities for medium-frequency melters are in the 800–1000 kW/ton range. For a given melt rate, the smaller furnaces available with medium-frequency units offer lower furnace losses through the refractory linings. In most cases, operating efficiencies are approximately 90%, which is 10% higher than that of heel melters.[5-7,11]

LINING MATERIALS

The most suitable lining material for induction furnaces melting ductile iron has been acid quartzite, which consists of 99% silica mixed with 0.5–2.4% boric acid or boron oxide. The refractory is vibrated in place behind a steel form conforming to the specific crucible dimensions. The amount of boric acid or boron oxide in the refractory affects the sintering point of the lining and is dictated by the operating temperature of the furnace. The higher the percentage of boron the lower the sintering point. Boron oxide is often being used over boric acid in formulations because of its lower moisture content.

The durability of the furnace crucible depends on the following factors:

- achieving maximum density during installation;
- developing a solid, continuous hotface by properly sintering the furnace lining;
- minimizing the amount of FeO and MnO contaminants introduced to the bath by using clean, dense scrap, keeping fines-removal systems in operation, and using silicon carbide to reduce these oxides before they can attack the silica lining;
- furnace stirring should not be excessive in order to avoid mechanical erosion of the lining.

Lining material requirements are from 2–5 lb/ton of iron melted, depending on the operating temperature, iron chemistry, amount of iron oxide present in the charge and the stirring action in the melter. Ductile iron producers will not realize the same long lining life as those producing gray iron, because of the higher carbon and lower silicon

TABLE 5-4. CORELESS INDUCTION FURNACE BASE IRON CHARGE FOR AS-CAST PEARLITIC DUCTILE IRON

MATERIALS	PERCENT	SILICON		CARBON		MANGANESE		SULPHUR		PHOSPHORUS	
		IN METAL	IN CHARGE	IN METAL	IN CHARGE	IN METAL	IN CHARGE	IN METAL	IN CHARGE	IN METAL	IN CHARGE
Returns	47.00	2.60	1.22	3.70	1.74	0.55	0.27	0.009	0.004	0.030	0.014
Steel Scrap	50.00	0.20	0.10	0.25	0.12	0.60	0.30	0.020	0.010	0.030	0.015
Graphite	2.31	—	—	98.50	2.27	—	—	0.050	0.001	—	—
Silicon Carbide (90%)	0.20	60.00	0.12	30.00	0.06	—	—	—	—	—	—
Totals	—	—	1.44	—	4.19	—	0.57	—	0.015	—	0.029
Melting Loss	—	10.00	–0.15	10.00	–0.42	10.00	–0.06	—	—	—	—
Estimated Composition	—	—	1.30	—	3.77	—	0.51	—	0.015	—	0.029

Table 5-5. Coreless Induction Furnace Base Iron Charge for As-Cast Ferritic Ductile Iron

Materials	Percent	Silicon		Carbon		Manganese		Sulphur		Phosphorus	
		In Metal	In Charge	In Metal	In Charge	In Metal	In Charge	In Metal	In Charge	In Metal	In Charge
Returns	50.00	2.60	1.30	3.70	1.85	0.23	0.115	0.008	0.002	0.030	0.015
Sorelmetal	30.00	0.09	0.03	4.25	1.27	0.010	0.003	0.020	0.006	0.022	0.006
Steel Scrap	19.07	0.20	0.04	0.20	0.04	0.600	0.115	0.020	0.004	0.030	0.006
Graphite	0.93	—	—	98.50	0.91	—	—	0.050	0.005	—	—
Totals	100.00	—	1.37	—	4.13	—	0.232	—	0.018	—	0.027
Melting Loss	—	10.00	−14.00	10.00	−41.00	10.00	−0.023	—	—	—	—
Estimated Composition	—	—	1.23	—	3.72	—	0.209	—	0.018	—	0.027

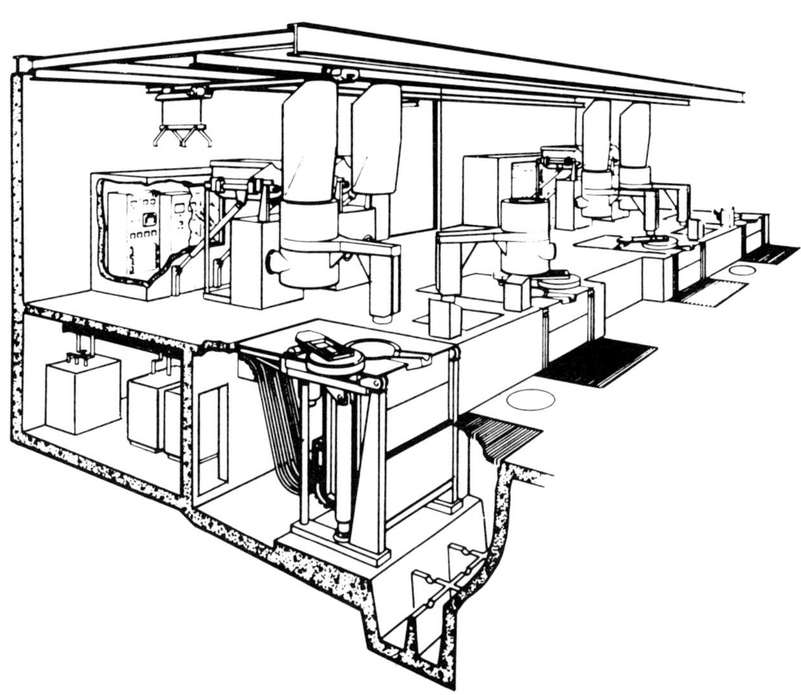

Fig. 5-8. A typical, multiple, coreless induction furnace installation.

contents of ductile base iron. The carbon and silicon in ductile iron are not in equilibrium and, as the temperature of the iron increases, the lining reacts with carbon to produce silicon and carbon dioxide ($SiO_2 + 2C = Si + 2CO$). The silica lining materials are also consumed in melting as the silica combines with iron oxide in the molten iron. The iron oxide comes from rust, or (possibly) overheated scrap. This results in the formation of iron silicate (Fe_2SiO_4) or fayalite.

Part of the iron silicate floats off as slag and is skimmed from the surface of the molten iron just prior to pouring off the furnace. The remaining fayalite can contribute to a buildup on the furnace walls in the "dead area" of the molten bath—between the upper and lower coils. This problem, however, can be considerably reduced by using silicon carbide grain as a part of the furnace charge. Silicon carbide can remove the FeO ($SiC + FeO = Fe + Si + CO$) to prevent the attack from occurring, or the fayalite (Fe_2SiO_4) after the attack has occurred. A silicon carbide addition of 5–10 lb/ton will satisfy this requirement.

PUSH-OUT LINING

Another advancement in induction furnace technology has been the push-out lining. This system incorporates a cast refractory plug, which is attached to a hydraulic ram when the lining is to be removed. The ram pushes the plug against the refractory lining, distributing the ejection forces evenly along the lining sidewalls and bottom, removing the entire worn lining in a matter of minutes. This method is much safer and exposes employees to less silica dust than the old method of chipping out a lining by hand, which can take hours. The increased efficiency, reduced furnace downtime and safety aspects will make the push-out lining a necessity in the future.[5]

■ DIRECT-ARC ELECTRIC FURNACE MELTING

ADVANTAGES AND DISADVANTAGES

The melting of ductile base irons in direct-arc furnaces can be done quite successfully. As with any melting method, there are advantages and disadvantages to be considered prior to selecting electric arc melting. The primary advantage of direct-arc melting is relatively low capital investment coupled with the flexibility that a batch process gives. Chemical analyses can be varied easily and the low oxidation losses afford the potential for excellent control of chemistry.

As with other melting processes, charge components determine the analysis of the iron produced. With careful selection of materials, sulfur content of the base iron can be held under 0.015% and the need to desulfurize can be eliminated.

After meltdown, the relative quiet of the bath allows the lighter oxide and other slag constituents to float to the top for skimming. This is an advantage over the constant mixing action of the coreless induction furnace.

While the direct-arc furnace is capable of producing high-quality, low-inclusion ductile base iron, its greatest disadvantage can be the cost of operation. Approximately 500 kWh are required to melt a ton of iron. Depending on geographic area, the cost of electrical energy can be quite high. Additionally, the cost premium for electricity use during peak daytime hours can be substantial.

Electrode consumption can be 12 or more pounds per ton of iron melted. Extended melting and holding times can increase energy and electrode costs even further. To take advantage of the potential increased melt quality, higher cost melt materials are required. In addition, considerable noise is generated during melting.

DIRECT-ARC FURNACE DESCRIPTION

The type of direct-arc furnaces used for melting ductile base irons are normally acid-lined furnaces of the type shown in Figures 5-10 and 5-11.

As shown, a direct-arc furnace is a refractory-lined bowl with three electrodes passing through the roof and suspended vertically so as to have the arcs in series. The electric current passes from one electrode through the charge or bath and out through the other electrode, generating heat through the arcing that occurs between the electrode and the charge or metal bath. The electrodes are carried on the structure, as shown and are spaced in relation to each other so that arcs do not occur between them. The hearth slopes to a pouring lip. The entire furnace is enclosed in a steel shell, which in turn is insulated from the furnace hearth and refractory lining.

REFRACTORY PRACTICE

High-volume melt operations benefit from the extensive use of water cooling in furnace sidewall panels, slag doors, door jams and roof rings. Fully water-cooled roofs are also available and are in use for high-volume steel-melting operations. Electrode holders are also water cooled. In addition to water cooling, carbon block refractories can also be used in critical areas, such as the tap hole. These practices—or some form of them—are recommended where furnace productivity, reduced refractory maintenance and uptime are important.

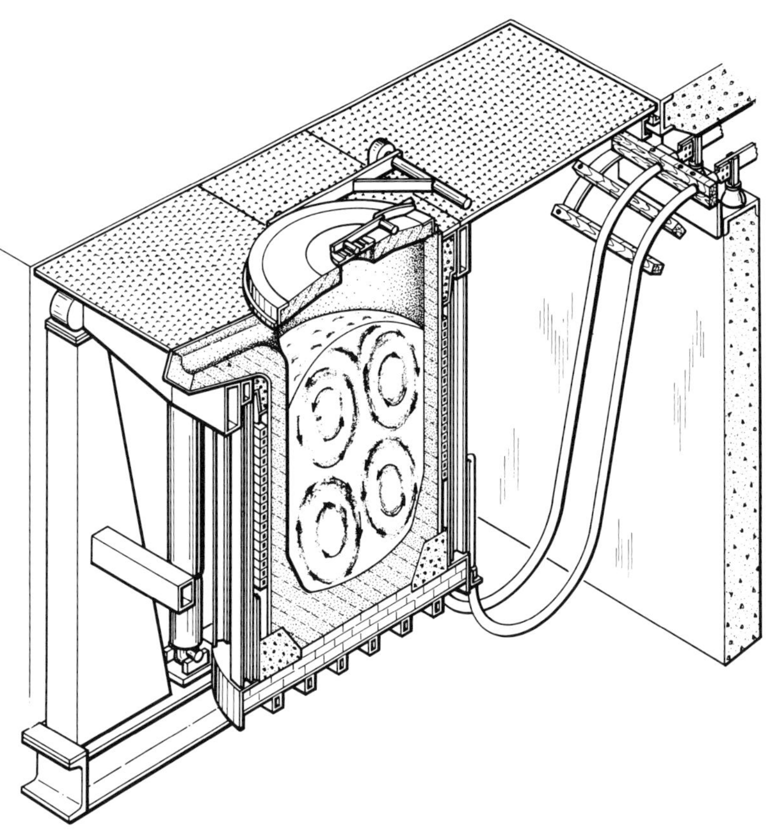

Fig. 5-9. Cross section of a typical coreless, line frequency, induction furnace.

Fig. 5-10. An electric arc furnace in operation.

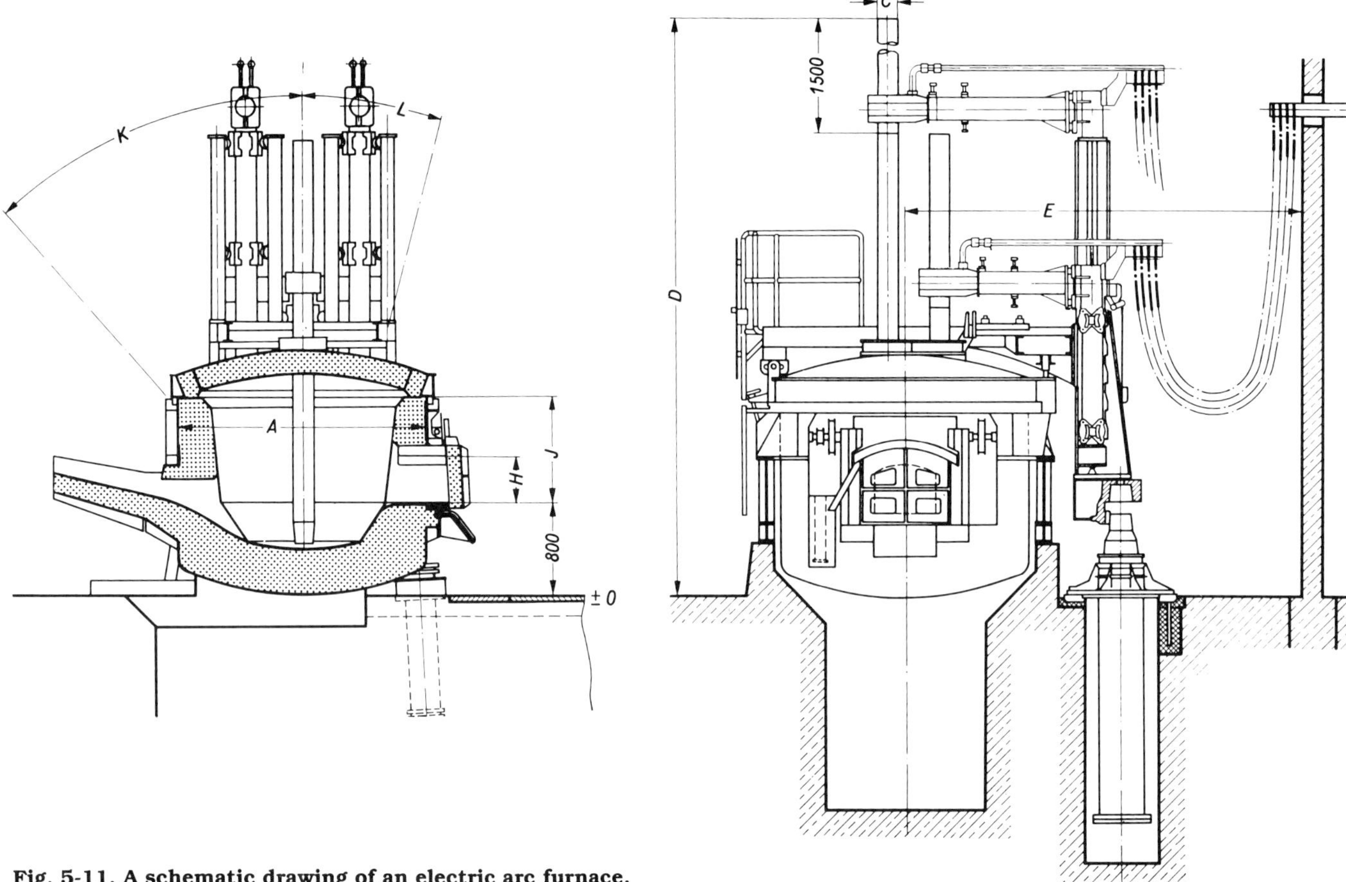

Fig. 5-11. A schematic drawing of an electric arc furnace.

For intermittent or limited operations, the following refractory practice should be adequate. Furnace linings will be siliceous or of an acid-type refractory. A completely new furnace lining will consist of one layer of heavy-duty firebrick laid next to the shell. These firebricks are cemented together with a high-temperature cement. Silica bricks are then laid in such a manner that they form a cup-shaped bottom. All silica bricks are laid loosely with dry silica sand brushed into the joints after each course is laid. This construction constitutes a lifetime bottom and is not removed when rammed bottom linings are replaced. Working bottoms in these furnaces are from 4 to 8 inches in depth, depending on the furnace size. They consist of rammed moistened material, which is a mixture of fireclay and ganister. A typical mixture is 90% silica ganister, 5% fireclay and 5% water.

Sidewalls are commonly constructed of silica or heavy-duty firebrick, which gives satisfactory service when furnaces are operated continuously. When operation is intermittent, as in the case of a jobbing foundry, mixed super-duty fireclay-ganister walls, rammed over heavy-duty firebrick, are used to minimize spalling.

Silica brick roofs cannot be used due to the cracking and spalling that occurs during heating and cooling in intermittent furnace operations. For this reason, many foundries use a furnace roof constructed from super-duty high-alumina brick or high-alumina rammed refractories. through dusting. The order of addition can be varied to accommodate differences in charge materials.

Charge Makeup

Two ferritic ductile iron sample charges are listed in Table 5-6. A sample pearlitic charge makeup is shown in Table 5-7. In order to keep the sulfur content at a low level, the carbon raiser and the pig iron should both be of low-sulfur composition. Turnings and borings should be used cautiously, because of the potential for sulfur- or lead-bearing free machining steels.

TABLE 5-6. TWO ELECTRIC ARC FURNACE BASE IRON CHARGES FOR AS-CAST FERRITIC DUCTILE IRON

(SAMPLE 1)

MATERIALS	PERCENT	SILICON		CARBON		MANGANESE		SULPHUR		PHOSPHORUS	
		IN METAL	IN CHARGE	IN METAL	IN CHARGE	IN METAL	IN CHARGE	IN METAL	IN CHARGE	IN METAL	IN CHARGE
Returns	50.00	2.65	1.32	3.75	1.87	0.22	0.11	0.006	0.003	0.030	0.005
Pig Iron (Low P)	20.00	0.18	0.03	4.10	0.82	0.00	—	0.016	0.003	0.030	0.006
Steel Scrap	28.32	0.12	0.03	0.25	0.07	0.60	0.17	0.020	0.005	0.020	0.005
Recarburizer*	1.68	—	—	98.50	1.66	—	—	0.050	0.001	—	—
Totals	—	—	1.38	—	4.42	—	0.28	—	0.012	—	0.026
Melting Loss	—	−10.00	−0.13	−15.00	−0.58	−10.00	−0.03	—	—	—	—
Estimated Composition	—	—	1.25	—	3.85	—	0.25	—	0.012	—	0.026

(SAMPLE 2)

MATERIALS	PERCENT	SILICON		CARBON		MANGANESE		SULPHUR		PHOSPHORUS	
		IN METAL	IN CHARGE	IN METAL	IN CHARGE	IN METAL	IN CHARGE	IN METAL	IN CHARGE	IN METAL	IN CHARGE
Returns	40.00	2.65	1.06	3.70	1.48	0.25	0.10	0.006	0.002	0.03	0.012
Pig Iron (Low P)	20.00	0.18	0.03	4.10	0.82	0.00	—	0.015	0.003	0.02	0.004
Steel Scrap	37.50	0.12	0.05	0.25	0.09	0.60	0.22	0.030	0.011	0.03	0.011
Sil Carbide (85%)	0.50	59.00	0.29	30.00	0.15	—	—	0.100	0.005	—	—
Recarburizer*	2.00	—	—	98.50	1.97	—	—	0.050	0.001	—	—
Totals	—	—	1.00	—	4.51	—	0.32	—	0.022	—	0.027
Melting Loss	—	−10.00	−0.15	−15.00	−0.67	−10.00	−0.10	—	—	—	—
Estimated Composition	—	—	1.28	—	3.84	—	0.22	—	0.022	—	0.027

*Carbon recovery (estimated): 20% loss on graphite; 10% loss on balance.

TABLE 5-7. ELECTRIC ARC FURNACE BASE IRON CHARGE FOR AS-CAST PEARLITIC DUCTILE IRON*

MATERIALS	PERCENT	SILICON		CARBON		MANGANESE		SULPHUR		PHOSPHORUS	
		In Metal	In Charge	In Metal	In Charge	In Metal	In Charge	In Metal	In Charge	In Metal	In Charge
Returns	50.00	2.65	1.32	3.75	1.87	0.56	0.28	0.006	0.003	0.03	0.015
Steel Scrap	47.58	0.12	—	0.25	0.12	0.65	0.31	0.030	0.014	0.03	0.014
Graphite	2.41	—	—	98.50	2.38	—	—	0.050	0.001	—	—
Totals	—	—	1.37	—	4.37	—	0.59	—	0.018	—	0.029
Melting Loss	—	−10.00	−0.12	−15.00	−0.57	−10.00	−0.05	—	—	—	—
Estimated Composition	—	—	1.25	—	3.80	—	0.54	—	0.012	—	0.029

*Add 0.5% copper for sections over 0.75 in.

Productivity of the melting operation and refractory maintenance can be greatly affected by the bulk density of the charge. A denser charge will give a larger heat size without having to add additional material after melt-down. Usually, however, high-density melt materials command a premium price. If water-cooled panels are used, a low-density charge may expose the refractory under the panels to the electrode arc and accelerate refractory loss. A recommended charge makeup would consist of thin, low-density steel on the bottom to cushion the charge when dropped into the furnace. Large castings and other returns are placed in the middle to avoid charge "cave-ins" and resulting electrode breakage, which could occur if large pieces are placed on top. Remaining materials are placed on top. Carbon raiser should be added to minimize loss.

MELTING PRACTICE

The melt cycle begins with the opening of the furnace roof. As an energy conservation measure, the roof should not be opened until the charge is ready to be placed in the furnace. Charging should be done as soon after tapping the previous heat as possible. That way, the heat from the furnace walls will be absorbed by the next charge.

Furnace rheostats should be adjusted to generate a long arc during initial meltdown. Energy radiated from the arc will be absorbed by the charge materials as the electrodes bore into the charge. After the charge has been melted to a flat bath, the length of the arc should be shortened to reduce energy loss to the sidewalls. Modern computerized furnace controls can control the arc length and electrode movement automatically. Computers can also control the amount of electricity used, so that the usage does not exceed the demand rate contracted for with the utility company.

The metal is heated to approximately 2750F (1510C). A sample is taken to check chemistry and the slag is skimmed from the bath. Adjustments to analysis can be made with the addition of carbon or other alloys. Electrodes should not be dipped into the bath to raise carbon because of their high cost. Also, metal should not be held in the arc furnace any longer than necessary. Heating under these conditions is inefficient and with extended holding, the silicon content of the bath increases and carbon decreases through the reduction of SiO_2.

Other key operating tips are: keep the electrodes vertical and parallel; and keep the electrode holders all at the same height.

BASIC DIRECT-ARC FURNACE MELTING

Base irons for ductile iron can be successfully melted in basic-lined arc furnaces. When a reducing basic slag is used in a melting practice, similar to those described in the section on acid-arc melting, the sulfur content of base irons can be reduced to as low as 0.004%. However, due to the heat time required and slag handling problems, basic melting is very seldom used in ductile iron production.

REFERENCES

CORELESS FURNACE

1. A.F. Spangler, ed.; *The Ductile Iron Process,* Miller & Co, p 189, (1978).

2. H. Colthurst, M. Donome; "An Introduction to the Metallurgical Aspects of Melting Gray, Malleable, and Ductile Base Irons in Coreless Induction Furnaces," *Current Information Report,* CR 8202 (1982).

3. J. Powell; "The Effective Control of Electricity Use in Melting and Holding Furnaces," BCIRA Report 1777.

4. R.G. Batson, T.A. Henry, D. B. Webster; "Energy Management in the Induction Melting of Iron," Center For Metals Production, Report no. 89-2 (Mar 1989).

5. R. Smith; "Trends in Electric Melting," *Foundry Management and Technology,* p 38 (Mar 1988).

6. R.Q. Sharpless; "Medium Frequency Induction Melting: It's Come of Age," *Electric Furnace Conference Proceedings.*

7. J.H. Mortimer; "Batch Melting: Advancing Induction Furnace Technology," *Modern Casting* (Sep 1987).

8. C.J. Edgerley; "Electrical Characteristics of Coreless Furnaces and Their Relevance to Practical Operating Conditions," BCIRA Report 1647.

9. J. Powell; "Electric Furnace Operation: A Survey of Performance and Practice," BCIRA Report 1364.

10. R.G. Batson, T.A. Henry, D.B. Webster; "Energy Management in the Induction Melting of Iron," Center For Metals Production, Report no. 89-2 (Mar 1989).

11. J.R. Smith; "The Use of Medium-Frequency Coreless Induction Furnaces in Iron Foundries," BCIRA Report 1728.

BIBLIOGRAPHY

GENERAL

American Foundrymen's Society, Inc. AFS–CMI Conference Proceedings: "Cupola Operation: State of the Art," (1980).

American Foundrymen's Society, Inc. *AFS Cupola Handbook.* Des Plaines, IL: American Foundrymen's Society, Inc. (1984).

Spangler, A.F., ed. *Ductile Iron Process.* Chicago: Miller & Co (1978).

Wanda, R.D and E. F. Dark. "60 Years of Cupola Literature." Physical Metallurgy Research Laboratories, Canada Centre for Minerals and Energy Technology.

U.S. Environmental Protection Agency. "Energy Conservation Techniques for the Iron Foundry Cupola." (1976).

CUPOLA

Abe, K. "On the Relationship Between Sulphur Pick–Up in Molten Cast Iron and Cupola Operating Conditions." *AFS Cast Metals Research Journal,* no. 11, (1975).

"AFS Congress on Cupola Productivity Divided Blast and Hot Blast." *Foundry Facts,* no. 35 (1975).

Alexander, A.P. "Experiences with Basic Refractories and Melting." *Ductile Iron Bulletin,* 6, INCO (1957).

American Foundrymen's Society, Inc. "Pollution–Free Gas Fired Cupola." *modern casting,* (Jan 1972).

Apthorp, N. "Natural Gas and Coke Mix Cuts Costs." *Foundry,* vol 91 (Apr 1963): 134–136.

Bardenheuer, P.W. "Melting of Cast Iron in a Gas Fired Cupola." *Casting Plant and Technology,* Issue no. 4/87.

Briggs, J. and E.J. Jago. "Iron Foundry Slags and Fluxes." *Foundry,* no. 104 (Nov 1976): 26–27; (Dec 1976): 101–104; no. 105 (Jan 1977): 116, 118, 120.

Burke, T.H. "Melting Ductile Iron in an Acid Cupola." *Foundry* (Jul 1962).

Carlson, R. "Evaluation of Cast Iron Melting in the Cupola, Arc and Induction Furnace, Foundry Melting." *modern casting,* no. 51,(1967): 99–104.

——— "Evaluation of Cast Iron Melting in the Cupola, Arc and Induction Furnace." *AFS Transactions,* vol 75. Des Plaines, IL: American Foundrymen's Society, Inc. (1967): 99–104.

Carter, S.F. "American Experiences with Basic Cupola Melting." *AFS Transactions,* vol 61. Des Plaines, IL: American Foundrymen's Society, Inc. (1953).

——— "Basic Cupola Operation." Ductile Iron Bulletin 7, INCO (Mar 1958).

——— "Slag Control." *Foundry Facts* (Mar 5–7, 1978).

Clow, S.C. "The Effect and Control of Sulfur in Iron." *AFS International Cast Metals Journal* (1979): 45–54.

Coates, R.B. and H.J. Leyshon. "A Method for the Economic Production of Nodular Graphite Cast Iron Based on the Use of an Acid Cupola and Electric Furnace." *BCIRA Journal,* vol 13 (1965): 469–475.

Dahlberg, H.R. and W.W. Levi. "Water Cooled Cupola Operation with a Basic Slag." *AFS Transactions,* vol 72. Des Plaines, IL: American Foundrymen's Society, Inc. (1964): 158–163.

Draper, A.B. and J.G. Sylvia. "Cupola Technology: Key to Reduced Melting Energy and Cost." *AFS Transactions,* vol 86. Des Plaines, IL: American Foundrymen's Society, Inc. (1978): 533–548.

Duca, W.J. "HCIF: The Electric Cupola." *AFS Transactions,* vol 80. Des Plaines, IL: American Foundrymen's Society, Inc. (1972): 13–16.

East, W.R. "GM's Central Foundry Division Installs First Plasma Melter at Defiance Facility." *Foundry* (Nov 1989).

Eastwood. D.H. and D.W.N. Pitts. "Desulphurization of Cupola Cast Iron by Tuyere Injection of Calcium Carbide." *Foundry Trade Journal,* vol 108 (1960): 99–102.

Fighs, S.V., W.N. Provis, B. Buckowski, W.J. Peck and A.D. Kamp. "Plasma–Fired Cupola: An Innovation in Iron Foundry Melting." *AFS Transactions.* Des Plaines, IL: American Foundrymen's Society, Inc. (1986).

Gopal, R., R.L. Kumar and P.N. Chakraborty. "Some Studies on Desulphurization in an Acid–lined Cupola." *British Foundryman,* vol 60 (1967): 480–482.

"Grede's Basic Cupola." *Foundry Facts* (Mar 8–9, 1978).Harper, J.D. "Operation of the Basic Cupola for the Production of Light Nodular Iron Castings 1956–66." *British Foundryman,* vol 60 (1967): 54–67, 413–415.

Heine, N.J. "Foundry Refractories: An Update." *Foundry,* no. 107, Part I (July 1979): 28, 30, 32, 36–37; Part II (Aug 1979): 86–88, 90, 92, 94.

Henderson, H.E. "Acid Cupola Melting for Ductile Iron." *AFS Transactions,* vol 67. Des Plaines, IL: American Foundrymen's Society, Inc. (1959).

————"Acid Cupola Melting for Ductile Iron." *AFS Transactions,* vol 67. Des Plaines, IL: American Foundrymen's Society, Inc. (1959): 661–668.

Isleib, C.R. "Ductile Iron Directly From the Cupola." Ductile Iron Bulletin 6, INCO (Jan 1957).

Katz, S. "Fluidization of Basic Slags and the Role of Calcium Fluoride, Fluorspar." *AFS Transactions,* vol 86. Des Plaines, IL: American Foundrymen's Society, Inc. (1978): 297–302.

Katz, S. and H.C. Rezeau. "Factors Governing Desulphurization in the Cupola." *modern casting* (1979): 89–92.

————"The Cupola Desulphurization Process." *AFS International Cast Metals Journal,* (1979): 9–18.

————"The Cupola Desulphurization Process." *AFS Transactions,* vol 87. Des Plaines, IL: American Foundrymen's Society, Inc. (1979): 367–376.

Katz, S. and V.R. Spironelol. "Effects of Charged Aluminum on Silicon Recovery and Desulfurization in an Iron Producing Cupola." *AFS Transactions.* Des Plaines, IL: American Foundrymen's Society, Inc. (1984).

KGT Giessereitechnik. "Cupola 2000 for Long Term Operation."

Kyle, J.K. "An Example of Basic Lining–less Hot Blast Cupola Operation and Control." *British Foundryman,* vol 60 (1967): 44–54, 418–419.

Langner, E.E., R. Carlson, J.A. Hamilton and W.J. Koepplin. "The Commercial Application of Oxygen Enrichment in Iron Cupolas." *AFS Transactions.* Des Plaines, IL: American Foundrymen's Society, Inc., vol 74 (1966): 150–155.

Lee, R.S. "Some Effects of Equipment, Material and Methods of Melting on Mechanical Properties of Ductile Iron." *AFS Transactions.* Des Plaines, IL: American Foundrymen's Society, Inc. (1975).

Leyshon, H.J. "A Note on the Possibility of Reducing the Sulphur Content of Cupola–Melted Iron by Increasing the Slag Volume." *BCIRA Journal,* vol 14 (1966): 761–763.

————."The Effect of Blast Temperature and Slag Basicity on The Performance of a Water–Cooled Cupola." *BCIRA Journal,* vol 13 (1965): 621–641.

Leyshon, H.J. and M.J. Selby. "Desulphurization of Iron by Injection of Calcium Carbide Into the Cupola." *BCIRA Journal,* no. 15 (1967): 108–114.

Leyshon, H.J. and M.R. Thibault. "Recent Work on Improved Cupola Performance." *AFS International Cast Metals Journal* (1978): 25–38.

Lloyd, A.T. "Selection and Application of Carbon Raising Materials." *AFS Transactions,* vol 82. Des Plaines, IL: American Foundrymen's Society, Inc. (1974): 229–234.

Lovett, M. and J. Sessanna. "Refractory Practice for Cupola Melting." *Foundry,* no. 101 (Sep 1973): 40–43.

Mrdjenovich, R., R.J. Warrick and D. Poggi. "Evaluation of Ilmenite as a Slag Fluidizer in Cupola Operations." *modern casting,* no. 65 (1975): 37–40.

Pugh, S. "Economic Evaluation of Melting Systems." *Foundry Trade Journal,* no. 142 (1977): 305–318.

Rollman, M.E. and D.J. Pusack. "Lining–less Watercooled Cupola Operating Experience." *AFS Transactions,* vol 68. Des Plaines, IL: American Foundrymen's Society, Inc. (1960): 489–494.

Rolseth, H.C. "The Effect of Oxygen Enrichment on Energy Consumption in Iron Cupola Melting." *AFS International Cast Metals Journal,* no. 2 (Sep 1977): 53–56.

Sale, J.H. "Operating Experience Using Shredded Automobile Scrap for the Production of Ductile Iron Pipe." *AFS Transactions,* vol 83. Des Plaines, IL: American Foundrymen's Society, Inc. (1975): 107–108.

Schaum, J.N. "Modernizing Your Iron Melting Department." *modern casting,* no. 68, (Sep 1978): 36–47.

Sheppard, H.T. "Cupola Slag Control." *AFS Transactions,* vol 75 no. 68 (1967): 735–740.

Spironello, V.R., N.W. Kilau and W.M. Mahan. "Auxiliary Fluxes in Cupola Operations." *AFS Transactions,* vol 86. Des Plaines, IL: American Foundrymen's Society, Inc. (1978): 563–572.

Taft, R.T. "Ten Years of Cokeless Melting." *British Foundryman,* no. 65 (1972): 321–328.

————."The Cokeless Cupola." *Foundry Trade Journal,* no. 137 (1974): 707–718.

Taft, R.T. and H.R. Perkins. "Cokeless Before the 1980s." *Foundry Trade Journal* (1978): 471–485.

Timmins, A.A. "An Analysis of Cupola Operation Performance." *BCIRA Journal,* vol 8 (1960): 247–265.

Tunder, S. and L. Hoehle. "Progress in Melting Techniques Applied to Spheroidal Graphite Iron." *Foundry Trade Journal,* no. 125 (1968): 945–951.

White, W. "Introduction to Basic Cupola Operation." Ductile Iron Bulletin 5, INCO (Jun 1956).

Williams, J.T. "Basic Cupola Melting." *AFS Transactions,* vol 67. Des Plaines, IL: American Foundrymen's Society, Inc. (1959).

———— "Basic Cupola Melting for Ductile Iron." *AFS Transactions,* vol 67. Des Plaines, IL: American Foundrymen's Society, Inc. (1959): 669–670.

SCRATA. "Exogenous Inclusions: Their Occurrence and Control." *Journal of Research,* no. 33.

Sommer, R.A. "Buildup in Channel Induction Furnaces." Gray Iron Research Institute, Special Report no. 81–13.

Stark, R.A. "Electric Furnace Plugging and Buildup." Gray Iron Research Institute Report no. 480, Mar 28 (1983).

———— "Refractory Buildup in Electric Induction Furnaces." Issued as an AFS Joint Committee Report, *AFS Transactions.* Des Plaines, IL: American Foundrymen's Society, Inc. 84–107 (1987).

———— "Refractory Buildup in Molten Metal Systems." Norton Company Internal Report (1977).

———— "Refractory Problems Peculiar to Induction Melting." *modern casting* (Mar 1985).

Sundberg, Y. "The Condition Diagram for Channel Furnaces and Its Practical Use." *ASEA Journal,* vol 53, no. 4–5 (1980).

Thielke, J. "Using Electric Furnaces to Store and Pour Treated Ductile Iron." *modern casting,* (Oct 1987).

Vertical Channel Melting

Becque, J.G. "Improving Induction Furnaces Refractory Life." *Foundry* (Apr 1986).

Dotsch, E. "Holding and Casting of Gray and Spheroidal Graphite Cast Iron." *Transactions of the BBC Conference for Induction Furnace Plants.* Dortmund (1986).

———— "Holding and Pouring of Magnesium–Treated Cast Iron." *modern casting* (Mar 1983).

Hengch, I. "Some Metallurgical Aspects of Producing Ductile Iron and Holding Treated Metal." George Fischer AG.

Hoff, J.L. "Channel Geometry: Key to Inductor Life." *Foundry* (Jun 1967).

Lilycrap. D.C. "Refractory Erosion in High Powered Channel Inductors." Electrowarme International 44, B 3 (June 1986).

Oberthur, A., Et al. "Report on Direct Casting of Nodular Graphite Cast Iron From A Pneumatic Pouring Furnace Without Using Any Intermediate Vessel." 7th International Junker Furnace Conference, Lecture A6 (1982).

Patron, R. "Concretions in Channel Induction Furnaces." *AFS International Cast Metals Journal.* Des Plaines, IL: American Foundrymen's Society, Inc. (June 1979).

Schlusselberger, R. Daubmeier and Grossman. "Long–Term Holding of Magnesium Treated Iron Melts." Austrian Foundry Institute.

Schroeder, K. "Operating Experience With 20–Ton and 50–Ton Induction Channel Furnaces." 6th International Junker Furnace Conference, Lecture no. 8 (1978).

Duplexing

American Foundrymen's Society, Inc. *AFS Cupola Handbook.* Des Plaines, IL: American Foundrymen's Society, Inc. (1984).

Calamari, G. and F. Abdel–Razek. "Induction Melting in the Foundry: The Italian Experience." *Foundry Trade Journal* (Jul 5, 1984).

Hennes, W. "Less Environmental Problems by Economic Iron Melting." *Giesserei–Praxis,* no. 17/18 (1985).

6

Selection of Charge Materials

John E. Foltz, Jr.

Intermet Foundries
Lynchburg, Virginia

◼ INTRODUCTION

In today's competitive global foundry business, with higher quality assurance criteria and improved design specifications, the foundry metallurgist and the melting department are under continual pressure to achieve higher degrees of process control. Greater use of computer-controlled machine tools, operating at high speeds, has resulted in increasing demands for consistency. In the past, engineers would specify mechanical or structural requirements for castings, but these requirements now include statistical capability limits. Statistical treatment of data, in effect, tightens the limits of accepted standards and, thus, increases the demand for higher levels of metallurgical control.

Casting properties and consistent machinability are controlled by the "big three" process control parameters: (1) metal composition; (2) inoculation; and (3) casting cooling rate. Each of these parameters must be consistently controlled on each and every heat of casting to ensure that customer requirements are met 100% of the time.

The selection of charge materials has a major effect on metal composition and, to some degree, on the required inoculation practices. There are many references in the literature stating the importance of charge material selection in controlling the five basic elements: carbon, silicon, manganese, sulfur and phosphorus. The increasing demands for consistency, and the statistical treatment of data, have expanded this basic list to include elements such as copper, chromium, nickel, tin, etc. Continually increasing customer requirements, many of which are statistically-based, has made the already difficult job of selecting charge material even more complex.

Fortunately, there are many sources in the literature that define the composition and recoveries of the various charge materials. In addition, there are many computer programs available to assist the metallurgist in the development of least-cost charge recipes. There are a number of reference and computer software sources, including the library and software services at American Foundrymen's Society.

◼ DEVELOPMENT OF INTERNAL REQUIREMENTS

The first step in the selection of charge materials is to review and list the customer's requirements and the casting design to establish the basic composition limits of the molten metal. These composition limits are further refined, based on the melting, treatment and inoculation practices, and on the predicted mold cooling time for the castings. Some of the composition limits will be spelled out in the customer requirements, but many will be established, based on experience and past practice.

Once the basic composition limits are defined, the next step is to consider any other internal constraints. For instance, if the lab testing equipment is limited, and the compositional range limits are narrow, high-risk, low-cost charge materials should be avoided.

Environmental issues also need to be included in the selection of charge material. Scrap steels painted with lead-base paints can cause serious issues in shops using coreless melters. By-product wastes leaving the plant as slags or baghouse dust must be tested using the U.S. EPA's Toxicity Characteristic Leaching Procedure (TCLP) to determine whether it is classified as "hazardous" or "nonhazardous" waste. (For further information, contact the American Foundrymen's Society's Environmental Department.)

The metallurgical requirements of certain jobs may be defined using statistical capability limits as to restrict or preclude the use of certain grades of charge materials, due to the lack of consistency in some low-cost materials. The size of the iron yard and the number of bin spaces available may limit the number of grades of materials that can be stored on-site.

Melting yield is an often overlooked, but very important, factor in the selection of charge materials. Melting yield is the ratio of the tons of molten metal output to the tons of charged metallic input. Grades of steel that contain large amounts of rusty material, thin material, or nonmetallics can produce low melt yield. Many of the negative results of melting low-yield materials are not immediately obvious. Slag and gas volumes can increase when melting low-yield materials and, thus, disposal costs will increase. If the composition of by-products are altered, there is a potential for environmental issues with the attendant costs. In many cases, low-yield materials contain higher levels of tramp elements. The resulting levels of tramp elements may force the use of a heat-treatment step, or the addition of a higher cost charge material, such as pig iron, to dilute the tramp levels.

Low melting yield can be the result of iron oxidation during melting. The potential for iron oxidation will increase as the surface area per unit weight of the scrap increases (i.e., thin steel scrap). Excessive iron oxidation typically leads to increased oxidation losses of manganese, silicon and carbon. Replacing these alloy losses with increased additions of ferroalloy, coke or carbon raiser is expensive.

When melting certain low-quality charge materials, combinations of the above effects can be observed. In these cases, melting yield can have a major impact on both the charge and final product costs.

■ EXTERNAL CONSIDERATIONS WHEN PURCHASING METALLIC CHARGE MATERIALS

Once the customer requirements and internal constraints have been considered, a list of possible charge materials can be developed based on a review of the external considerations, e.g., availability, delivery, composition, consistency, bulk density, sizing, and economics. Dealing with reputable, cooperative scrap dealers in your geographic area can eliminate many problems with off-specification material, late delivery, improperly sized material, etc. The development of internal charge material specifications is highly recommended. Visits to scrap dealers, inviting scrap dealers to visit you, and joint reviews of your internal specifications are excellent methods to prevent shipment of nonconforming charge material.

Freight rates can have a major impact on the delivered cost of charge materials, so scrap dealers in your geographic area should be investigated as primary sources. Your unloading equipment, iron yard size and rail siding availability will determine shipping methods and frequency.

Composition and consistency of the charge materials will be dictated by internal requirements, as discussed earlier. The bulk density and sizing of the charge materials must also be considered. Acceptable sizing of charge materials should be matched to the charge handling, preheating and melting equipment. Oversize materials can cause downtime and safety concerns due to bridging and blockage. Undersize material can cause problems if the fine materials are sucked into the baghouse system. Materials with low bulk density can slow down melt rates, due to an inability to get a full charge into the charge bucket, and thus necessitate excessive repeat trips to charge the melter. Repeat trips cause excessive heat loss in electric melters, due to the large amounts of heat lost when the lid is opened for charging. Low-density materials can adversely effect combustion in the cupola by restricting air flow through the stack. Low bulk-density materials can generate excessive slag in both cupolas and coreless melters.

The plant metallurgist must not only conform to all the requirements of the customer, but he must also meet the plant manager's goal to accomplish conformance at the lowest cost. The purchase of charge materials is consistent with market theory in that the least desirable materials are the lowest cost materials. As discussed above, the lower cost charge materials can have inconsistent composition, nonferrous parts, coatings, heavy rust, excessive dirt, low melting yield, etc. Selection of charge materials based solely on melt cost can lead to poor decisions. A better method is to consider the impact on the final product costs when selecting charge materials.

TYPICAL METALLIC CHARGE MATERIALS

Once all the selection criteria are listed and reviewed, the metallurgist can begin selecting charge materials based on charge recipes (see Chapter 5). There is computer software available from many sources that calculate least-cost charges. To develop these recipes, the composition and recovery of each of the possible charge materials must be known. In addition to the basic elements C, Si, Mn, S and P, the residual or tramp elements must also be considered. There are numerous sources of this information, including the Institute of Scrap Iron & Steel (ISIS), the *AFS Cupola Handbook,* numerous BCIRA articles and ASM material property handbooks.

STEEL SCRAP

Steel scrap typically makes up 40–60% of a ductile charge for casting producers and up to 80–95% for pipe producers. Steel scrap is used for inexpensive iron units and can be purchased in many grades. The ISIS listing of scrap steel grades can be found in the Appendix. Typical compositions of steel scrap are shown in Table 6-1. Metallurgical and purchasing departments must be constantly aware of subversive and detrimental elements that can be introduced from steel scrap. Chromium and copper levels need to be watched in "auto grade" steels. Electric furnace produced steels will have higher tramp elements than open hearth steels, due to hot metal dilution in the open hearth furnaces. Typically, the lower cost grades of steel scrap have a higher risk of tramp element contamination. It is a good practice to inspect each incoming load of steel. Be on the outlook for proper grade, correct sizing, plated materials, nonferrous parts, alloy steels, cast irons, excessive dirt and rust, etc. If you purchase more than one grade of steel, take care to set up good unloading and iron yard bin marking procedures.

DUCTILE IRON REMELT (RETURNS)

Ductile iron remelt consists of sprues, gates and scrap castings from ductile iron heats. If grades other than ductile iron are produced, special care must be taken to keep the ductile returns separated in the iron yard. If various grades of ductile iron are produced, it can be cost effective to separate ductile returns by silicon and alloy levels. Some shops have successfully used color coding of hoppers and bins to aid in returns segregation.

PIG IRON

Pig iron is the cleanest (lowest level of tramp elements) source of iron units of known composition for a ductile charge. However, pig iron is one of the highest unit-cost charge materials. When evaluating charge cost, consider-

ation must be given to end product cost. The use of pig iron may reduce heat treatment requirements, increase melt rate, or aid in meeting statistically established standards of acceptance. If other process costs can be eliminated by using pig iron, then a higher charge cost may produce a lower end-product cost. Pig iron used in ductile charges is typically a low-phosphorous, low-sulfur grade. The alloy levels of C, Si and Mn need to be selected based on the internal requirements. The elements Ni, Cu, Cr, Ti, V and Al can vary from the various suppliers of pig iron and, thus, must also be considered. Typical pig iron compositions are shown in Table 6-2.

PURCHASED CAST SCRAP

Many scrap dealers make a practice of accumulating cast scrap from varied sources in one pile. Purchasing cast ductile scrap for charge material can lead to unpredictable composition and the possibility of sulfur and phosphorus contamination from gray iron. If you have a consistent source of purchased ductile cast scrap, it can be a potentially cost-effective charge material, providing you are aware of the pitfalls and effectively monitor the phosphorous and sulfur levels of your melt. Table 6-3 shows typical composition of cast iron scrap.

DIRECT REDUCED IRON (DRI)

Although it is not widely used, DRI can be a successful ductile iron charge material. The major advantage of DRI is that it contains extremely low levels of all the common tramp elements. Disadvantages of DRI include availability, heavy slag formation, extra fuel requirements to melt, and oxidation of charge Si and Mn.

OTHERS MATERIALS

There is a wide variety of other charge material, such as borings, turnings, cans, runouts, etc., which have not been popular materials for ductile charges due to high levels of tramp elements or excessive slag generation. However, with the advent of newer refining technology, such as the plasma arc, many of these materials may become economically feasible.

METALLURGICAL EFFECTS OF TRAMP ELEMENTS

The metallurgical effects of elements have been well documented in many references, including the *AFS Cupola Handbook,* the *Iron Casting Handbook,* various BCIRA reports, and wall charts. Metallurgical effects of various elements were discussed in detail in Chapter 4.

TABLE 6-1. TYPICAL ANALYSES OF THE MOST COMMON STEEL SCRAP GRADES

Steel Scrap*	% of Usage	% Cu	% Ni	% Cr	% Sn	%Mo	% P	% S
Punchings and Plate Scrap	13	0.09	0.06	0.06	0.015	0.010	0.015	0.025
Electric Furnace Bundles	8	0.10	0.07	0.07	0.020	0.010	0.020	0.025
Railroad Wheels, Rails, Splice Bars, Tie Plates & Spikes	3	0.09	0.07	0.07	0.015	0.010	0.050	0.050
Plate and Structural	17	0.13	0.09	0.09	0.025	0.015	0.025	0.025
No. 1 Railroad Heavy Melting	3	0.18	0.15	0.10	0.020	0.020	0.020	0.040
No. 1 Industrial & Dealer Heavy Melting	5	0.25	0.09	0.10	0.025	0.030	0.020	0.040
Railroad Car Sides	2	0.30	0.10	0.10	0.025	0.030	0.020	0.050
Shredded Scrap (No forward power trains)	7	0.22	0.11	0.14	0.020	0.020	0.025	0.040
Foundry Steel	15	0.35	0.14	0.20	0.035	0.035	0.030	0.050
No. 2 Industrial & Dealer Heavy Melting	5	0.55	0.20	0.18	0.040	0.040	0.030	0.070
No. 2 Dealer Bundles	5	0.50	0.10	0.18	0.100	0.030	0.030	0.090
Briquetted Turnings & Borings	10	0.20	0.40	0.40	0.015	0.150	0.030	0.080
Flashings, Alloy Steels & Miscellaneous	7	(See Text)**						
All Steel Scrap Used in Iron Melting	100%							

*Analytical samples extracted from a homogeneous melt of a large, representative batch, e.g., 20 tons or more.

**Source: *Cupola Handbook*, American Foundrymen's Society, Inc., Des Plaines, Illinois (1984).

TABLE 6-2. TYPICAL ANALYSES OF LOW PHOSPHORUS METALS SELECTED FOR DUCTILE IRON PRODUCTION (%)

Element	Source "A" F-1 & F-10	Source "B"	Source "C"*
Carbon	4.30 (Ave)	4.25	4.20
Silicon	0.18	1.35	1.16
Sulfur	0.015 (F-1) (ave) 0.006 (F-10)**	0.010	0.006
Phosphorus	0.015	0.035	0.028
Manganese	0.005	0.12	0.11
Trace Elements by Spectrographic Analysis Alloying Elements			
Nickel	0.067	0.030	0.030
Copper	0.018	0.010	0.028
Graphite Promoting Elements			
Titanium	0.006	0.030	0.050
Aluminum	0.003	0.001	0.002
Carbide and Pearlite Forming Elements			
Chromium	0.029	0.030	0.030
Vanadium	0.020	0.020	0.015
Magnesium	Nil	Nil	Nil

*low-phosphorus pig iron
**low-sulfur grade, 0.010% max

In addition to the metallurgical effects on ductile irons, tramp elements must be evaluated for their environmental impact, in accordance with permitted by-product levels. Tramp elements may be classified into groups according to the manner in which they are recovered in the molten iron:

1. Elements such as zinc, cadmium and, to some extent, lead oxidize to a white irritating smoke. These oxide dusts can cause sludge or baghouse dust to be classified as a high-cost disposable by-product. Employees in the melt area must be protected from these oxide dusts.

2. Elements such as lead and bismuth, which are partially soluble in iron, can be trapped by vigorous stirring just before solidification. Lead, if not volatilized, has a well known tendency to settle to the bottom of a bath, penetrate the refractory at a joint or imperfection, and even penetrate the bottom of the vessel, causing a runout.

3. Aluminum, boron, vanadium, zirconium and titanium are soluble in iron, but are highly oxidizing. Varying concentrations are slagged off during melting and treatment. These elements can effect slag disposal, depending on permits.

4. Chromium and tellurium are soluble in both molten iron and slag, and become partitioned between the two phases. Approximately one-third of the charged chromium will be removed in the slag, with the balance going to the molten iron.

5. Elements that have solubility in molten iron and low oxidizing tendencies include: copper, nickel, tin, molybdenum, columbium, arsenic, antimony, tungsten and cobalt. Once these elements are recovered in the iron, their levels can be lowered only by dilution.

■ CUPOLA FUELS AND FLUXES

The composition and sizing parameters for fuels and fluxes were discussed in Chapter 5. As with the other charge materials, both metallurgical and by-product effects must be considered in the charge material selection decision. The development of internal specifications and the practice of inspecting incoming materials are highly recommended. Suppliers of fuels and fluxes should be required to alert the customer of any process changes, including changes in raw material deposits or sources.

■ CHARGE ALLOYING AGENTS

The most common charge alloying agents for ductile iron charges include carbon, silicon, manganese, copper and nickel. Cupola melters typically require more Si and Mn as charge alloys than do electric melters, due to the oxidation of Si and Mn in the cupola. Electric melters depend on additions of carburizers to achieve desired carbon aims, whereas cupola melters pick up a majority of the required carbon units from the coke used for fuel.

The development of internal specifications recommended for metallic charge materials, fuels and fluxes is also recommended for charge alloys.

Charge alloys are discussed in detail in chapter 7 of the *AFS Cupola Handbook*, 5th edition. Selection parameters for charge alloys include composition, sizing, recovery and economics. Following are some of the commonly used charge alloying elements.

CARBON

Artificial graphites and low-sulfur petroleum coke are the

TABLE 6-3. TYPICAL ANALYSES OF SCRAP IRON CASTINGS

Types of Castings	T.C.	Si	Mn	P	S	Ni	Cu	Cr	Mo
Automobile Engine Blocks	3.30	2.10	0.60	0.12	0.10			0.30	
	3.25	2.00	0.65	0.12	0.10	0.75		0.50	
Truck & Tractor Engine Blocks	3.25	2.00	0.65	0.12	0.10			0.40	0.40
	3.25	2.00	0.65	0.12	0.10		0.75	0.40	
Automobile Brake Drums	3.40	2.00	0.70	0.20	0.10				
Automobile Clutch Plates	3.25	2.20	0.60	0.20	0.10				
Automobile Crankshafts									
Pre-1960: plain carbon, alloy steel or alloy cast irons	2.80	2.50	0.80	0.15	0.10	0.50		0.50	0.50
Post 1960: largely ductile iron	4.00	2.80	0.90	0.05	0.01				
Automobile Connecting Rods									
Pearlitic malleable iron	2.40	1.40	0.40	0.05	0.09				
Some ductile iron	4.00	2.80	0.90	0.05	0.01				
Farm Machinery Castings									
Light	3.40	2.40	0.55	0.60	0.10				
Medium	3.30	2.20	0.60	0.50	0.12				
Heavy	3.15	1.80	0.65	0.35	0.12				
Misc. Machinery Castings									
Light	3.50	2.50	0.50	0.30	0.10				
Medium	3.30	2.00	0.60	0.30	0.12				
Heavy	3.15	1.50	0.65	0.25	0.12				
Very Heavy	3.10	1.25	0.70	0.20	0.14				
Pipe, Water									
Pit Cast	3.50	1.60	0.50	0.65	0.10				
DeLevaud	3.60	1.50	0.40	0.50	0.10				
Sand Spun	3.60	1.95	0.60	0.20	0.10				
Stove Plate	3.50	2.40	0.50	0.80	0.12				
Radiators	3.50	2.30	0.50	0.50	0.12				
Grate Bars	3.25	2.00	0.60	0.40	0.10				
Chilled Railroad Car Wheels	3.50	0.55	0.50	0.30	0.12				
Chilled Plow Shares	3.40	1.40	0.60	0.40	0.12				
Chilled Mold Boards	3.50	0.80	0.60	0.30	0.12				
Chilled Glass Molds	3.50	2.20	0.60	0.20	0.10	1.00		0.30	
White Iron Bearings (Agr.)	3.40	0.75	0.70	0.25	0.12		1.00	1.00	
White Iron Crusher Plates	3.30	0.60	0.60	0.15	0.10	4.50		1.50	
Light to Medium Malleable Iron	2.50	1.40	0.45	0.15	0.14				
Heavy Section Malleable Iron	2.35	0.95	0.40	0.15	0.12				

most popular carbon raisers for ductile charges, due to their low sulfur levels and high percent recovery. Calcined petroleum cokes, natural graphites, coke breezes, etc., which contain high sulfur levels, are typically added only when desulfurization takes place prior to ductile treatment. Sizing and composition need to be evaluated to determine the optimum parameters for maximum recovery. Typically, the higher the carbon content and the lower the ash content, the better the recovery. Both the upper and lower screen sizing can have an effect on recovery, de-

pending on the degree of mixing, metal composition and temperature, pollution hoods, etc. Silicon carbide can be used as a source of carbon if there are also additional silicon units required.

SILICON

Common sources of charge Si include lump ferrosilicon, pig iron, silicon carbide (SiC) grain (for electric melting),

SiC briquettes (for cupola melting) and briquetted Si fines (for cupola melting). When using lump ferrosilicon, specify "No boron" on your orders, as some manufacturers produce FeSi-containing boron for producers of malleable iron. Silicon levels in pig iron can be specified according to required final Si levels and the Si units in the balance of the charge.

Silicon carbide, which is a deoxidizer, has been reported to provide a "pretreatment" inoculating effect on irons. It is composed of 70% Si and 30% carbon. Alloys can be purchased containing varying percentages of SiC. Electric melters commonly use 90–92% SiC grain. Cupola melters often select 20–80% SiC in a briquetted form.

Si briquettes—another source of charge Si—can be purchased with varying levels of Si. They are typically produced from screened by-products of Si products. It is recommended that you require your supplier to alert you to any raw material changes made in the production of Si briquettes supplied to you.

Cupola melters using briquetted materials should select suppliers who will guarantee consistent, certified composition. Briquetted material is not recommended for electric melting shops, due to the difficulty in drying briquetted material. Dealing with reputable suppliers, who will provide certification of composition with each shipment of briquetted alloy, is highly recommended.

Flux requirements may have to be adjusted by cupola melters using briquetted products bonded with cement binders. If adjustments to the flux are overlooked, potential problems can arise with slagging, carbon pickup, silicon and manganese oxidation rates, refractory erosion, etc. These factors should be considered in evaluating alloy performance.

Manganese

Generally, the final manganese target analysis for ductile iron is relatively low, with the result that the manganese levels in many melts will have to be diluted by charges of clean iron units (such as pig iron). Should additional manganese be required, the most commonly used charge alloy is ferromanganese. Electric melters often use lump FeMn; cupola melters often use FeMn briquettes.

Nickel

Nickel is a good alloying element for increasing the mechanical properties of ductile irons without adversely affecting the machinability of the castings. However, the high market price of nickel limits its use to special applications. One growing application for nickel is in austempered ductile irons, where it is one of the preferred elements.

Sources for nickel as a charge alloying element are electrolytic nickel squares, nickel shot and nickel pellets.

Copper

One of the most widely used ductile alloying elements is copper. The best source of copper is 99.99% pure electrolytic copper. Users of secondary copper (wire croppings, etc.) should beware of contamination of subversive elements such as lead, antimony, and arsenic.

The recovery of copper from the various charge alloys varies depending on the composition and sizing of the alloy, type of melting unit, combustion conditions (cupolas), mixing conditions (electric melters), charge makeup, etc. Any changes in the above variables can effect alloy recovery to the point of having a notable effect on charge cost. The selection of a charge alloy should be based on a carefully controlled performance evaluation. When evaluating an alloy performance trial make sure to consider all the elements in the charge alloy. For example, allow credit for the carbon units in SiC and the Fe units in ferroalloys.

■ SUMMARY

Customer demands for continual quality improvement and the statistical treatment of data have forced foundry metallurgists into tighter levels of process control. Primary metallurgical control areas include molten metal composition, inoculation, and casting cooling rates. The selection of charge materials is a key step in the control of metal composition. Both internal and external requirements need to be considered in the selection of charge materials. Internal requirements include customer requirements, casting design, charging, melting and molding equipment, environmental issues, melt yield, and economics. External requirements include availability, delivery, composition, bulk density and sizing.

The development of internal material specifications and the inspection of incoming charge material is highly recommended. The selection of charge material should be based on experience or the results of carefully monitored performance evaluations.

■ BIBLIOGRAPHY

American Foundrymen's Society, Inc. *The Cupola and Its Operation.* Des Plaines, IL: American Foundrymen's Society, Inc. (2d ed., 1954; 3d ed., 1965; 5th ed., 1985).

American Society for Metals. "Properties and Selection of Metals." *Metals Handbook,* 8th ed., vol 1, Metals Park, OH: American Society for Metals (1961).

BCIRA. "Cupola Metallic Charge Materials." Broadsheet 61 (1972).

Burlingame, R.D. "Metallics for Cupola Melting." *AFS Cupola Handbook,* 5th ed., Des Plaines, IL: American Foundrymen's Society, Inc. (1984): 65–92.

Engineering Material Handbook, 1st ed. New York: McGraw-Hill Book Co. (1958).

Graham, J.L. Personal communications to author.

Henderson, H.E. "Use of DRI in Foundries." *Direct Reduced Iron, Technology and Economics of Production and Use.* New York: AIME (1980): 120.

Institute of Scrap Iron and Steel Inc. *ISIS Handbook,* Washington: ISIS (Nov 1982).

Kotschi, R.M. and C.R. Loper. "Foundrymen's Guide to Quality Iron." Baird Corporation (1988).

Krause, D.E. and T. Barlow. *Cupola Operator's. Manual,* Columbus, OH: Gray Iron Research Institute, Inc., (1946): 168–194.

Lloyd, A.T. "Selection and Application of Carbon Raising Materials." *AFS Transactions,* vol 82, Des Plaines, IL: American Foundrymen's Society, Inc. (1974): 229.

Sale, J.H. "Operating Experience Using Shredded Automobile Scrap for the Production of Ductile Iron Pipe." *AFS Transactions,* vol 83, Des Plaines, IL: American Foundrymen's Society, Inc. (1975): 107–108.

Spengler, A.F., ed. *The Ductile Iron Process.* Chicago: Miller & Co. (1978).

Thomas, W.A. Personal communications to author.

Wallace, J.F., G.F. Ruff and R.C. Helmink. "Further Studies on Ferrous Metal from Municipal Wastes as Charge Material for Cast Iron." *AFS Transactions,* vol 83, Des Plaines, IL: American Foundrymen's Society, Inc. (1975): 467–478.

Warda, R.D., R.K. Buhr. "A Detailed study of Cupola Emissions." *AFS Transactions,* vol 81, Des Plaines, IL: American Foundrymen's Society, Inc. (1973): 467–478.

7

Desulfurization

Donald B. Craig

*Elkem Metals Co.
Niagara Falls, New York*

■ INTRODUCTION

The production of high-quality ductile iron castings requires a reliable source of low-sulfur base iron. This low sulfur content is desired in order to improve process control, to minimize the amount of alloy required to nodulize the iron, and to reduce the incidence of dross defects in the castings. A low level of sulfur in the base iron improves the efficiency during treatment and provides a more consistent residual magnesium in the treated iron.

Low-sulfur irons can be produced by a variety of processes, the most common of which include electric melting of low-sulfur materials, basic-slag cupola practice and acid-slag cupola practice with external desulfurizing. Careful selection of the charge materials for an electric furnace has proven to be an effective and reliable method of producing a low-sulfur base iron.

Desulfurization is necessary when the cupola is used as the source of molten iron. Coke, used as both fuel and carbon source in the cupola, contains sulfur, which dissolves readily into the molten iron. This sulfur can be removed from the iron within the cupola by using a basic slag process. Alternatively, when an acid slag process is preferred, the sulfur can be removed from the iron using external desulfurization. The individual ductile iron treatment process will determine the desired level of sulfur required in the base iron, but generally a sulfur content ranging from 0.006 to 0.02% is acceptable for most treatment processes. The capability of each desulfurizer and desulfurizing technique to achieve this goal will be discussed in this chapter.

■ DESULFURIZERS

SODIUM OXIDE

Historically, sodium oxide (Na_2O) has been used to control the sulfur content of cast iron. A common practice in the manufacture of gray iron pipe was to add sodium oxide to the molten iron to keep the sulfur within the desired range of 0.06–0.08%.[1] Active Na_2O can be obtained by adding either soda ash (Na_2CO_3) or sodium hydroxide (NaOH) to the molten iron. The sodium oxide was typically added to the iron as it flowed in the launder from the cupola to the treatment vessel. Due to the low density and low melting point of Na_2O, the slag was a liquid phase that separated readily from the molten iron. Sodium oxide slags are very fluid and highly reactive with most refractories. Basic, 90% alumina, or carbon-based refractories are recommended for systems using Na_2O for desulfurization.

Sodium oxide reacts with the dissolved sulfur in the iron to form sodium sulfide (Na_2S), which is also a liquid phase at iron-making temperatures. Typically, an addition of 2 to 3% soda ash was required to reduce the sulfur in the iron from 0.1% to 0.02%. The addition of graphite to the slag improves the desulfurizing effectiveness of soda ash. An increase in either the metal temperature or the interfacial area between slag and metal also improves the desulfurizing effectiveness of sodium oxide.

Desulfurization with sodium oxide has several drawbacks, which has limited its usefulness in the foundry industry. Soda ash (Na_2CO_3), when mixed with molten iron produces obnoxious fumes, that can create a hazardous environment for both operators and foundry equipment. The highly reactive nature of Na_2O causes heavy refractory wear, and sodium oxide slags have difficulty in reducing the sulfur content of the iron consistently below 0.02%.

CALCIUM CARBIDE

Calcium carbide (CaC_2) is the most popular and most widely used desulfurizer within the foundry industry. This popularity was created by calcium carbide's strong desulfurizing ability, process flexibility and ease of use. Calcium carbide can consistently reduce the sulfur in iron to a level less than 0.01%. It has been used effectively in a variety of desulfurization techniques, each using a different method of agitation. These factors provided the foundryman with an effective and robust method of removing sulfur from cupola iron.

The melting point of CaC_2 is well above the temperature range required to produce ductile iron. Thus, the calcium carbide desulfurizer remains solid in the liquid iron. The much lower density of the calcium carbide causes the desulfurizer to float to the top of the molten iron. Agitation is required to mix the calcium carbide into the liquid iron. A study by Talballa, Trojan, Bigelow and Flinn[2] demonstrated that CaC_2 dissociates into calcium and graphite at the interface with the liquid iron. The calcium reacts with the sulfur dissolved in the liquid iron to form calcium sulfide (CaS). The graphite forms a film at the interface. The study indicated that the thickness of each of these layers increases as the reaction proceeds, thereby slowing down the desulfurization reaction.

An increase in the surface area, achieved by reducing the particle size, generally increases the efficiency of the desulfurization reaction. Typically, the particle size of calcium carbide falls within the 8 mesh to 60 mesh (2.36 mm to 0.25 mm) range. An addition of 0.4–1.0% of calcium carbide is usually required to reduce the sulfur in the cupola iron from 0.1% to a level of less than 0.01%. An increase in both temperature and agitation will also improve the efficiency of desulfurization with calcium carbide.

This buildup of the reaction products on the surface of the calcium carbide particles, combined with inefficient us-

age of the desulfurizer, has created a significant challenge for calcium carbide. If the CaC_2 is not totally consumed during desulfurization, the resultant slag can pose a serious environmental problem. Calcium carbide, when mixed with water, forms acetylene gas, which is highly flammable. Mixtures of acetylene and air containing between 2% and 85% acetylene are explosive. Therefore, the desulfurization slags from foundries using CaC_2 as a desulfurizer have come under the scrutiny of the environmental agencies. The manufacturers of calcium carbide have responded to this challenge by developing a new generation of CaC_2-based desulfurizers. These new desulfurizers, when efficiently used, significantly lower the residual calcium carbide retained in the slag, thus reducing the environmental hazard from the desulfurizing slag.

CALCIUM OXIDE

Calcium oxide (CaO), or lime, has been used as a desulfurizer in the steel industry, but has not yet found extensive use as a desulfurizer in the iron foundry. The high melting point of lime creates a solid-liquid interface for desulfurization. As in calcium carbide, the surface area of the lime particles, metal temperature, agitation and desulfurizing time all affect the efficiency of the desulfurizer. Talballa, Trojan, Bigelow and Flinn[2] demonstrated, however, that calcium oxide unlike calcium carbide, does not dissociate into calcium and oxygen at the interface with liquid iron. Rather, the CaO reacts with the sulfur dissolved in the liquid iron to form CaS and $CaFeO_2$, which accumulate at the interface.[2] This accumulation of reaction products at the interface retards further desulfurization by that lime particle. Initially, lime provides excellent desulfurization, but the accumulation of reaction products on the surface of the lime particles rapidly reduces the rate of desulfurization, thereby limiting the effectiveness of lime as a desulfurizer.

Studies by Katz and Landefeld[3] and by Coon[4] indicate that an addition of 5–10% fluorspar (CaF_2) to the lime can increase the desulfurizing effectiveness of the lime. The fluorspar fluxes the reaction products, CaS and $CaFeO_2$, away from the liquid-iron interface, thus providing a fresh CaO surface for additional desulfurization. This fluxing action allows lime's initial high rate of desulfurization[2] to continue over a longer period of time, thereby allowing the lime-fluorspar (CaO-CaF_2) mixture to reduce the sulfur content of the iron to the desired level of less than 0.01%. Five percent fluorspar and 95% lime was determined to be the optimum mixture.[3,4]

Typically, a 1.3–2.0% addition of CaO-CaF_2 is required for desulfurization, as compared to the 0.4–1.0% addition of CaC_2 required for desulfurization. The mixture of limefluorspar creates higher carbon and silicon losses from the liquid iron than does calcium carbide.[4]* The addition of

5% graphite increases the desulfurizing efficiency of lime fluorspar. Katz and Landefeld[3] also demonstrated that the efficiency of the lime-fluorspar desulfurizer is dependent on the thickness of the slag covering the bath of iron. The desulfurizer, lime-fluorspar, also tends to be sensitive to both agitation and to the length of time that the iron remains in the ladle.

***[Editors' Note: Recent industrial experience indicates that only 1.1–1.8% CaF$_2$ is necessary at the higher temperatures associated with continuous desulfurization. Carbon and silicon losses are lower than when 5% fluorspar is used.]**

MAGNESIUM

The boiling point of magnesium is well below the temperature range required to produce ductile iron castings. Thus, magnesium vaporizes when added to molten iron. Desulfurization occurs when magnesium vapor encounters the dissolved sulfur in the liquid iron. The introduction of pure magnesium into molten iron has proven to be quite difficult, due to the volatility of magnesium. Thus, specialized equipment and processes (such as a converter and plunging bell) have been developed to add pure magnesium to molten iron. These processes are discussed in Chapter 9.

While the addition of magnesium to molten iron is commonplace in the production of ductile iron, the use of magnesium to desulfurize cupola iron has not been adopted to any large extent. A correlation between the sulfur content of the base iron and the number of dross defects found in the casting has been established.[4] This relationship indicates that, as the sulfur content in the base iron increases, the number of dross defects found in the castings also increases. Thus, the treatment of high-sulfur base irons with magnesium tends to increase the incidence of dross defects found in the castings. Therefore, foundries have found it more economical to desulfurize the cupola iron prior to the treatment of the iron with magnesium.

▬ DESULFURIZING PROCESSES

BASIC-SLAG CUPOLA

The basic-slag cupola was used extensively to produce the high-carbon, low-sulfur iron required for the manufacture of ductile iron. In fact, the basic-slag cupola was once the preferred desulfurizing system.[1] Improvements in external desulfurization, combined with several inherent process difficulties, have caused a shift away from the basic-slag cupola.

The basic desulfurizing slag was created by the addition of flux materials to the charge. The flux addition was often a combination of limestone, dolomite and fluorspar. A charge typically contained 10–12% flux, which created a slag with a basicity ratio in excess of 2.0.[5] The basic slag tends to promote carbon absorption in the iron, thus, permitting a higher percentage of steel scrap to be used in the charge. The sulfur content in the iron from a basic-slag cupola typically ranged from 0.01 to 0.035%, which was suitable for most ductile iron treatment methods.

Silicon losses from the metallic charge are much higher for basic-slag cupolas than for acid-slag cupolas.[1] The very high slag-melting temperature, combined with the large volume of slag generated during melting, tends to make the composition of the iron from a basic-slag cupola difficult to control. A basic-slag cupola tends to have lower melting capacity than a similar acid-slag cupola, since additional fuel is required to melt the large volume of nonmetallic material required to create the basic slag. These factors provided the foundries with the incentive to convert from basic to acid melting with external desulfurization.

INJECTION

Injection was at one time used rather extensively for desulfurization, but currently is used only on a limited basis in a few, relatively large batch applications.[5] The desulfurizing reagent is injected into the bath of molten iron via a refractory or graphite lance (Fig. 7-1). An inert gas, usually

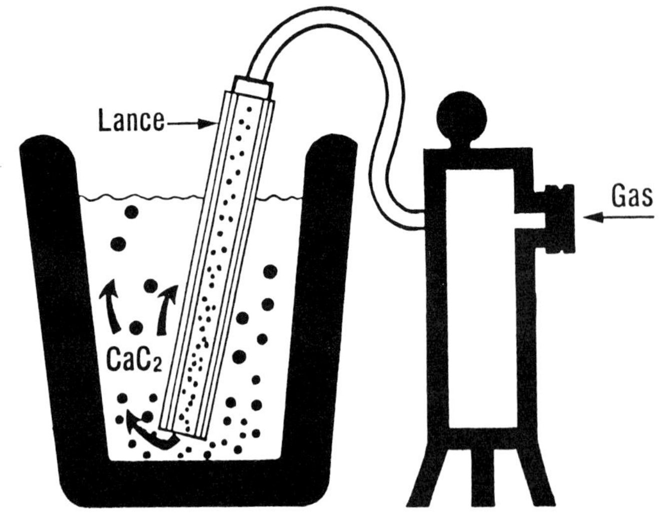

Fig. 7-1. Diagram of the injection desulfurization process.

nitrogen, carries the reagent through the lance and into the molten iron. The lance enters the bath of molten iron at a 70–80° angle to allow for better dispersion of gas bubbles carrying the desulfurizer, and to minimize erosion at the tip of the lance. The inert carrier gas also provides some desirable agitation within the iron bath. The desulfurizing reagent typically consists of finely ground particles of calcium carbide (30 mesh × Down).[6] These particles are carried to the surface of the ladle within the bubbles of inert gas. The desulfurization reaction occurs between the solid CaC_2 particles and the liquid iron at the bubble interface. The slag particles within the bubbles then agglomerate into a surface slag once the bubbles reach the surface. The reagent particles are contained within relatively large gas bubbles. This tends to insulate the reagent from the liquid iron, which inhibits the good surface contact required for efficient desulfurization.[1]

The desulfurizing efficiency of injection is controlled by the injection rate, reagent particle size and amount of agitation in the bath.[7] An addition of 1.0–1.5% CaC_2 can reduce the sulfur content of the iron from 0.1% to 0.01–0.015%. The large volume of gas needed to inject the reagent tends to cause a large temperature loss during desulfurization. The refractory or graphite lances have a limited life; additionally, the cost of the injection tubes and the large volume of carrier gas required must be considered when evaluating the efficiency of the system.

SHAKING LADLE

The shaking ladle was developed in Sweden and has been widely use in Europe, but it has found limited application in North America. The ladle does not, in fact, shake. Rather, it rotates quietly and smoothly in a circular, eccentric motion about the vertical axis, creating a wave that swirls around inside the refractory-lined ladle (Fig. 7-2). This wave action provides the agitation necessary to mix the desulfurizer into the molten iron. The mixing action is determined by the amount of liquid iron in the ladle, the particle size of the desulfurizer, the number of revolutions per minute and the eccentricity of revolution of the ladle.

The shaking ladle is generally lined with a 70% alumina refractory. Iron is poured into the ladle until it is one-third full. The desulfurizer is usually added while the ladle is being filled. Once filled, the ladle is then rotated for 8–10 minutes at a rotational speed of 50–55 rpm. A 0.5–1.0% addition of the desulfurizer, usually calcium carbide, is required to reduce the sulfur content of the iron from 0.1% to the desired range of 0.009–0.015%. After the mixing cycle is complete, the liquid iron is decanted from the ladle. The ladle is then inverted to dump out the desulfurizing slag.

Desulfurization using a shaking ladle requires a large volume of iron—usually 5–10 tons (4.5–8 tonnes) per treatment—to be efficient. The temperature loss for the shaking

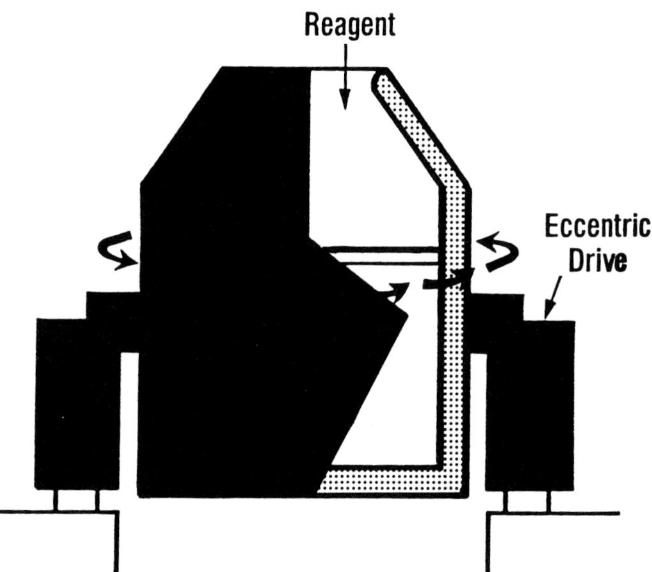

Fig. 7-2. The shaking ladle desulfurization process.

ladle is reported to be 50–80F (28–44C).[1] The shaking ladle process has a high capital cost and requires a significant amount of maintenance, but is an effective method of removing sulfur from molten iron.

RHEINSTAHL STIRRER

The agitation required for effective desulfurization is provided by a rotating refractory paddle in a Rheinstahl stirrer. This paddle, shaped like an inverted "T," is connected to a variable-speed electric motor that is mounted on top of the cover of the ladle (Fig. 7-3). The rotation of the refractory paddle creates circular movement within the molten iron. This stirring action mixes the desulfurizing reagent into the metal bath.

The ladle is usually lined with 70% alumina refractory. Liquid iron is poured into the ladle and the preheated stirrer is placed into the molten iron. The refractory paddle is rotated at speeds between 30 and 100 rpm.[6] The desulfurizing reagent is then introduced into the ladle through a hole in the cover. An addition of 0.7–1.0% of the desulfurizer, usually calcium carbide, is required to reduce the sulfur in the iron from 0.1% to the desired range of 0.009–0.014%. To achieve good desulfurization, 6–10 minutes of stirring is required. The Rheinstahl stirrer generally requires between 5 and 20 tons (4.5–18 tonnes) of iron to maintain sufficient temperature within the iron.[6]

143

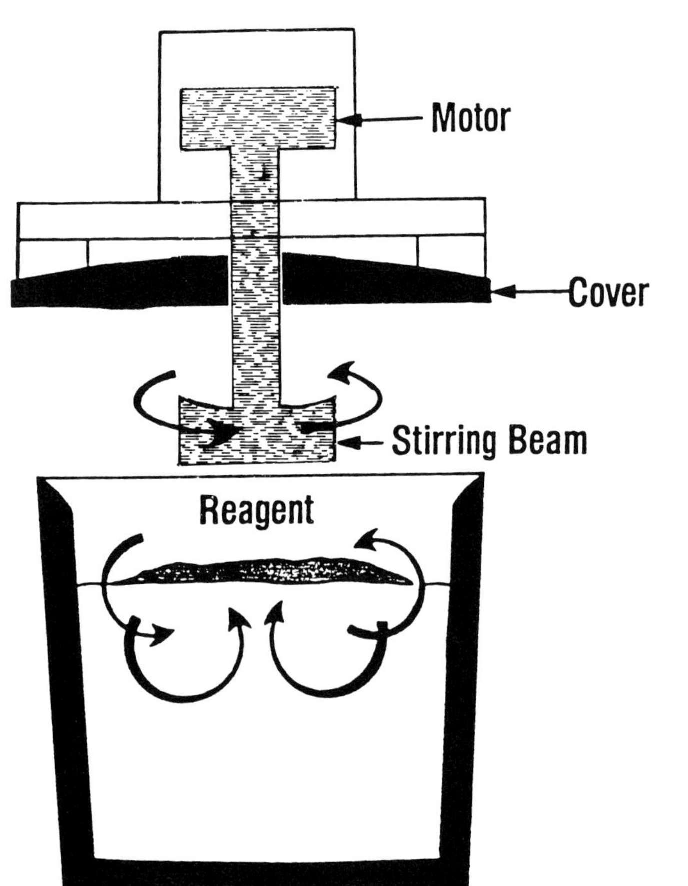

Fig. 7-3. Diagram of a rotating refractory paddle in the Rheinstahl stirrer process.

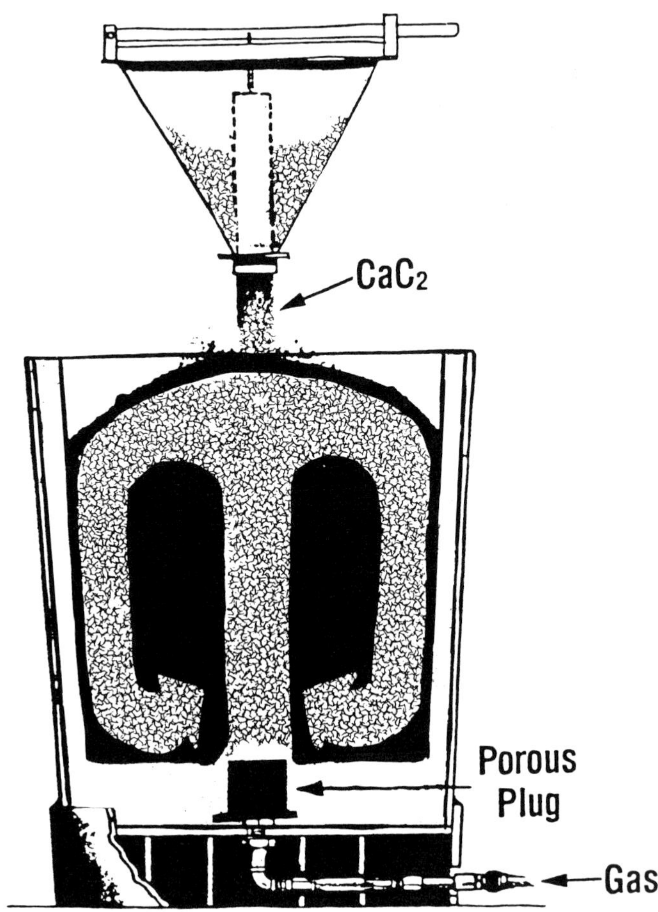

Fig. 7-4. The batch porous plug desulfurization process.

Porous Plug

In the porous plug desulfurization process, a porous refractory plug is placed in the bottom of the ladle. An inert gas, usually nitrogen, is passed through the interstices of the refractory, producing a stream of tiny bubbles that rise to the surface of the ladle. This finely disseminated stream of bubbles creates a pronounced meniscus at the top of the ladle and promotes circulation of the iron within the ladle. The iron is drawn upward in the stream of bubbles and downward along the walls of the ladle (Fig. 7-4). This stirring action draws the desulfurizer from the surface of the iron into the metal bath. The porous plug desulfurization practice can be used as a continuous or a batch process.

Batch porous plug applications range in size from 800 pounds to 30 tons (0.4–27 tonnes).[5] The ladle is usually lined with a 50–70% alumina refractory. For small ladles, a single plug located in the center of the ladle bottom provides sufficient agitation within the ladle. Larger ladles usually require three plugs, spaced 120° apart and located at the mid radius of the ladle, to provide the necessary agitation (Fig. 7-6). To avoid damage to the plug, the bottom of the ladle should be contoured to allow any residual liquid iron to drain away from the surface of the plug. The flow of inert gas is started through the refractory plug prior to the addition of liquid iron. The level of slag in the ladle should be 6–8 inches below the rim of the ladle. This freeboard is required to minimize spillage of iron during stirring. The desulfurizing reagent, usually calcium carbide, is weighed and added to the surface of the ladle (Fig. 7-4). An addition of 0.8–1.5% of desulfurizer is required to reduce the sulfur content of the iron from 0.1% to 0.006–0.015%. The particle size of the desulfurizer should be within the 8 mesh to 60 mesh (2.36 mm to 0.25 mm) range for efficient sulfur removal. Stirring time for the batch porous plug technique is usually 3–5 minutes.

The continuous porous plug desulfurizing technique must be sized to accommodate the melting rate of the cupola. This technique requires that the iron remain in the ladle for 4–8 minutes. The optimum dwell time for the iron in the ladle is approximately 6 minutes.

The desulfurizing unit has a modified teapot design with a slag notch located opposite the ladle spout (Fig. 7-5). A single plug is typically used for desulfurizing ladles treating small volumes of liquid iron. This plug is usually located one inch (25 mm) off the center of the bottom of the ladle. For volumes greater than 20 tons (18 tonnes) per hour,[5] three plugs are recommended. These plugs are spaced 120° apart and are located on the mid radius of the bottom of the ladle (Fig. 7-6). Each plug, whether in a single or multi-plug application, must have its own control system to control the flow of nitrogen through the plug. The gas system should be capable of delivering approximately 5–8 ft^3 (0.14–0.23 m^3) of nitrogen per minute to each plug. The control system must be able to control the gas pressure within the range of 15–60 psi (103–414 kPa). A gas flow of 7 ft^3 (0.20 m^3) under 15 psi (103 kPa) of pressure is typical for most porous plug applications. Higher

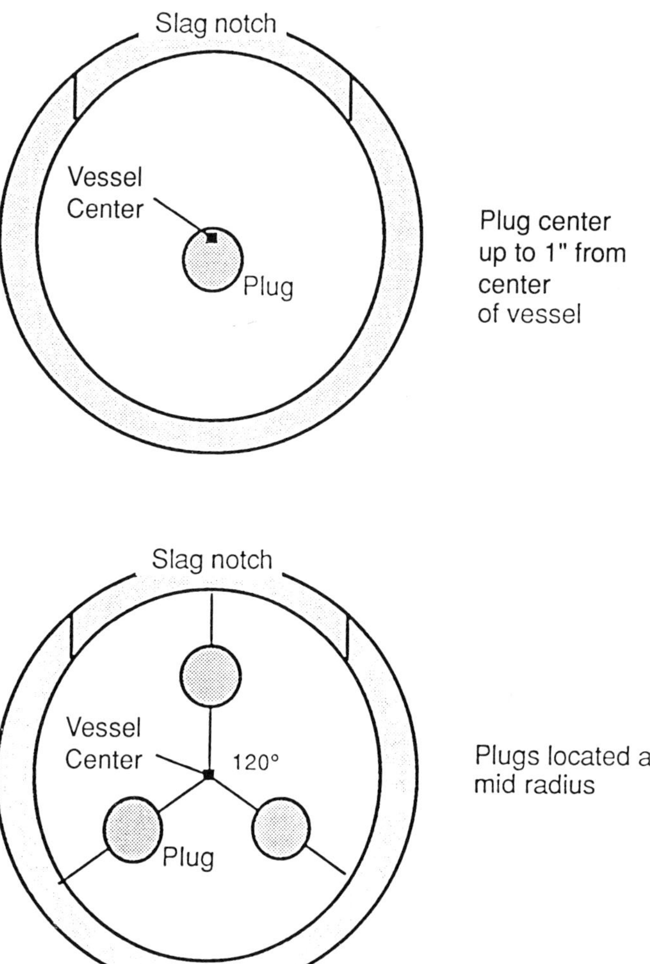

Fig. 7-6. Optimal ladle location for single and multiple porous plugs.

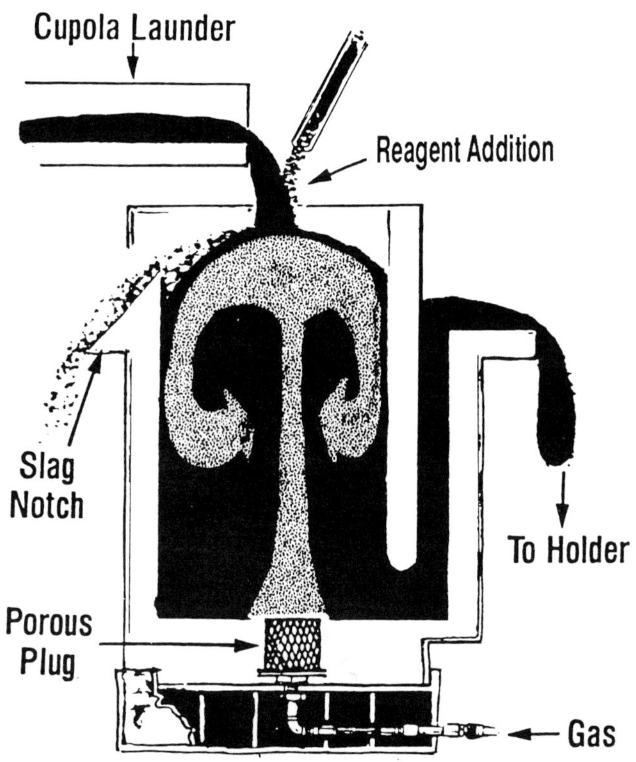

Fig. 7-5. Diagram of the continuous porous plug process.

pressures and/or lower flow rates indicate blockage in the plug and possible failure.

The desulfurizing ladle should be located under the cupola launder, such that the liquid iron flowing out of the launder impinges upon the bath of iron directly over the stream of bubbles from a porous plug. The desulfurizing reagent must be added to the desulfurizing ladle in a manner that allows the stream of desulfurizer particles to enter the bath of iron simultaneously with the stream of iron from the launder. This allows the fresh desulfurizer to be drawn into the bath, which increases the efficiency of the desulfurization reaction. The dispensing mechanism, usually a screw conveyor, should have a variable speed control system that can be calibrated to allow the accurate metering of the reagent into the desulfurizing ladle.

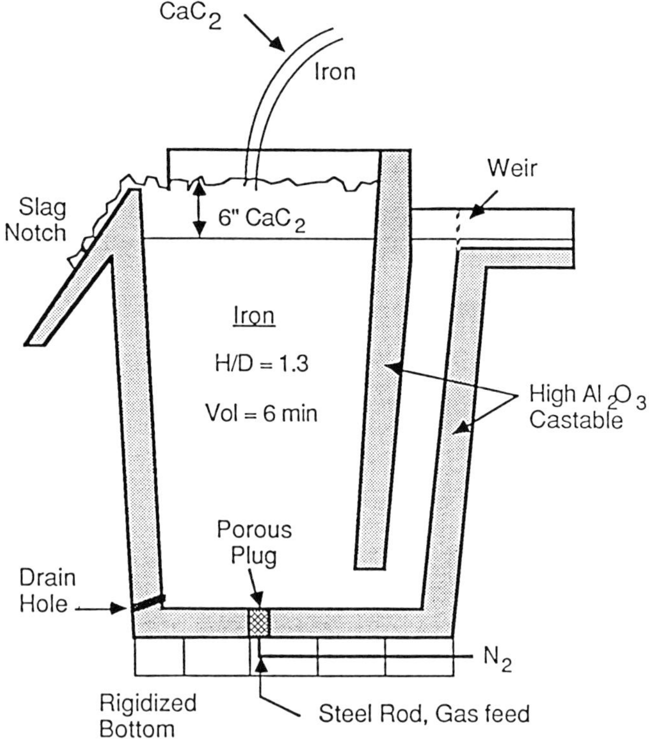

Fig. 7-7. Key features of a modified teapot desulfurizing ladle.

notch should be very steep, and the notch should not have a section thickness greater than two inches. This will minimize any slag buildup on the slag notch. The rim of the vessel should be six inches higher than the slag notch to minimize metal spillage and allow for vigorous stirring. This combination of dimensions will create a top slag on the surface of the ladle that ranges from 6–10 inches thick. A dry, granular and free-flowing slag is required for optimum desulfurizing efficiency. The slag notch must be kept free from any slag buildup and the gas flow should be adjusted so that the slag flows freely over the notch.

***[Ed.N.: In 1991 the (several) foundries employing continuous porous plug desulfurization, with the use of lime as a reagent, report increased addition rates as compared to calcium carbide, as well as the need for certain modifications in dwell unit design parameters. The main differences are the need for multiple porous plugs, where one is currently used, or four plugs where three are currently used. A longer dwell time (increased vessel capacity) of 10–12 minutes is required; and the height of the slag notch has to be raised to allow a thicker slag blanket to develop.]**

For optimum efficiency, it is important that the desulfurization ladle operator have the ability to make minute adjustments in the feed rate of desulfurizer flowing to the ladle. This allows the operator to compensate for fluctuations in the volume of iron treated and the sulfur content of the iron. The control mechanism must have precise control of the desulfurizer within the addition range of 0.3–1.5% of the melt rate. At start-up, higher feed rates of the desulfurizer (up to 3% of the melt rate) are required to build up the slag volume. The system should also be designed to allow for a periodic check of the flow of reagent to the ladle.

The nitrogen flow is started through the refractory plug prior to the flow of iron into the ladle. As in the batch process, an individual gas control system for each plug is desirable for optimum desulfurization. The desulfurizer should be metered into the ladle once the flow of iron begins. An addition rate 0.3–0.8% of the desulfurizer is required to reduce the sulfur content of the iron from 0.1% to the desired range of 0.006–0.015%.

The level of iron in the ladle is controlled by the height of the spout (Fig. 7-7). A slag notch located six inches higher than the spout creates the slag blanket necessary for efficient desulfurization.* The slag notch must encompass approximately 100° of the circumference to allow for continuous self-slagging. The discharge chute of the slag

■ REFERENCES

1. S.G. Clow; "The Effect and Control of Sulfur in Iron," *AFS Transactions*, vol 87, p 401 (1979).

2. M. Talballa, P.K. Trojan, W.C. Bigelow, R.A. Flinn; "Mechanism of Desulfurization of Liquid Iron Carbon Alloy with Solid CaC_2 and CaO," *AFS Transactions*, vol 84, p 775 (1976).

3. S. Katz, G.F. Landefeld; "Plant Studies of Continuous Desulfurization with $CaO-CaF_2-C$," *AFS Transactions*, vol 93, p 215 (1985).

4. P.M. Coon; "The Development and Industrial Application of Lime-Fluorspar Mixtures for the Desulfurization of Cast Iron," *AFS Transactions*, vol 88, p 471 (1980).

5. American Foundrymen's Society, Inc. "Desulfurization," *Ductile Iron Molten Metal Processing*, (1986).

6. A.F. Spengler, ed.; *Ductile Iron Process*, pp 211–220 (1976).

7. E. Campomunes, R. Goller; "External Desulfurization in Cast Iron by Injection," *AFS Transactions*, vol 84, p 137 (1976).

8. S. Katz, H. Rezeau; "The Cupola Desulfurization Process," *AFS Transactions*, vol 87, p 367 (1979).

9. W.A. Henning; "Efficiency in Desulfurizing Practices: Committee 5-L Report," *AFS Transactions*, vol 94, p 815 (1986).

10. P.K. Trojan, W.N. Bargeron, M. Talballa, R.A. Flinn; "Sulfur Removal From Liquid High Carbon by Various Slags and by Calcium Carbide," *AFS Transactions*, vol 80, p 291 (1972).

11. G. McGlothlin; Continuous Desulfurization of Iron in High Production Foundries by the Porous Plug Process, *AFS Transactions*, vol 85, p 5 (1970).

12. W.N. Bargeron, P.K. Trojan, R.A. Flinn; "Desulfurization of Molten Iron Droplets While Floating on Slag Baths," *AFS Transactions*, vol 78, p 117 (1970).

13. P.K. Trojan, M. Talballa, R.A. Flinn; "Desulfurization of Iron Droplets During Floating and Immersion in Slags and by Solid Desulfurizers," *AFS Transactions*, vol 79, p 415 (1971).

14. W.N. Bargeron, P.K. Trojan, R.A. Flinn; "The Kinetics of Sulfur Transport between Slag and Molten Iron Droplets," *AFS Transactions*, vol 77, p 303 (1969).

Treatment Alloys and Materials

Jack Klein

SKW Metals & Alloys, Inc.
Niagara Falls, New York

■ INTRODUCTION

The variety of commercially available treatment alloys and materials that the foundryman can use to produce ductile iron is extensive. Intertwined in the evolution of nodulizing products is the evolution of nodulizing processes, which continue to strive to maximize productivity, stabilize and maximize recovery of nodulizing elements, and upgrade the quality of the ductile iron produced. However, ductile iron processes have tended to develop around certain known or suspected characteristics of an individual element or a combination of elements beneficial to producing ductile iron. This chapter will examine those chemical elements that have been considered to be important to the nodulization of ductile iron. Some of the major commercial forms in which these elements are available to the ductile iron prducer will also be considered. Where appropriate, differentiation will be attempted among those factors involved in the choice of a nodulizing material, although final selection of a nodulizing product is usually best determined through a controlled foundry trial.

■ NODULIZING ELEMENTS

MAGNESIUM

In commercial practice, while other elements are capable of producing spheroidization, magnesium is most commonly used. However, use of magnesium is not without problems. It has limited solubility in iron, and a boiling point of 2025F (1107C), far below normal iron treatment temperatures of 2700–2750F (1482–1510C). After treatment, the magnesium content of the iron decreases or "fades," making a carefully timed holding/pouring regimen imperative.

Additionally, magnesium is a carbide stabilizer and is not effective in neutralizing elements deleterious to spheroidization of graphite (i.e., lead, bismuth, titanium, antimony, and arsenic, which promote vermicular and flake graphite).

The extreme tendency for magnesium to volatilize can be somewhat controlled through various processes: use of a deep pocket or a lid for the treatment ladle, a generous application of a ferrous covering material for the magnesium-containing alloy, dilution of the magnesium vapor with an inert gaseous phase via magnesium particle injection, cored wire injection to make use of the ferrostatic head pressure of the iron, or use of a pressure or plunging process.

However, it is also possible to control the rate of reaction of magnesium by chemical alloying of magnesium with other elements. As an example, calcium has the observed effect that, when alloyed with magnesium, the volatility of the liquid iron-magnesium reaction is reduced, reducing the magnesium fade rate. The role of calcium will be discussed later in more detail.

But while the various design processes just mentioned can reduce the negative effects of magnesium volatility, magnesium treatment alone can still result in difficulties stemming from carbide formation and a lower than desired nodule count—even when the "desired" magnesium content in the iron is attained. To minimize these negative effects, foundries have added cerium and other rare earth elements.

CERIUM AND OTHER RARE EARTH ELEMENTS

Rare earth elements play a rather complex role when used in ductile iron. They can nodulize on their own, neutralize deleterious elements, and provide nucleation sites for graphite. For commercial applications, two types of rare earth additions are used. Rare earth additions can either be of a mischmetal or low-cerium rare earth type or of a high-cerium rare earth type. Compositions for these two types of additions are shown in Table 8-1. Morrogh[1] and later Mickelson and Merrill[2] proved that ductile iron could be produced using cerium-rich or mischmetal type rare earths. However, production results indicated the time between treatment and pouring had to be very short or nodules deteriorated and carbides formed. Church[3] reported on the effects of combining magnesium and rare earths. Rice and Malizio[4] reported the beneficial effects of cerium as a nodulizing material, particularly in light-section castings. Lalich,[5] on the other hand, found that low-cerium rare earths were more beneficial than high-cerium rare earths in maximizing nodule count in ductile irons. While there is controversy over whether high-cerium or low-cerium rare earths are of greatest benefit to ductile iron, it is not questioned that rare earths are beneficial in the nodulization of cast iron, and that the magnesium content of the iron can be reduced when rare earths are present.

Rare earths also have the ability to neutralize the deleterious effects of subversive elements (such as bismuth, titanium, antimony, lead and arsenic) which form vermicular or flake graphite. Work by McCluhan[6] indicates all of the lighter rare-earth elements of cerium, lanthanum, and

TABLE 8-1. ANALYSES OF RARE EARTH SOURCES

	Approximate Analyses, %	
	Mischmetal, Bastnasite or Low-Cerium Rare Earths	Cerium Concentrate or High-Cerium Rare Earths
Cerium	50	90
Lanthanum	33	5
Neodymium	12	2
Praseodymium	4	1
Other Rare Earths	1	2

praseodymium are essentially equal in controlling subversive elements.

Not unlike magnesium, the cerium or total rare earth content in the iron must be controlled. It is typically accepted that, in lighter sections, rare earths should total 0.017% when magnesium is used in combination with either high-cerium or low-cerium rare-earth elements. Too low of an addition (<0.009%) will not provide a benefit, while too high an addition (>0.025%) will promote carbides.

It has been reported by Lalich[8] that cerium, neodymium and praseodymium promote carbides to a greater degree than lanthanum when added to iron in large percentages (>0.05%) without magnesium or calcium present. Due to its expense, however, lanthanum by itself or in combination with magnesium has never been widely used commercially.

Yttrium is yet another of the rare earths that have proved[9] to be singly effective as a nodulizing element, but its higher cost prevents widespread use.

CALCIUM

Calcium has also been reported to be an effective nodulizing agent, particularly by the Japanese.[10] Indeed, some Japanese and European foundries rely on greater concentrations of calcium in their nodulizers than U.S. foundries. It is claimed that high levels of calcium in the nodulizer assist in pouring carbide-free, thin-section castings. While it is accepted that irons treated with calcium-free nodulizers and postinoculants tend to exhibit poor microstructures, this has not led to widespread use in the U.S. due to fear of excessive slag generation.

Some Japanese foundries contend that calcium-treated irons are cleaner than magnesium-treated irons. However, it has been proven that, in automatic pouring furnaces, low calcium and low aluminum levels in the treatment alloy may be required to minimize slag buildup both in the furnace and around the pouring nozzle. Nonetheless, some U.S. foundries are employing higher calcium nodulizers in order to improve casting quality, and reduce the rate of magnesium fade.

TREATMENT ALLOYS (NODULIZERS)

While the discussion on the role of elements other than magnesium in producing ductile iron is interesting (if not confusing and controversial), magnesium is the primary element used in producing ductile iron. Essentially, and for purposes of our discussion, magnesium additions can be classified in one of two ways: *elemental* and *alloyed.*

ELEMENTAL MAGNESIUM ADDITIONS

Processes that utilize elemental magnesium have one common denominator: they rely on a physical method of introducing a relatively pure form of magnesium into the iron in such a way as to minimize the escape of magnesium from the treatment vessel. There are many variations to the methods employed.

One widely used technique is the converter method, which traps the metallic magnesium in a reaction chamber at the bottom of the treatment vessel. The form of magnesium used is typically commercially-pure magnesium ingots. Some processes, such as the Kubota process, utilize a plunging technique to hold the metallic magnesium deep below the iron surface. Usually, the treatment vessel is pressurized with air to minimize the escape of magnesium vapor from the iron, improving magnesium's efficiency.

In the MAP process all but one corner of a slab of metallic magnesium coated with a refractory material is plunged into the molten iron. The refractory material helps to control the release of the magnesium. Magnesium efficiencies in all these processes have been reported as high as 50–70%, depending on treatment vessel size and treatment temperature (larger treatment vessels and lower treatment temperatures improve magnesium efficiencies).

Treatment with metallic magnesium has been reported to have the following benefits:

- for those foundries that hold and pour treated ductile iron, the absence of calcium and aluminum during treatment minimizes slag buildup in the holding furnace and pouring nozzle;
- partial or complete desulfurization can be accomplished in the same treatment vessel used for magnesium treatment;
- no silicon is added during treatment.

The disadvantages frequently listed include:

- the need for larger treatment vessels to take full advantage of the process;
- the absence of elements such as cerium, calcium and aluminum in the iron can lead to a need for high amounts of late-stage additions of graphite nucleating (or subversive element neutralizing) alloys to produce the desired graphite structure in the iron;
- the capital investment costs.

Although powdered metallic magnesium is far less commonly used, it can be introduced in a variety of ways. These include pneumatic injection deep into a treatment ladle, or through a slide-gate, or ladle additions by means of a cored wire.[11,12]

Another approach to introducing elemental magnesium into iron is to dilute the magnesium with other materials. Examples of this would be Mag-Coke, which consists of pieces of coke, the pores of which are impregnated with the magnesium. This product contains 43.0–45.5% magnesium. However, these dilution methods are used in very few foundries.

ALLOYED MAGNESIUM ADDITIONS

The most widely used nodulizing materials are those that alloy magnesium with other elements—elements that have been determined to be beneficial to the formation of nodular graphite or that help to control reactivity and, hence, magnesium release. These materials can be classified as either *nickel-based* or *silicon-based* magnesium alloys.

Nickel-Based Magnesium Alloys

Table 8-2 represents the four types of nickel-based magnesium alloys available today.

Alloy 1 has been used for the commercial production of ductile iron since 1949. It can be used effectively in a wide variety of processes from a simple sandwich technique to plunging. Magnesium recoveries are typically in the range of 50–70% as a result of using this alloy. The nickel added by alloy 1 acts a graphitizer as well as a pearlite stabilizer and helps to minimize variations in mechanical properties between thin and thick sections of castings.

Alloy 2 was developed as a more economical alternative to alloy 1. While the silicon in the alloy reduces alloy costs, there is some sacrifice in magnesium recovery compared to alloy 1.

Alloys 3 and 4 offer significant changes from alloys 1 and 2. Both these alloys have a density greater than that of liquid iron and sink when added to liquid iron. Magnesium treatment with these alloys can be accomplished by dropping the alloys into a ladle of molten iron. The lower magnesium content of alloys 3 and 4, compared to 1 and 2, result in greater magnesium recovery efficiencies.

Silicon-Based Magnesium Alloys (MgFeSi)

Silicon-based alloys are the most commonly used treatment alloys in the U.S., treating in excess of 3,000,000 metric tons (MT) of ductile iron per annum. The most common silicon-based treatment alloy is magnesium ferrosilicon (MgFeSi). There are a variety of producers of these alloys in the U.S. and a seemingly endless variety of chemical compositions of MgFeSi alloys available to the ductile iron producer. It is recommended that the individual foundry discuss with the supplier, in detail, which composition would be best for its needs. Some of the general principles in the design and selection of MgFeSi alloys will be explained next. It is also advised that aluminum levels in MgFeSi nodulizers used in automatic holding/pouring operations be kept on the low side. Aluminum can also contribute to the production of slag not beneficial to the proper functioning of an automatic holding/pouring operation.

Table 8-3 represents a typical line of MgFeSi nodulizers currently available. The critical elements in the design of MgFeSi are magnesium, calcium, aluminum, cerium and/or total rare earths, and, of course, silicon. Each of these elements may need to be considered in the selection of a MgFeSi nodulizer based on the given application or process.

Magnesium content is probably the most straightforward of the various elements. It has been proven that, as the magnesium content of the alloy decreases, the recovery of magnesium increases.[13,14] While use of 9% Mg alloys were at one time very typical, their use has declined in favor of the improved recovery afforded by 6% Mg alloy.

Process constraints may also dictate the level of reaction volatility permitted. The various processes used include: flow-through, which typically uses a 3–4% Mg alloy; in-the-mold, which uses a 4.5–6.5% Mg alloy; or the teapot spout tundish process, for which a 5% Mg (or lower) content is sometimes recommended. Alloys containing 9%

Mg are still occasionally used in foundries, typically because of constraints on silicon analysis.

Foundries using magnesium-ferrosilicon should be aware that the analysis and reporting of magnesium content may vary between suppliers. This can lead to differences in performance between an alloy of the same reported chemical composition supplied by two different producers. In some cases, this difference can be 0.5% Mg or greater, which can lead to magnesium undertreatment or overtreatment if these differences are not recognized by the foundry.*

[*Editors' Note: The author is referring to the controversy surrounding the reporting of "total Mg" or "effective Mg" content.]

Calcium content in U.S.-produced MgFeSi alloys have almost always been at the 1% level. While 1% Ca is still the most typical level in MgFeSi, alloys are now produced with calcium levels from 0.3% (or lower) to 2.0% (or higher). Calcium's principal functions in MgFeSi are to act as a nucleant for graphite precipitation and to moderate the rate of the magnesium reaction.

Also, the presence of calcium in foundry practice has been shown to reduce the magnesium fade rate. While some foundries find this beneficial, others (particularly foundries trying to hold and pour treated ductile iron) find calcium to be a problem because the calcium oxide (CaO), which is formed in favor of magnesium oxide (MgO) (calcium is reducing to MgO), tends to produce slag build-up in holding furnaces and small pouring nozzles. As a result, low-calcium MgFeSi alloys were developed. However, some U.S. foundries use MgFeSi alloys with calcium contents in excess of 1%. In reality, this technology is nothing new, as some MgFeSi alloys used in Europe and Japan have contained on the order of 2% Ca for years. Some of the reported benefits of higher calcium alloys are:

- reduced carbides, particularly in thin sections;

TABLE 8-2. NICKEL BASE NODULIZERS

Designation	Analyses, %					
	Mg	Si	C	Fe	Ni	Normal Form
Alloy 1	13–16	—	2.0	—	bal	2-in. × 8M crushed
Alloy 2	13–16	26–33	—	5 max	bal	2-in. × 8M crushed
Alloy 3	4.2–4.8	—	2.0 max	—	bal	3 & 15 lb pigs
Alloy 4	4.0–4.5	—	2.5 max	32–36	bal	3 & 15 lb pigs

TABLE 8-3. MAGNESIUM-FERROSILICON NODULIZERS

	Typical Analysis, %*						
Designation	Si	Mg	Ce	TRE**	Ca	Al	Typical Sizing
5	43–48	5–6.5	—	—	0.8–1.3	1.20 max	1.25 x 0.25 in.
5 (0.3 Ce)	43–48	5–6.5	0.25–0.40	—	0.8–1.3	1.20 max	1.25 x 0.25 in.
5 (0.5 Ce)	43–48	5–6.5	0.50–0.75	—	0.8–1.3	1.20 max	1.25 x 0.25 in.
5 (1.0 Ce)	43–48	5–6.5	0.95–1.20	—	0.8–1.3	1.20 max	1.25 x 0.25 in.
5 (0.5 TRE)	43–48	5–6.5	0.30–0.45	0.50–0.75	0.8–1.3	1.20 max	1.25 x 0.25 in.
5 (1.0 TRE)	43–48	5–6.5	0.45–0.60	0.75–1.10	0.8–1.3	1.20 max	1.25 x 0.25 in.
5 (1.75 TRE)	43–48	5–6.5	0.85–1.0	1.5–2.0	0.8–1.3	1.20 max	1.25 x 0.25 in.
5 (high Ca)	43–48	5–6.5	0.3–0.6	0.6–1.0	1.5–2.25	1.20 max	1 in. x 8M
9	43–48	8.5–10.0	—	—	0.8–1.5	1.20 max	1.25 x 0.25 in.
9 (Ce)	43–48	8.5–10.0	0.5–0.75	—	0.8–1.5	1.20 max	1.25 x 0.25 in.
9 (TRE)	43–48	8.5–10.0	0.35–0.50	0.60–1.0	0.8–1.5	1.20 max	1.25 x 0.25 in.
3	43–48	3.0–4.0	—	—	0.8–1.3	1.20 max	1.25 x 0.25 in.
3 (TRE)	43–48	3.0–4.0	—	1.50–2.0	0.8–1.3	1.20 max	1.25 x 0.25 in.
3 (Ce)	43–48	3.0–4.0	1.50–2.5	—	0.8–1.3	1.20 max	1.25 x 0.25 in.
In-the-Mold	43–48	5.5–6.5	0.3–0.5	0.35–0.60	0.35–0.60	1.20 max	5 x 18M
Flow-Through	43–48	3.0–4.0	—	1.5–2.0	0.8–1.3	1.20 max	3/8 in. x 12M

*Analyses reflect a wide range of specifications and may vary among producers.

**Total rare earth content is a calculated figure, which may vary among producers.

- reduced flare and fume, due to the tendency of the increased calcium to reduce the volatility of the magnesium reaction;
- improved magnesium recoveries, due to suppression of the magnesium reaction;
- reduced magnesium fade rate.

When using higher calcium alloys in foundry practice, careful attention should be given to deslagging practices. Some foundries have found the use of fluxing compounds to be beneficial when high calcium levels are present in their treatment alloy.

Cerium and Oher Rare Earths

Much has been said in other sections of this handbook concerning the role of cerium and other rare earth elements. Magnesium-ferrosilicon alloys can be produced using either a high-cerium or low-cerium (mischmetal) addition. The question of "which type is better" is controversial, not only among MgFeSi producers, but among foundries themselves. The final selection will have to be determined by actual foundry experience.

The most widely used alloys, however, are those containing up to 1% Ce or 1% total rare earths (TRE). Typically, foundries making heavy-section castings will use MgFeSi nodulizers with a minimum of cerium or rare earth in order to minimize variation in nodule size, shape, count and distribution within the iron matrix. However, some foundries producing extremely thin-section castings prefer cerium and total rare earth additions on the order of 0.5% in the alloy, claiming that carbides can become a problem if rare earth or cerium levels are too high. Then again, there are some producers of thin-section ductile iron castings using 3–4% Mg alloys with 1.5–2.0% rare earth content. Each foundry should carefully review the various options available with its MgFeSi supplier(s), and make a final decision based on their own carefully controlled foundry tests.

A minimum *aluminum content* of 0.5% in postinoculants and nodulizers has been deemed desirable by some foundries to promote nucleation of graphite. However, there may be overriding concerns of having too much aluminum present in the iron, causing hydrogen pinholing problems or slag related problems in foundries that hold and pour ductile iron. Certainly, with the advent of late-stage inoculation processes, the 0.5–1.0% aluminum content typically found in magnesium-ferrosilicon may not be necessary to the production of quality ductile iron. It is best to consult with your MgFeSi supplier to discuss your options.

The *silicon content* in magnesium-ferrosilicon tends to be in the 41–48% Si range for a 6% Mg alloy. Process limitations in the production of magnesium-ferrosilicon prevent production of MgFeSi with less than 40% Si. Magnesium-ferrosilicon is available in the 40–43% Si range, if a foundry is interested in utilizing more returns, but these products are usually sold at a premium.

Sizes of Alloy Product

Before leaving this section on silicon-based (MgFeSi) alloys, there are a few guidelines on selection of sizes of alloy product for ladle or tundish processes. Typical sizes for MgFeSi products are listed in Table 8-3. The actual size a foundry may best utilize is usually determined by ladle size and calcium content of the alloy. As ladle treatment size increases, a size larger than the standards indicated may be desirable, and as the calcium content of the alloy increases, the size of alloy that is used typically decreases.

As with any general guidelines there are exceptions. Some foundries have found it to their economic benefit to utilize the undersized MgFeSi generated in the crushing of MgFeSi alloy over the typical sizes indicated in Table 8-3.

Cast MgSiFeCa Block

Another form of alloyed magnesium used in the U.S. is a cast MgSiFeCa block, which is plunged into the ladle to produce ductile iron. It is known that aluminum can cause pinholing in ductile iron. Excessive calcium and aluminum levels can also have a negative effect on controlling excessive slag buildup in the holding furnace or pouring nozzle. With an approximate composition of 31% Mg, 2% Ca, 53% Si, and 1% Al, its use is favored by shops preferring to minimize silicon, calcium and aluminum additions.

■ NODULIZER SELECTION

There is a wide variety of products available for use by the foundryman to make ductile iron and only a few of the major magnesium-bearing products used today have been mentioned. The intent was not to provide a buyer's guide, but to provide some of the background necessary to make an informed selection of a magnesium treatment product for a specific situation. Obviously, both the chemical limitations of the iron and the process limitations of the foundry are the two major factors that will need to be considered.

Chemical requirements will be influenced by end product specifications, base metal limitations and magnesium residual requirements. End product specifications may affect nodulizer selection in a number of ways. Certainly, cost can be a consideration based on the nickel requirement of the alloy. The silicon requirements of the casting may dictate the choice of either an elemental magnesium product if base silicon levels are high, or MgFeSi additions if the base metal silicon is low relative to casting specifications. The residual magnesium content required for producing heavy section castings is greater than that of ductile iron pipe or thin-section castings, which may influence nodulizer selection.

Process conditions can vary widely from foundry to foundry. Many factors within a foundry can affect the selection of nodulizers, including pollution control requirements, variation in base sulfur levels, capital costs, shop logistics and metal transfer limitations, types of slag generated (which may affect pouring capabilities), degree of automation desired, treatment temperatures, variation in casting section thickness, required utilization of returns, etc.

■ REFERENCES

1. H. Morrogh; "Production of Nodular Graphite Structures in Gray Iron," *AFS Transactions*, vol 56, p 72 (1948).

2. R.I. Mickelson, T.W. Merrill; "Experiments in Making Cerium Ductile Iron," *AFS Transactions*, vol 76, p 289 (1968).

3. N.L. Church; "Magnesium-Cerium Combination in Cast Iron," AFS Transactions, vol 81, p 301 (1973).

4. M.A. Rice, A.B. Malizio; "The Use of Multiple Nodulizing Elements in Making Ductile Iron Pipe," AFS Transactions, vol 82, p 15 (1974).

5. M.J. Lalich; "Effective Use of Rare Earths in Magnesium Treated Ductile Cast Irons," *AFS Transactions*, vol 82, p 441 (1974).

6. T.K. McCluhan; "Nodulizing Materials," Quality Ductile Iron Production—Today and Tomorrow, *Proceedings of Joint AFS-DIS Conference* (Oct 14-16, 1975).

7. "The Effect of Cerium on Carbides in Thin Section Ductile Iron," *AFS Transactions*, vol 75, p 372 (1967).

8. A.F. Spengler, ed; *The Ductile Iron Process*, p 226 (1978). [In discussion of reference 5.]

9. J.J. Kaater, Et al.; "Yttrium Nodular Iron," *Foundry*, vol 90, p 52 (Jan 1962).

10. T. Kusakawa; "Role of Calcium in the Production of CC Iron," 27th International Casting Congress, Vienna, Austria (1961).

11. M.C. Ashton, Et al.; "The Use of Magnesium Wire Injection for the Production of Nodular Iron," *Modern Casting*, p 55 (May 1975).

12. H. Bruckmann, J. Klein; "Research Testifies to Cored Wire's Applicability in Ductile Iron Production," *Modern Casting*, p 27 (Dec 1989).

13. P.K. Trojan, R.A. Flinn; "Fundamentals of Magnesium Addition to Ductile Iron," *S.A.E. Transactions*, vol 73 (1965).

14. R.A. Clark, T.K. McCluhan; "Influence of Magnesium Content on the Nodulizing Efficiency of Magnesium Ferrosilicon Alloys," *AFS Transactions*, vol 73 p 442 (1965).

9

Magnesium Treatment Methods

William A. Henning

Miller and Company
Chicago, Illinois

■ INTRODUCTION

The addition of spheroidizing agents to molten iron is probably the most important single step in the production of ductile iron. Magnesium is used exclusively as the spheroidizing element in commercial practices, and it is frequently added with cerium and other rare earth-containing materials. Magnesium and its alloys have low vaporization temperatures and, consequently, their addition to molten iron must be done with extreme care.

The object of the founding business is to produce the casting quality required at the lowest possible cost. Efficiency of the treatment alloy is, therefore, one of the most important considerations, and yet it is probably the most inconsistent factor in the entire process. The list of variables that influence magnesium recovery is long and includes: metal temperature, type and size of nodulizing material, quantity of metal being treated, rate of tapping, and treatment method. The recovery figures cited for each method are only meant to represent typical experiences or reported results. It is understood that higher or lower values are possible, depending on how well individual foundries are able to control their operation. This chapter will focus mainly on those treatment methods that have proven to be the most commercially successful, although lesser used methods will also be described.

It has been demonstrated that the treatment and pouring ladle refractory composition has an influence on final residual magnesium levels. Magnesium loss is aggravated when silica (SiO_2) ladle linings are used, due to the tendency to form magnesium silicate ($MgSiO_3$), which may end up in the castings as stringy-type dross inclusions. Magnesium loss is minimized by using alumina (Al_2O_3) or magnesite (MgO) refractories. Increased temperature losses associated with the use of these more conductive linings may be offset by the use of insulating backup materials.[1]

■ MAGNESIUM RECOVERY

Although numerous variations of formulas for calculating magnesium recovery have been published, their major basic difference concerns whether or not credit is given for the amount of sulfur removed from the base metal by magnesium in the treating alloy. According to atomic weights, desulfurization with magnesium requires 0.76 lb of magnesium to remove one pound of sulfur, forming magnesium sulfide. However, for all practical purposes, it is easier to assume that they will combine on an equal weight basis. A simplified formula for calculating magnesium recovery thus becomes:

magnesium recovered = $(Mg_R + S_B) / (Mg_A \times 100)$

where

Mg_R = magnesium residual of treated iron

S_B = base sulfur

Mg_A = magnesium added.

This formula was more popular in early ductile iron days, when base sulfur levels tended to be higher than they are today. Now that sulfur levels in the furnace or holder are very often in the 0.010% area, it has become well accepted to simplify the formula even more by deleting the sulfur credit:

$$\% \text{ Mg recovered} = Mg_R / (Mg_A \times 100)$$

When discussing magnesium recovery in the balance of this chapter, the latter formula is the one being used. However, if one desires to calculate the amount of magnesium to be added for a new ductile iron practice, it would be better to take the sulfur into account:

$$M_A = (Mg_R + S_B) / (\% \text{ Mg required} \times 100)$$

OPEN-LADLE, POUR-OVER PROCESS

This treatment technique is the simplest to use, which probably accounted for its early popularity. In this method, a measured amount of spheroidizing alloy is placed on the bottom of an open, heated ladle, after which a given quantity of base iron at the required treatment temperature is added as rapidly as possible (Fig. 9-1).

If the nodulizing alloy in use contains 5–6% magnesium, normal recovery of that magnesium will be 20–30%. Magnesium recovery is sensitive to many factors, primarily because magnesium has such a low vaporization temperature and low solubility in iron. For these reasons, a deep bath is desired to prolong the contact between the iron and the magnesium alloy and, thus, increasing the efficiency of the spheroidizing element. Therefore, the ladle depth should be two-and-one-half to three times the diameter, and the metal should be tapped into the ladle rapidly. The tapping rate should approach 15 seconds per ton. Ladle design should allow at least 12 inches of freeboard, as splashing is quite common. Denser spheroidizing alloys will produce better magnesium recoveries than the lighter alloys when using the pour-over method.

SANDWICH METHOD

The sandwich method is a direct descendant of the open ladle process, retaining its basic ladle configuration and

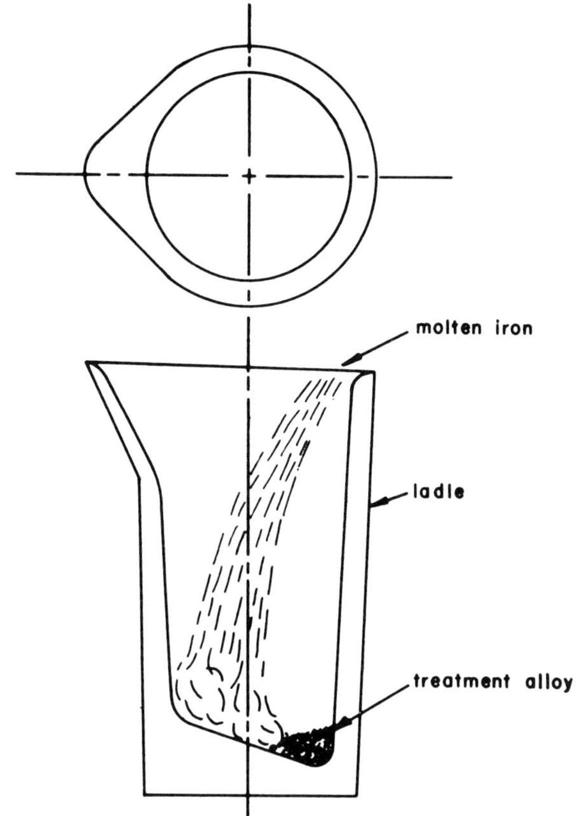

Fig. 9-1. Schematic of the open ladle transfer for treating ductile iron.

advantage of process simplicity. The principal difference lies in the ladle bottom design, wherein a pocket is provided to hold the treatment alloy and cover material (Fig. 9-2). The term "sandwich" derives from the layers of nodulizing alloy and cover material.

Magnesium recoveries can be improved to 40–50%, because the major part of the reaction is confined to the bottom of the ladle and takes place at a reduced rate—probably as a result of reduced base-metal temperature in the area of the alloy and cover material—as the cover material melts. In effect, overtreated, colder iron at the bottom of the treating ladle is diluted by the hotter iron above it, resulting in improved magnesium efficiency.

The most popular cover materials are clean steel punchings, or 50% ferrosilicon if the process is tolerant to this extra silicon addition. Steel punchings should be no thinner than one-sixteenth of an inch, and should be flowable in nature so as to form a relatively dense cover. If 50% ferrosilicon is used, it should be sized approximately 1-in. × down. Many foundries have reported an optimum cover addition of 2%. If steel is being used, an added temperature loss of about 50F (28C), and loss of about 0.07% carbon should be expected. Other cover materials used include ductile iron splash, resin bonded sand, calcium carbide, or silica sand, but none of these have ever gained wide acceptance.

Just as with the pour-over process, rapid tap times approaching 15 seconds per ton should be used, and weighing of the alloy and the molten metal is highly recommended. The sandwich process is easily adaptable to the full range of ladle sizes.[2]

■ TUNDISH COVER LADLE

In 1978, it was estimated that 75–80% of the ductile iron produced in the U.S. was nodulized using either the pour-over or sandwich method. There were no tundish ladles in commercial use. According to an AFS survey completed in 1988,[3] the tundish ladle is now used in 40% of the U.S. foundries producing ductile iron. This dramatic growth was a result of magnesium recoveries as high as 60–70%, and an accompanying reduction in flare and smoke generated during treatment.

The tundish cover (Fig. 9-3) operates on the simple premise that, by limiting the amount of available oxygen during the reaction of molten iron with the magnesium alloy, two important and related consequences follow:

1. Less available oxygen results in a significant reduction in the amount of magnesium oxide fume generated. That which does form is restricted in its exit and tends to deposit on surfaces inside the covered ladle. Consequently, very little fume escapes into the foundry atmosphere.
2. Since less magnesium is lost by oxidation, tundish cover treatments are more effective and economical when compared to the previously discussed ladle treatment techniques.[4]

Early designs featured a removable cover, set on the ladle after placement of the alloy and cover metal, and removed after treating to permit pouring of the nodulized metal. In spite of the inherent economic and environmental advantages, foundries were slow to adopt the tundish ladle due to the difficulty of moving and removing the lid with every nodulizing treatment. This major drawback was overcome by semi-permanently fixing the lid to the ladle top by means of simple wedge-clamps or bolts, and a mortar joint. Treated iron is simply poured from the ladle through the filling orifice. With this type of ladle design, there is a tendency toward slag build-up in the area of the filling/pouring hole. Alloy is placed through a separate opening in the lid, which is subsequently sealed with a steel plug or cap prior to treating (Fig. 9-4). Further design

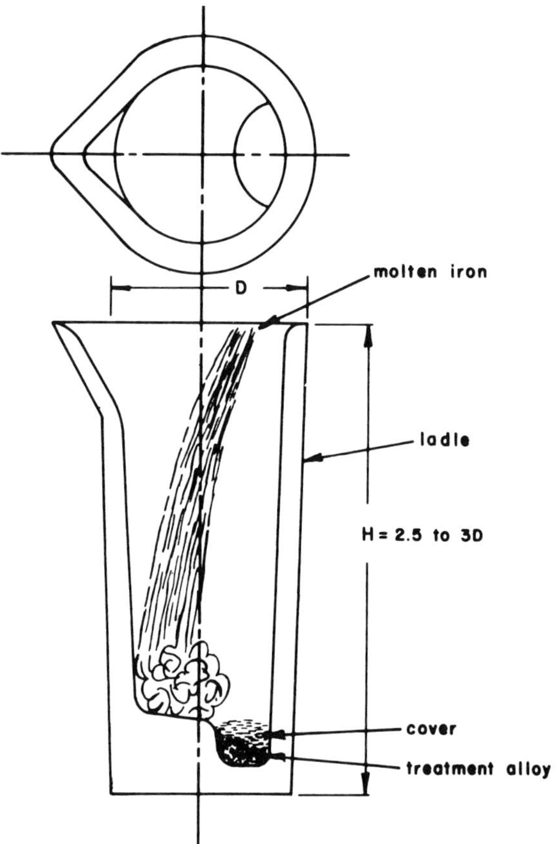

Fig. 9-2. Schematic of the sandwich method for treating ductile iron.

improvement involved relocating the alloy chute to a position 180° from the filling orifice, and lining it with refractory to serve also as the pouring spout. This change allows treatment slag to exit the ladle more readily through this larger opening, rather than be held back to build up in the ladle interior around the relatively small filling orifice.

Most covered ladle designs incorporate the use of a divider wall in the bottom, but a step or pocket ladle has also been used very successfully. The use of some type of sandwich cover material over the alloy is still advised, though generally in smaller quantities than used in the sandwich ladle. Because the bolted-on ladle cover results in an airtight seal, some means of venting is required to relieve pressures generated during the treatment and to prevent blowback from the tundish orifice. There are several methods available to accomplish this, but a one-half inch hole in the alloy chute cap is normally adequate. A metal weigh scale is absolutely necessary because the level of iron in the ladle is not visible with the cover in place.

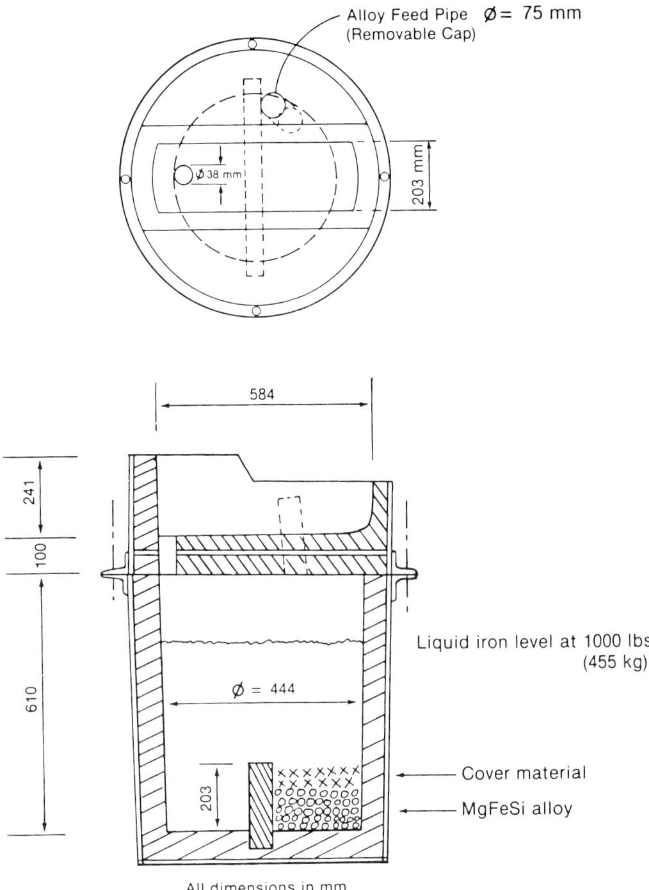

Fig. 9-3. Plan and schematic cross section of tundish cover ladle.

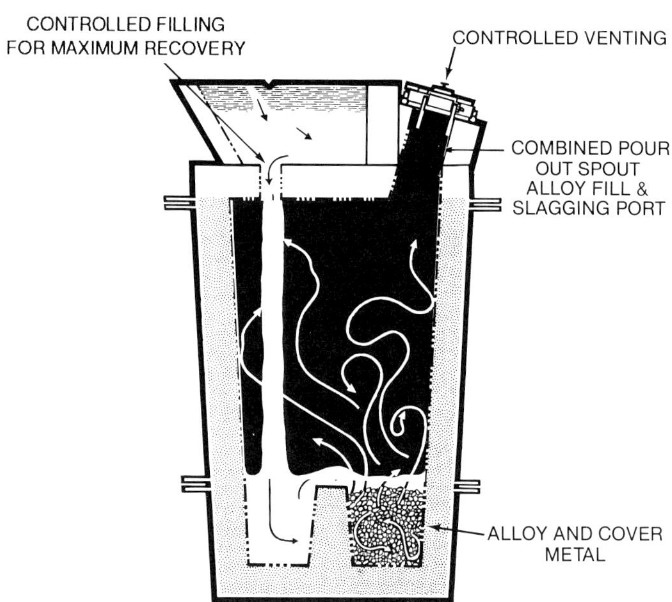

Fig. 9-4. Improved tundish ladle design.

A modification of the tundish cover ladle can be seen in the teapot type design (Fig. 9-5). The teapot ladle combines several of the features previously described except that a tundish is no longer used; metal enters the treatment ladle through the enlarged opening of a teapot spout. After placing the alloy, the cover cap—normally employing a fiber seal—is closed and clamped shut. With alloys containing no more than about 5% magnesium, it is possible to operate this ladle without venting. Magnesium efficiencies are comparable to the standard design. Due to the relatively large area of refractory in the teapot spout, slag buildup is a potential problem if base sulfur levels exceed 0.02%.

Tundish cover ladle treatment offers improved and more consistent magnesium recovery, virtual elimination of flare, about 90% fume reduction, reduced metal splashing, and reduced carbon and temperature losses. The principal disadvantage is cover construction and maintenance.

POROUS PLUG PROCESS

This technique found early popularity because desulfurization and nodulization could be performed in the same ladle in a single cycle. Magnesium recoveries averaged 30–35%. Conversions to electric melting and development of the tundish ladle practice have caused porous plug nodulizing to lose favor, and it is best used where it is still convenient to combine the desulfurization and nodulizing steps.

In this process, intense stirring or rolling action of the bath is created by the introduction of multitudes of tiny inert gas bubbles through a porous plug in the bottom of the ladle (Fig. 9-6). This causes the magnesium alloy to be pulled beneath the top surface of the molten metal in the ladle. Bubbling times for a 5% magnesium-ferrosilicon alloy range between 15 and 30 seconds, but excess bubbling time reduces magnesium recovery due to oxidation. Postinoculant can also be added during the last few seconds of the bubbling cycle. Should some degree of carburizing be needed, maximum carbon recoveries from graphitic materials are obtained after about one minute of bubbling. About 10°F (6°C) additional temperature loss will accompany each 0.1% carbon addition.

Temperature loss associated with the gas bubbling is a disadvantage of the process. To ensure turbulence over the entire surface of the metal, multiple porous plugs will be required in ladles of more than about 6000 pound capacity.

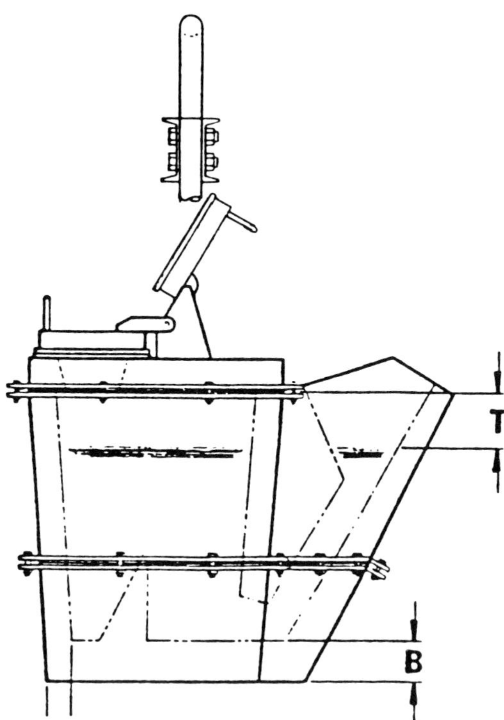

Fig. 9-5. Modified pressure-sealed ductile iron treating ladle.

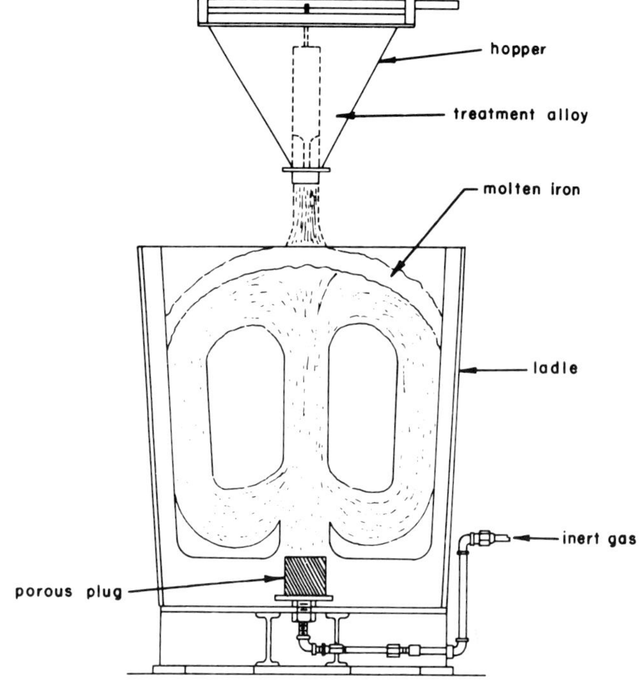

Fig. 9-6. Schematic of porous-plug agitation method for treating ductile iron.

Nitrogen is the most widely used gas, and will require a flow of 5–7 cubic feet per minute, per plug. Because of the degree of temperature loss, the process is best suited for ladles in excess of 2000 pounds.

IN-THE-MOLD TREATMENT

As high-production foundries moved to eliminate manual pouring in favor of automatic pouring, the introduction of in-the-mold treatment (Fig. 9-7) permitted holding of untreated, low-sulfur base iron in vessels used in conjunction with the automatic process, thus eliminating buildup associated with moving treated ductile iron through receivers, inductors, or nozzles.

Other advantages include:

- elimination of smoke and treatment effluent, as these reaction products are absorbed by the mold;
- magnesium recoveries of 70–80% (although 100% recovery has been reported);
- elimination of postinoculation;
- attainment of base silicon levels as high as 2.0% as a result of the lower alloy inputs;

- reduction in metal temperature losses compared to other ductile iron processes involving several ladle transfers.

The actual process involves incorporating a special reaction chamber within the runner and gating system of the mold. Nodulizing alloy, usually about one percent of the mold weight, is placed within the chamber, and the magnesium treatment occurs during the pouring process.

Gating systems normally used in the process consist of a single inlet and outlet, to and from the reaction chamber. The system is based on the following relationships:

A = Choke area or casting ingate area in sq. in.

Area B (Runner) = A + 0.10A

Area C (Chamber Outlet) = A + 0.12A

Area D (Chamber Inlet) = A + 0.30

Area E (Downsprue) = A + 0.30A Min.

The design of the reaction chamber is based on a pouring rate of pounds per second and a surface area of the chamber measured in square inches. Alloy sizing and pouring temperatures must remain constant in this process. The steps involved in reaction chamber design are as follows:

1. Determine the total poured weight of the mold. This includes the casting, gating and risers.

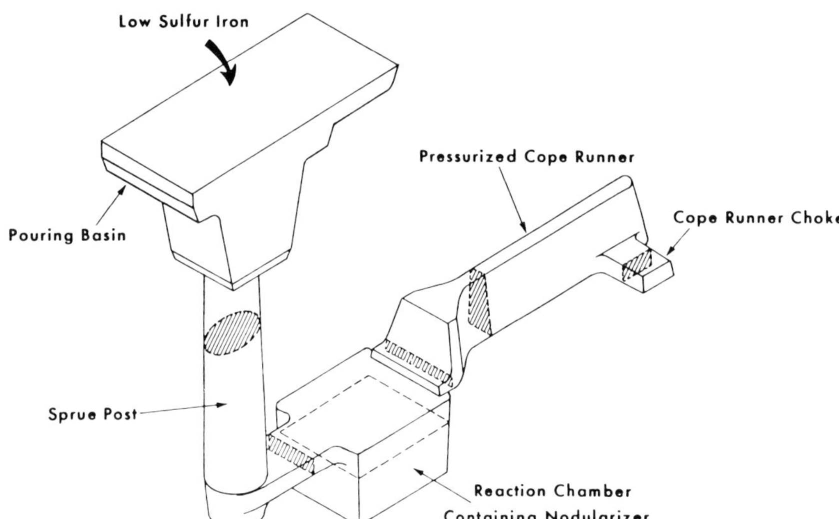

Fig. 9-7. Typical features of mold nodulizing gating arrangement.

Casting yield for the process is in the range of 40–60% for castings up to 100 pounds.

2. Determine the pouring time in seconds from the Standard AFS "Pouring Rate Table."
3. Calculate pouring rate in pounds per second.
4. Calculate the reaction chamber area in square inches using the solution factor equation.

solution factor = pouring rate (lb/sec) ÷ chamber area

The solution factor is a constant and usually equals 0.7–0.9 (for 0.04% final magnesium, use 0.8 as the constant).

reaction chamber area (sq. in.) = pouring rate (lb/sec) ÷ solution factor

5. Calculate the amount of magnesium alloy required (one percent by weight will result in a 0.04% final magnesium).

magnesium alloy weight = poured weight of casting × 0.01.

6. Calculate volume of magnesium alloy. (Density of 5 M x 18 M (4 mm x 1 mm) magnesium-ferrosilicon alloy is 1.117 ounces per cubic inch, or 0.073 pounds per cubic inch).

alloy volume = alloy weight (oz) ÷ 1.17 *or* alloy weight (lb) ÷ 0.073

7. Calculate the alloy depth using (6) and (4).

alloy depth = alloy volume ÷ reaction chamber area

8. Calculate reaction chamber depth.

chamber depth = alloy depth + reaction space (0.6 in. min.) + inlet depth (varies)

It is absolutely essential that base sulfur content be at 0.01% or below to minimize the chance of inclusion defects. Pouring times must be carefully controlled to ensure that alloy is being dissolved throughout the entire filling of the mold. Unacceptable graphite shape at gate areas can result from slow filling of the mold.[5]

The disadvantage of lower casting yield is apparent because the reaction chamber(s) is a part of each runner system. It is believed that there are practical casting sizes, geometry, and casting quantity restrictions for this process. The increased possibility of slag and other reaction products entering the casting is a further disadvantage. There probably is a greater need for testing the nodularity of every casting produced, due to the fact that each mold is a separate treatment.

EXTERNAL STREAM TREATMENT

External stream treatment (Fig. 9-8) employs the same treating principle as in mold nodulizing, only the reaction chamber is external to the mold. The dimensions of the chamber and the ratio of inlet hole to outlet hole diameters are critical to achieve successful, consistent results. Because the reaction chamber is of predetermined dimensions, wide variations in sulfur levels cannot be tolerated.

Magnesium recoveries of 60–70% are reported. Initially, alloy with only about 3.5% magnesium was used to avoid pre-ignition in the chamber. This meant that the normal lower magnesium alloy with somewhat restrictive sizing may negatively impact the overall economics, but external stream treatment has performed very well, particularly where there are plant layout problems associated with incorporation of other treatment techniques.

■ PLUNGING METHOD

Plunging was the first important improvement in ductile iron processing techniques after development of the pour-over process and has been in commercial use since the 1950s. While it can be used with all forms of nodulizing materials, its popularity has generally been based on using more concentrated magnesium sources.

With the plunging method, a bell-shaped refractory device containing the spheroidizing material is submerged deeply into the ladle of molten iron (Fig. 9-9). This plunging and holding of the nodulizing material below the melt surface prevents the material from floating and, hence, usually results in an overall increase in recovery of the magnesium. The spheroidizing material is usually held in the plunger by means of a sheet metal can or other suitable container.

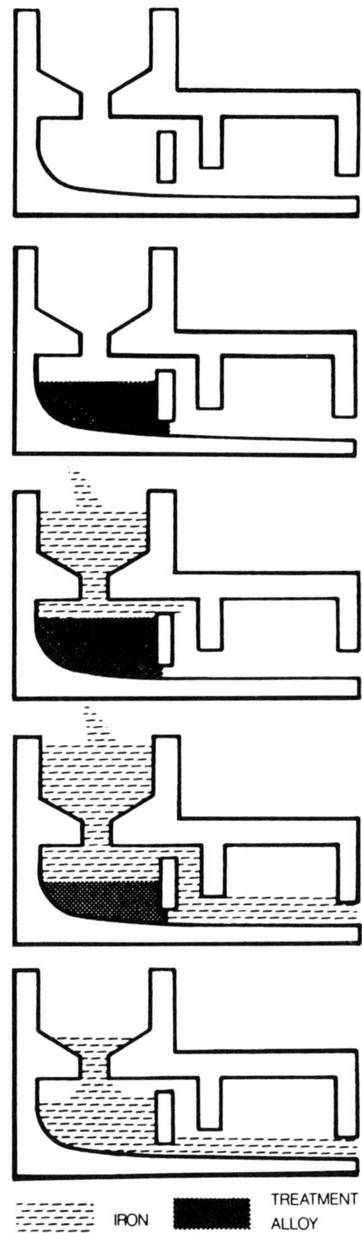

Fig. 9-8. Schematic of flow-through nodulizing process.

alloy addition was about 2%, depending upon treatment weight, temperature, and sulfur content. More recently, 5–6% magnesium alloys have been tried, with higher calcium levels to control the reactivity. Addition rates are reduced accordingly. Alloy sizing is normally 3/8-in. x 16M, to allow easy placement in the reaction chamber.

Chamber design innovations have made the process more suitable for use in production foundries. The need for

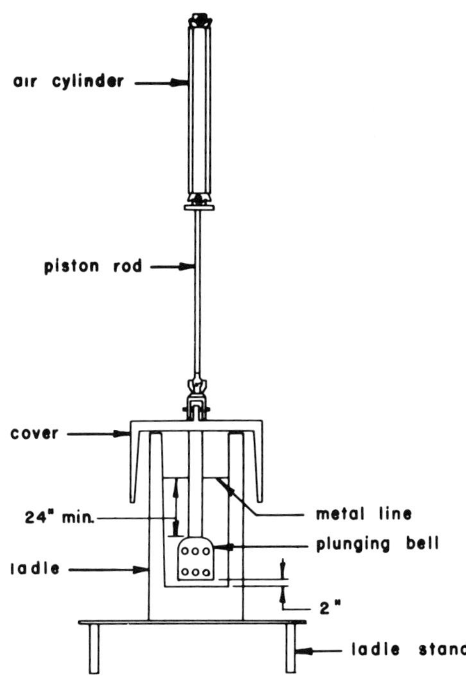

Fig. 9-9. Schematic of plunging method for treating ductile iron.

161

Recovery of magnesium from plunging is high and can easily exceed 50% when using 5–6% magnesium-ferrosilicon alloys. Because of this high recovery, the plunging method has the advantage of requiring less spheroidizing material than the open ladle or sandwich methods. This results in less smoke and fume, creates less slag with cleaner metal, and reduces alloy cost. The added cost of the plunger bell assembly and labor, however, may offset the reduced alloy cost. Another advantage is that this method does not necessarily require a completely empty ladle for each treatment as do the open ladle and sandwich processes.

A tall, slender ladle is required to allow for plunger displacement and the boiling action that takes place in the ladle. A metal column height-to-diameter ratio of two is recommended. Particularly when plunging 30% or higher magnesium alloys, a 12- to 20-inch deep skirt should be attached to the cover to better contain any ejected metal or sparks.

Plunging bells are subjected to extremely severe operating conditions. Heat shock, impact and erosion (the higher the magnesium content, the shorter will be the life of the bell) are all involved. The bell is an expendable item and must be replaced when it fails. Bell life expectancy is dependent on the care with which it is used, as well as the type of material being plunged. Bells are available from commercial sources in a variety of materials: graphite, high-alumina clay-graphite refractories, or in some cases, high-alumina refractory only. Strength of the refractory materials may be improved by the inclusion of stainless steel needles.[6]

The primary disadvantage is a greater temperature loss, due to the mass of the cold plunger assembly. The process does not lend itself to variable treatment quantities or batch sizes. In addition, the cycle time is usually slightly longer than the other methods mentioned.

■ CORED WIRE TREATMENT

The addition of magnesium to liquid iron by means of a cored wire was first mentioned in the literature in 1975. Later experimentation gave reproducible results, but gained virtually no commercial acceptance, due to a variety of problems: equipment used to feed the wire was not reliable, and arbors used to store wire created problems during high-speed feeding. The cost of encapsulating pure magnesium wire in a steel sheath was high (Fig. 9-10).

The success in the 1980s of cored wire for steel ladle metallurgy applications rekindled interest in ductile iron wire nodulizing. Wire feeding machines were improved, as was the design of static payoff devices. By using powders in the wire, other materials, such as calcium carbide, ferro-silicon and/or rare earths, can be blended with the magnesium. As the magnesium concentration in the wire is reduced, recovery and consistency of the element improves. Some capital investment is necessary, but not to the extent as in most pure magnesium applications.[7]

Initial acceptance of this technique has been more rapid in Europe. Advantages are said to include high metal cleanliness, applicability of inert gas shielding, minimal smoke evolution, reduced temperature losses, simultaneous or sequential addition of postinoculating agent, and the potential for automation. The disadvantage is the cost of the cored wire.

■ MAP PROCESS

The MAP process is a variation of plunging that involves the use of refractory-coated magnesium billets or ingot. Hardened concrete nails are driven into the corner of preweighed blocks of magnesium. A double loop of 14-gauge wire is then attached. The magnesium blocks are coated with three individual air-dried layers of a refractory slurry, dip coated in such a way as to have about one-half to three-quarters of an inch of the corner nearest to the nail exposed, without any refractory coating. The second refractory coating is dipped three-quarters of an inch below the first coat, and the third coat is three-quarters of an inch below the second coat. When dipped, the ingots must be clean and free from grease or oil. The refractory coating must not peel or crack on flat surfaces or corners, or the reaction rate between the molten iron and magnesium will accelerate, and an explosion may occur.

One commercial foundry employing the MAP process casts its own bells and sleeves with 3000F (1650C) high-alumina castable containing 3% stainless steel needles. The bells are cast in a steel mandrel with a steel core and wood plugs to form the vent holes. Stems are also cast in steel molds. The bell and stem are assembled using a threaded rod, and the nut inside the bell is later covered with refractory. Prior to use in production, the assembly is lowered into a treatment ladle and gas-fired for eight hours. During use, it is mandatory that the vent holes in the bell be kept open and free of slag buildup to ensure the magnesium vapor will be homogeneous throughout the treatment, contributing to high and consistent magnesium recovery. It is not unusual for a bell and stem assembly to last 140 heats.[8]

The MAP process has found its best application in the production of ductile iron pressure pipe. Recovery of the pure magnesium can be as high as 50% in a pipe foundry due to the five to seven ton batch size, low-sulfur base iron, and plunging temperature of only about 2650F (1450C). Magnesium recoveries will drop dramatically as batch size decreases, sulfur increases, or treating temperature increases.

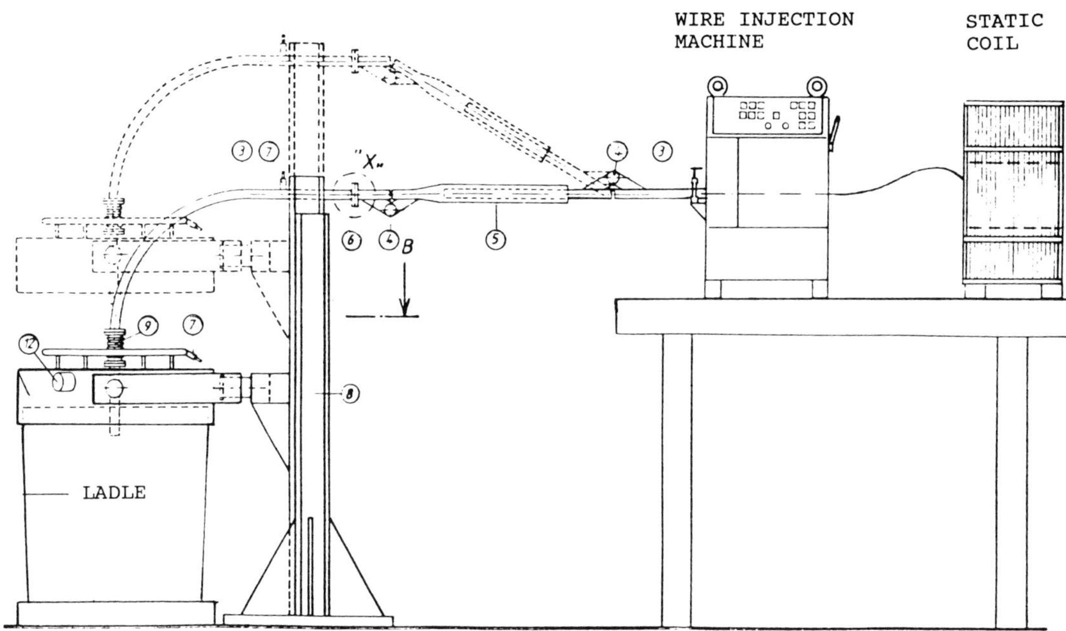

Fig. 9-10. Schematic of an automated wire injection station.

TILTING REACTOR METHOD

This method was developed in Europe, as has been the case with several other techniques employing complex equipment and metallic magnesium. The process has also remained more popular in Europe than in the U.S. European foundries have not developed desulfurization techniques to the same level of efficiency as those in the U.S. Also, European magnesium-fer-rosilicon alloys have historically been higher in cost and lower in quality than U.S. alloys, both factors favoring the use of pure magnesium processes in Europe.

The reactor is a refractory-lined, cylindrical vessel ca-pable of almost 180° rotation on a stationary axis. A perforated reaction-chamber plate is located as shown in Figure 9-11, serving to form a chamber for the pure mag-nesium pieces.

Initially, the converter is filled with base iron while in a horizontal position, so that the reaction chamber is located above the molten iron to prevent contact between the pure magnesium and the base iron. After a predetermined amount of magnesium is charged into the reaction cham-ber, the breech plug is clamped in place, and the molten metal charging cover is closed hydraulically. Treatment oc-curs as the reactor is rotated to the vertical position. The reaction rate is controlled by the size of the holes in the re-action chamber plate, with total treating time varying from one to two minutes, depending upon batch size. After the magnesium reaction is complete, the charge door is opened, and the converter is rotated to the pouring posi-tion. Treated iron is then poured from the vessel into transfer or pouring ladles. Preparing the reactor for the next treatment involves removal of the breech plug, and cleaning slag from the reaction chamber and reaction plate holes.

Magnesium recoveries of 50–60% have been reported. Normal cycle time is seven to twelve minutes, dependent on the foundry layout for metal handling. Because of the total absence of inoculating-type elements introduced dur-ing nodulizing, the iron will require heavier postinoculation than irons treated with magnesium-ferrosilicon. Careful maintenance of the charge door, reaction plate and breech plug areas is imperative.

Advantages include the ability to obtain good recovery with the use of relatively low-cost metallic magnesium, the ability to simultaneously add carbon, and the capability of treating high-sulfur base irons without a separate desulfu-rization step.

PRESSURE LADLE

The pressure ladle technique involves overtreating approxi-mately one-fourth of the total amount of iron desired with magnesium, and subsequently combining this overtreated

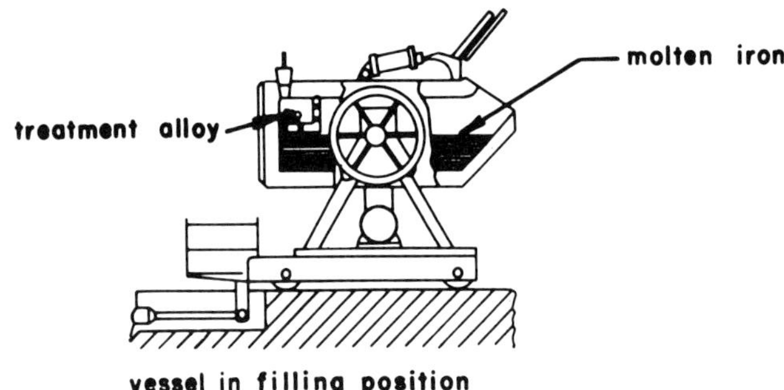

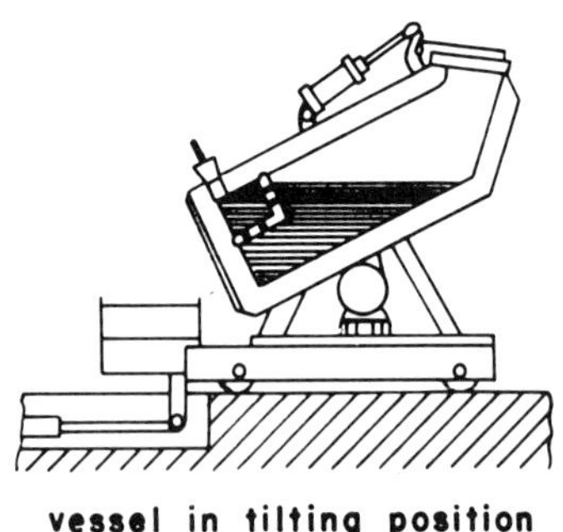

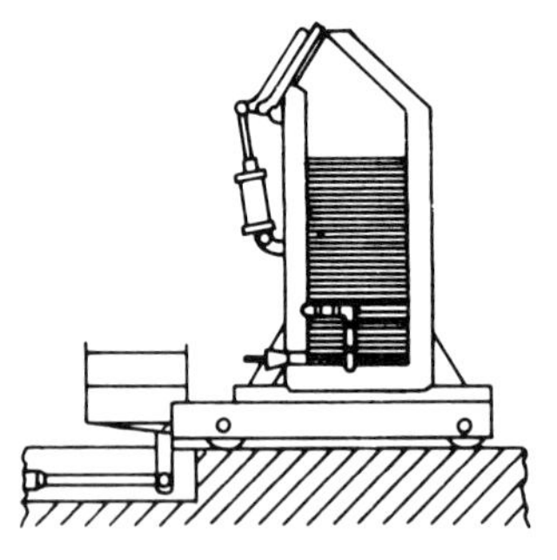

Fig. 9-11. Schematic of the tiltable ladle for treating ductile iron.

metal with the appropriate quantity of untreated iron to attain the desired magnesium level. The primary purpose of this overtreatment/dilution is to reduce temperature losses associated with metal handling. Engineering and operational problems are also minimized by reducing the size of the pressure treatment vessel and its associated equipment to a minimum.

Blocks of pure magnesium are used as the magnesium source with this method. These blocks of magnesium are bolted to steel rods, which can be suspended above the iron in a fixed position in the chamber pressure head. The ladle itself is the pressure vessel (Fig. 9-12).

To begin the treatment, base iron is poured into the pressure treatment ladle. This special ladle is then moved under and held against the fixed pressure chamber head. A pressure-tight seal between the fixed head and the ladle is accomplished with a rubber gasket. The pressure of the atmosphere above the iron in the pressure vessel is increased to 15–40 psig, depending upon iron temperature. The block of pure magnesium is then immersed beneath the surface of the iron and held near the bottom of the ladle until it is totally consumed. Pressure inside the ladle will build to 60–110 psig immediately. In approximately one minute the pressure inside the chamber will return to the prepressurized level. At this point the magnesium treatment is complete, and the remaining pressure is released. The overtreated and untreated iron are then combined to produce the final ductile iron quantity and quality desired.

The primary advantage of this system is the use of relatively inexpensive pure magnesium. Magnesium recoveries range from 60–90%, depending upon iron temperature and the desired residual magnesium level. There is no silicon addition associated with the actual magnesium treatment.

This method can tolerate some sulfur in the base iron, if the level is low and consistent. When overtreating, the sulfur content must be considered, because the untreated iron must be desulfurized with the magnesium in the overtreated iron. The solubility of magnesium in iron is quite small.

Disadvantages of the method include the cost of the rather complex equipment, as well as maintenance and operation. The iron temperature loss during the treatment process must also be considered. Heavy postinoculation will be necessary to produce iron with good metallurgical properties.

The pressure ladle method is not widely used, even though it will produce good quality ductile iron. This method is more suited to casting operations that require very large quantities of ductile iron that must be poured in a short period of time.[9]

MISCELLANEOUS METHODS

Since the early days of ductile iron production, many other techniques have been tried in an effort to promote more efficient and consistent use of magnesium during treatment. Although there may have been in-depth interest in some of these processes at various times, many have faded from use in commercial practice and do not merit lengthy discussion. A brief listing of processes in this category would include:

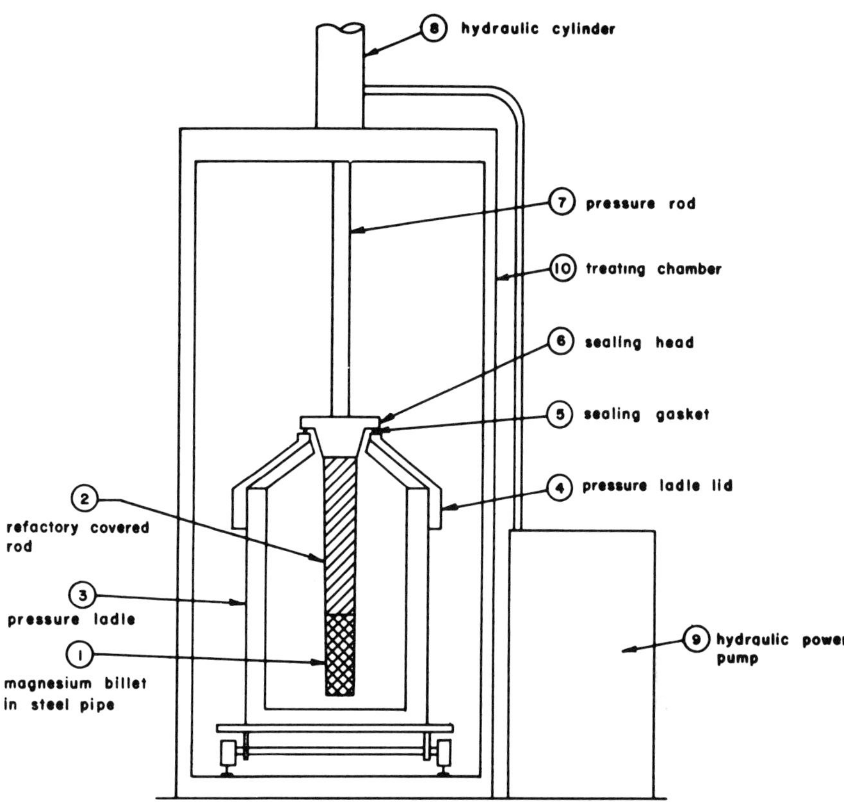

Fig. 9-12. Schematic of sealed ladle for treating ductile iron.

- injection of powdered or granular magnesium (discontinued due to high temperature losses and erratic results);
- shelf and/or detachable bottom ladles designed for use with magnesium-coke (discontinued as magnesium-coke was no longer produced);
- the "T-nock" process, which gained only very limited acceptance.

SELECTION OF A TREATMENT METHOD

The choice of the treatment method for an individual foundry involves a great many factors. Priorities should be given to the most important factors after honest and unbiased consideration of the circumstances of each individual situation. Criteria involved in selecting a treatment method include the following.

Base iron. Type of melting equipment and raw material availability may result in base iron sulfur or silicon levels, for example, that would preclude certain treatment methods and materials.

Types of castings being poured. Thin-sectioned castings requiring higher pouring temperatures could dictate the use of the most rapid treatment method available.

Cycle time. Batch size limitations must be considered in view of cycle time or holding time after treatment.

Batch size. Very small treatments preclude a method such as plunging, for example. Individual plant capabilities may dictate the batch size within rather narrow limits.

Simplicity. Lack of experienced personnel and sophisticated support equipment may influence the selection process.

Mechanization possibilities. Highly repetitive operations involving manual labor may be easily mechanized during certain treatment methods.

Reproducibility. Some treatment methods have innate variability that would prevent them from meeting certain quality control requirements.

Physical plant limitations. Layout and available headroom may eliminate certain techniques from consideration.

Need for inoculation. Inoculation after spheroidizing may dictate the need for reladling and, consequently, have some bearing on the selection process.

Environmental considerations. Pyrotechnics and fume evolution during treatment can vary widely with some of the methods described.

Cost. Although important to all foundrymen, cost becomes a particularly important criteria in high-production plants. Some treatment methods allow the use of lower-cost nodulizing materials.

◼ REFERENCES

1. American Foundrymen's Society; *Ductile Iron Molten Metal Processing,* 2nd ed., pp 29–45 (1986).

2. W.A. Henning; "Sandwich Technique for Producing Ductile Iron," *Technical Bulletin #54,* Ductile Iron Society, pp 34–39 (Dec 1978).

3. D.W. Brewster et al.; "Survey on Ductile Iron Practice," *AFS Transactions,* vol 97, pp 79–84 (1989).

4. R.D. Forrest; *The Progress and Industrial Acceptance of the Covered Ladle for Ductile Iron Production* (N.p., n.d.).

5. A.F. Spangler, ed.; *The Ductile Iron Process,* Miller and Company, pp 237–286 (1986).

6. D. Matter, "Nodularizing Methods," *Quality Ductile Iron Production—Today and Tomorrow,* pp 126–142 (1975).

7. H. Mair et al.; "Advancement of Cored Wire Applications," originally presented at 1989 Infacon, New Orleans, LA.

8. J.R. Vest; private communication to author (June 1989).

9. C.R. Kern; private communication to author (June 1989).

10

Postinoculation

Harvey E. Henderson

Foundry Consultant
Concord, Virginia

■■■ WHY INOCULATE DUCTILE IRON?

Even before the invention of ductile iron in 1948, inoculation was a well established practice in the manufacture of gray iron castings. Ladle additions and inoculation mechanisms have, perhaps, been the most carefully guarded and misunderstood secrets in the whole field of iron casting. Many important improvements in engineering properties are obtained by adding a small amount of inoculating alloy to the ladle, rather than adding a similar quantity of regular alloy directly to the furnace.

In gray and ductile cast iron, there is a delicate balance between the reactions:

(1)   Liquid → iron carbide + austenite

(2)   Liquid → graphite + austenite

either of which may occur on freezing. If reaction (1) occurs, the resulting product is hard and unmachinable, whereas the structure resulting from reaction (2) possesses excellent machinability.

The carbide-forming tendency can be greatly reduced by an addition of 0.10–1.00% of an inoculant which may be 45–85% ferrosilicon* containing 0.5–1.25% aluminum and 0.50–1.0% calcium. The use of 75% ferrosilicon is predominant in ductile iron inoculation.

> ***[Editors' Note: 85% ferrosilicon has not been available in the U.S. for about 25 years, but there is reportedly some currently being produced in the People's Republic of China.]**

Inoculation is a necessary step in the production of ductile iron. Uninoculated, pure magnesium-treated iron is almost always completely carbidic, and precipitated graphite is compacted and not always spheroidal. Inoculation, therefore, should be carried out with care and precision. Magnesium-ferrosilicon alloys contain nucleating elements, such as calcium and cerium; irons treated with them will tend to be considerably less carbidic and, thus, easier to inoculate.

Suitable inoculation provides sufficient nucleation centers for solidification to take place in the "graphite-plus-austenite system." In general, the greater the degree of nucleation in ductile cast iron, the greater the number, and the smaller and more uniform in size and shape will be the graphite spheroids. There will also be less tendency for carbides to precipitate during solidification.

Figures 10-1, 10-2 and 10-3 show the effect of inoculation on the structure of ductile cast iron. The iron shown in Figure 10-1 was not inoculated and shows mostly primary carbides in the microstructure. In Figure 10-2, the iron was under-inoculated and still shows considerable primary carbide. Properly inoculated ductile iron contains no primary iron carbide in the microstructure, as shown in Figure 10-3.

The effect of inoculation is at a maximum immediately after the addition of the inoculant and is not permanent. The loss of inoculating effect, which occurs over time without noticeable change in the composition of the metal

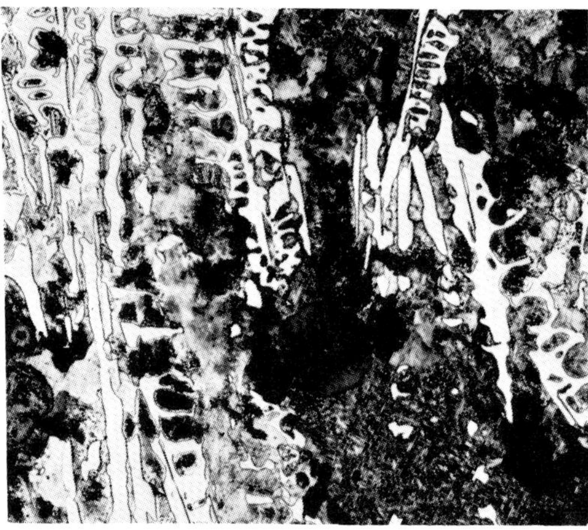

Fig. 10-1. Uninoculated ductile iron; $\times 250$, etched.

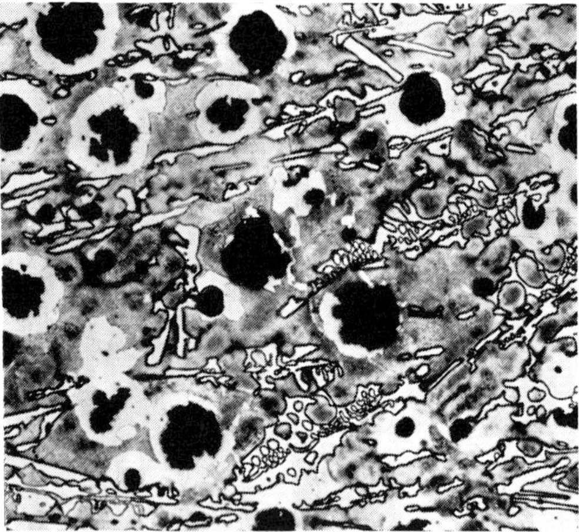

Fig. 10-2. Underinoculated ductile iron; $\times 100$, etched.

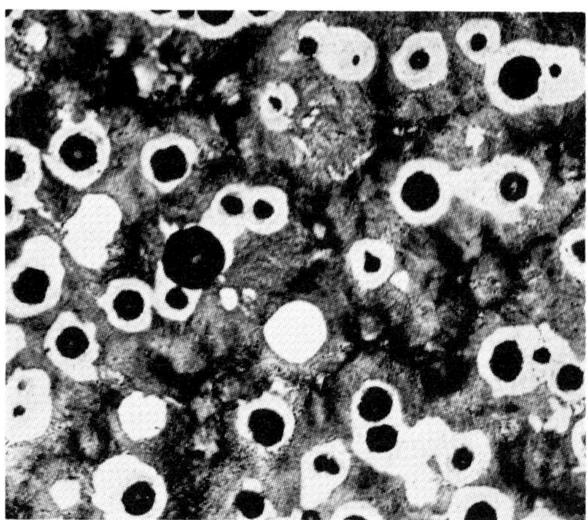

Fig. 10-3. Properly inoculated ductile iron; $\times 250$, etched.

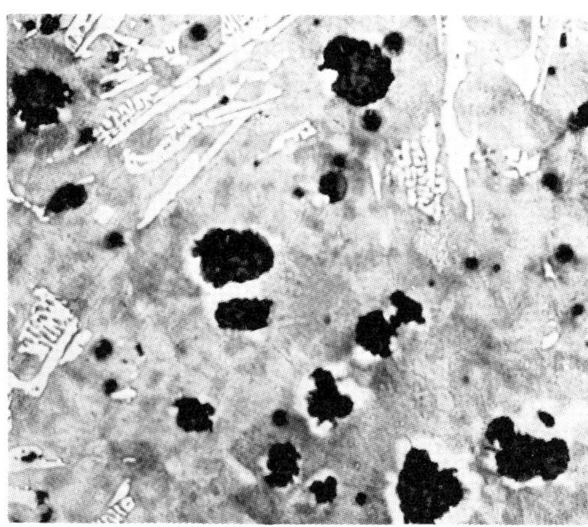

Fig. 10-4. Faded inoculation; note carbides and irregular shape of nodules; $\times 250$, etched.

(fade) may be seen in Figure 10-4. Note the presence of primary carbides and the deterioration in the shape of the nodules.

■ WHY CARBIDES FORM

The formation of primary carbides is most dependent on the section size of the casting; therefore, the factors influencing the occurrence of carbides in ductile iron are generally shown for various section sizes. The effects of pouring temperature and carbon equivalent (carbon + 1/3 silicon) for various section sizes are shown in Figure 10-5.

The microstructure in a 1/8-inch section poured at 2400F (1316C) showed 20% carbides at 4.39% carbon equivalent, and only 2% carbides at 4.60% carbon equivalent.

It is desirable—economically and for minimum dross—to keep the base sulfur as low as possible. However, as shown in Figure 10-6, the carbide formation is much greater when the base sulfur is under 0.02%. Actual industry experience has shown increased propensity for carbides when base sulfur levels are less than 0.007%.*

***[Ed. N.: This whole issue may be related to the tendency to overtreat with magnesium at the low sulfur levels. Good structures can be produced in low-sulfur base irons by using very low magnesium residuals.]**

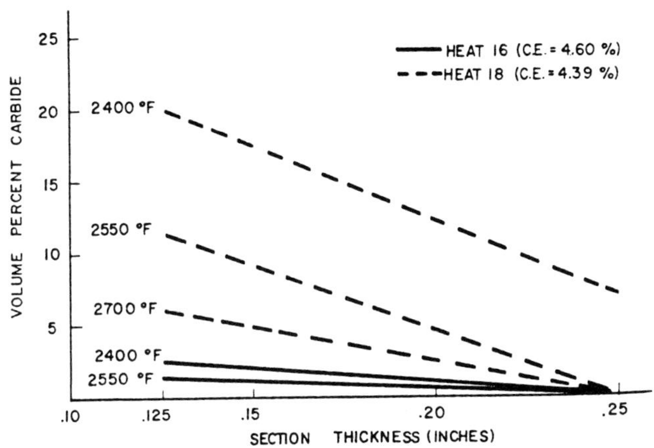

Fig. 10-5. Effect of section size, pouring temperature and carbon equivalent on the volume percent carbide in cerium-bearing ductile iron postinoculated with 0.31% silicon.[1]

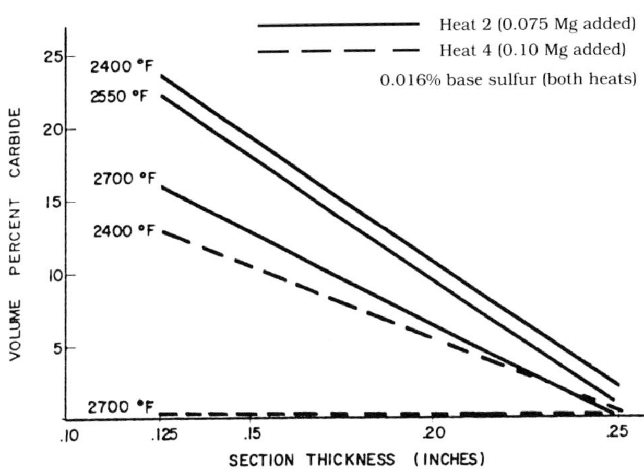

Fig. 10-7. Effect of section size, pouring temperature for 0.075 and 0.10 Mg additions on the volume percent carbide in cerium-bearing ductile iron postinoculated with 0.31% silicon.[1]

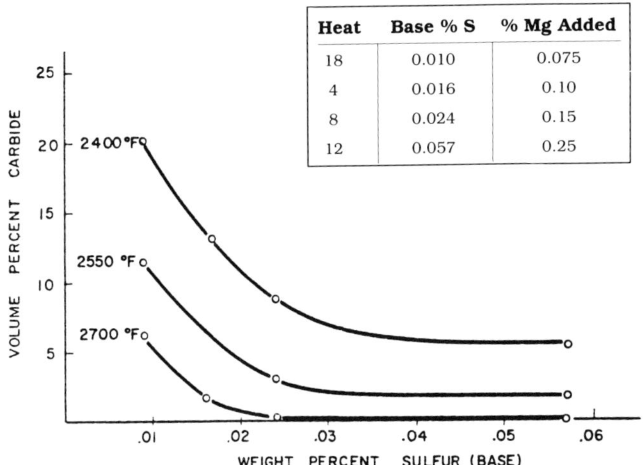

Fig. 10-6. Effect of pour temperature and base sulfur level on the volume percent carbide in cerium-bearing ductile iron postinoculated with 0.31% silicon.[1]

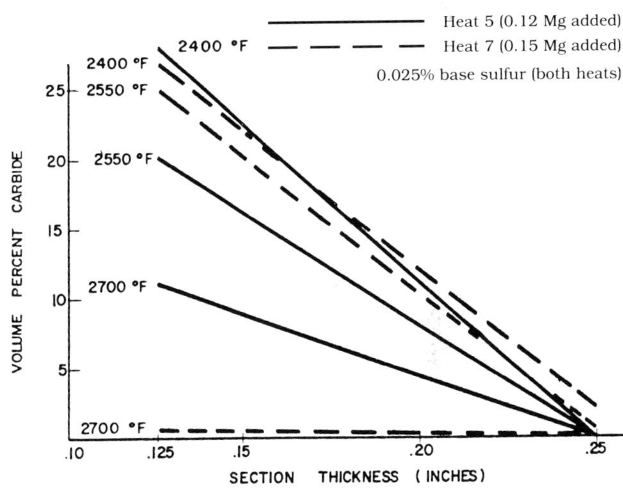

Fig. 10-8. Effect of section size, pouring temperature, for 0.12 and 0.15 Mg additions on the volume percent carbide in cerium-bearing ductile iron postinoculated with 0.31% silicon.[1]

Figures 10-7 and 10-8 show that higher magnesium additions reduce the amount of carbide for all pouring temperatures up to a residual magnesium level of 0.05%. Residual magnesium contents over 0.05% may increase carbides, reduce nodule count and adversely affect graphite shape.

■ CHOICE OF INOCULANTS

A large number of inoculants with high silicon content are available for ladle or late inoculation. They are based on ferrosilicon alloys containing 45–75% silicon. All inoculants based on ferrosilicon will contain one or more of the

TABLE 10-1. COMPOSITIONS OF SOME FERROSILICON-BASED INOCULANTS FOR DUCTILE IRON

Type of Inoculant	Si%	Al%	Ca%	Ba%	Sr%	Zr%	Mn%	Mg%	RE%
75% FeSi, Ca bearing	75.0	0.6–1.25	0.6–1.0						
FeSi-Ba	60–65	1.0	0.8	0.8		6	6		
FeSi-Ba	60–65	0.5–1.7	1.0	9–11					
FeSi-Ba	60–65	1.5	2.0	5–6			9–10		
FeSi-Ba	70–75	0.8–1.20	0.8–1.20	1.75–2.25					
FeSi-Zr	80	1.5–2.5	2.5			1.5			
FeSi-Sr*	75	<0.5	<0.1		0.8				
FeSi-Sr*	45–50	<0.5	<0.1		0.8				
FeSi-Ce	45	0.5	0.5						13 (10% Ce)
FeSi-Ce	45	0.5	0.5						3.5 (3% Ce)
45% FeSi	45–50	0.8	0.8						
45% FeSi-Mg	45–50	0.8	0.8					1.25	
FeSi-La	75	1.5							2.0–2.5 La

*Refer to Editors' Note on Sr additions, this chapter.

minor elements—aluminum, calcium, strontium,* barium, magnesium, zirconium, cerium, or other rare earth elements—to stimulate their inoculating effects. Typical compositions of some commercially available inoculants are shown in Table 10-1.

> *[Ed. N.: Strontium-bearing alloys are not normally recommended for ductile iron, except where there is a total absence of calcium and cerium in the nodulizing system (e.g., pure magnesium).]

The most commonly used silicon-based inoculant is 75% ferrosilicon containing 0.5–1.25% aluminum and 0.6–1.0% calcium. These elements are generally present, at varying percentages, in the raw materials used to produce the ferrosilicon. Final chemistry adjustments can be made through ladle metallurgy; the levels controlled at consistent amounts.

The main reasons for inoculating ductile iron are to prevent the formation of primary carbides, to improve the nodule shape, and to increase the nodule count. A ladle addition of 0.5% inoculation-grade ferrosilicon usually gives adequate nucleation to satisfy the requirements, while additions as high as 1% may be required for thin sections. If chill formation still persists, the choice of a more effective, special inoculant containing additional minor elements must be made.*

> *[Ed. N.: The expected performance differentials of proprietary inoculants in ductile irons are greatly diminished as compared to their use in gray iron. Careful evaluation on the shop floor is recommended.]

Particularly in thin sections, carbide formation is reduced and the nodule count is increased if cerium is added. Figure 10-9 shows the most effective range of addition to be 0.0045–0.008%. Inoculants containing cerium and other rare earths are more effective than those containing only cerium.*

> *[Ed. N.: The literature will verify that cerium and rare earth additions made as part of the nodulizing alloy are equally as effective as adding them in the inoculant.]

METHODS OF ADDING INOCULANTS

The best method of ladle inoculating ductile iron consists of adding the inoculant to the stream during reladling. The amount of inoculant required depends on the composition of the iron, the thickness and solidification rate of the casting, and the composition of the inoculant.

Additions of inoculant up to 1.0% of inoculating-grade ferrosilicon are often used in light section castings to minimize carbide formation. In the case of castings with heavy sections, the addition may be as low as 0.25% to keep the carbon equivalent below 4.50% and prevent carbon flotation.

The microstructure of an adequately ladle-inoculated ductile iron in a 1/2-in. section may be seen in Figure 10-10. A "golden rule" for ladle inoculation is to select the inoculant most suited to the purpose, and then use the minimum amount possible to achieve the required result.

The term "late inoculation" includes any method of adding the inoculant to the metal stream as the casting is poured, or within the mold cavity itself. It is possible to produce greater levels of inoculation by these methods with much less inoculant than is necessary with ladle inoculation. The amount of inoculant required for late inoculation is a maximum of 0.20% of the metal poured, but additions as low as 0.02–0.05% are often adequate. The microstructure of a 1/4-in. section, using ladle inoculation, is shown in Figure 10-11. With a small amount of additional late inoculation, the primary carbides are eliminated and the nodule count is increased greatly, as shown in Figure 10-12. This is also shown in Figure 10-13, where one gram of inoculant added to the three-pound casting gave similarly improved results. With late inoculation, the main problem is to obtain uniform distribution throughout the casting, rather than to select the most effective inoculant. Figure 10-14 shows non-uniform structure, which may

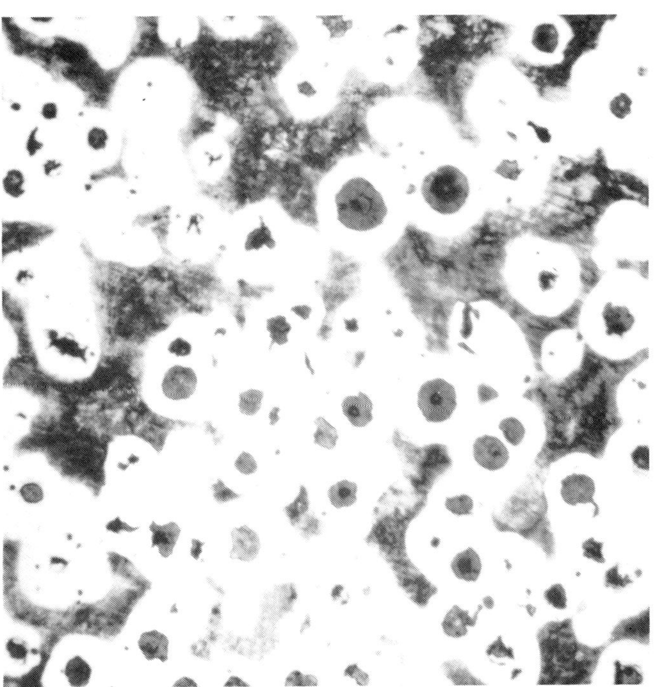

Fig. 10-10. Adequately postinoculated ductile iron; as-cast, nital etched, ×100, 1/2-inch section.[2]

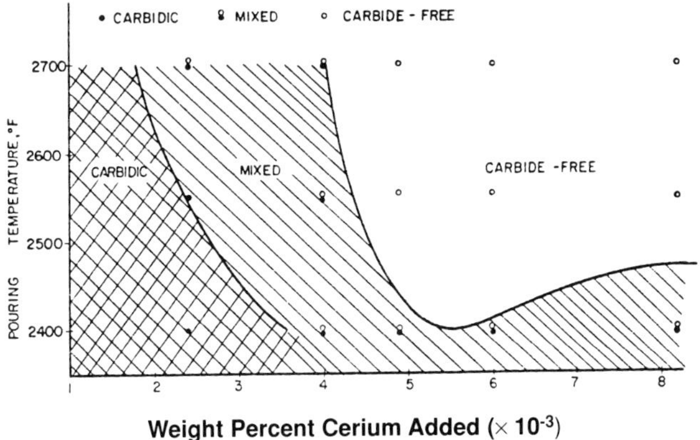

Fig. 10-9. Range of cerium effectiveness in ductile iron postinoculated with 0.31% silicon.[1]

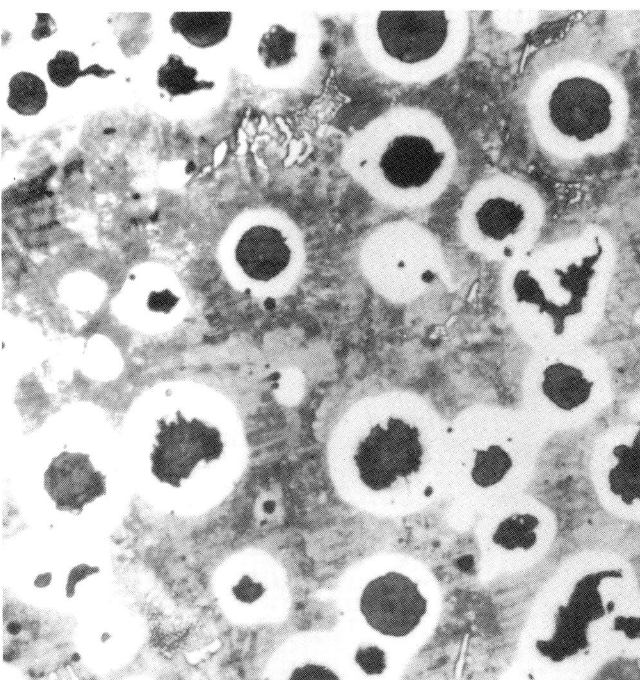

Fig. 10-11. Microstructure of 1/4-inch section as-cast ductile iron, ladle inoculated with 1% of 75% ferrosilicon; nital etched, ×100.[2]

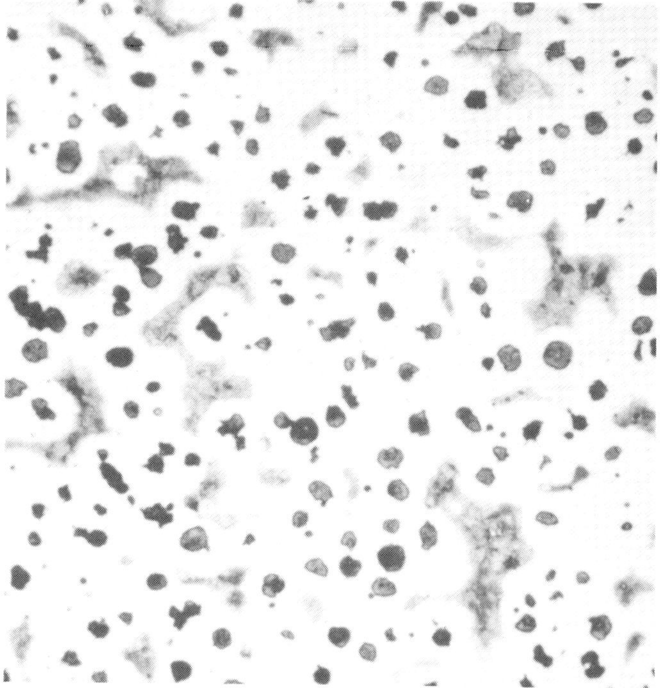

Fig. 10-12. Microstructure of 1/4-inch section as-cast ductile iron, ladle inoculated with 1% of 75% ferrosilicon plus mold inoculation; nital etched, ×100.[2]

occur with in-the-mold inoculation. In-the-mold magnesium treatment accomplishes both the nodulizing and inoculation required. The disadvantages are the risk of nonuniformity and dross inclusions.

The newest methods of late inoculation of ductile iron are late metal stream inoculation and wire inoculation.

Late metal stream inoculation provides the advantages of very uniform inoculation, automatic operation, correct proportioning of the inoculant addition in relation to the weight of metal cast into each mold, and the use of commercially available inoculating materials, which only have to be sized appropriately before use. Microstructural comparison of the effect of ladle and metal stream inoculation upon the structure of 2-mm (5/64-in.) sections of ductile iron may be seen in Figure 10-15. Variation in nodule count for the same additions is shown in Fig. 10-16.

Wire inoculation consists of two key elements: (1) an inoculant-cored wire; and (2) a wire feeding device with suitable controls for maintaining a preset wire feed rate. Figures 10-17 and 10-18 show the microstructure of ductile iron casting uninoculated, and wire inoculated with 0.067% inoculating grade ferrosilicon, in a 1/4-in. section and in a 1/2-in. section. It may be seen that the wire inoculation has effectively removed the primary carbides. Encapsulation of inoculant in the wire is expensive, and addition rates are limited to the rate at which the steel sheath will melt in the metal stream. Wire inoculation is not practiced widely in commercial use.

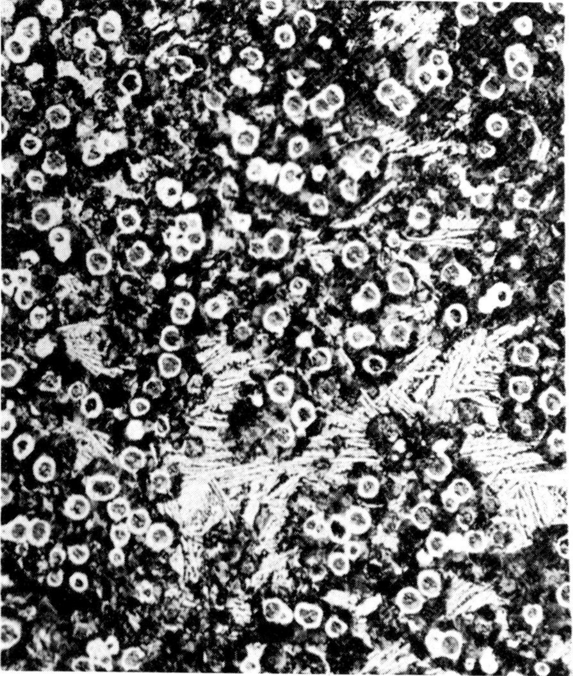

Fig. 10-13. Microstructure in 1/4-inch arm of a 3-lb production casting with no addition (left) and one gram of 85% FeSi added to mold downsprue (right); picral etch, ×100.[3]

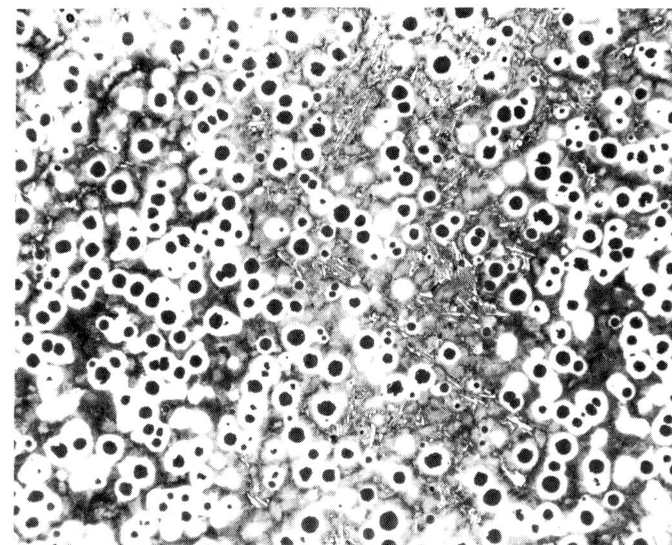

Fig. 10-14. Non-uniform inoculation in a thin casting; 4% picral etched, ×100. (The inoculant was added in the mold.)[4]

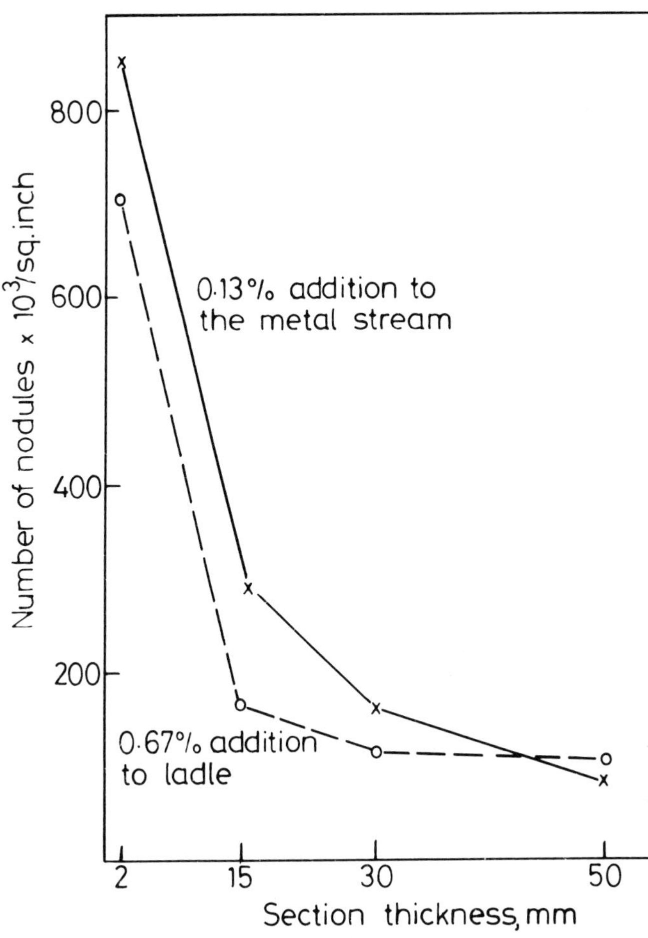

Fig. 10-16. Nodular numbers in ferrosilicon-magnesium treated irons inoculated with foundry grade ferrosilicon (75%) silicon) added either to the ladle within 1 min of casting, or to the metal stream.[4] (© BCIRA 1979, reprinted with permission)

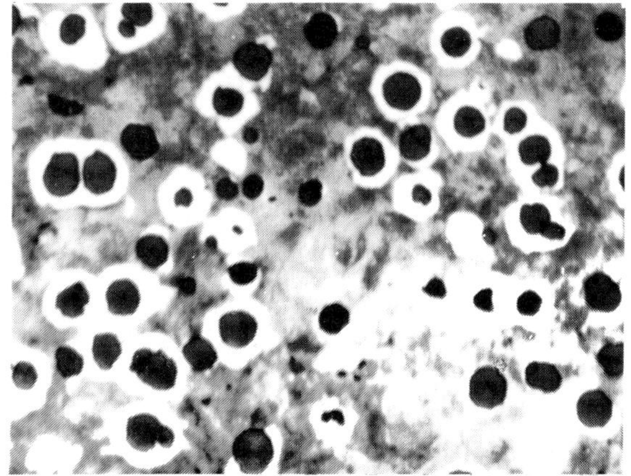

(a) 0.13% addition to metal stream

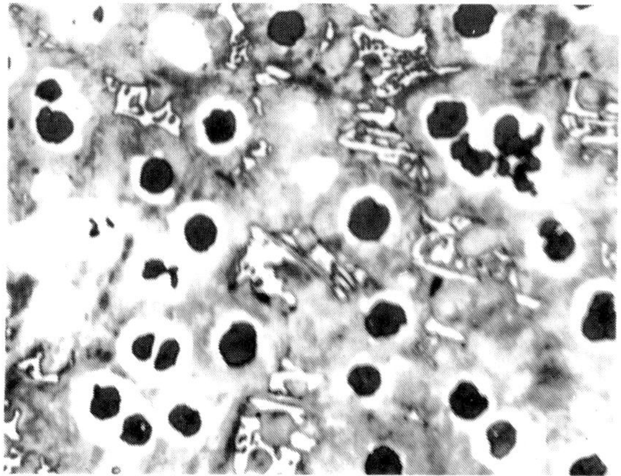

(b) 0.67% addition to ladle

Fig. 10-15. Effect of ladle and metal stream inoculation upon the structure of 2-mm (5/64-inch) sections of irons treated with ferrosilicon-magnesium and inoculated with foundry grade ferrosilicon; ×500, 4% picral etched.[4]

FADING OF INOCULATION

The effect of inoculation is at a maximum immediately after the addition of an inoculant. In the presence of small amounts of cerium, higher nodule numbers are obtained, and rates of fading are reduced. In ductile irons, the fading of inoculants causes a reduction in the number of centers from which growth of the eutectic occurs. This increases undercooling during eutectic solidification and the likelihood of formation of primary carbides. The decrease in nodule number is often accompanied by a deterioration in the shape of the nodules, as shown in Figure 10-19a. There will be a loss of magnesium at the same time of the inoculant fading. If the magnesium content does not become too low, re-inoculation will increase the number of nodules and will restore the shape of the nodules to their original form, as shown in Figure 10-19b.

CONTROLS REQUIRED IN INOCULATION

Selection of a source of an inoculant is based on the reliability of the producer to control the chemical analysis, cleanliness of the alloy, size distribution, protection of the material during shipment, and turnover of material to maintain fresh stock. Ferrosilicon can become oxidized on the surface, due to long-term storage and humidity, without being obvious. The oxide coating can contribute to an increase in slag or dross, especially in the smaller sizes that are used for late inoculation.

The amount of inoculant added to the downsprue, or to the stream as it enters the mold, is very small and must be clean, and non-oxidized. It must be added in such a manner that it becomes well mixed into the iron before the metal enters the casting cavity. All inoculants must be free of moisture and fine, combustible material that may flash

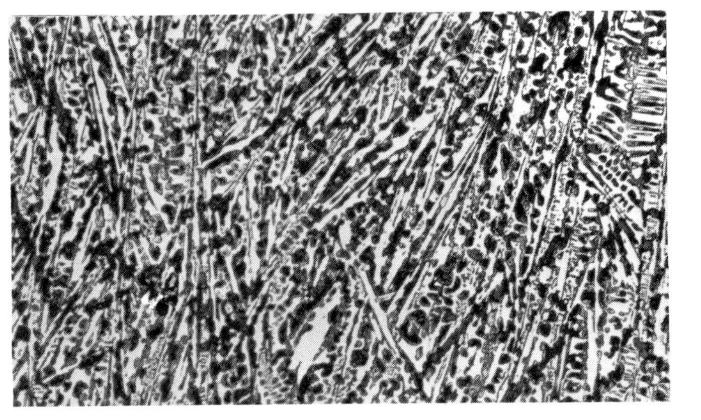

uninoculated

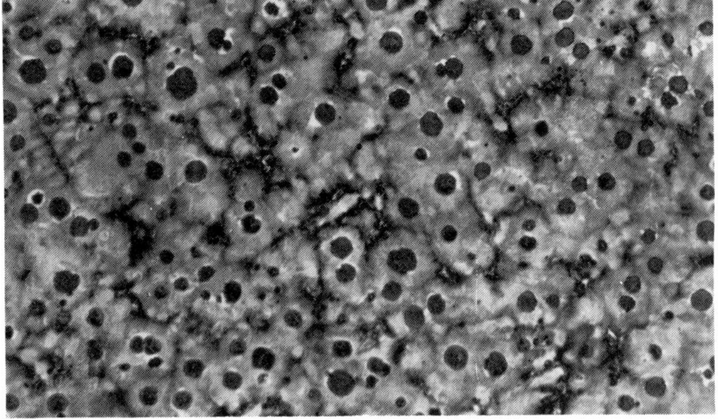

inoculated

Fig. 10-17. Microstructure of nodular iron casting, NiMg treatment alloy, wire inoculated with 0.067% FeSi, 1/4-inch sections.[5]

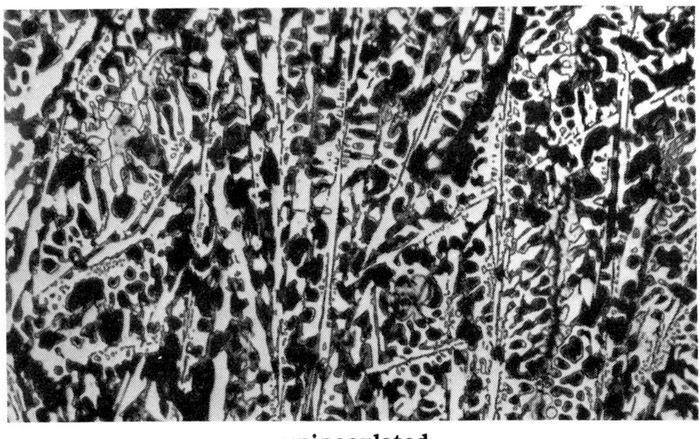

uninoculated

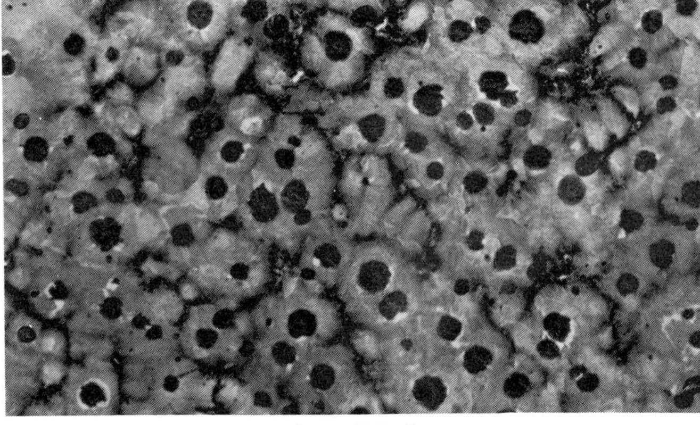

inoculated

Fig. 10-18. Microstructure of nodular iron casting, NiMg treatment alloy, wire inoculated with 0.067% FeSi, 1/2-inch sections.[5]

or blow and become a safety hazard. All material brought into the foundry from a cold storage *must remain sealed* until it reaches room temperature to prevent sweating and formation of moisture on the alloy. The size of the inoculant must be controlled for the method of addition. It must be small enough to be mixed into the iron in the time it takes to fill the pouring ladle.

The method of addition to the pouring ladle must be controlled for uniformity and consistency. The inoculant is added to the pouring ladle after the iron is about two inches deep in the bottom of the ladle. It is added so that it contacts the metal stream and at a rate that all inoculant is added before the ladle is half full. The metal stream must be wide enough and fast enough to thoroughly mix the alloy as the ladle fills. If a teapot ladle is used, the iron in the spout may not become inoculated and should be pigged or poured back into the furnace. Pouring ladles should be

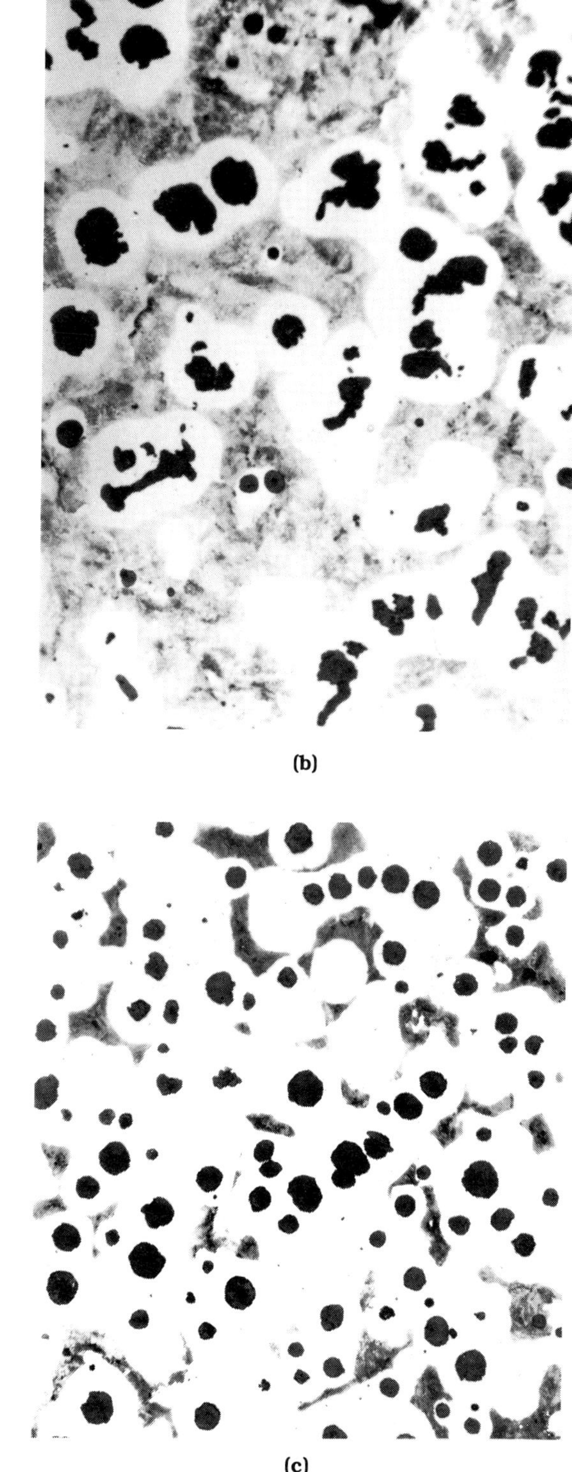

(b)

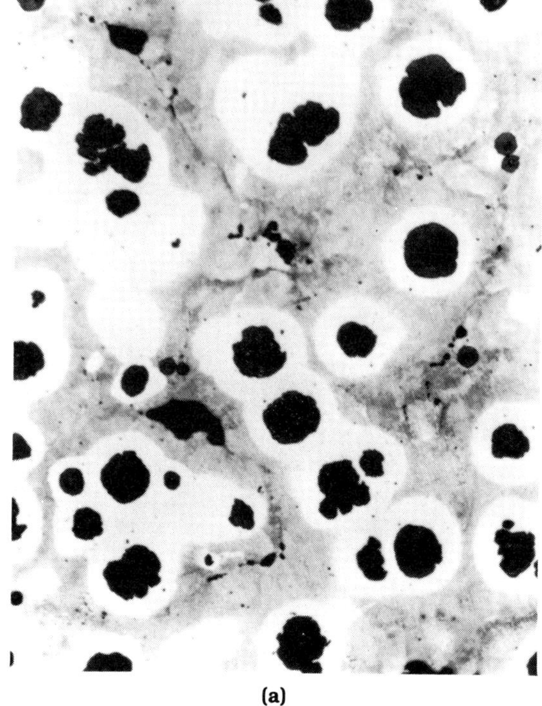

(a)

(c)

Fig. 10-19. **Structures of 1-3/4-inch square bars of ductile iron showing the effect of fading of inoculants upon nodule number and shape: (a) cast 2-1/2 minutes after inoculation; (b) cast 20 min after inoculation; (c) re-inoculated shortly before casting, but 20 minutes after initial inoculation. picral (4%) etched; ×100.**[6]

emptied of previous iron left over from pouring, because that iron is cold and contains oxides that reduce the effectiveness of the inoculant, as well as producing cold iron with resulting slag and dross in the castings.

Improper selection and method of using inoculants may contribute to microshrinkage, intercellular segregation and carbides. Process control procedures must include monitoring the effectiveness of inoculation by observing microstructures of test pieces of the casting. Monitoring frequency is dependent upon the criticality of the application and requirements of the quality control program. Any change in inoculant material or addition method must be thoroughly evaluated before a final change in procedure is made.

REFERENCES

1. W.J. Evans, S.F. Carter, Jr., J. Wallace; "Factors Influencing the Occurence of Carbides in Thin Sections of Ductile Iron," *AFS Transactions*, vol 89, pp 299, 302, 304-305, 315 (1981).

2. V.H. Patterson; "What Innoculation Accomplishes (Why Innoculate?)," *Proceedings of the AFS-CMI Conference: Modern Inoculation Practices for Gray and Ductile Iron*, pp 29-31 (1979).

3. W.J. Dell, R.J. Christ; "Chill Elimination in Ductile Iron by Mold Inoculation," *AFS Transactions*, vol 72, p 409 (1964).

4. G.F. Sergeant; "Late Metal Stream Inoculation—BCIRA Developments," *Proceedings of the AFS-CMI Conference: Modern Inoculation Practices for Gray and Ductile Iron*, pp 250, 258-259 (1979).

5. D. Sanders; "Wire Inoculation—A State of the Art Report," *Proceedings of the AFS-CMI Conference: Modern Inoculation Practices for Gray and Ductile Iron*, p 227-228 (1979).

6. A.G. Fuller; "Fading of Inoculants," *Proceedings of the AFS-CMI Conference: Modern Inoculation Practices for Gray and Ductile Iron* , p 167 (1979).

BIBLIOGRAPHY

Dawson, J.V. "Choice of Inoculants and Why Small Additions," *Proceedings of the AFS-CMI Conference: Modern Inoculation Practices for Gray and Ductile Iron*, Des Plaines, Illinois: American Foundrymen's Society, Inc. (1979).

Dell, W.J. and R.J. Christ. "Chill Elimination in Ductile Iron by Mold Inoculation," *Proceedings of the AFS-CMI Conference: Modern Inoculation Practices for Gray and Ductile Iron*, Des Plaines, Illinois: American Foundrymen's Society, Inc. (1979).

Evans, W.J., S.F. Carter, Jr. and J.F. Wallace. "Factors Influencing the Occurrence of Carbides in Thin Sections of Ductile Iron," *AFS Transactions*, vol 89, Des Plaines, Illinois: American Foundrymen's Society, Inc. (1981).

Flinn, R.A. "Fundamentals of Metal Casting," Reading, Massachusetts: Addison-Wesley Publishing Co. (1963).

Fuller, A.G. "Fading of Inoculants," *Proceedings of the AFS-CMI Conference: Modern Inoculation Practices for Gray and Ductile Iron*, Des Plaines, Illinois: American Foundrymen's Society, Inc. (1979).

Jenkins, L.R. "Inoculation," *Ductile Iron News*, Issue No. 2, Mountainside, New Jersey: Ductile Iron Society (1989).

Patterson, V.H. "What Inoculation Accomplishes (Why Inoculate?)," *Proceedings of the AFS-CMI Conference: Modern Inoculation Practices for Gray and Ductile Iron*, Des Plaines, Illinois: American Foundrymen's Society, Inc. (1979).

Sanders, S.D. "Wire Inoculation," *Proceedings of the AFS-CMI Conference: Modern Inoculation Practices for Gray and Ductile Iron*, Des Plaines, Illinois: American Foundrymen's Society, Inc. (1979).

Sergeant, C.F. "Late Metal Stream Inoculation," *Proceedings of the AFS-CMI Conference: Modern Inoculation Practices for Gray and Ductile Iron*, Des Plaines, Illinois: American Foundrymen's Society, Inc. (1979).

Stefanescu, D.M., R.J. Warrick, L.R. Jenkins, G. Chen and F. Martines. "Influence of the Chemical Analysis of Alloys on the Nodule Count of Ductile Iron," *AFS Transactions* , vol 93, Des Plaines, Illinois: American Foundrymen's Society, Inc.(1985).

Wallace, J.F., P. Du, H-Q Su, R.J. Warrick and L.R. Jenkins. "The Influence of Foundry Variables on Nodule Count in Ductile Iron," *AFS Transactions*, vol 93, Des Plaines, Illinois: American Foundrymen's Society, Inc.(1985).

11

Foundry Practices

Hugh Kind

Robert Neuman

Nick Wukovich

FOSECO, Inc.
Cleveland, Ohio

■ INTRODUCTION

The demand for ductile iron castings with close dimensional tolerances and a high degree of soundness cannot be met without controlling the individual processes that collectively result in the finished product. These include: pattern design, metal handling, molding, gating practices, pouring rates, effective sprue heights, correct choke areas, use of pouring basins, design of runners and ingates, selection and use of filters, and risering practices. An attempt will be made in this chapter to review these processes, providing some insight into the basic knowledge necessary to consistently produce castings that meet the stringent demands of today's marketplace.

■ PATTERN DESIGN: SHRINKAGE

In order to build patterns that will yield castings of acceptable dimensional capability, some knowledge of pattern shrinkage for ductile iron is required. In particular, it should be realized that the matrix structures of ductile iron will affect the amount of pattern shrinkage encountered during the casting process.

Pattern shrinkage for castings that have a predominantly pearlitic matrix will be 0.15 inch per foot, while annealing these same castings to achieve a ferritic matrix will result in a growth of approximately 0.05 inch per foot. Castings that have a ferritic as-cast structure will have no pattern shrinkage, while castings with a carbidic as-cast matrix will shrink 0.20 inch per foot. Full annealing of a component with a carbidic as-cast matrix will result in a growth of 0.30 inch per foot.

When producing castings of varying section thickness, care must be taken to insure that the matrix structure is consistent throughout the casting. Emphasis should be placed on the control of metal chemistry and inoculation practice, so that variations in matrix structure can be minimized.

METAL HANDLING

If the goal of producing high quality ductile iron castings is to be attained, the foundryman must have some degree of control over the metal handling aspects of the production process. It is possible to design gating systems for ductile iron castings that will effectively eliminate dross and non-metallic inclusions, if the amount of reaction products introduced into the mold can be limited to some predictable level. For this reason, the condition of treatment and the pouring vessels themselves should be monitored closely.

The design of risers to feed these castings cannot be effective if there are wide variations in the shrinkage characteristics of the iron produced. These variations can be reduced by controlling factors such as metal composition, magnesium treatment, pouring temperature, and inoculation.

METAL CLEANLINESS

The quantity of magnesium or alloys containing magnesium used to produce ductile iron is directly related to the amount of oxide slag and dross continuously formed on the melt's surface in the furnace and in the treatment ladle. This dross is detrimental—ending up as inclusions in the casting as well as producing excessive buildup in the ladle—which will affect the efficiency of the nodulization and ladle inoculation phases of the process. The magnitude of the problem is directly increased as the base metal sulfur increases.

When transferring treated iron from the treatment ladle to pouring ladles, care must be taken to minimize turbulence and exposure of the iron to the atmosphere. The pouring ladle should be filled rapidly to prevent oxidation and temperature loss.

LADLES

Clean ladles are essential to insure consistent results. Treatment and pouring ladles that have been allowed to

build up with slag will adversely affect the magnesium recovery during treatment and magnesium fading during pouring. These slags are rich in sulfur, and some of the sulfur will revert into the iron and tie up a portion of the magnesium, which can result in insufficient magnesium being available to produce the required nodularity. Teapot pouring ladles are often recommended to minimize slag entering the mold. The use of ladle covers helps to minimize temperature losses.

The ladle refractory used should be neutral, high-alumina or basic. Silica, fireclay or other acid refractories can react with the magnesium sulfide-rich basic slag and cause a reversion to vermicular or flake-graphite structure in the treated iron.

POURING TEMPERATURES

The pouring temperature of ductile iron is very important, since it is a metal that is readily oxidized. When the pouring temperature falls below 2550F (1399C), the volume of oxidation products increases significantly. Many founders who continue their practice of pouring molds as long as the metal will flow, will find themselves producing ductile iron castings with inclusions. One approach to help insure that the desired pouring temperature of 2550–2600F (1399–1425C) is maintained is to preheat all the ladles. Pouring temperatures should only be high enough to avoid misruns, pinholes, and slag defects. Excessive pouring temperature can dramatically increase shrinkage defects caused by mold wall movement.

FLUXES AND SLAG COAGULANTS

The addition of a flux to the metal during the inoculation transfer helps cleanse the metal of existing slag and suppresses the further oxidation of the metal. Small additions of this prepared flux are sufficient. Depending on the amount of metal treated, the addition may range from 0.02–0.10%. For large quantities of metal, or where large molds are poured, the addition of the flux to the pouring basin is most effective.

These fluxes are, for the most part, based on the use of fluorides such as calcium fluoride, sodium fluoride or cryolite. Fluxing agents assist in maintaining clean, consistent ladle volume, and in the rapid flotation of oxides that are in the molten iron. Slag coagulants are often used to assist in a skimming. Floating out or agglomerating the oxides before pouring the molds reduces the chance of entrapping slag inclusions in the castings.

It is important to note that a flux should never be applied to the bare refractory in a ladle. Flux should be added to a heel of metal in the ladle as it is filling, or perhaps on top of the nodulizing alloy in the treatment ladle. Since fluxes are designed to combine with and lower the melting point of slags, they will also have the same effect on refractory if they are allowed to concentrate on the bare refractory surface prior to filling the ladle with iron.

■ MOLDING

Producing castings of consistently high quality requires a high degree of control in the molding process. Particular attention should be paid to variables that affect mold wall rigidity. During solidification, ductile iron undergoes a volumetric expansion, resulting in increased pressure on the mold surface. If the mold wall yields significantly, the resulting volumetric expansion of the casting may exceed the capability of the risers to provide an adequate supply of feed metal, resulting in internal porosity in the casting.

Green sand molds are particularly susceptible to this phenomenon when mold hardness levels are not at least 95 psi. If a penetration-type gauge is used to measure mold strength, a minimum level of 35 psi should be maintained. These minimum hardness levels are often overlooked by molding personnel, who choose to reduce squeeze pressure as a means of improving the stripping characteristics of the pattern, when faced with fluctuating sand properties, inadequately maintained molding equipment, or worn tooling.

Sand properties should be monitored regularly to reduce variations in mold hardness, mold permeability, and moisture content. In addition, excess moisture above the temper point of the sand contributes to dross and pinhole defects in the castings produced.

Molds produced with dry sand, shell sand, or chemically-bonded sand are generally much more rigid than green sand molds, and are, therefore, less likely to be affected by mold wall movement. Some producers have reported that there are binders that become plastic when molten metal enters the mold, resulting in low mold permeability. If this occurs, the castings produced will be susceptible to gas-related defects.

Tight-fitting jackets or flasks should be used so that the mold walls can be supported as a safeguard against mold wall movement. The molds should be clamped or weighted properly to prevent mold separation during and after pouring.

■ GATING PRACTICES

Proper design of the gating system plays an integral part in the process of producing sound castings. Pattern layout should incorporate positioning of casting impressions to improve directional solidification and provide adequate room for the gating system.

A properly designed gating system should accomplish the following tasks:

- prevent slag and other nonmetallics from entering the casting cavity;
- introduce metal into the mold cavity at a predetermined velocity and with a minimum amount of turbulence;
- introduce metal into mold cavities in a manner conducive to controlled directional solidification.

Gating systems used for producing ductile iron differ from gray iron gating systems because of the high levels of magnesium and silicon in ductile iron, which increase the possibility of dross formation. For this reason, turbulence must be minimized in ductile iron gating systems. Ductile iron systems usually incorporate the use of large runners, which reduce the metal velocity and provide time for the slag to be separated from the molten metal, before entering the mold cavity. While this is effective in reducing nonmetallic inclusions in the castings produced, mold yield is reduced.

■ POURING RATES

There are many factors to consider when making a decision on the desired pouring rate for a given gating system. One of the most important factors to be considered is that metal velocity must be minimized, so that turbulence can be reduced as much as possible. This reduction in turbulence will insure that separation of the nonmetallics occurs in the runner system.

Time constraints are often dictated by molding machine cycle times in today's automated foundries. When this occurs, ideal pouring times may not be allowable if productivity is to be maximized.

Another factor that should be considered in green sand systems is the ability of the sand to resist erosion when the gating system is being filled. Considerable research has been conducted to determine the limitations placed on metal velocity by molding sands. One such formula, derived for use with synthetic sands bonded with a combination of western and southern bentonites, is:

$$PT = (0.83 + 1.025\,T)\,\sqrt{CW}$$

where

$$PT = \text{Pouring time (sec)}$$

$$T = \text{Minimum casting wall thickness (in.)}$$

$$CW = \text{Casting weight (lb)}$$

or

$$PT = (1.23 + .06\,T)\,\sqrt{CW}$$

where

PT = Pouring time (sec)

T = Minimum casting wall thickness (mm)

CW = Casting weight (kg)

Table 11-1 gives pouring time values for castings up to 1000 pounds in weight with minimum section thickness of 1/8–1-1/2 inch. Similarly, Table 11-2 gives this same information for castings up to 500 kg in weight with minimum section thickness ranging from 3.0–38.0 mm. Note that these formulas are based on pouring temperatures in the range of 2500–2600F (1370–1425C).

TABLE 11-1. POURING TIME FOR DUCTILE IRON IN SECONDS; 2550–2600F POURING TEMPERATURE

Casting Weight (lb)	Thickness (inches)									
	1/8	1/4	3/8	1/2	5/8	3/4	7/8	1	1-1/4	1-1/2
10	3.1	3.4	3.8	4.3	4.7	5.1	5.5	5.9	6.7	7.5
20	4.3	4.9	5.4	6.0	6.6	7.2	7.7	8.3	9.4	10.6
30	5.3	6.0	6.7	7.4	8.1	8.8	9.5	10.2	11.6	13.0
40	6.1	6.9	7.7	8.5	9.3	10.1	10.9	11.7	13.4	15.0
50	6.8	7.7	8.6	9.5	10.4	11.3	12.2	13.1	14.9	16.7
60	7.4	8.4	9.4	10.4	11.4	12.4	13.4	14.4	16.4	18.4
70	8.0	9.1	10.2	11.2	12.3	13.4	14.5	15.5	17.7	19.8
80	8.6	9.7	10.9	12.0	13.2	14.3	15.5	16.6	18.9	21.2
90	9.1	10.3	11.5	12.7	14.0	15.2	16.4	17.6	20.0	22.5
100	9.6	10.9	12.1	13.4	14.7	16.0	17.3	18.6	21.1	23.7
120	10.5	11.9	13.3	14.7	16.1	17.5	18.9	20.3	23.1	25.9
140	11.3	12.9	14.4	15.9	17.4	18.9	20.4	22.0	25.0	28.0
160	12.1	13.7	15.4	17.0	18.6	20.2	21.8	23.5	26.7	30.0
180	12.9	14.6	16.3	18.0	19.7	21.5	23.2	24.9	28.3	31.8
200	13.6	15.4	17.2	19.0	20.8	22.6	24.4	26.2	29.9	33.5
250	15.2	17.2	19.2	21.2	23.3	25.3	27.3	29.3	33.4	37.4
300	16.6	18.8	21.0	23.3	25.5	27.7	29.9	32.1	36.6	41.0
350	17.9	20.3	22.7	25.1	27.5	29.9	32.3	34.7	39.5	44.3
400	19.2	21.7	24.3	26.9	29.4	32.0	34.5	37.1	42.2	47.4
450	20.3	23.0	25.8	28.5	31.2	33.9	36.6	39.4	44.8	50.2
500	21.4	24.3	27.2	30.0	32.9	35.8	38.6	41.5	47.2	52.9
600	23.5	26.6	29.8	32.9	36.0	39.2	42.3	45.4	51.7	58.0
700	25.4	28.7	32.1	35.5	38.9	42.3	45.7	49.1	55.9	62.6
800	27.1	30.7	34.4	38.0	41.6	45.2	48.8	52.5	59.7	67.0
900	28.7	32.6	36.4	40.3	44.1	48.0	51.8	55.7	63.3	71.0
1000	30.3	34.4	38.4	42.5	46.5	50.6	54.6	58.7	66.8	74.9

TABLE 11-2. POURING TIME FOR DUCTILE IRON IN SECONDS; 1370–1425C POURING TEMPERATURE

Casting Weight (kg)	Thickness (millimeters)									
	3	6	9	12	15	18	21	24	27	30
5	3.2	3.6	4.0	4.4	4.8	5.2	5.6	6.0	6.4	6.8
10	4.5	5.0	5.6	6.2	6.7	7.3	7.9	8.4	9.0	9.6
15	5.5	6.2	6.9	7.6	8.3	9.0	9.6	10.3	11.0	11.7
20	6.3	7.1	7.9	8.7	9.5	10.3	11.2	11.9	12.8	13.6
25	7.1	8.0	8.9	9.8	10.7	11.6	12.5	13.4	14.3	15.2
30	7.7	8.7	9.7	10.7	11.7	12.7	13.6	14.6	15.6	16.6
35	8.3	9.4	10.5	11.5	12.6	13.7	14.7	15.8	16.9	17.9
40	8.9	10.1	11.2	12.3	13.5	14.6	15.8	16.9	18.0	19.2
45	9.5	10.7	11.9	13.1	14.3	15.5	16.7	17.9	19.1	20.3
50	10.0	11.2	12.5	13.8	15.1	16.3	17.6	18.9	20.2	21.4
60	10.9	12.3	13.7	15.1	16.5	17.9	19.3	20.7	22.1	23.5
70	11.8	13.3	14.8	16.3	17.8	19.3	20.8	22.3	23.8	25.4
80	12.6	14.2	15.8	17.4	19.1	20.7	22.3	23.9	25.5	27.1
90	13.4	15.1	16.8	18.5	20.2	21.9	23.6	25.3	27.0	28.8
100	14.1	15.9	17.7	19.5	21.3	23.1	24.9	26.7	28.5	30.3
120	15.5	17.4	19.4	21.4	23.3	25.3	27.3	29.3	31.2	33.2
140	16.7	18.8	20.9	23.1	25.2	27.3	29.5	31.6	33.7	35.9
160	17.8	20.1	22.4	24.7	26.9	29.2	31.5	33.8	36.1	38.3
180	18.9	21.3	23.8	26.2	28.6	31.0	33.4	35.8	38.2	40.7
200	19.9	22.5	25.0	27.6	30.1	32.7	35.2	37.8	40.3	42.9
250	22.3	25.1	28.0	30.8	33.7	36.5	39.4	42.2	45.1	47.9
300	24.4	27.5	30.7	33.8	36.9	40.0	43.1	46.3	49.4	52.5
350	26.4	29.8	33.1	36.5	39.9	43.2	46.6	50.0	53.3	56.7
400	28.2	31.8	35.4	39.0	42.6	46.2	49.8	53.4	57.0	60.6
450	29.9	33.7	37.6	41.4	45.2	49.0	52.8	56.6	60.5	64.3
500	31.5	35.6	39.6	43.6	47.6	51.7	55.7	59.7	63.7	67.8

◼ EFFECTIVE SPRUE HEIGHT

The velocity of the metal entering the mold cavity through an ingate varies according to the level of the metal in the mold cavity, unless the entire cavity is positioned below the ingate (Fig. 11-1). For this reason, an average ferrostatic head can be calculated, which will reduce the complexity of the formula used to calculate the choke area in the runner system.

The formula for calculating the effective sprue height is as follows:

$$ESH = H \, (P^2/2C)$$

where

ESH = Effective sprue height

H = Distance from choke to the top of the sprue or pouring basin

P = Height of the casting above choke

C = Height of the casting in the mold

Figure 11-2 illustrates the terms used in calculating the effective sprue height. Figure 11-3 illustrates a special case where P = C.

CHOKE AREA

The minimum cross-sectional area in the gating system, which governs the rate of filling of the mold, is called the *choke*. In the case of a pressurized system, the choke is the ingate that connects the runner to the casting (or possibly the riser). If multiple ingates are used, the choke is the sum of the cross-sectional areas of all ingates. The following formulas, simplified to accommodate friction losses, can be used in the calculation of choke cross-sectional area:

$$CA = (0.33 \times WP) / (PT \times \sqrt{ESH})$$

where

CA = Choke area (in.^{2})

WP = Poured weight of casting (lb). (This includes casting weight plus the estimated weight of connected risers.)

PT = Pouring time (sec)

ESH = Effective sprue height (in.)

or

$$CA = (11.89 \times WP) / (PT \times \sqrt{ESH})$$

CA = Choke area (cm^2)

WP = Poured weight (kg)

PT = Pour time (sec)

ESH = Effective sprue height (cm)

POURING BASINS

The pouring basin is an important part of the gating system, which, unfortunately, is often overlooked by those attempting to design a functional gating system. The basin must have sufficient volume to allow the iron pourer to keep it full during the pouring process. This insures that the dross, created by metal treatment and transfer, will not be allowed to enter the downsprue.

The depth of the pouring basin should be three to four times the diameter of the downsprue, so that the shock of the iron falling from the pouring vessel to the mold can be absorbed. The width of the basin should allow the iron pourer to keep the basin full, without splashing iron on top of the mold.

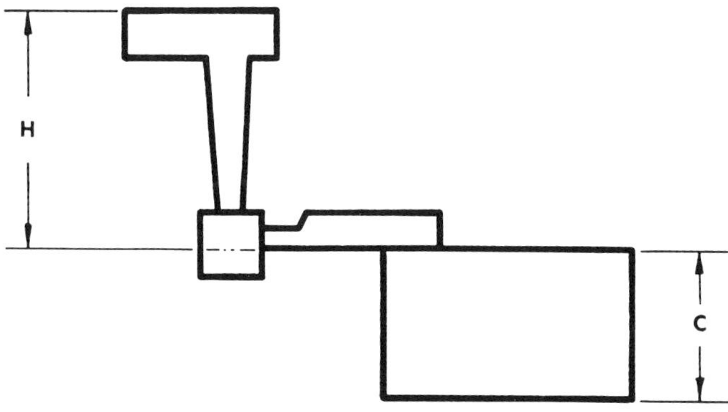

Fig. 11-1. A mold with ingates at the top of the casting cavity illustrates the one case where the ferrostatic head is constant.

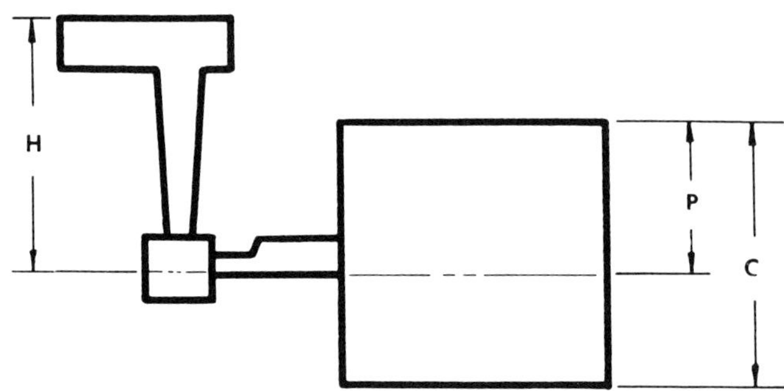

Fig. 11-2. The terms used in calculating the effective sprue height are illustrated.

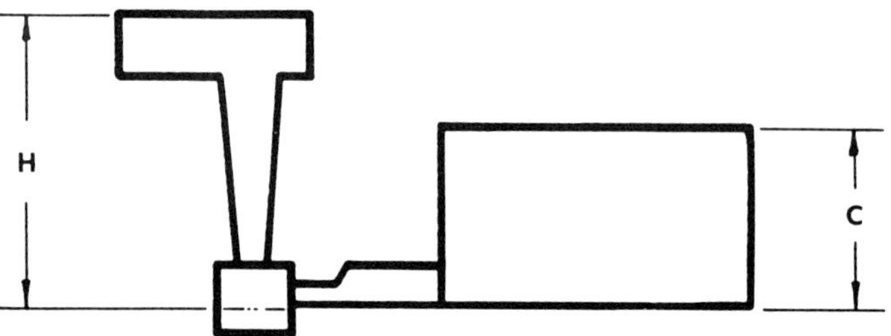

Fig. 11-3. A special case in calculating the effective sprue height, where (P) = (C).

Whenever possible, conical pouring basins should be avoided, since they allow a vortex effect, which introduces air into the downsprue. Figure 11-4 illustrates the design of a shelf-type pouring basin, which has proven effective for ductile iron gating systems.

■ RUNNERS AND INGATES

A moderately pressurized gating system having a sprue:runner:ingate ratio of approximately 4 : 8 : 3 or 2:4:1 has consistently produced quality castings in all size ranges. The downsprue, which connects the pouring basin to the runner system, should be tapered approximately 3°.

The reason for this is that the cross section of the liquid metal stream tends to decrease as the metal falls from the pouring basin to the bottom of the downsprue. If the downsprue is not tapered, a vacuum will be created between the metal stream and the walls of the downsprue, causing air to flow from the mold into the metal stream. This condition will increase the amount of dross present in the metal stream and, subsequently, in the castings produced.

In vertically parted systems, where several runners or ingates branch off of the downsprue at different levels, the downsprue cross section can be reduced in a step-like manner. By making sure that the cross-sectional area of the downsprue is 10–20% greater than the combined areas of the downstream chokes, the amount of time necessary to pressurize the gating system can be reduced. This type of downsprue is often effective in reducing penetration defects in the cavities located in the lower portion of the mold.

Using a well at the base of the downsprue that is approximately twice the runner depth and twice the downsprue diameter (at the base) will reduce the turbulence at the junction of the downsprue and the runner. This will allow the runner to be filled properly, and will reduce erosion at the bottom of the downsprue.

The runners should have height-to-width ratios of 2:1, and be larger than runners used in gray iron gating systems, so that metal velocity can be reduced. Reducing metal velocity will, in turn, decrease the turbulence in the runner, particularly when directional changes are made. When directional changes are incorporated into the runner system, these areas should be scrutinized in an effort to reduce turbulence. Sharp corners should be avoided, since they are susceptible to erosion. If bends are necessary, the radius should be as large as is practical. Ingates should be located away from the radius whenever possible.

When the liquid metal reaches the end of the runner, it tends to lap over itself and flow against the oncoming stream of metal. This turbulence is detrimental to the process of metal-slag separation, which should occur in the runner. The effects of this phenomenon can be reduced by tapering the end of the runner. If possible, the runner should extend well beyond the last ingate connection, so that turbulent metal does not flow into the ingate.

The ingate dimensions recommended for ductile iron require a width-to-height ratio of 4:1. This improves the effectiveness of the ingate in trapping nonmetallics, and will also facilitate removal of the casting from the gating system upon shakeout. The connection between the ingate and the runner should be perpendicular, to insure that the ingate remains choked during the entire pouring process.

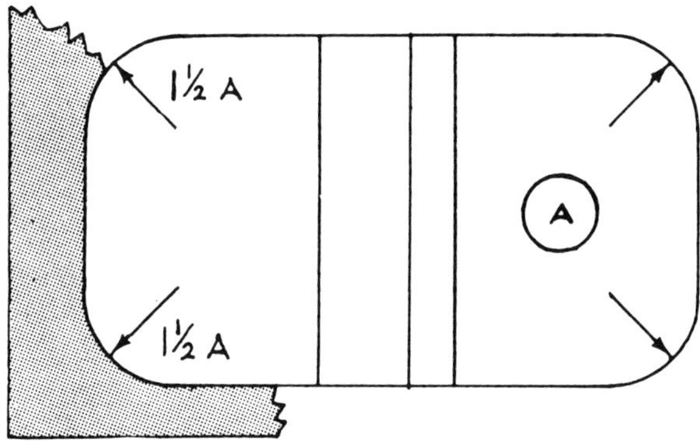

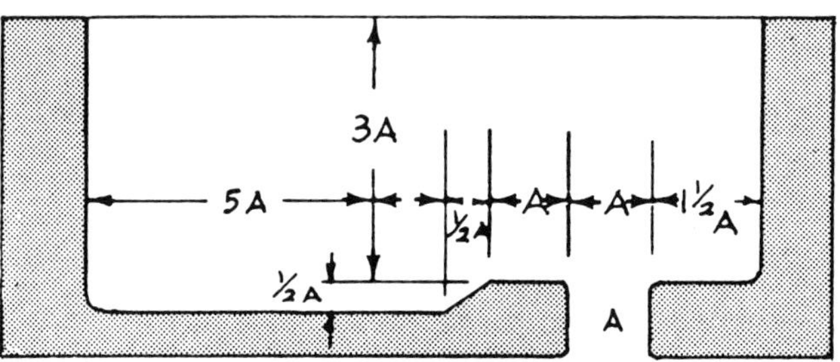

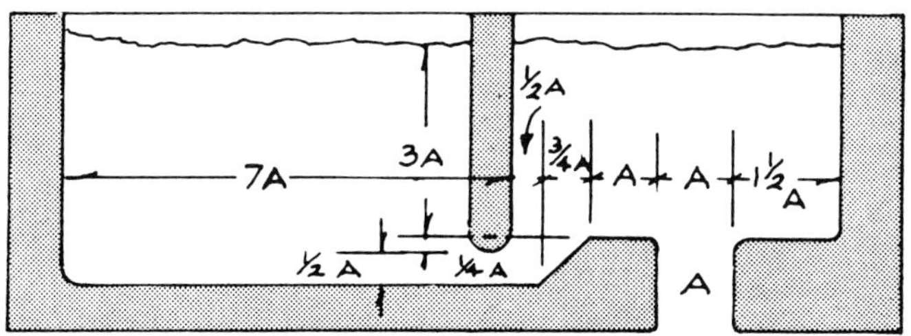

Fig. 11-4. The design of a shelf-type pouring basin is shown.

FILTER APPLICATION

The use of ceramic filters in ductile iron casting applications has expanded rapidly in recent years, with users obtaining a number of technical and economic benefits. Of primary importance are the reduction in inclusion-related defects that result in unacceptable castings either before or after machining, and the reduction in size and weight of conventional gating systems. Both of these benefits are a result of the efficient removal of slag, dross, and other non-metallics from the molten iron with a correctly designed ceramic filter.

The conventional approach used to prevent inclusion-related defects has been through gating systems designed to separate nonmetallic inclusions from the molten iron. This requires a gating system of adequate cross section and length to allow the lower-density nonmetallic materials to float and adhere to the mold surface. In many cases, runner extensions, skimmer cores, and elaborate swirl bobs are used to assist this process. Unfortunately the conventional approach to gating system design does not always provide adequate casting quality and often sacrifices mold yield.

The use of a ceramic filter in a conventionally designed gating system can be an effective method of reducing inclusion-related defects, but obtaining optimum filtration effectiveness and reduced gating system size requires a gating system designed specifically for ceramic filter use. The next section discusses the methods of filter application and gating system design that provide optimum technical and economic benefits based on laboratory and industrial experience.

Ceramic Filter Types

Cellular Ceramic Filters

Cellular ceramic filters are available in a range of sizes and cell geometries (described in cells/in.^{2}). The cell geometry required for a given application depends on the metal type and metal cleanliness. Normally 0.080 in. cell size is recommended for ductile iron (Fig. 11-5).

Ceramic Foam Filters

Ceramic foam filters are produced in a wide variety of sizes and different porosities (described in pores per linear inch, or ppi). Normally 10 ppi filters are recommended for ductile iron (Fig. 11-6).

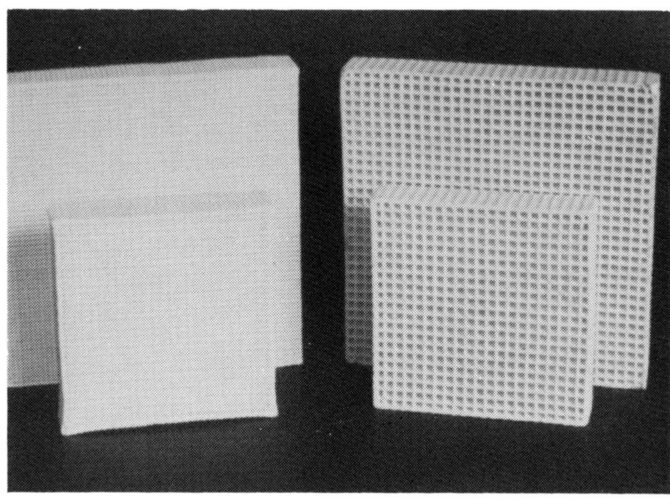

Fig. 11-5. Cellular ceramic filters.

Guidelines for Filter Application

A correctly designed gating system for ceramic filter use should provide the following:

- simple filter placement;
- consistent mold filling time;
- optimum filtration effectiveness;
- minimum erosion/turbulence past the filter;
- minimum gating system size.

In order to accomplish these objectives, it is necessary to determine the size and number of filters required for a given application, and to locate these in a correctly designed gating system.

Fig. 11-6. Ceramic foam filters are pictured.

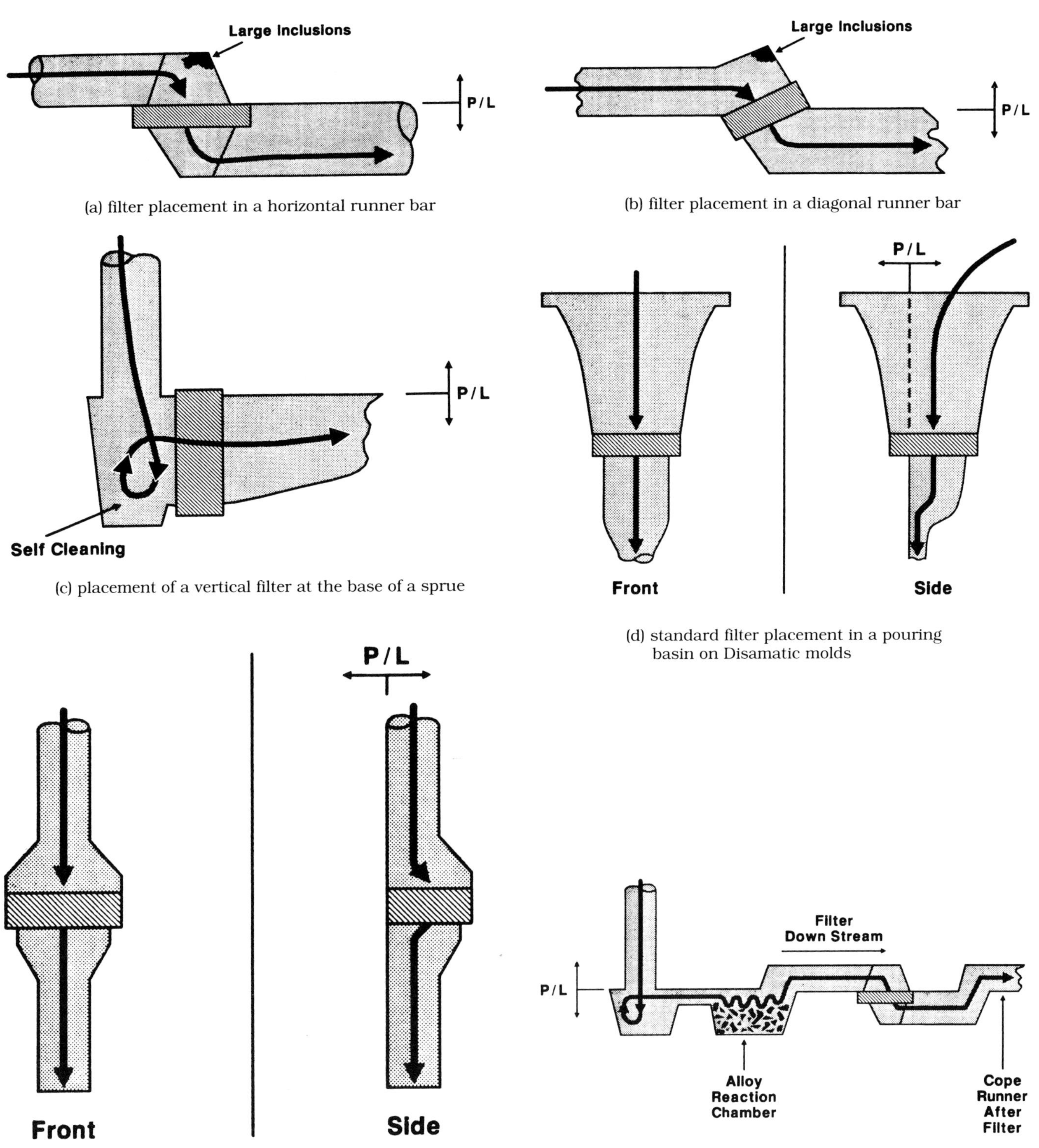

Fig. 11-7. Illustrations of various locations and positions of filters.

Filter Location and Position

The location and position of a filter are influenced by the molding method, the pattern layout, nodulizing treatment, and inoculation practice used. Several examples are shown in Figures 11-7a–f.

Filter Frontal Area: Choke Ratio

The size and number of ceramic filters required for a given application is determined by the pouring rate and the total amount of metal to be passed. To ensure that the filter does not affect the pouring rate, ratios for the amount of filter area required to permit the desired volume of metal to pass (filter frontal area) and the smallest cross section (choke) in the gating system have been developed. For ductile iron, the filter frontal area to choke ratio is 4–6:1.

Using this guideline, a filtered gating system should fill at or near the same rate as an unfiltered system. Ceramic filters should not be used to control the pouring rate, since the varying amounts of material trapped on a filter, from mold to mold, can result in variations in pouring times. Variable pouring times cause serious problems when nodulizing alloys and inoculants are used in the gating system, since the metallurgical properties of the casting can be affected. Variable pouring times can also cause serious problems in foundries with automatic pouring devices.

Gating System Design

In horizontally parted molds, filter gating systems designed similar to those shown in Figure 11-8a and 11-8b have proven particularly successful in minimizing the runner cross section and providing suitable mold filling rates. These systems are nonpressurized, since the cross-sectional area of the ingates is greater than the choke area located either at the sprue or after the filter. Placing the gating system choke before the ingates has proven to be beneficial and differs substantially from the commonly used pressurized gating systems found in many iron foundries (such as the 4:8:3 system).

A gating system for the mold nodulizing process is a two-part system encompassing what might be termed "production" and "distribution." In all cases, a pressurized gating system is required to accomplish the production of ductile iron. After that phase, a standard 4:8:3 system is suggested for distribution, with the choke area considered as part of the sprue area in the 4:8:3 calculation. Using this technique, it is possible to "produce and distribute" ductile iron in the mold.

Designing a gating system using ceramic filters in vertically parted molds is normally a compromise between how the filter must be located, and the room available for the gating system. The optimum gating system design would place the filter at the bottom of the mold below the casting cavities. This allows the mold to fill from the bottom up and ensures that the filter will be surrounded by metal prior to filling the casting's cavities. Unfortunately, designing a gating system in this manner is not always practical, and successful filter applications have used a variety of filter locations and gating techniques.

Filter Blockage

The quantity of ductile iron that will pass through a given size filter depends on the cleanliness of the iron. Those factors that will adversely affect cleanliness are:

    1. *High Residual Magnesium Levels*—In general, as residual magnesium levels increase, the quantity

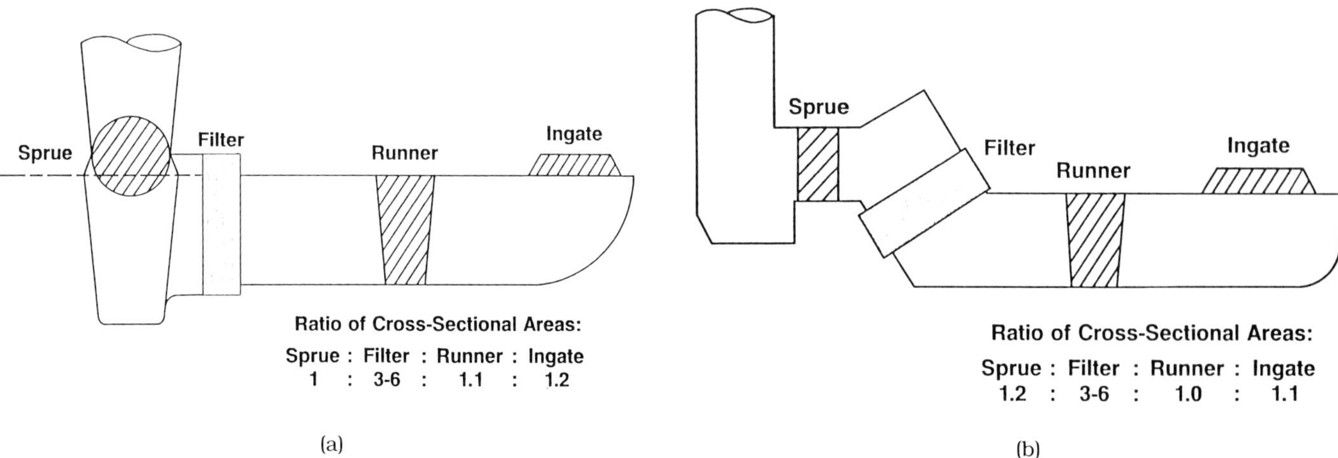

Fig. 11-8. Application of the filter frontal area to choke ratio guideline is shown for two different filtering gating designs.

of ductile iron that will pass through a given size filter decreases.

2. *High Sulfur Levels*—As base sulfur levels increase, the quantity of ductile iron that will pass through a given size filter decreases. This can be attributed to the increased level of magnesium sulfides present in the iron, which are retained by a ceramic filter.

3. *Pouring Temperatures*—The higher the pouring temperature, the greater the quantity of ductile iron that will pass through a given size filter. Although no direct relationship between filter blockage and pouring temperature has been established, it is well documented that the level of nonmetallics present increases with decreasing pouring temperatures. This has a significant effect on filter blockage, particularly at temperatures below 2500F (1371C).

4. *Treatment Alloys and Practices*—Metal cleanliness is greatly affected by the type of nodulizing alloy and the treatment practice employed. In general, the cleaner the nodulizing alloy (i.e., low in alloy slag and oxide content) and the lower the silicon content of the alloy used, the higher the degree of metal cleanliness. Treatment practices that result in high magnesium recoveries usually result in a higher level of metal cleanliness. An exception, however, is the in-the-mold nodulizing process, which can provide very high magnesium recoveries, while contributing a substantial volume of reaction products in the gating system.

5. *Inoculant Types and Practices*—The type of inoculant used and the method with which it is applied may negatively affect metal cleanliness, causing filter blockage. This should be considered, particularly if evidence of oxidized or undissolved inoculants is found on ceramic filters.

Metal flow prior to blockage will be in the range of 15 to 30 pounds of iron per square inch of filter area, an important factor in selecting the size and number of filters required for a given application.

Benefits of Filtration

Ceramic filters have been proven to provide the following benefits:

- reduced inclusion-related scrap (both before and after machining);
- simplified gating system design;
- improved casting machinability;
- reduced machine stock allowances;
- improved casting appearance.

RISERING

EFFECTS OF FOUNDRY PROCESSES ON RISERING

Successful use of risers to produce quality ductile iron castings of high soundness, requires that melting, inoculation, molding, and rigging must be controlled.

Control of those variables related to carbon equivalent (CE) that affect feeding may be summarized as follows:

- as CE is decreased, the feed requirement increases;
- as carbon content increases, the graphite available for expansion increases, and the feed metal required decreases;
- as the silicon increases above 1.8% (for a 4.6% CE iron), the feed requirement can increase;
- when the total carbon plus 1/7 silicon is equal to or greater than 3.9, it is reported that little or no shrink occurs;
- when CE is above 4.55%, carbon flotation is possible in heavy sections.

Figures 11-9 and 11-10 show these relationships.

The metal, prior to pouring, should be inoculated. Because of reabsorption of nucleation sites, time is both an asset and a deterrent to good risering practices. When the metal has been inoculated, no more than 15 minutes should occur before pouring. Beyond this time, both the proper graphite shape as well as the casting's feeding requirements become questionable.

The molding methods also have a contributing effect on risering efficiency, especially mold hardness. Because of low mold hardness, a riser can become ineffective in eliminating shrinkage, resulting in increased riser requirements

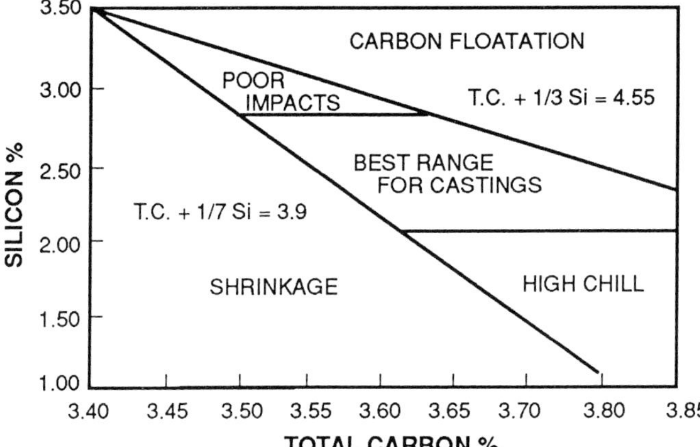

Fig. 11-9. The effect of composition on shrinkage and flotation in ductile iron. (Reynolds, Maitre and Taylor)

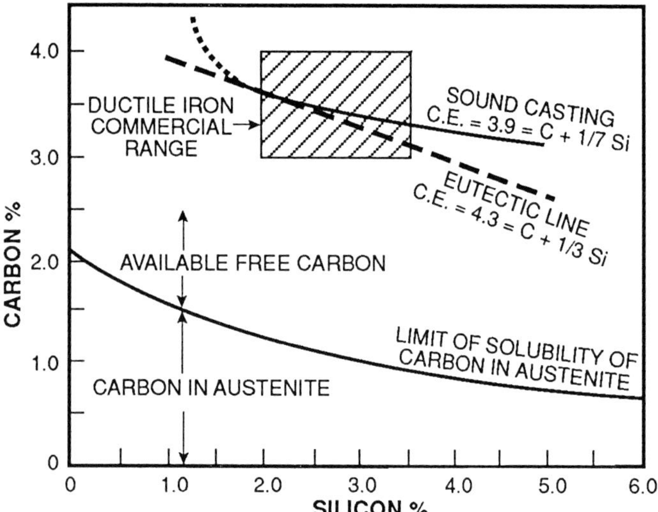

Fig. 11-10. Relationship of carbon and silicon that will remain in solution at various carbon and silicon levels.

for ductile iron castings. Large risers are often required for castings produced in low-pressure green sand molds, while the same castings made in high-pressure green sand or CO_2 molds will require risers that are relatively small. Figure 11-11 shows the effect.

The effect of green sand and nobake sand on mold dilation is shown in Figure 11-12. Molds with a hardness of 95 and over have minimum mold wall movement and also require the smallest risers. The rigging engineer should consider the foundry process in his shop when rigging a new casting or trying to correct a problem casting.

VOLUME CHANGES

Before proceeding with the risering of ductile iron, an understanding of the volume changes (liquid contraction from 2650F (1454C) occurring as the casting solidifies should be considered.

Typically, ductile iron castings are poured with 300–500°F (149–260°C) of superheat, i.e., heavy sections poured at about 2400F (1316C) and thin sections 2550–2650F (1399–1454C). The importance of this temperature is related to the volumetric changes that occur during solidification from pouring through the liquid state, and final graphitization and solidification.

In the liquid state, the volumetric change, or contraction, is about 0.4% to 0.8% per one hundred degrees (F) of superheat, depending on the CE (for 500°F [149°C] this will be 2.0%–4.0%). The second volumetric change occurs during the time the metal is solidifying. This is known as

austenitic contraction and varies from 0.25–2.00%, depending on the carbon equivalent (i.e., 4.5–4.3% CE).

The final volumetric change that will occur is during graphite precipitation. If inoculation is correct and there is little or no absorption of the nuclei, an expansion of about 3.4% will occur.

If all of the volumetric changes are totaled, we can see that there should be about +1.75 to +2.25% expansion for a 4.3% CE iron. Then why should the foundryman be concerned with a riser? Because the mold also contributes a volumetric change. Mold wall movement will contribute a 2–4% volume increase to the feeding requirement, and this will be a negative—caused by pressure exerted by the expansion of the casting, and burn-out of the sand in contact with the casting.

For a 4.3% CE casting, these volumetric changes when the liquid contracts from 2650F (1454C) may be summarized as follows:

0.4% × 5.5	=	–2.20%
Austenitic contraction	=	–2.00%
Mold wall movement	=	–2.00%
Graphite expansion	=	4.00%*
Total	=	–2.20%

***[Editors' Note: Various references cite this figure as much as 6–10%.]**

A comparison of carbon equivalent and volumetric change is shown in Figure 11-13.

Therefore, as molds become very rigid while pouring temperatures are held about the same, a thin casting may not require risering; or, if carbon equivalents are high, elimination of risers is possible.

THE RISERING OF DUCTILE IRON

Risering of ductile iron is done in two ways, either with the conventional sand riser, or with insulating or exothermic/insulating sleeves. Some foundries are finding that sleeves offer both an improvement in quality (i.e., sound castings) and yield.

There are several methods used for risering. The simplest method is to use 30–40% of the casting volume for a sand riser, or 15% of the casting volume for a sleeved riser. A second method is to use a shape factor, while a third method involves the use of a computer program that covers most of the variables discussed in the previous sections (i.e., mold hardness, pouring temperature, and chemistry).

Method 1 needs no description; method 3 is beyond the scope of this section, since the computer program makes all of the calculations and also makes the recommendations for risering requirements.

Method 2, the shape method, will be reviewed. The steps are as follows:

1. Calculate the weight or volume of the casting.

2. Locate the hot spots on the casting where risers should be attached (unless a side riser will be used and dimensioned to control the liquid feed into the casting).

3. Decide whether a top riser or a side riser is to be used.

4. If a side riser is to be used, calculate the surface area of the casting. This area will be divided into the casting volume to determine the side neck dimension. This is the casting modulus (M_c). It will be used with Figure 11-14 to obtain the neck modulus (M_N).

5. Measure the length, width and height of the casting in the position it will be cast. Then use the following formula:

$$f = (L + W) / T$$

where

f = Factor Number

L = Length

W = Width

T = Thickness or Height

This gives a factor number.

6. Look up the factor number on Figure 11-15. This gives the percent of total riser weight or volume.

7. Decide how many risers to use. Portion the riser weight or volume into each section to be risered. Repeat Step 6 for individual sections of the casting.

8. If a top riser is used, make a breaker core based on the riser (metal) diameter (D), i.e., opening 40–50% of D and thickness 10% of D. If a side riser is used, calculate the side riser dimensions (Fig. 11-14 and Example 2).

9. Check hot spots where top risers are attached to make sure that bleed back into the riser does not cause a shrink (Fig. 11-16). (See Example 1 for breaker core problems that might result.)

■ RISERING EXAMPLES

The following examples of method 2 are shown: a plate and a bar casting, along with problems that may result in shrinkage due to the riser contacting the casting when the riser is slightly undersized, and one method of correction.

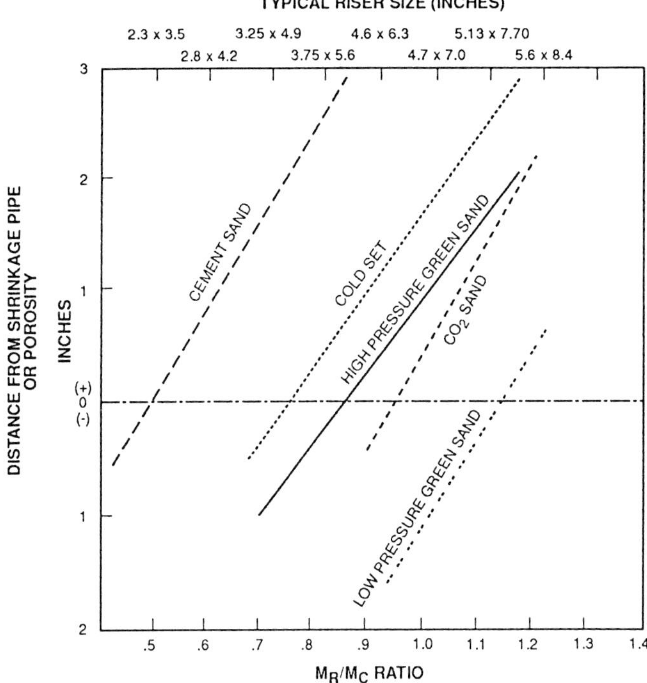

Fig. 11-11. Effect of molding materials upon casting soundness with a top-risered, chunky casting of a modulus of 1.0 in. (Gough and Morgan)

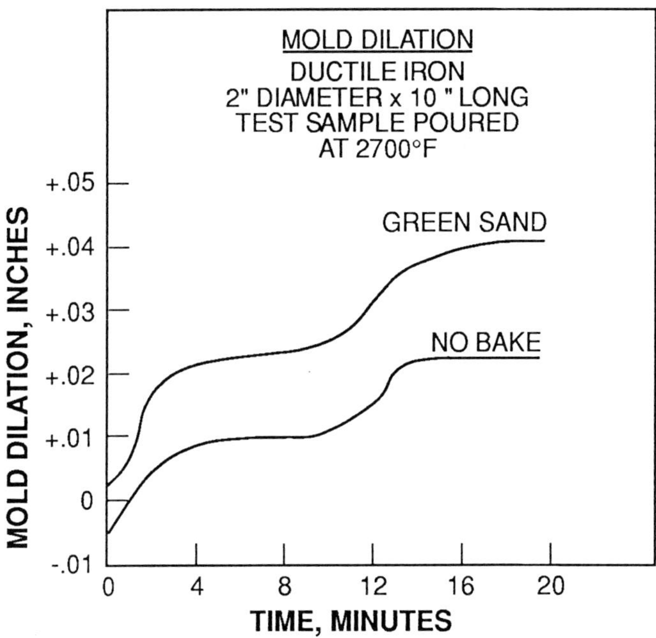

Fig. 11-12. The effect of molding sand on mold dilation for ductile iron.

189

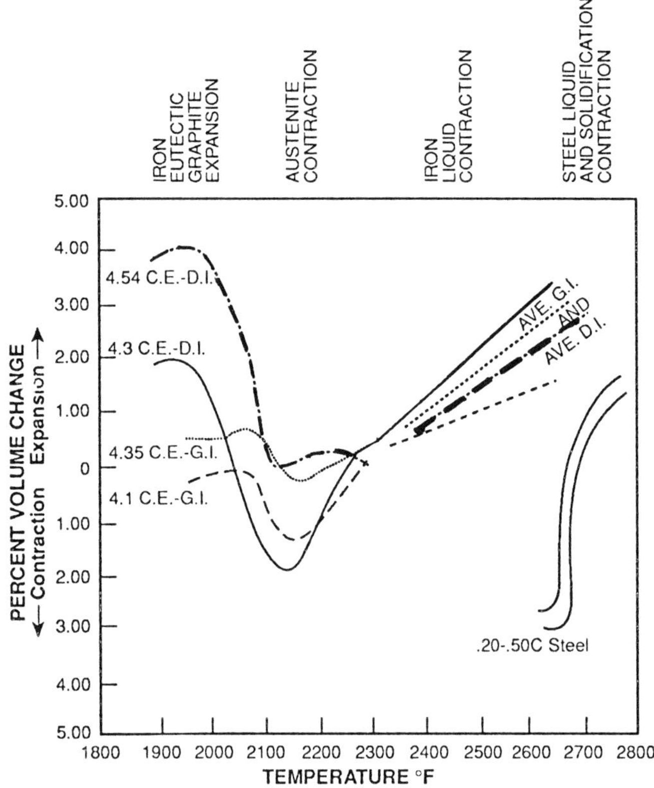

Fig. 11-13. Volumetric change on cooling from liquid to solid for ductile iron, gray iron and steel.

EXAMPLE 1: PLATE CASTING

(Using Method 2—shape factor)*

Step 1

$L \times W \times T$ = weight

20 in. $\times$ 10 in. $\times$ 0.625 in. = 125 in.^{3}

0.254 lb/in.^{3} $\times$ 125 in.^{3} = 31.75 lb

Step 2—Not Applicable

Step 3

Decide on top or side risering: this casting will be top risered.

―――――――――

*Refer to Figure 11-17.

Step 4—N/A, top risered.

Steps 5 and 6

Factor number (using Fig. 11-15)

$f = (L + W\ /\ T) = (20 + 10\ /\ 0.625) = 48$

From Figure 11-14: 5.3% for sleeved riser; 20.0% for sand riser.

Step 7

Use one riser.

Sleeved riser: about 1.68 pounds, 2 in. dia $\times$ 2 in. high.

Sand riser: about 6.38 pounds, 3 in. dia $\times$ 3.75 in. high.

Step 8

Breaker core for top riser (Fig. 11-18).

Sleeved riser: opening 0.8 in. dia $\times$ 0.25 in. thick

Sand riser: opening 1.20 in. dia $\times$ 0.30 in. thick

(Use 64 AFS or finer sand for breaker core).

Step 9

Make a layout of the casting-riser junction, and check the hot spot with the riser attached to locate its center (back purging might occur into the riser). The center of the hot spot will determine where shrinkage will occur (e.g., in the casting—shrink in casting; in the riser above the contact—no shrink in casting; at the contact—shrink at the contact). Sometimes a change in the core thickness or hole opening will correct this condition (Fig. 11-19).

Problems at Riser Contacts

One of the checks needed, since this is a plate section, is for the possibility of back-purging due to the casting-riser junction. Figure 11-16 gives the pictorial description of the back purging phenomenon.

Breaker Cores

The center of the hot spot is a candidate for a shrink if the riser diameter is not large enough to bring the hot spot center above the casting-riser interface. Figure 11-20 shows the effect of a properly and improperly sized riser contact due to the hot spot. One way to correct this problem is to reduce the contact size or to break up the hot spot by increasing the thickness of the core (Fig. 11-19).

The opening in the breaker core can be calculated as follows:

Sleeved Riser: $D = (1.13\ \text{to}\ 1.45) \times M_c$ (Ref. 2)

Sand Riser: $D = (1.72\ \text{to}\ 2.20) \times M_c$ (Ref. 3)

Side Risers

When a side riser is used, the size of the neck becomes critical. Most foundrymen tend to increase the riser size to keep the neck open. The effect of a large riser is to superheat the sand around the neck and slow down its cooling, allowing liquid feed to the casting. Breaker cores can be used on necks for side risers the same as on top risers, allowing easy riser removal. The effect of neck dimensions on a plate casting is shown in Figures 11-21 and 11-22.

EXAMPLE 2: BAR
(Using Method 2—shape factor)*

Step 1

$$L \times W \times T \times d = \text{weight}$$

$$10 \text{ in.} \times 2.5 \text{ in.} \times 2 \text{ in.} \times 0.254 \text{ lb/in.}^3 = 12.7 \text{ lb}$$

Step 2

The side riser will be dimensioned to control the liquid metal fed into the casting, preventing hot spots.

Step 3

This will be a side riser.

Step 4

[Refer to Step 8.]

Steps 5 and 6

Factor number (using Fig. 11-15):

$$f = (L + W) / T = (20 + 2.5) / 2 = 6.25.$$

(From Figure 11-15: 12% of casting weight for sleeved riser; 40% of casting weight for sand riser.)

Step 7

Use one riser. For a 1.53 lb sleeved riser, the sleeve dimensions should be 2 in. dia $\times$ 2 in. high; for a 5.08 lb sand riser, the sleave dimensions should be 2.5 in. dia. $\times$ 3.75 in. high.

Step 8

Calculation for neck connection of side riser.
Area:

$$\text{Ends} = 2 \text{ in.} \times 2.5 \text{ in.} \times 2 = 40 \text{ in.}^2$$

$$\text{Top and Bottom} = 2.5 \text{ in.} \times 10 \text{ in.} \times 2 = 50 \text{ in.}^2$$

$$\text{Total} = 100 \text{ in.}^2$$

The modulus of the casting is:

$$M_c = (\text{VOLUME/AREA}) = (50 \text{ in.}^3 / 100 \text{ in.}^2) = 0.50 \text{ in.}$$

Using 0.95 to 1.00 $\times M_c$ for neck modulus (Fig. 11-22), the neck modulus is:

$$M_N = 0.95 \times 0.5 \text{ in.} = 0.475 \text{ in.}$$

Using Figure 11-14, the dimensions of the neck can be approximated. In this case 1.88 in. $\times$ 2 in. will be used, giving an actual neck modulus of:

$$M_N = (W \times T) / 2 (W + T)$$

$$= (1.88 \text{ in.} \times 2 \text{ in.} / 2 (1.88 \text{ in.} + 2 \text{ in.}) = 0.484 \text{ in.}$$

Since the neck will contact the casting on one side and the risers on the other, the two contact faces are not included. (If the neck is over one inch in length, the length of the neck should be included when calculating the surface area). The riser base width will be the outside diameter of the sleeve or the diameter of the sand riser. The height of the riser base will be as high as the neck, with a base extension of 1/4 to 1/2 inches into the drag.

Step 9

Since this is a side riser and the neck has been dimensioned to control the liquid feed into the casting, the hot spot exercise will not be needed.

Because of the extra metal required for necks and bases, side risered castings will have a lower yield than top risered castings. The cleaning and riser removal play a large part in the choice of how a casting will be risered.

---

*Refer to Figure 11-23.

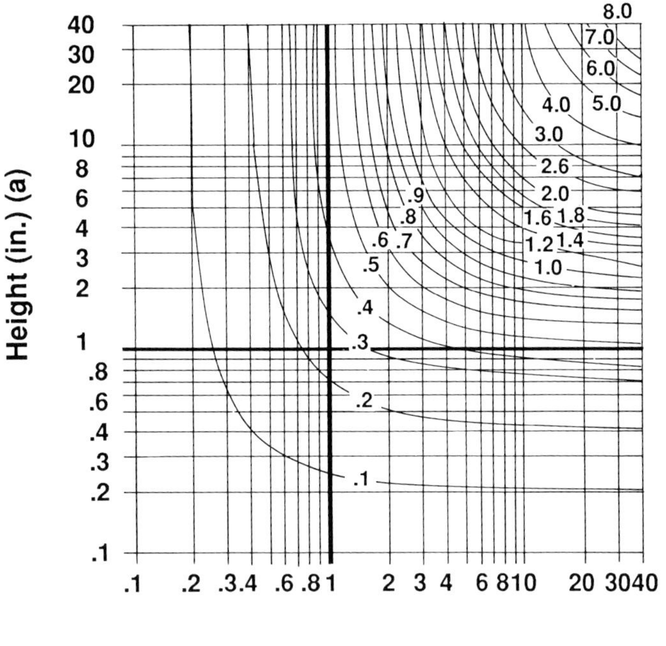

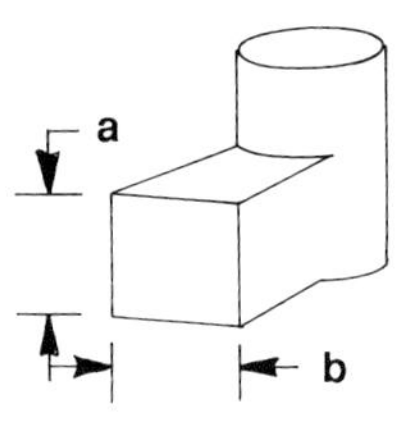

Example: 4x4 neck

$$M_n = \frac{a \times b}{2a + 2b}$$

$$= \frac{4 \times 4}{2(4) + 2(4)}$$

$$= \frac{16}{16}$$

$$= 1.0''$$

Fig. 11-14. Side riser chart for modulus and dimensions are shown. (Wlodawer)

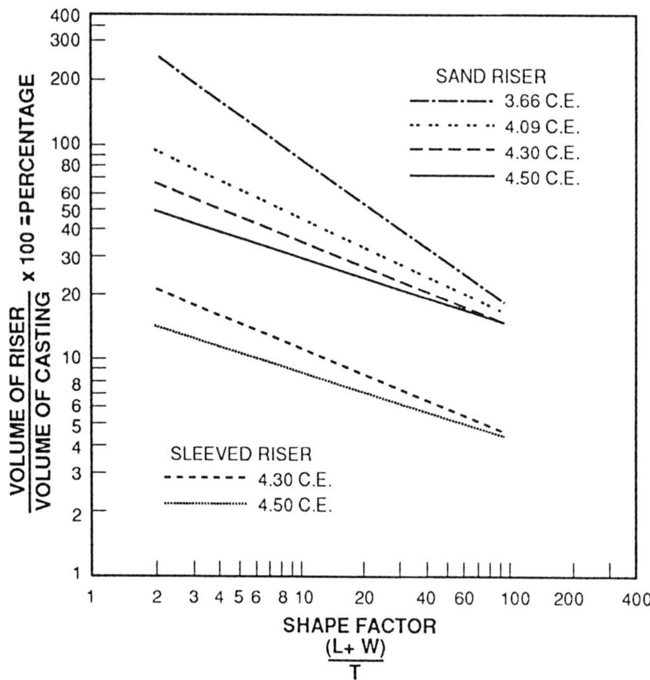

Fig. 11-15. The percent of total riser weight or volume can be determined by correlating the shape factor for castings with different carbon equivalents.

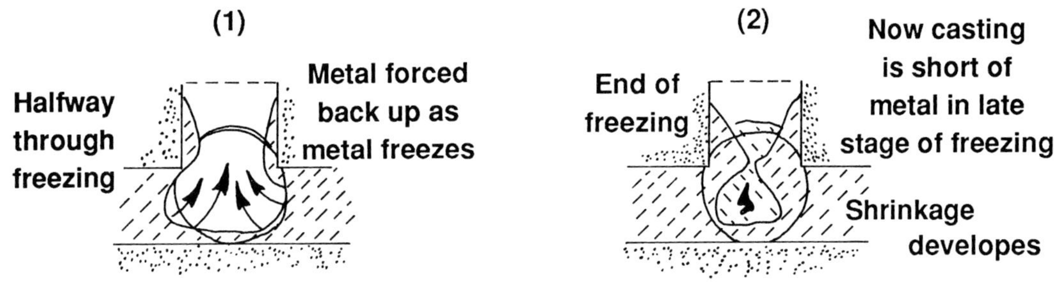

Fig. 11-16. The effect of a hot spot on shrinkage due to back purging in a top riser is shown.

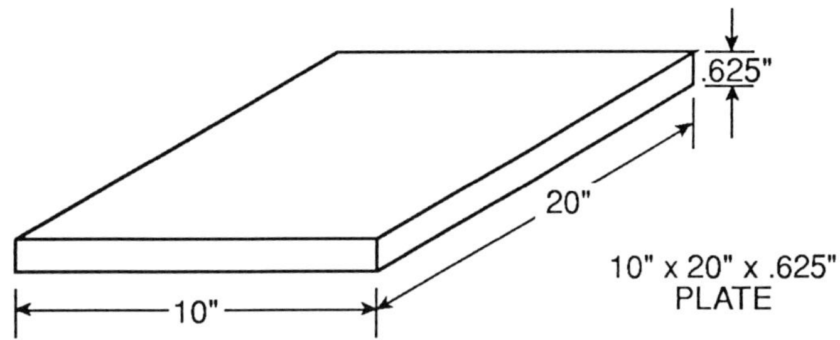

Fig. 11-17. Dimensions of plate used as an example of the "shape method" to calculate a ductile iron riser.

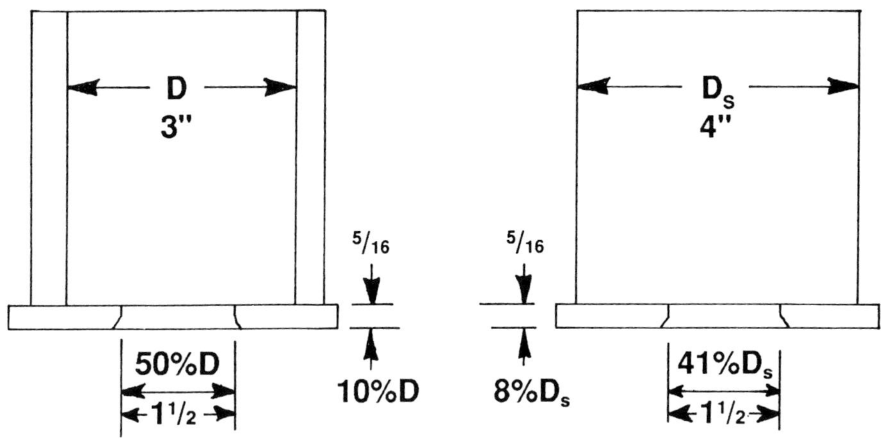

Fig. 11-18. Calculation of a breaker core for a top riser: examples with different riser diameters, same breaker. If hole or thickness is varied, the "rule of thumb" is suggested.

	Standard Dimensions		"Rule of Thumb"	
	Thickness	Hole	Thickness	Hole
Steel	10% D	50% D	T	5T
Grey Iron	10% D	40% D	T	4T
Ductile	10% D	40–50% T	T	45T

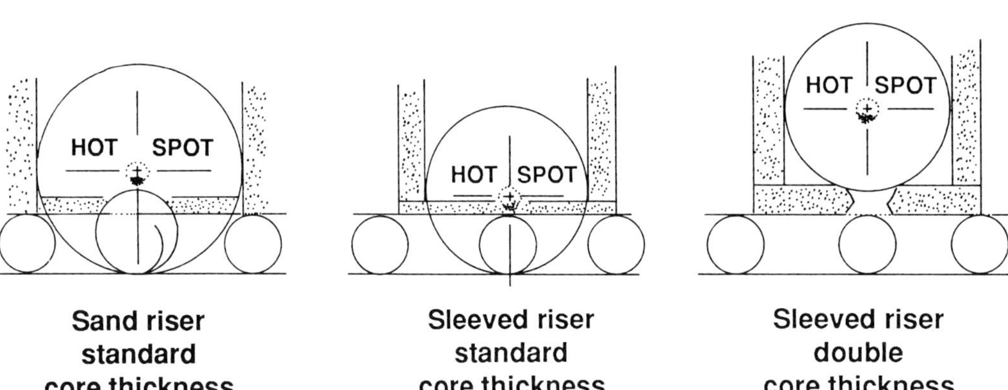

Fig. 11-19. Effect of riser size and breaker core thickness on both the hot spot center and shrinkage in a top-risered casting.

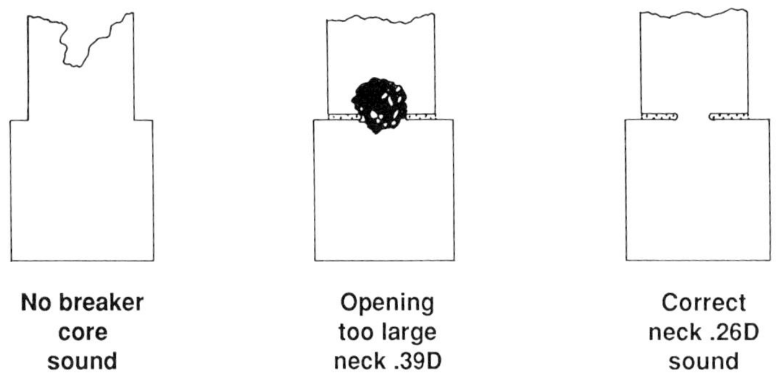

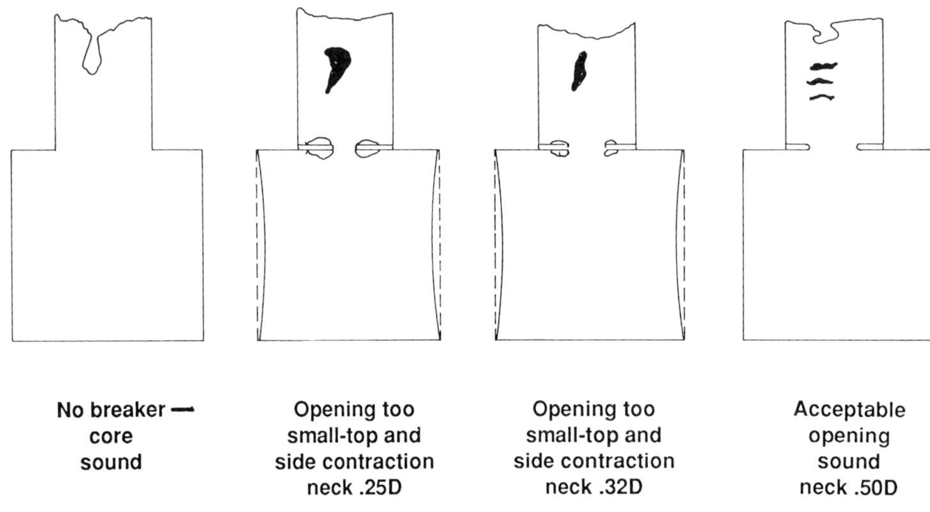

Fig. 11-20. The effect of properly and improperly sized riser contact on breaker core opening.
(Gough and Morgan)

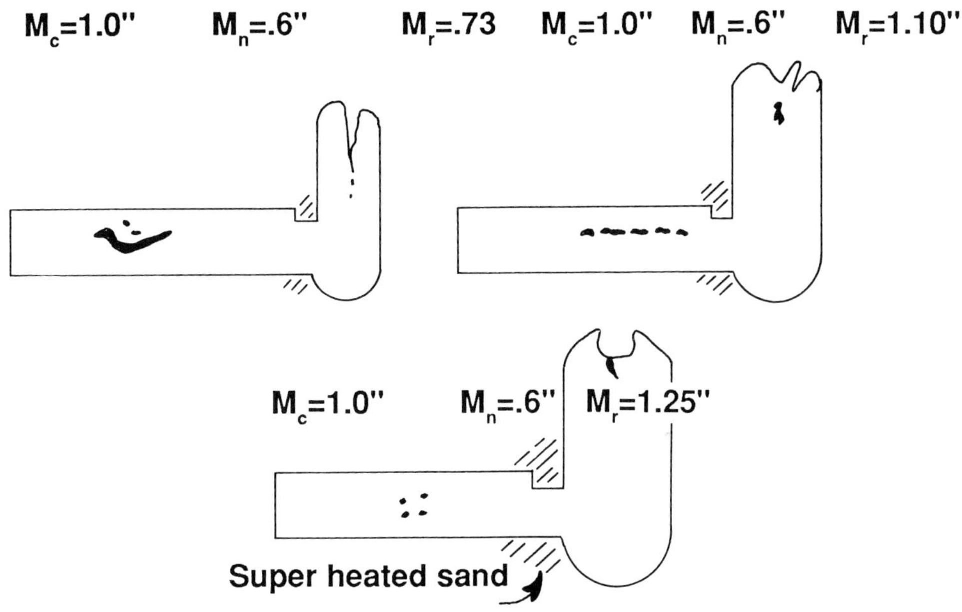

Fig. 11-21. The effect of changing riser size to superheat the neck area, resulting in improved feeding.

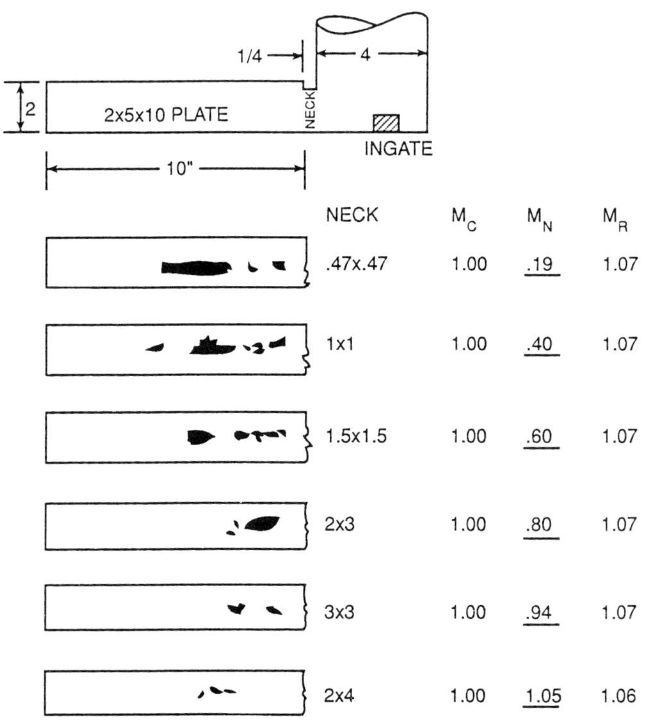

NECK	M_C	M_N	M_R
.47x.47	1.00	.19	1.07
1x1	1.00	.40	1.07
1.5x1.5	1.00	.60	1.07
2x3	1.00	.80	1.07
3x3	1.00	.94	1.07
2x4	1.00	1.05	1.06

Fig. 11-22. Various shrinkage defects produced in a test casting with various neck sizes in an adequately risered casting.

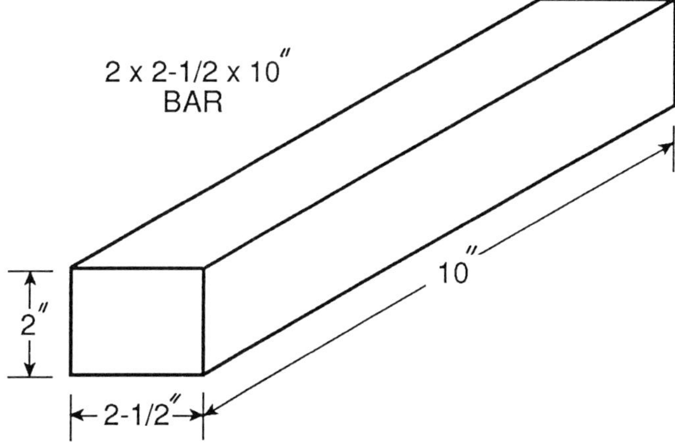

Fig. 11-23. The dimensions of the bar casting (Example 2) are shown.

■ BIBLIOGRAPHY

Bates, C.E. and B. Patterson. "Volumetric Changes During the Freezing of Hypoeutectic Ductile Iron," *AFS Transactions*, vol 87. Des Plaines, IL: American Foundrymen's Society, Inc. (1979).

Bates, C.E., R.H. McSwain and G.L. Oliner. "Volumetric Changes During Freezing of Ductile Cast Iron," *AFS Transactions*, vol 85. Des Plaines, IL: American Foundrymen's Society, Inc. (1977).

Bertolino, M.F. and J.F. Wallace. "Influence of Metal Factors, Molding Sand Composition, and Casting Shape on Mold Cavity Enlargement," *AFS Research Report.* Des Plaines, IL: American Foundrymen's Society, Inc. (1968).

Day, J.P. and H. Kind. "The Development and Application of Cellular Ceramic Filters for Gray and Ductile Iron," *AFS Transactions*, vol 92. Des Plaines, IL: American Foundrymen's Society, Inc. (1984).

Gough, M.J. and J. Morgan. "Feeding Ductile Iron Castings: Some Recent Experiments," *AFS Transactions*, vol 84. Des Plaines, IL: American Foundrymen's Society, Inc. (1976).

Karsay, S.I. *Ductile Iron III: Gating and Risering.* Montreal: Q.I.T. Fer et Titane Inc. (1981).

Kind, H. "Filter Application and Gating System Design for Iron Castings," *AFS Ferrous Filtration Conference Proceedings.* Des Plaines, IL: American Foundrymen's Society, Inc. (1986).

Spangler, A.F., ed. *The Ductile Iron Process.* Chicago: Miller and Company (1984).

Reynolds, C., J. Maitre and H. Taylor. "Feed Metal Requirements for Ductile Iron Castings," *AFS Transactions*, vol 66. Des Plaines, IL: American Foundrymen's Society, Inc.(1958).

Ruddle, R.W. "Gating and Risering of Iron Castings," AFS Chapter Talk (1978).

White, R.W. "Gating and Risering of Ductile Iron Castings," *Foundry Management and Technology* (Feb/Mar 1960).

Wlodawer, R. *Directional Solidification of Steel Castings.* Pergamon Press, (1966).

Wukovich, N. and R. Neuman. "Risering Ductile Iron," unpublished (June 1, 1984).

12

Influence of Section Size on Microstructure and Mechanical Properties

Al Alagarsamy

Grede Foundries, Inc.
Milwaukee, Wisconsin

■ INTRODUCTION

Ductile iron is section sensitive, though not to the same degree as gray iron. Mechanical properties of ductile irons are influenced by matrix microstructure, which, in turn, is influenced by the section thickness of the casting and its chemical analysis. Section thickness mainly affects solidification rate (time required for complete solidification from the beginning to the end of solidification) and casting cooling rate, through the austenite transformation range. In heavier sections, because of extended solidification times, magnesium fading will also be affected. These two factors— solidification rate and cooling rate—influence nodule count, presence of carbides, and the amount of ferrite and pearlite present. There have been many investigations in the past to evaluate and quantify the effect of section size on the matrix and mechanical properties.[1] Most of the work was done using laboratory-produced iron, which may not represent commercially produced irons. However, we can still learn from the general conclusions of these studies pertaining to alloy effects on matrix formation and relative cooling rates of different section sizes.

■ EFFECT OF SECTION SIZE ON SOLIDIFICATION RATES

Solidification time is a function of volume and surface area of the casting, mold heat extraction rate, pouring temperature, and the thermal properties of the solidifying metal. For a given molding medium and solidifying metal, solidification time (t) depends on the volume-to-surface area ratio (V:SA). Figure 12-1 shows the effect of different shapes and sizes and their solidification times. The smaller the V:SA ratio, the faster the solidification rates. Thin-section castings have small V:SA ratios and, consequently, have very short freezing times.

Solidification starts from the outside walls and proceeds towards the center. External corners solidify first at a rapid rate, and as solidification progresses, the molding medium is heated, reducing the heat extraction rate and leading to slower solidification at the center of the casting. When a small amount of molding medium is surrounded by fairly large sections of metal, as in small cores in heavy-section castings, the molding mass is heated to a high temperature, thus reducing heat transfer from the solidifying

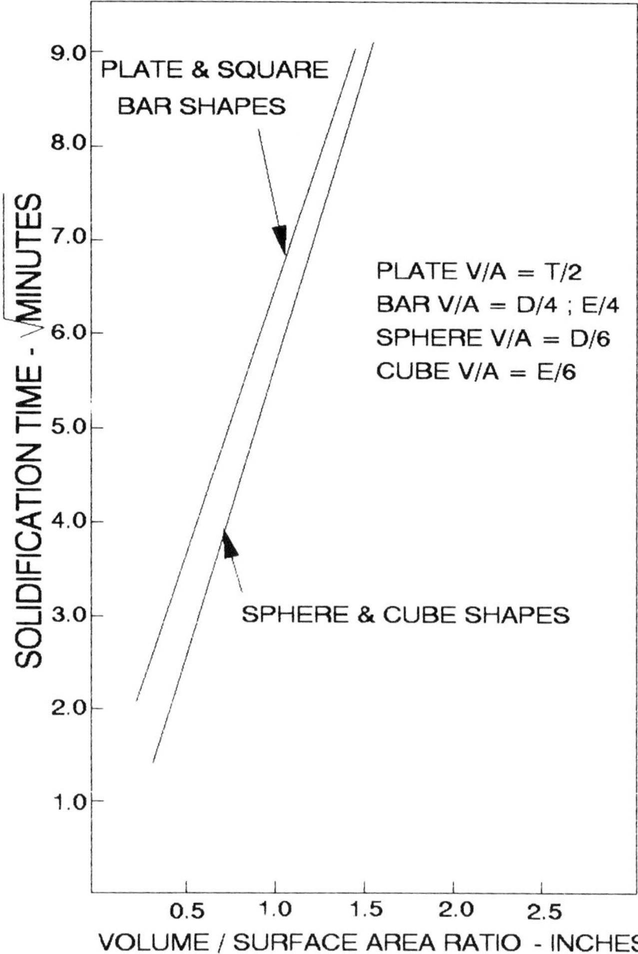

Fig. 12-1. Solidification time vs. shapes and sizes.

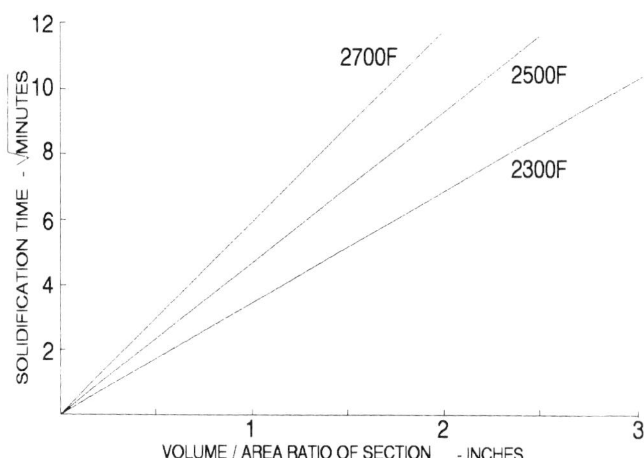

Fig. 12-2. Effect of pouring temperature on solidification time of castings made in green sand molds.

forming tendency. Nucleation rate is increased with increased undercooling, which results in increased nodule count if carbide formation can be suppressed. Figure 12-3 shows the effect of solidification times on nodule count.

Thermal capacity and conductivity of mold material influences the heat transfer rate and, thus, the solidification rate. Solidification rates are much faster in metal molds, especially in copper molds. The nodule count of ductile iron made in a metal mold is extremely high (500-1000 mm^2). As the section thickness increases, solidification rate decreases, resulting in lower nodule counts. Slow cooling rates promote segregation of elements like manganese, chromium, molybdenum, etc., towards last freezing areas and tend to stabilize carbides and promote microporosity. These intercellular carbides

metal. Cubes and spherical-shaped castings solidify faster than plates and cylindrical shapes for the same V:SA ratios. This is due to the fact that there is an increasing amount of molding medium—in the case of cubes and spheres—as the distance from the casting increases, thus providing a larger capacity for heat extraction.

Figure 12-2 shows the effect of pouring temperature on solidification times. When metal is poured at higher temperatures, it takes longer to solidify. There is an increased amount of heat to be removed before the metal starts solidifying, and the molding medium is heated to a higher temperature, thus reducing the heat transfer rate.

Ductile iron undercools before the bulk of the eutectic solidifies. The amount of undercooling increases with cooling rate, when all other metallurgical factors are kept constant. When the eutectic undercooling temperature is low (below 2080F/1138C), iron is prone to chill carbide formation. Late effective inoculation decreases the chill-

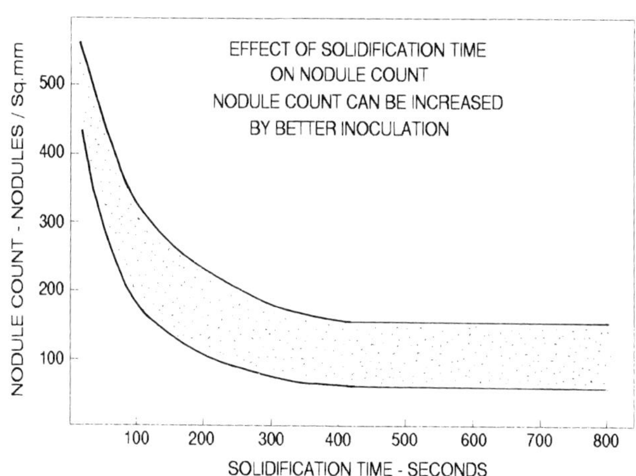

Fig. 12-3. Nodule count vs. solidification time.

and associated porosity are detrimental to dynamic properties, such as fracture toughness and fatigue endurance limit. In very heavy-section castings, chunk graphite may be formed, especially at the center of the castings. Chunk graphite formation can be minimized by the proper balance of flake graphite-promoting elements such as antimony, lead, tin, arsenic, bismuth, etc., and chunk graphite-promoting elements such as silicon, cerium, and nickel.[9] When the metal is relatively pure (very low levels of deleterious elements), cerium and silicon levels should be kept low in heavy-section castings.

Cooling Rate

For a given chemistry and nodule count, the casting cooling rate through the austenite transformation range determines the final matrix structure. Faster cooling rates, which are associated with thin sections, promote pearlite formation, whereas slower cooling rates favor ferrite formation. Generally, faster solidification rates are accompanied by faster cooling rates through austenite transformation.

Besides the section thickness, the total mass of the casting—relative to the amount of molding medium (sand:metal ratio) and the nature of molding medium—influences the cooling rate differently than the solidification rate. When the total weight of metal poured is increased, molding materials are heated to a higher temperature as the metal solidifies and cools to the transformation temperature range. This slows down the cooling rate, thus providing longer times for ferrite formation. And molding methods, such as "shell" and "lost foam," result in slower cooling rates, even though solidification rates may be high with

these processes. Schematic cooling curves are shown in Figure 12-4 with different combinations of solidification and cooling rates, and expected microstructures.

Nodule count plays an important role in controlling the final matrix. For the same cooling rate, increasing amounts of nodules provide increasing centers for ferrite nucleation and growth, thus resulting in increasing amount of ferrite. For this reason, late and multiple inoculation steps are employed to increase the nodule count and ensure the maximum amount of ferrite. But the conflicting effects of section thickness must be recognized. Thinner sections generally promote high nodule counts and, thereby, have a tendency to increase the amount of ferrite. But they also increase the cooling rate, thus reducing the ferrite content. The final matrix structure depends on nodule count, cooling rate (section thickness), and chemistry (carbon equivalent and other alloy factors).

▬ MECHANICAL PROPERTIES

Mechanical properties are influenced by microstructure, which, in turn, is influenced to some extent by the section thickness of the casting. Very thin sections solidify very fast, thus increasing the nodule count. When the undercooling is very pronounced, chill carbides are formed at the outer edges. The presence of carbides reduces elongation and increases hardness. But with efficient late inoculation, carbides can be minimized, even in 0.25-inch-section castings. Even when carbides are present at the time of solidification, they anneal out in the mold if the cooling rate after solidification is slow enough and the silicon level is high enough.

Mechanical properties are also influenced by alloying elements present in the section thickness under consideration. When very small amounts of pearlite stabilizers are present, predominantly ferritic structures are obtained. Higher silicon levels also favor ferrite formation. Copper, manganese, tin, etc., stabilize pearlite, increasing the casting's strength.

For heavier-section castings, copper alone may not be enough to ensure a high percentage of pearlite. In addition to copper, small amounts of tin can be added to achieve a high percentage of pearlite. When there are large differences in the cooling rate between different sections of the casting, the addition of tin is advantageous. However, tin additions, above the level necessary for pearlite stabilization (0.08%), is not advisable, because intercellular flake graphite may be formed. Tin also promotes embrittlement of the grain boundary areas, thus reducing elongation and raising impact transition temperature. Nickel is not as strong as copper in promoting pearlite, but is used in combination with copper in heavier sections to achieve higher strength.

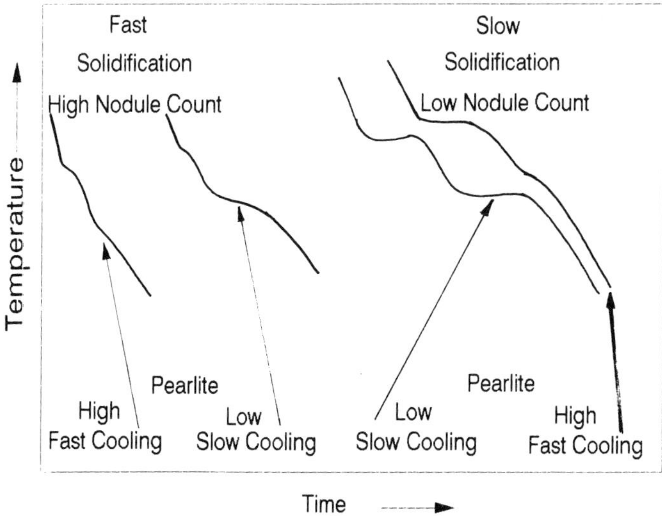

Fig. 12-4. Schematics of cooling curves of ductile iron.

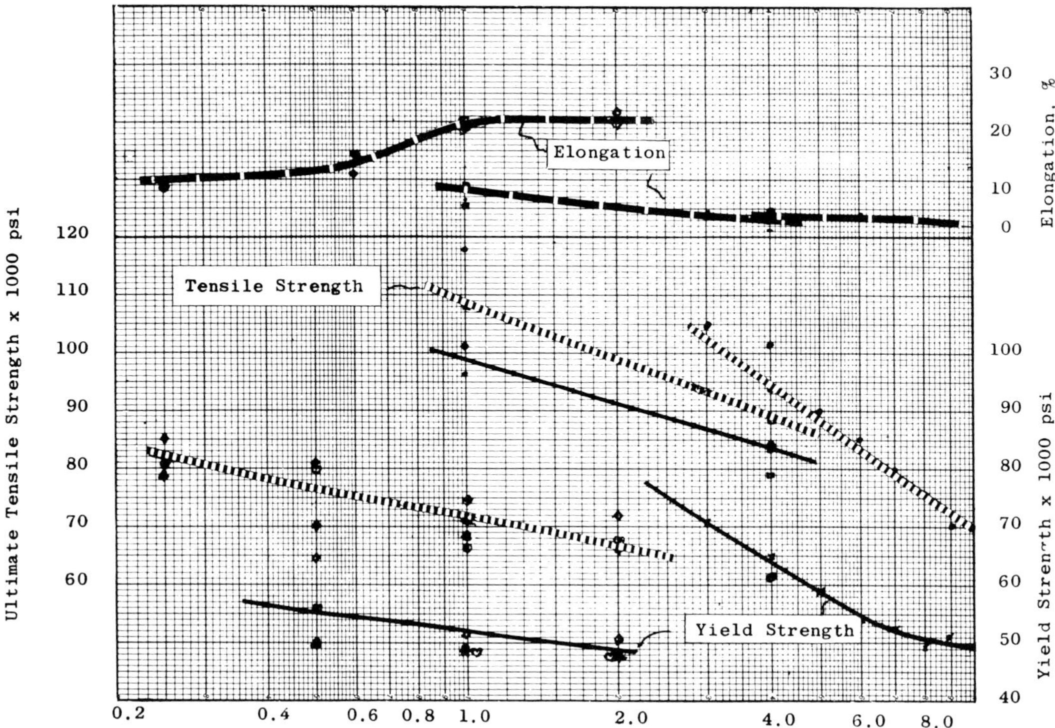

Fig. 12-5. Effect of section thickness on mechanical properties of as-cast ductile irons.

Figure 12-5 shows the relationship between mechanical properties and section thickness. The information was derived from several sources and, thus, represents different chemistries.[5] From this figure, it can be seen that, generally, as-cast tensile and yield strengths drop as the section thickness increases. Elongation increases with section thickness up to about two inches, and then gradually decreases.

In the as-cast condition, with increasing section thickness, ferrite content increases as a result of slower cooling rates. This accounts for the variation in the properties. In heavier sections, slow cooling rates ensure a ferritic matrix, resulting in lower strengths and higher elongation. When segregating elements, such as manganese, chromium, molybdenum, etc., are present, slow solidification rates magnify the segregation effect because of fewer grain boundaries and result in cell boundary carbides. These carbides have the effect of lowering elongation, even when the matrix is predominantly ferritic. Tool life decreases when there are grain boundary carbides. For these reasons, manganese is not generally used for pearlite stabilization. By increasing nodule count (maintaining at least 100 nodules/mm^2), the effect of segregation can be minimized.

With good foundry practice of melting, treating, and pouring the effect of varying cooling rates on mechanical properties can be minimized, at least up to 2-inch sections, as shown in Table 12-1. These results are derived from Y-blocks made in green sand molds. Iron was melted in an electric furnace, treated with a pure magnesium process, and poured from a pressure pour furnace. Iron was inoculated, while pouring, using in-stream inoculation.

There is a tendency for lower strengths with increasing section thickness, in both pearlitic and ferritic castings. These differences can be minimized with annealing for ferritic castings and normalizing for pearlitic castings. Variation in properties up to a 6-inch section of ductile iron alloyed with 0.8%Cu and 1%Ni is shown in Figure 12-6.[6] Even after adding considerable amounts of alloys (Cu, Ni, Mo and Mn combinations), a fully pearlitic structure was not ensured in heavier sections.[6]

Fatigue endurance limit is also affected by section thickness. As the section size increases, nodule count decreases, increasing grain boundary segregation, which, in turn, decreases fatigue endurance limit. This decrease in fatigue properties is more pronounced than in tensile properties.[8]

Table 12-1. Mechanical Properties of As-Cast and Subcritically Annealed Ductile Irons of Different Section Sizes

Iron	"Y" Block Size	Tensile Strength (KSI)	Yield Strength (KSI)	Elongation Percent	Hardness (BHN)	Nodule Count
As-Cast						
A	5/8 in.	75.8	48.3	18.5	170	200
A	1 in.	74.3	48.5	17.1	176	125
A	2 in.	70.1	46.1	19.7	170	100
B	5/8 in.	115	66.5	8.5	244	200
B	1 in.	110	63	8.3	255	125
B	2 in.	117.5	67	9.8	262	125
Subcritical-Annealed Properties						
A	5/8 in.	66.5	46.6	22	146	
A	1 in.	64.3	45.3	23.5	160	
A	2 in.	63.3	44.3	21	154	
B	5/8 in.	66.8	48.2	20.5	149	
B	1 in.	66.4	48.1	19.7	163	
B	2 in.	62.8	47.1	18.75	159	

Chemistry (%)							
Iron	Si	Mn	Cu	Cr	Mg	S	C
A	2.35	0.3	0.18	0.04	0.038	0.006	3.65
B	2.35	0.3	0.55	0.04	0.038	0.006	3.65

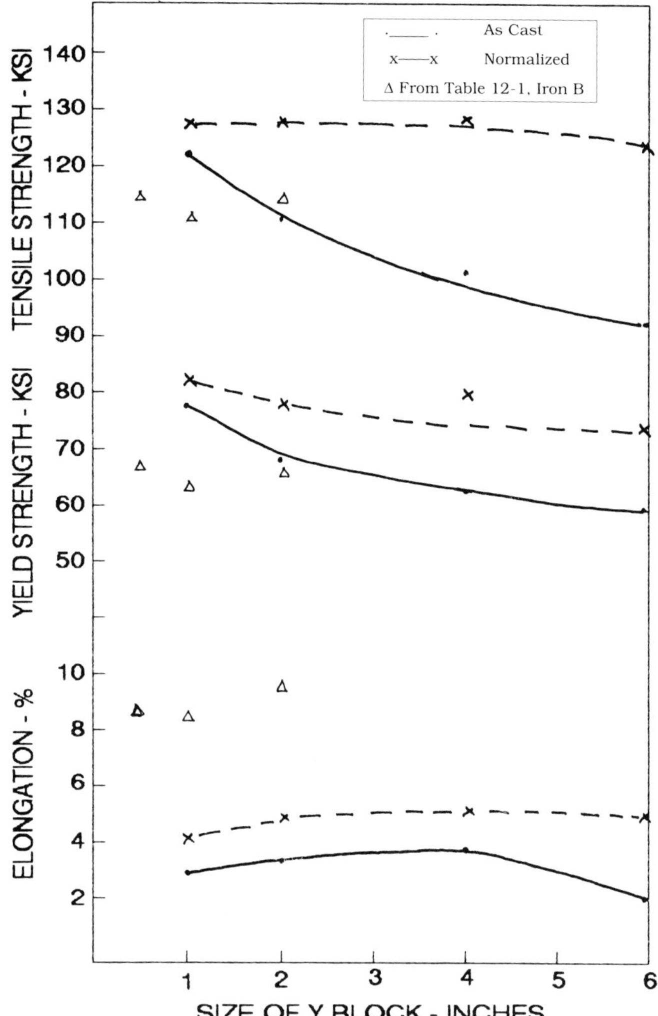

Fig. 12-6. Mechanical properties of heavy-section castings with pearlitic matrix.

Fig. 12-7. Recommended carbon equivalents for different section thicknesses.

bide-promoting elements like manganese, chromium, vanadium, etc., should be maintained as low as possible.

Elements that lower graphite eutectic and raise carbide eutectic temperatures are detrimental for the production of as-cast thin-section castings. Because of the propensity to chill formation, late and effective inoculation is necessary for the successful production of thin-section castings. To produce ferritic castings in the as-cast condition, all pearlite-stabilizing elements, such as copper and nickel, should be controlled to very low levels, along with the carbide stabilizing and segregating elements. Silicon levels are kept high (above 2.5%) to facilitate formation of ferrite and to minimize carbide formation.

THIN-SECTION CASTINGS

Careful consideration should be given in producing very thin-section castings of less than 0.25 inch in thickness. Carbon equivalent should be selected according to the section thickness. Figure 12-7 shows the relationship between section thickness and recommended carbon equivalents. Higher carbon equivalents are needed to prevent chill carbide formation in thin-section castings. Very high carbon equivalent irons may reduce fluidity and may result in misruns. Pouring temperatures need to be higher (above 2600F/1427C) for predominantly thin and rangy castings. Magnesium residual should be maintained at low levels (0.025–0.035%) to prevent carbide formation. Other car-

HEAVY-SECTION CASTINGS

As already discussed, heavy-section castings solidify slowly, resulting in lower nodule count. With fewer nodules and fewer cells, cell boundary segregation is very pronounced. Since these elements are also pearlite and carbide stabilizers, cell boundary areas will have stable carbides and residual pearlite. This reduces the ductility of heavy-section castings. For this reason, all forward-segregating elements should be controlled to very low levels to ensure maximum ductility.

Cerium is known to promote chunk graphite in heavy sections. Cerium-free magnesium alloys or the pure magnesium process should be used in producing heavy-section castings.

Carbon equivalent control is also important, as slow solidification enhances carbon flotation. Carbon flotation is detrimental to mechanical properties in the areas where flotation occurs. Either eutectic or slightly hypoeutectic compositions are preferred for heavy-section castings. When large risers are used in the production of intermediate sections, carbon flotation can occur in the risers, preventing piping, and making the risers ineffective in feeding castings. When carbon flotation occurs, the lower sections of the castings will show an analysis to near-eutectic composition.

Magnesium residuals should be maintained at a higher level (0.045–0.055%) for heavy-section castings in order to avoid chunk graphite formation and other degenerate graphite forms. Since the nodule count will tend to be lower, late and efficient inoculation is also essential for heavy-section castings. With higher nodule counts, many of the problems associated with slow cooling can be minimized.

The carbon equivalent of iron should be selected according to the predominant section thickness of the castings poured. To minimize porosity in heavy-section castings, maintaining a silicon below 2.25% is recommended. With lower silicon, it is easier to stabilize pearlite. Even with lower silicon, a combination of pearlite stabilizing elements may need to be used to ensure a high percentage of pearlite in the as-cast condition.

For the production of very heavy castings, where solidification times can extend over a period of hours, additional cooling in the mold is a must to eliminate chunk graphite formation and to maintain a decent nodule count. A foundry in Germany uses water-circulating pipes, embedded in the molds, in the production of ductile iron nuclear casks weighing over 170 tons. Without this additional cooling, it will take days to produce a casting, and the mechanical properties obtained from trepanned bars will not pass the minimum requirements.

■ REFERENCES

1. D.R. Askeland, S.S. Gupta; "Effect of Nodule Count and Cooling Rate on the Matrix of Nodular Cast Iron," *AFS Transactions*, vol 83, pp 313–320 (1975).

2. C.W. Briggs; "Solidification of Steel Castings," *AFS Transactions*, vol 68, pp 157-168 (1960).

3. E. Compomanes; "Effect of Minute Additions of Antimony on Structure and Properties of Ductile Iron," *AFS Transactions*, vol 71, pp 57–62 (1971).

4. T. Watmough, W.F. Shaw, F.C. Bock; "Combined Effects of Selected Elements on the Properties of Ductile Iron," *AFS Transactions*, vol 79, pp 225–246 (1971).

5. A.F. Spangler, ed; *The Ductile Iron Process*, Miller and Company, p 326 (1976).

6. K. Reifferscheid; "Characteristic Strength Values of Ductile Iron With Completely Pearlitic Matrix in the 1–6" Wall Thickness Range," *AFS International Cast Metals Journal*, pp 30–36 (Dec 1978).

7. M.J. Lalich, C.R. Loper; "Describing the Eutectoid Transformation in Hyper-Eutectic Ductile Cast Irons," *AFS Transactions*, vol 51, pp 238–245 (1973).

8. K.B. Palmer; "Effect of Cast Section Size on Fatigue Properties and the Prevention of Corrosion Fatigue of Nodular Irons," BCIRA Conference Paper, pp 201–212 (N.d.).

9. S.I. Karsay; *Ductile Iron Production Practices*, American Foundrymen's Society, Inc. (1975).

13

Heat Treatment

Bela V. Kovacs

AFC Technical Center
Livonia, Michigan

■ INTRODUCTION

A great deal of early published data on ductile iron evaluated products made from either raw materials or processes that have since been modified and improved. New molding processes, nodulizing alloys, postinoculation techniques and melting practices have made ductile iron a consistently reliable engineering material.

Standard grades of ferritic/pearlitic ductile iron can now be routinely produced by foundries without subsequent heat treatment. But when employed on these ferritic/pearlitic grades, heat treatment is used to: (1) correct a unique, out-of-spec condition or, (2) assure consistent quality in a less-than-capable casting process, or (3) to stress relieve or normalize castings that will be subsequently machined or surface hardened.

The first part of this chapter will deal with proven heat treatment techniques for producing the standard grades of ductile iron. Later, nonconventional grades of heat treated irons will be discussed. These include quench-and-tempered irons, surface-hardened irons, austempered irons, and austenitic irons. In these processes, the subsequent heat treatment is an integral part of the casting process. The heat treatment produces strength, wear resistance, or high temperature properties not available in the conventional ferritic/pearlitic grades.

In order to understand the heat treating process in ductile iron, it is important to discuss the metallurgy involved. As in all ferrous alloys, the amount of carbon in the matrix determines the final hardness of that matrix. The other alloying elements effect the "hardenability" of the alloy, but have little effect on the hardness. In steel, the amount of carbon in the matrix is fixed (i.e., a 1050 steel has 0.5% carbon at any temperature, unless it is intentionally carburized). In ductile cast iron, the carbon nodules provide "sinks" and sources for carbon, while the silicon in the matrix improves the "mobility" of the carbon. Therefore, unlike steel, the carbon content in the ductile iron matrix is a function of the silicon content and the thermal history of the casting. Figures 13-1a–d show the effect of thermal history on the final microstructure. In this chapter the various heat treatments and their resultant structures and properties will be reviewed.

■ MOLD COOLING

Ductile irons solidify with the small graphite spheroids encased in austenite. As the casting cools to the eutectoid temperature, the matrix essentially has eutectoid carbon

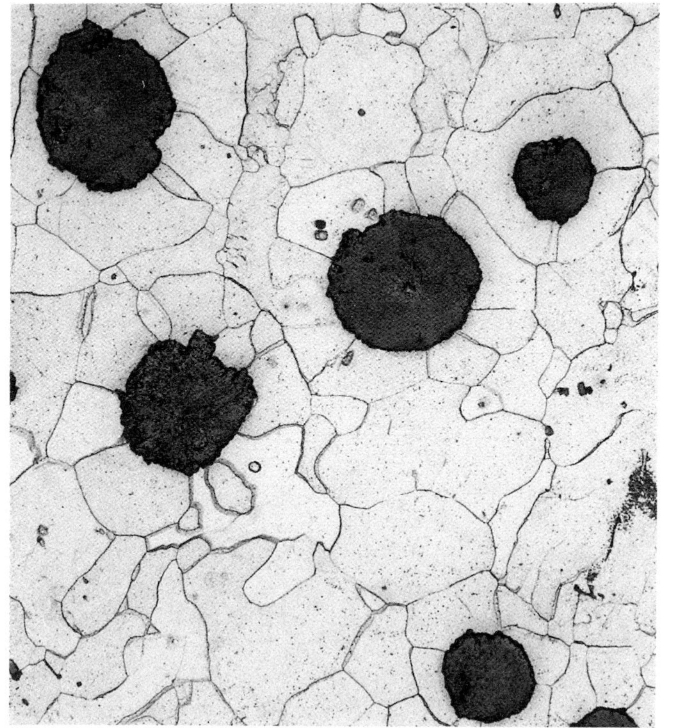

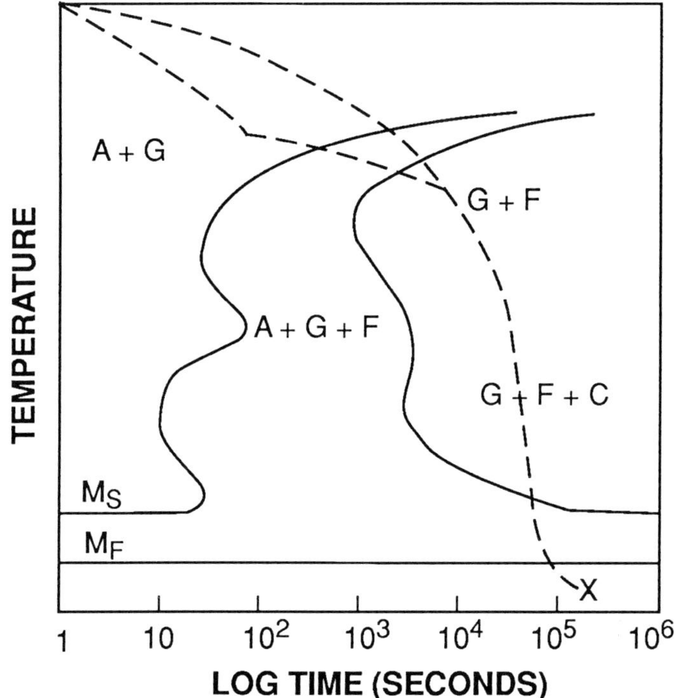

Fig. 13-1(a). The cooling cycle and the resultant microstructure for a ferritic grade of iron; ×400.

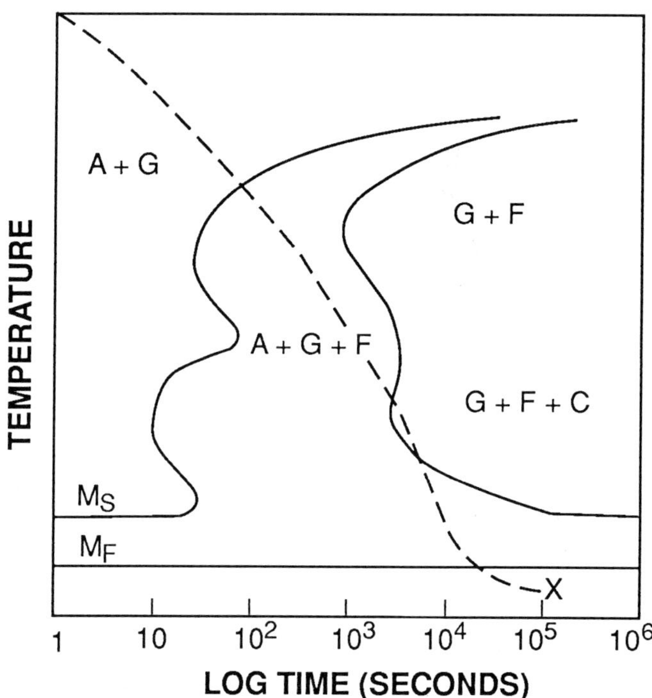

Fig. 13-1(b). The cooling cycle and the resultant microstructure for a pearlitic grade of iron; ×400.

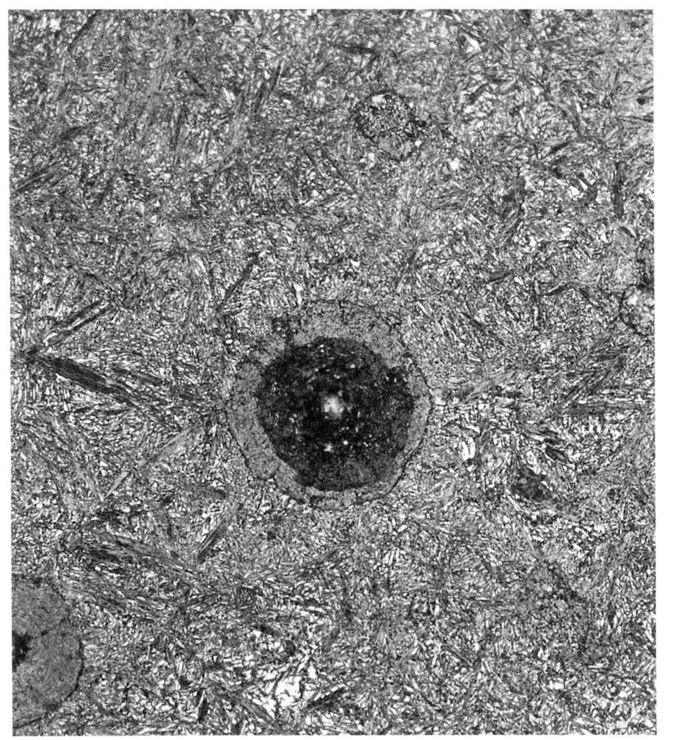

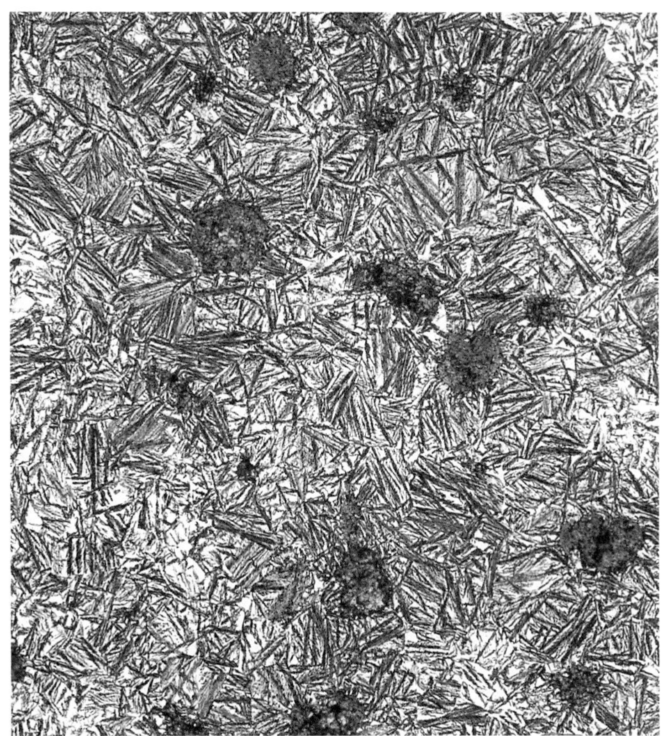

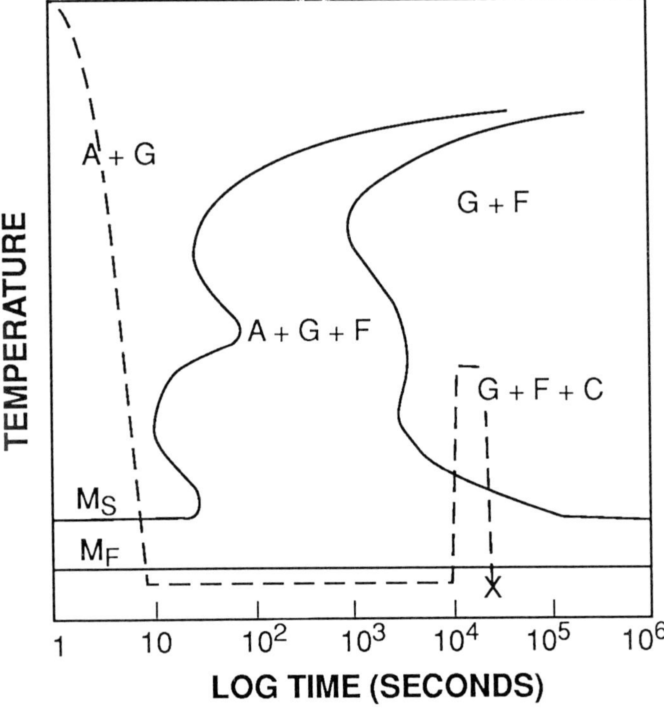

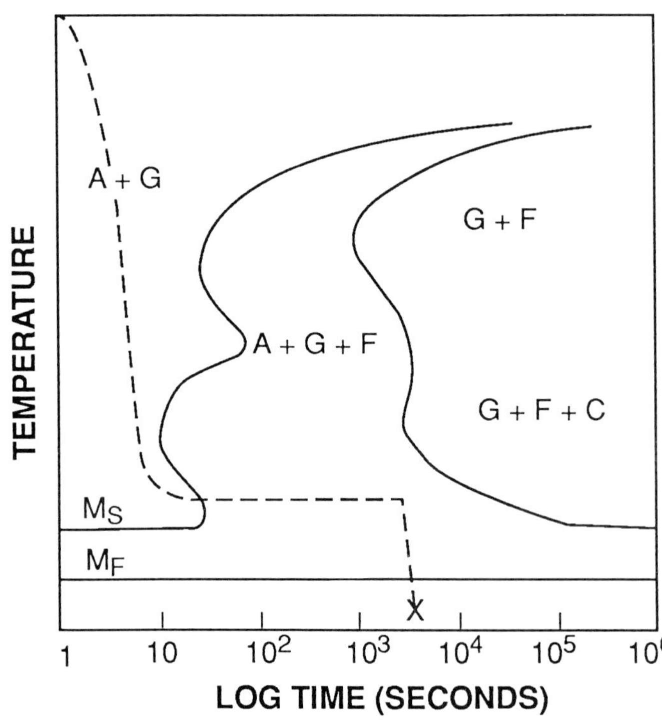

Fig. 13-1(c). The quenching and tempering (700F [371C]) cycle and the resulting microstructure for martensitic grade; ×400.

Fig. 13-1(d). A typical austempering treatment (675F [357C]) and the resultant microstructure; ×400.

concentration (i.e., approximately 0.7%). If subsequent cooling is slow, as in heavy-section castings, the carbon has time to migrate into the nodules, leaving a largely ferritic matrix (i.e., less than 0.02% carbon). If subsequent cooling is rapid, as in castings with thin sections, the carbon does not have time to migrate into the nodules and, as the casting cools beyond the lower critical range, the matrix transforms to pearlite (alternate platelets of iron carbide and ferrite).

The as-cast structure is largely determined by the cooling rate (which is largely influenced by the section size) and the chemical composition of the casting. The resultant ferrite-to-pearlite ratio (or presence of acicular structures) in the casting is a function of the time the casting is above the critical temperature range and the rate of cooling through the lower critical range. More specifically, the resultant structure is determined by the ability of carbon to diffuse from the matrix into the nodules. This diffusion process is controlled by the elapsed time above the critical temperature *and* by the presence or absence of carbon diffusion barriers. Certain elements have the tendency to segregate in the area immediately adjacent to the graphite nodules, forming diffusion barriers that hinder the migration of carbon into the nodule. The most significant of these elements are antimony, tin and copper. The presence of these elements will significantly affect the pearlite content in an as-cast ductile iron. Therefore, the as-cast microstructure can be consistently reproduced by controlling the casting chemistry and the shakeout process. Figure 13-2 shows the variation in mechanical properties available in as-cast ductile iron.

STRESS RELIEVING

Castings with variable section sizes can be stress relieved to minimize or eliminate cracking in service. (It should be noted that, in service, cracking is less prevalent in ductile iron than in gray iron.) Stress relieving can also minimize distortion during subsequent machining.

Recommended stress relieving cycle(s):
- *Castings with regular analyses:*
  1050–1100F (565–593C) for one hour per inch of section, *plus* one hour. Cool uniformly to 700F (371C).
- *Alloyed and intricate castings:*
  1150–1250F (621–676C) for one hour per inch of section, *plus* one hour. Cool uniformly to 700F (371C).
- *Ni-resist Castings:*
  1150–1250F (621–676C) for one hour per inch of section, *plus* one hour. This should be done after rough machining. (See Chapter 3, "Production of Ductile Ni-resist," for additional details.)

At 900F (402C), 60% of the stresses are relieved without measurable softening. At stress relieving temperatures (at or below 1100F [593C]), no measurable softening will occur. At temperatures above 1100F (593C), some softening (50–70 points BHN) may occur with a corresponding loss in strength.

SUBCRITICAL FERRITIZING

Pearlitic and ferritic/pearlitic grades of ductile iron can be fully ferritized by a subcritical annealing process. Castings should be heated to 1200–1300F (649–704C) and held for five hours, *plus* one hour per inch of section. Cool uniformly. *(Note: This treatment will not break down primary carbides.)*

CARBIDE BREAKDOWN

The breakdown of iron carbides in ductile iron is usually fast, due to the high silicon content and the presence of graphite spheroids acting as nuclei. (These same factors make it easy to produce carbide-free ductile iron as cast with the proper chemical balance and inoculation techniques.)

To break down primary carbides in ductile iron castings with normal chemistries, heat the castings to 1650F (899C) and hold for one to three hours at this temperature.

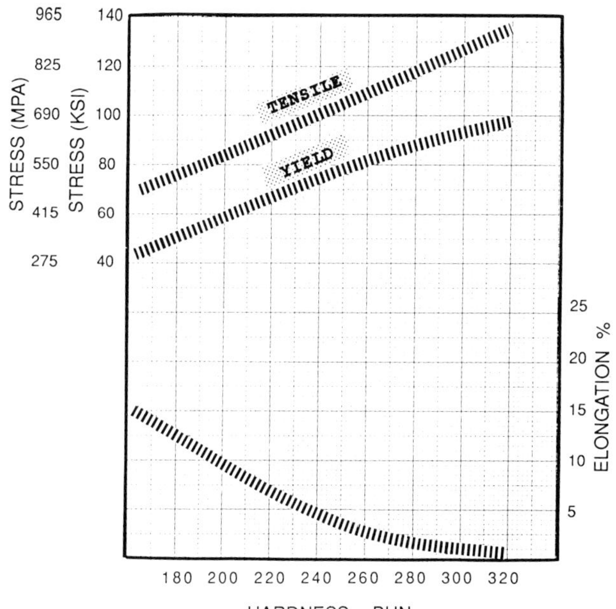

Fig. 13-2. Variation in mechanical properties available in as-cast ductile iron.

Figure 13-3 shows the results of annealing 1/4-inch plates that were uninoculated and heavily carbidic as-cast at a series of times and temperatures. These plates contained 3.15% C, 2.66% Si, 0.40% Mn, 0.15% P, 0.10% Cu and 0.04% Mg. In commercial practice, most carbides will be completely decomposed in less time than indicated in Figure 13-3. One to two hours at 1650F (899C) is usually sufficient.

First-stage annealing at lower temperatures, for instance at 1475F (802C), has been shown to help develop a fine-grained ferrite structure at room temperature. Fine-grained structures have somewhat greater impact toughness and retain toughness to lower temperatures than coarse-grained. However, the decomposition rate for primary carbides at these lower temperatures is very slow, (i.e., 10–20 hours at temperature is required). If maximum toughness at lower temperatures is required, it is more practical to break down the carbides at 1650F (899C), cool to room temperature, and then re-austenitize at 1475F (802C) for grain refinement. The net gain would probably be a 30–40F (17–22C) lower ductile/brittle transition temperature. From Figure 13-4, it is apparent that lowering the silicon content might be a more practical and economical method of lowering the transition temperature.

Certain carbide stabilizing elements, notably chromium, form carbides that are very difficult, if not impossible, to decompose. For instance, an iron with 0.26% chromium has been found to result in primary carbides that could not be broken down after two 20-hour treatments at 1700F (927C). The resulting structure, after pearlite breakdown, was a ferritic/carbidic matrix with only 5% elongation. As little as 0.05% chromium can double the annealing time necessary to break down carbides. In all cases, where a tough ferritic iron is required, chromium content must be kept low, preferably below 0.05%. Other carbide stabilizers include molybdenum, copper (at more than about 1%), manganese, boron, vanadium, zinc, tungsten and tin.

■ PEARLITE BREAKDOWN

In ductile irons, pearlite begins to decompose in a range from 1100–1400F (593–760C). More specifically, that temperature for a specific iron can be approximated by the following formula:

$$\text{Critical Temp. (F)} = 1350 + 50(\%Si) + 40(\%Mo) - 30(\%Mn) - 25(\%Ni) - 10(\%Cu)$$

At 1100F (593C) pearlite decomposition begins at a nearly undetectable rate. As the lower critical temperature is approached, the pearlite decomposition rate increases rapidly.

Because of the normal segregation of alloys and impurities common in all ferrous alloys, the pearlite present varies in its stability as well as its decomposition temperature. In the great bulk of ductile irons in the 3.2–3.8% C, 1.0–3.9% Si, 0–0.50% Mn, 0–3.0% Ni, 0.025–0.08% Mg

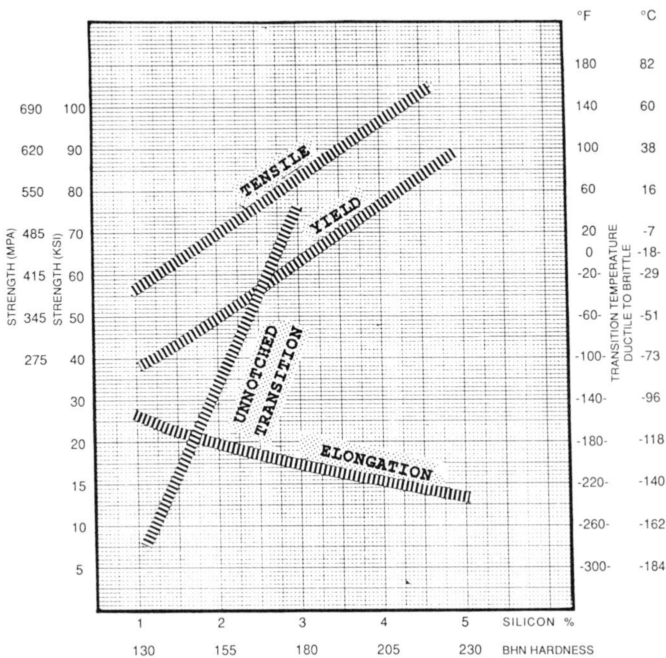

Fig. 13-3. Effect of time on temperature for decomposition of primary cementite.

Fig. 13-4. Influence of silicon on the mechanical properties of ferritic ductile iron.

composition range, at least 90% of the pearlite will be decomposed if held at 1275F (690C) for five hours following austenitizing.

The second-stage holding temperature should be raised 25°F (14°C) if the castings are cooled below 1200F (649C) between first-stage annealing (austenitizing and homogenizing) and second-stage annealing, (pearlite breakdown), or if first-stage annealing is omitted. This is because the austenite-to-pearlite transformation temperature upon heating is about 25F (14C) higher than the transformation temperature upon cooling.

It has been suggested, at times, that first stage annealing, or austenitizing, might be eliminated if carbides were known to be absent from the casting. However, it is now known that prior austenitizing is necessary if ferritic ductile iron castings are to retain their toughness to low temperatures. A loss of as much as half the impact strength and a rise of as much as 50°F (28°C) in transition temperature can be the result of omitting prior austenitization.

Ferritization is sometimes achieved by controlled cooling from the austenitizing temperature rather than by arresting for several hours at 1275F (690C). Ferrite starts to form directly from austenite at temperatures below 1475F (802C) in low-alloy irons. Therefore, a short, controlled, cool ferritizing cycle can sometimes be used, and in clean irons, a structure of 90–95% ferrite will result. The maximum rate of cooling over the critical range for several iron compositions is shown in Table 13-1.

Higher silicon content (i.e., 2.4–2.6%) might allow somewhat higher cooling rates to be used. Ferritization by a controlled cooling cycle could be of economic advantage for very large castings or large furnace loads.

Several alloying elements slow or prevent pearlite breakdown in ductile iron. The most common are phosphorus, copper, tin, arsenic and antimony.

Phosphorus, present in almost all irons, segregates at the grain boundaries. At levels above 0.10% it stabilizes pearlite in the surrounding areas. For maximum toughness, phosphorus is usually held to below 0.05%. At that level, it does not have a significant effect on pearlite breakdown.

Copper is a potent pearlite stabilizer, especially in low-silicon and low-nickel ductile irons. Irons containing 2.5–2.8% silicon, 0.3% manganese, and up to 1% nickel can tolerate up to 0.40% copper with the expectation that they can be annealed to 90% ferrite. However, the presence of copper may require double the time for second-stage annealing. Higher levels of copper, or copper in the presence of additional manganese, chromium, or molybdenum, can result in an iron that is practically impossible to ferritize.

Very small amounts of tin, arsenic or antimony are very powerful pearlite stabilizers. These were discussed in Chapter 4.

■ INFLUENCE OF MASS ON HEAT TREATMENT

HEATING

Heavy, thick castings, or tightly-packed piles of castings, take longer to reach temperature throughout the load than thin castings or lighter loads. One hour per inch of casting section, or equivalent section in the case of thin, tightly-packed castings, is ample time to get the center of the casting or load up to the furnace temperature. Often 30 minutes per inch of section suffices.

Difficulty may arise in estimating the heating requirements of an equivalent section, or the time needed for a pile or dense load of castings. The best solution for this is to arrange the castings as desired, and then sample a thorough cross section of the load for consistency. If the properties throughout the load are uniform, establish a procedure for loading the castings in the same pattern each time that particular casting is processed.

COOLING

Mass is important in cooling, too. It is obvious that in mold or air cooling, the center of the load will cool more slowly than the outside. Whether the difference will be significant in any case must be determined by sampling the load to assure uniform properties throughout. This is especially important in air-hardening heat treatments.

■ ANNEALING

FERRITIZING: GRADES 65-45-12 AND 60-40-18

A full anneal, or softening, produces maximum ductility and is a two-temperature treatment. In the first stage, the casting is heated above the critical (to about 1650F/899C) to dissolve any primary carbides present. The castings are held at this temperature for one hour per inch of section, *plus* one hour. As was discussed earlier in this section, alloy additions can retard the transformation process and can increase or decrease the critical temperature. At homogenization temperatures above 1650F (899C) carbides will decompose rapidly, but warpage may increase. At temperatures below 1650F (899C) warpage will be minimized, but carbide decomposition will take longer.

The second stage of the annealing process can be accomplished by either of two methods:

TABLE 13-1. THE MAXIMUM RATE OF COOLING OVER THE CRITICAL RANGE FOR SEVERAL IRON COMPOSITIONS

Alloy Composition (%)*			Max. Cooling Rate from 1475–1200F (802–649C)	
Si	Ni	Mn	(°F/hr)	(°C/hr)
2.0–2.2	0	0.10	110	61
2.0–2.2	0	0.30	85	47
2.0–2.2	0	0.50	40	22
2.0–2.2	1.0	0.10	70	39
2.0–2.2	1.0	0.30	40	22
2.0–2.2	1.0	0.50	15	8
2.0–2.2	1.5	0.10	40	22
2.0–2.2	1.5	0.30	15	8
2.0–2.2	1.5	0.50	Not adaptable to controlled cooling cycle	

*Note: All irons low in copper and phosphorus.

1. Furnace cool to 1275F (700C), hold for five hours plus one hour per inch of section, then cool to room temperature uniformly.
2. Furnace cool to 1200F (649C) at a maximum cooling rate of 35F (19C) per hour between 1450F (700C) and 1200F (649C). Cool to room temperature uniformly.

Some Ni-resist ductile irons have excessive amounts of carbide present. These irons may require annealing temperatures of 1700F (927C) to 1900F (1038C) for five hours. Furnace or air cool to break down the carbides. Grade D-2 is air cooled for maximum austenite stability. (See Chapter 3 for additional details.)

NORMALIZING

In this heat treatment, temperatures of 1600–1700F (871–927C) are used to austenitize and break down any carbides present. In the case of castings that are free of carbide, the castings should be heated to temperature and held for two hours per inch of section. Air cool to produce Grade 100-70-03 pearlitic ductile iron. In some cases, particularly when heavy-section castings or dense furnace loads are involved, it may be necessary to fan cool the castings to meet the hardness requirements.

STEP NORMALIZE

The initial part of a step normalizing treatment is the same as normalizing. However, the castings are furnace cooled to between 1450 and 1375F (788 and 745C), held for three hours and then air cooled in a gentle air blast. Usually this type of heat treatment is used to produce 80-55-06 grade ductile iron in which both pearlite and ferrite are present.

NORMALIZE AND TEMPER

Where optimum toughness and impact resistance along with relatively high tensile properties are required, normalizing is followed by tempering. This consists of reheating castings to temperatures of 800–1200F (427–649C) and holding at temperature for two hours for each inch of cross section. These temperatures are varied within the above range in order to bring casting hardness values within the specification limits (Fig. 13-5).

TEMPER EMBRITTLEMENT

It has been found that ferritic ductile irons of normally good toughness can be embrittled by quenching from 800–950F (427–510C). (By definition, embrittlement is raising the brittle-ductile fracture transition temperature.)

Castings are commonly air cooled after tempering, but in some cases, as in galvanizing cycles, they are cooled quite rapidly from this embrittling range. The embrittlement increases with longer times at temperature.

Low-phosphorus (0.05%), low-silicon (2%) irons exhibit as much as a 20–40°F (11–22°C) rise in transition temperature when quenched from the tempering temperature. High-phosphorus (0.08–0.16%), high-silicon (2.7%) irons can experience up to a 150°F (83°C) rise in transition temperature when quenched from the aforementioned tempering range. This embrittlement is similar to the effect noticed in malleable iron and steel, although steels embrittle when slowly cooled from tempering temperature.

QUENCHING AND TEMPERING

Hard, strong, acicular and lamellar structures are often developed by quenching in oil, salt or lead and then tempering to the desired hardness. A great variety of quench-and-temper heat treatments can be used to obtain high strength and hardness. Most small- to medium-size

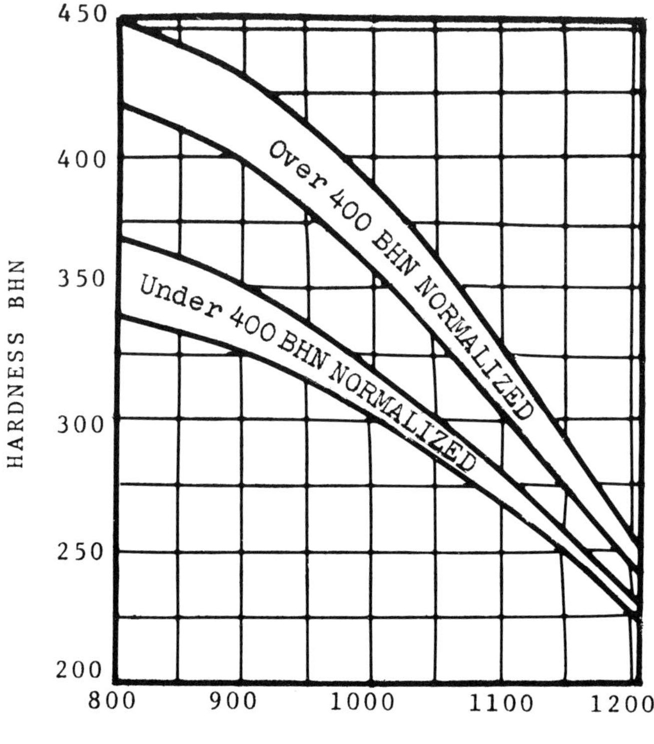

Fig. 13-5. Hardness of normalized ductile iron tempered at various temperatures.

castings can be easily heat treated without additional alloying in conventional equipment. Larger, heavier castings may require specialized heat treating equipment or alloying to respond to heat treatment.

As in normalizing, the primary aim in a quench-and-temper heat treatment is to saturate the austenite with carbon and keep it there. Oil, salt and lead quenching are more severe than air quenching, and permit even low-alloy irons to be quenched to high strength and hardness levels. Castings can usually be quenched and tempered to a narrower hardness range than is possible by normalizing and tempering. Thus, the type of heat treatment used becomes an economic as well as a technical decision.

Maximum as-quenched *hardness* is determined by the carbon content in the austenite, which is a function of the austenitizing temperature and the silicon content of the iron. The *hardenability* of the iron is a function of the carbon content and the presence of other alloying elements, such as copper, molybdenum, nickel and manganese. The resultant hardness is a function of the carbon and alloy content of the iron and the quench severity of the heat treat system used. The severity of the quench is determined by the quenching media, the degree of agitation in the bath, and the mass of the piece, or load, being quenched.

At 1650F (899C), most 2–3% silicon ductile irons will be fully austenitized. After one to two hours at temperature the austenite will be saturated with carbon and will be at its optimum condition for the highest as-quenched hardness.

Lower austenitizing temperatures are often used to minimize cracking and distortion in complicated or irregular shaped castings. The minimum practical austempering temperature will be a function of the chemical composition of the iron. Using a lower austenitizing temperature will result in a lower as-quenched hardness.

Among the alloying elements commonly used to increase hardenability in ductile irons, molybdenum is the most potent. It is followed, in order of decreasing potency, by manganese, nickel and copper. The carbon content in the austenite is inversely proportional to the silicon content. Therefore, the silicon content has an effect on both the hardenability *and* the ultimate hardness of the casting. Figure 13-6 shows the relative effect of various alloys on hardenability for several ductile irons.

As-quenched hardness levels are dependent on the circulation of the quenching medium or the agitation of the piece being quenched (see Tables 13-2 and 13-3). Effective quenching apparatus should include some means of inducing controlled agitation to the quench "bath." To avoid cracks due to stresses induced during quench transformation, castings should be tempered immediately after quenching.

Tempered hardness depends on the as-quenched hardness level, alloy content, tempering time and tempering temperature. Because of the varying chemical compositions of irons within a given grade, precise tempering curves or formulae, such as those available for steels, have not been derived for ductile irons. An average curve, such as Figure 13-7, should be used as a first approximation. With close control of the iron analysis and the heat treating process, a narrower control range can be expected and the heat treater can develop a much more accurate tempering curve for the application.

■ AUSTEMPERING

Austempered ductile iron (ADI) is an alloyed and heat-treated ductile cast iron. The base chemical composition and the casting process of ADI are similar to those of conventional ductile iron. Alloying elements are added, however, for one reason only: to promote heat treatability. The commonly used alloying elements are molybdenum, nickel, and copper. These elements improve hardenability, but have little effect on the physical properties in an austempered casting.

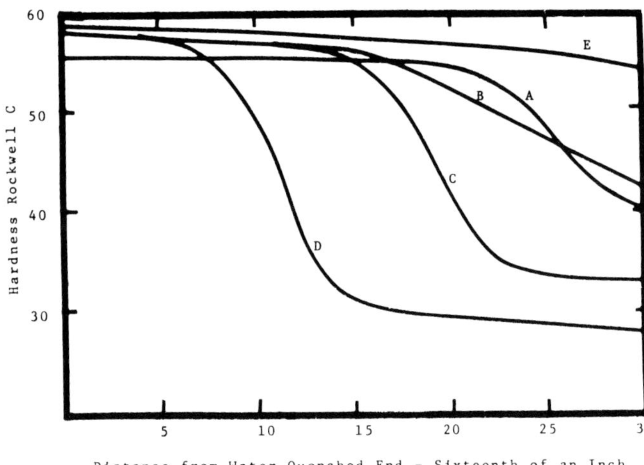

Chemical Analysis of Test Samples in Fig. 13-6					
Sample	%C	%Si	%Mn	%Ni	%Mo
A	3.27	2.35	0.45	2.56	—
B	3.27	2.38	—	1.02	0.50
C	3.27	2.35	0.03	2.45	—
D	3.58	2.26	0.03	0.99	—
E	3.45	2.56	0.28	3.92	0.57

Fig. 13-6. The Jominy hardenability curves for several ductile irons.

Austempering involves a two-step heat treat cycle (Fig. 13-8):

1. First, the casting is fully austenitized at a temperature range of 1550–1750F (843–954C). Full austenitization means that all the matrix of the casting is transformed to FCC (face centered cubic) austenite, and this austenite is saturated with carbon.

2. In the second step, the part is rapidly quenched to a temperature range of 460-750F (843–954C), and isothermally reacted for 1/2 to 4 hours. This phase of the heat treatment is called austempering. During austempering, the austenitic matrix decomposes to a unique ausferritic texture of acicular ferrite and a stable, high-carbon austenite. The austempering temperature determines the time duration of the reaction, the morphology of the phases in the matrix, and the physical properties. After austempering, the casting is air-cooled to room temperature before a significant amount of bainite is formed.

TABLE 13-2. QUENCH SEVERITY COMPARISON FOR COMMON QUENCHANTS

Condition	Air	Oil	Water	Brine
Still	0.02	0.25–0.30	0.90–1.0	2.0
Mild Circulation	—	0.30–0.35	1.0–1.1	2.0–2.2
Moderate Circulation	—	0.35–0.40	1.2–1.3	—
Good Circulation	0.10	0.40–0.50	1.4–1.5	—
Strong Circulation	—	0.50–0.80	1.6–2.0	—
Violent Circulation	0.12	0.80–1.1	4.0	5.0

TABLE 13-3. APPROXIMATE QUENCH SEVERITY COMPARISON FOR SALT QUENCHES.

Condition	360F (182C)	700F (371C)
Still and Dry	0.15–0.20	0.15
Agitated and Dry	0.25–0.35	0.20–0.25
Agitated with 1/2% Water	0.40–0.50	0.30–0.40
Agitated with 2% Water	0.50–0.60	0.50–0.60*
Agitated with 10% Water	0.90–1.3*	not possible

*Requires special quenching apparatus.

It should be noted that the ausferritic matrix develops at the bainitic reaction temperatures; however, it is not bainite. The reaction is terminated before bainite forms. Bainite is detrimental to ADI.

ADI has five ASTM standard grades. These grades are described by two ASTM standards: A897-90 and A897M-90. In the first one, the properties are listed in English units; in the second one, in metric units. The mechanical properties in the two standards are shown in Tables 13-4 and 13-5.

The effect of the austempering temperature on the microstructure is shown in Figures 13-9 and 13-10. At a low temperature (500F [260C]), the matrix structure is much finer than at high temperature (700F [370C]). The effect of the austempering temperature on the yield and tensile strengths is shown in Figure 13-11.

The recommended chemical composition for ADI is shown in Table 13-6. Note that the Mn level in ADI is lower than in ductile iron. The reason for this is that Mn, at levels higher than 0.35%, segregates in the casting, causing uneven carbon solubility. During heat treatment, the reaction is nonuniform when the carbon distribution is uneven; therefore, the mechanical properties are varied throughout the casting.

Nodule count also has a marked effect on the mechanical properties of ADI. High nodule count limits the solute

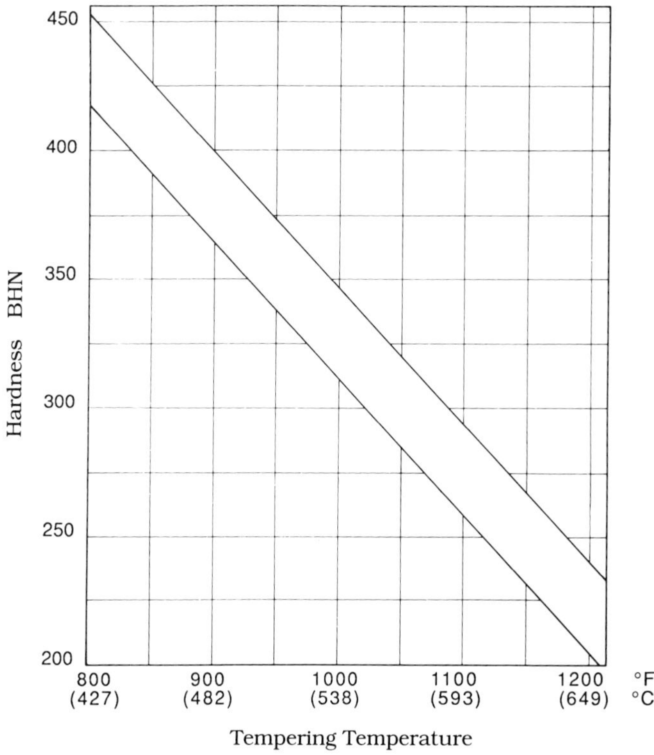

Fig. 13-7. Hardness of oil-quenched ductile iron tempered for one hour (as-quenched over 500 BHN).

segregation, making the solute distribution and strength properties more uniform. In general, the nodule count should be higher than 100 nodules per mm².

The ADI process for castings should be tightly controlled. Although the chemical composition itself is not that critical, the consistency is! Variation in the chemical composition from heat to heat causes diverse response during heat treating, leading to a variation in the observed mechanical properties. There are two casting defects that are particularly detrimental to ADI and should be avoided: shrinkage holes and eutectic carbides.

TABLE 13-4. THE FIVE ASTM STANDARD ADI GRADES (A897-90)†

Grade	Tensile Strength (KSI)	Yield Strength (KSI)	Elongation (%)*	Impact Energy** (ft-lbs)	Typical Hardness (BHN)
1	125	80	10	75	269–321
2	150	100	7	60	302–363
3	175	125	4	45	341–444
4	200	155	1	25	388–477
5	230	185	—	—	444–555

*minimum values

**unnotched Charpy bars tested at 72 ± 7 °F

†Reprinted, with permission, from the *Annual Book of ASTM Standards*, copyright American Society for Testing and Materials.

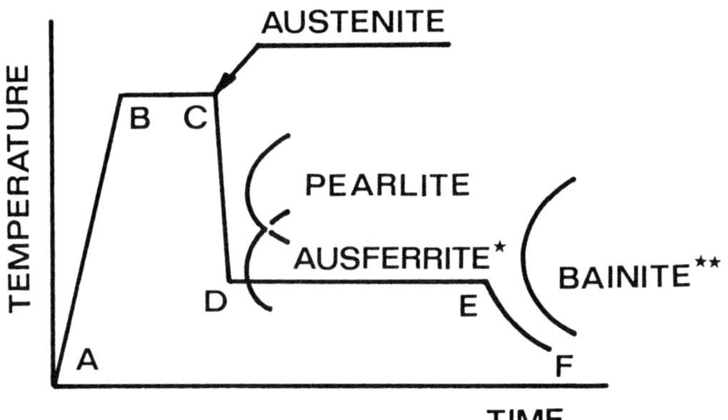

Fig. 13-8. The two-step austempering process.

TABLE 13-5. THE FIVE ASTM STANDARD ADI GRADES (A897M-90)†

Grade	Tensile Strength (MPa)	Yield Strength (MPa)	Elongation (%)*	Impact Energy** (Joules)	Typical Hardness (BHN)
1	850	550	10	100	269-321
2	1050	700	7	80	302–363
3	1200	850	4	60	341–444
4	1400	1100	1	35	388–477
5	1600	1300	—	—	444–555

*minimum values

**unnotched Charpy bars tested at 22 ± 4 °C

†Reprinted, with permission, from the *Annual Book of ASTM Standards*, copyright American Society for Testing and Materials.

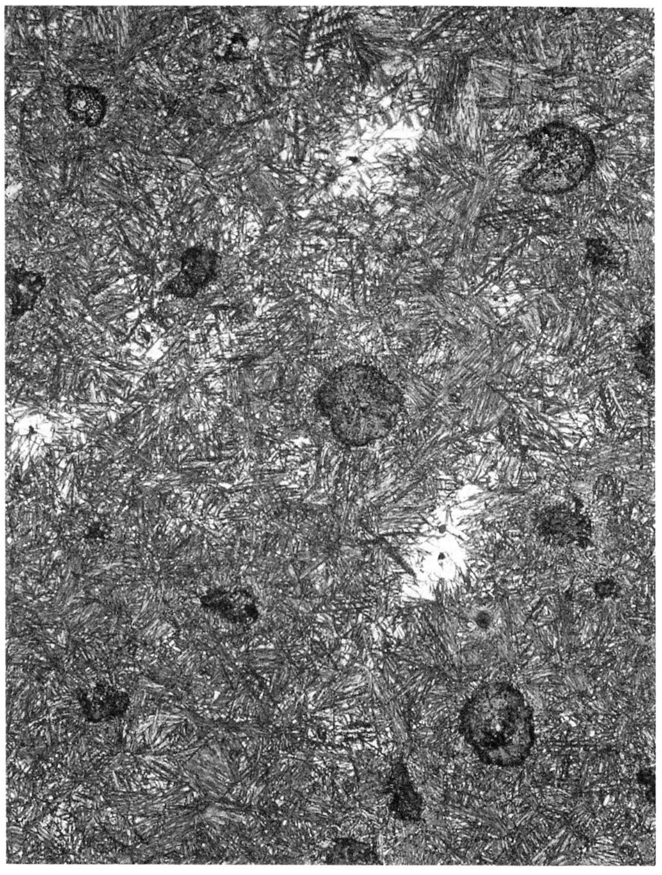

Fig. 13-9. Photomicrograph of heat treated Grade 5 ductile iron at 500F (260C); ×400.

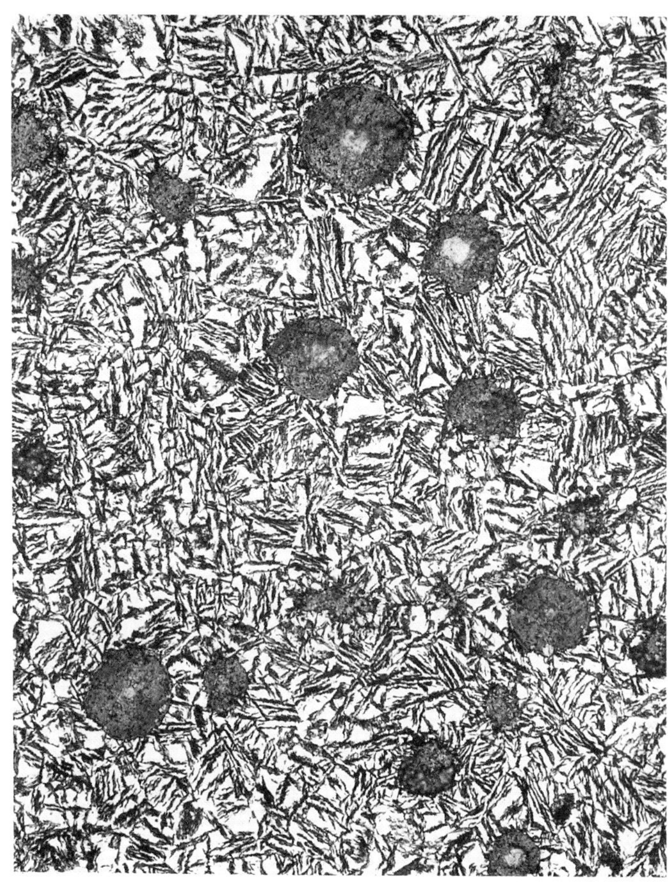

Fig. 13-10. Photomicrograph of heat treated Grade 1 ductile iron at 700F (371C); ×400.

▬ SURFACE HARDENING

Ductile iron can readily be surface hardened by induction, flame or other surface hardening techniques to a surface hardness value of over 60 Rc. The resulting castings will have a hard, wear-resistant case with a core that retains good ductility and impact resistance.

The existence of about 10%—by volume—graphite in most ductile irons means that, in hardness testing, the indenter will probably rest on some graphite, contributing almost nothing to the hardness reading. The ductile iron matrix is, therefore, actually somewhat harder than the reading indicates. For instance, a 50 Rc ductile iron is about equivalent in matrix hardness to a steel of 60–62 Rc.

INDUCTION HARDENING

Experimental work and successful production applications show that ductile iron can be surface hardened by induction heating to provide good case hardness with a wide variation of core properties. The operation is not overly sensitive and, while prior structure and composition have an influence, the hardening temperature can be adjusted for successful surface hardening of all but substantially ferritic structures.

The selection of a suitable structure for induction surface hardening of ductile iron should be guided primarily by the mechanical properties desired. Where the ductility desired (as represented by percent elongation) does not exceed 3–6%, a pearlitic as-cast or normalized structure is

TABLE 13-6. RECOMMENDED CHEMICAL COMPOSITION FOR ADI CASTINGS

Base Metal	
CE (Carbon Equivalent)	4.30–4.60
C	3.40–3.80
Si	2.20–2.60
Mn	0.35 max. (or 0.60 max. for section size under 1/2 in.)
S	0.02 max.
P	0.04 max.
Alloy Additions (if required)	
Cu	0.20–0.80
Ni	0.20–2.00
Mo	0.10–0.30
Residuals	
Al	0.050 max.
Cr	0.100 max.
Ti	0.040 max.
Sb	0.002 max.
Sn	0.020 max.

suitable, and provides immediate response to induction hardening.

When intermediate core ductility is required (8–12% elongation), the fact that the matrix has considerable ferrite becomes important, and the induction-hardening temperature may have to be increased to provide suitable response. Completely ferritized structures exhibiting maximum ductility (i.e., 18% elongation or more) respond very slowly to induction heating and are generally not suitable for induction surface hardening.

Although not in widespread use as of this writing, some work has been done on induction surface hardening of austempered ductile irons having 7–12% elongation. Since the ADI structure is fully saturated with carbon, it responds rapidly to induction heating.

An induction-hardening temperature of 1700F (927C) appears to be most satisfactory to minimize the influence of prior structure. Excessive hardening temperatures provide a danger of retained austenite with a loss of hardness, while lower hardening temperatures often fail to achieve maximum surface hardness.

The case depth depends not only upon the metallurgical variables, such as prior structure, composition and hardening temperature, but also upon the characteristics of the induction heating apparatus. Both the power available in the high frequency converter and the frequency itself may have an important influence. Minimum case depths are produced with converters that operate at high frequency and have sufficient power available to bring the surface to the hardening temperature in a minimum of time. This minimizes the conduction of heat to the interior of the part. But the depth of case hardening will also depend upon the design of the heating coil, the relative positioning of the part, the quenchant and the quenching fixture. If progressive hardening is desired, localized heating and quenching are utilized.

There is no practical limitation on the maximum case depth obtainable, except perhaps the limitation of hardenability in very large sections. The practical minimum case depth in ductile iron will approach 0.020 in. where minimum core ductility is required, but will be 0.030–0.040 in. where a ductile, ferritic core is required.

FLAME HARDENING

Ductile iron responds well to flame hardening. Like induction-hardened ductile iron, flame-hardened ductile iron demonstrates wear resistance superior to pearlitic ductile iron and better than many steels. The flame hardening technique used is similar to that used with alloy steel of medium hardenability.

Initial structure has the same influence on a casting's response to flame hardening as to induction hardening. A fully pearlitic or acicular structure, in which the matrix is saturated with carbon, will surface harden most readily, while a ferritic matrix will require a slower flame travel to get sufficient carbon in solution for the quench.

Pearlitic and acicular irons respond well to both self-quenching and water quenching after flame heating. However, an essentially ferritic matrix will not respond fully to the self-quenching method.

Care should be taken that all, or most, of the internal stresses are removed from the castings prior to flame hardening. If the castings are in the as-cast condition and are predominantly pearlitic, they should be given a fairly high temperature stress relief for a relatively short time (i.e., 1100–1200F (593–649C) for one hour per inch of casting section. If this is not done, and if certain casting sections are under restraint, cracks may develop under the hardened case.

Following flame hardening, if subsequent liquid quenching is employed, the castings should be transferred, before cooling to room temperature, to a bath or oven at 300–400F (149–204C) for a minimum of one hour, and then cooled in air for stress relief. This will reduce the high interface stresses between the hardened case and the softer core to a harmless level without appreciable reduction in surface hardness. It is generally not necessary to employ this stress relieving technique to material hardened by the self-quench method.

With proper technique and control of surface temperature (usually 1550–1650F/843–899C), the following ranges of surface hardness can be expected in commercial processing of ductile iron castings:

- Ferritic ductile, flame heated with a water quench, Rc 35–45;
- Pearlitic/ferritic ductile, stress relieved, flame heated and self quenched, Rc 40–45;
- Pearlitic/ferritic ductile, stress relieved, flame heated and water quenched, Rc 50–55;
- Pearlitic ductile, stress relieved, flame heated and water quenched, Rc 50–62.

Ultimately, the final hardness will be determined by the heating temperature and time, the amount of dissolved carbon in the microstructure prior to hardening, and the rate of quench. Values as high as Rc 64 have been reported on properly austenitized and water quenched castings. This compares with a maximum quenched hardness of Rc 65 for steels with carbon contents above 0.55%.

HEAT TREATMENT FOR AUSTENITIC DUCTILE IRONS

Austenitic ductile irons, by definition, are austenitic at room temperature. Therefore, no conventional quench transformation process can be used to harden the matrix. Strengths can be improved, however, by an oil or water quenching from 1700–1850F (927–1010C), since some carbon redissolves in the austenite, which is prevented from precipitation by the rapid cooling. No significant amount of stress is induced by quenching. For austenitic ductile irons to be used at elevated temperatures, stress relieving or preconditioning at a temperature of approximately 50F (28C) above the service temperature is necessary to avoid warpage and stress cracking in service. This technique is successfully used in many gas turbine and turbocharger castings.

MISCELLANEOUS GROWTH

A pearlitic structure, when annealed to ferrite, grows about 0.05 inches per foot. Since solidification shrinkage for the pearlitic grade is 0.15 inches per foot, the pattern shrinkage conforms generally to the 0.10 inches per foot exhibited by gray iron. Light castings, solidifying white (carbidic), grow about 0.20 inches per foot when poured under moderate restraint and annealed. Thus, the annealed light-section ductile casting is larger than the pattern dimensions.

EFFECT OF MOLD COOLING RATE AND HEAT TREATMENTS ON PROPERTIES

The effect of the rate of mold cooling is illustrated in Table 13-7, for ductile irons with and without nickel. Microstructures of the as-cast metal are shown in Figures 13-12 to 13-15. The fully annealed structures are shown in Figures 13-16 to 13-19 and the normalized structure of the same irons in Figures 13-20 and 13-21. Note, particularly, the pearlite stabilizing influence of nickel. While some of these data do not seem to match in total, it does offer very interesting comparisons.

SUMMARY

The heat treatments previously described and the resultant properties are summarized in Table 13-8.

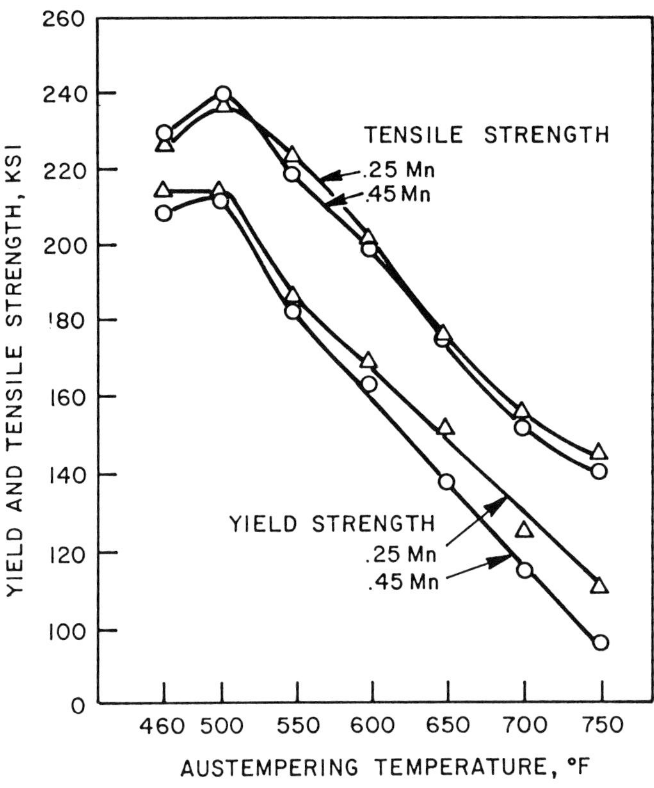

Fig. 13-11. The effect of the austempering temperature on the yield and tensile strength of ADI. (Abstracted from Report No. GRI-89/0017, The Gas Research Institute, reprinted with permission.)

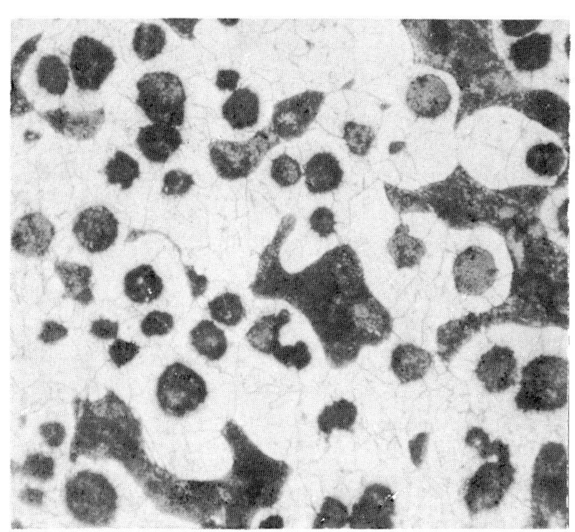

Fig. 13-13. K-2, as cast, nital etched; ×150.

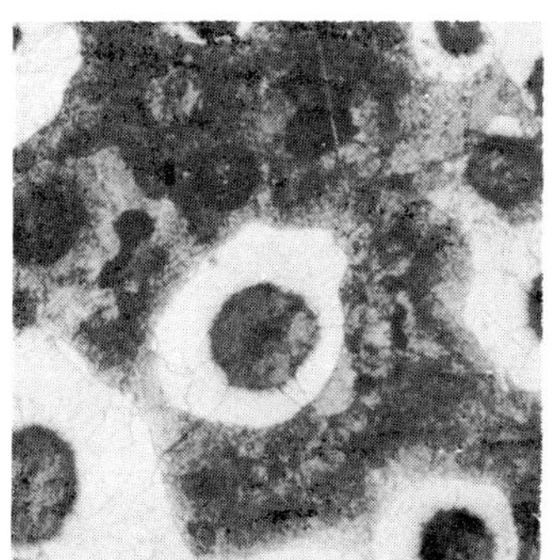

Fig. 13-12. K-1, as cast, nital etched; ×150.

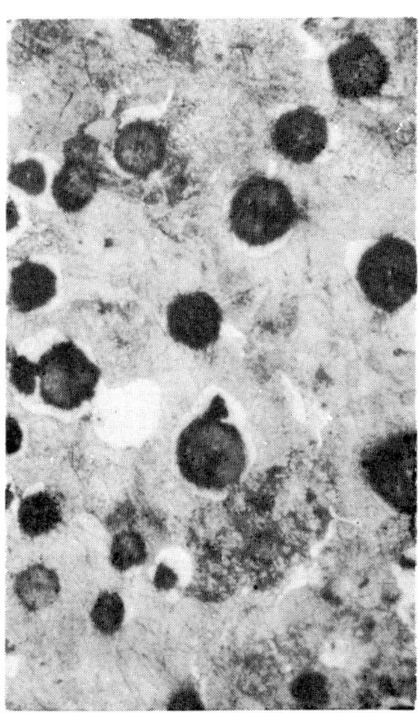

Fig. 13-14. K-4, as cast, nital etched; ×150.

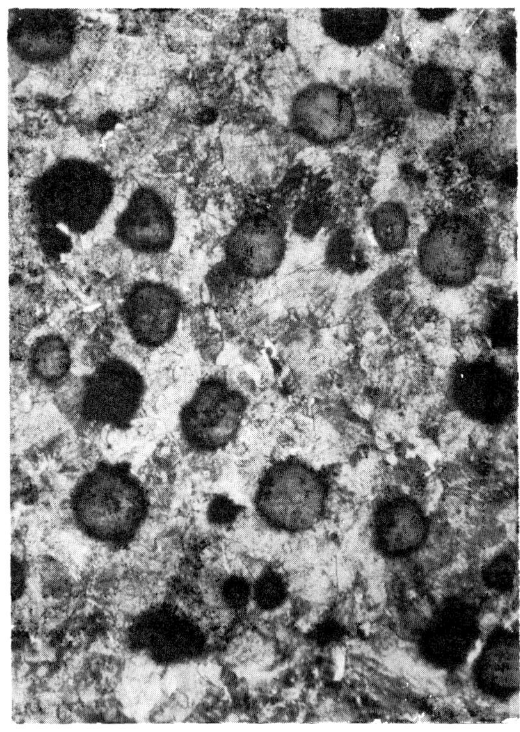

Fig. 13-15. K-5, as cast, nital etched; ×150.

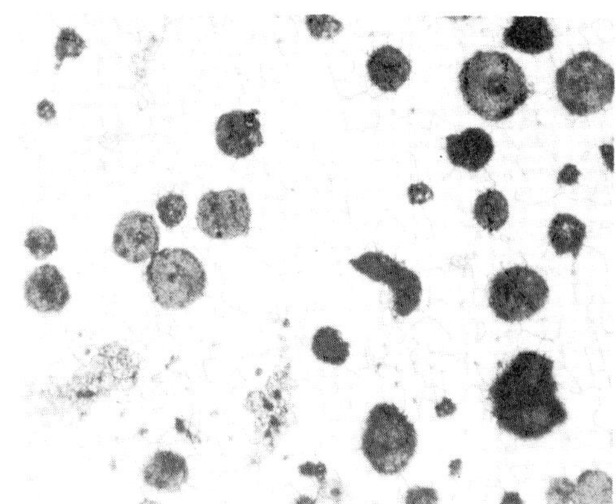

Fig. 13-17. K-2, annealed, nital etched; ×150.

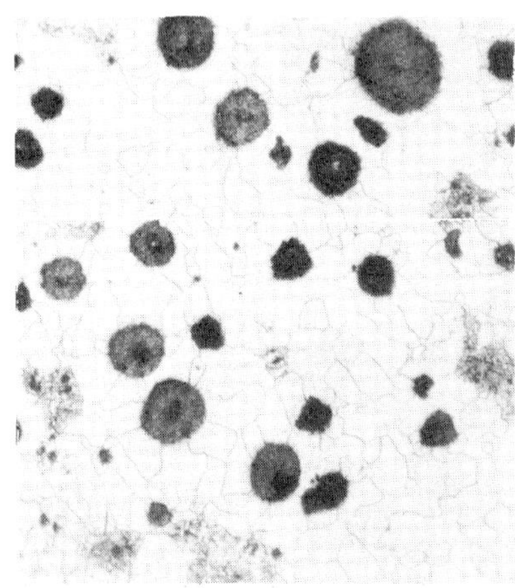

Fig. 13-16. K-1, annealed, nital etched; ×150.

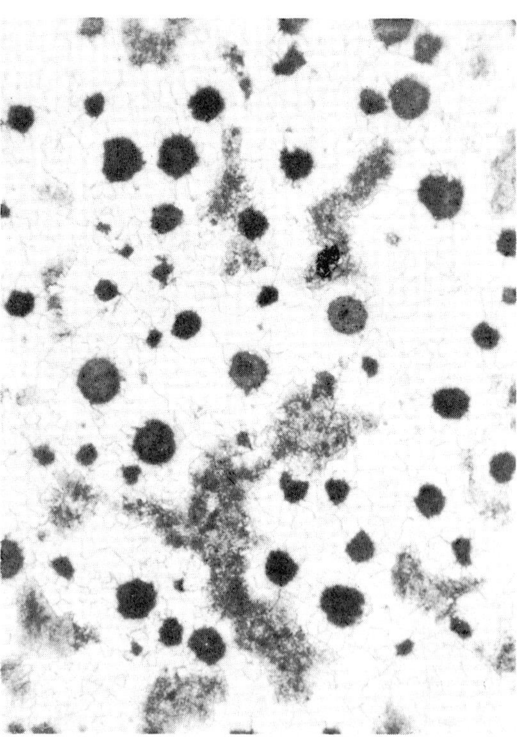

Fig. 13-18. K-4, annealed, nital etched; ×150.

Table 13-7. Effect of Cooling Rate and Heat Treatment on Mechanical Properties

Melt	K-1	K-2	K-3	K-4	K-5
Carbon %	3.50	3.50	3.55	3.52	3.32
Silicon %	2.15	2.15	2.01	2.03	1.69
Sulfur %	0.021	0.021	0.022	0.021	0.017
Phosphorus %	0.020	0.020	0.017	0.019	0.019
Manganese %	0.25	0.25	0.33	0.26	0.25
Magnesium %	0.066	0.066	0.053	0.048	0.065
Nickel %	Tr	Tr	Tr	1.70	1.50
As Cast (keel blocks)					
Cooling Rate	very slow	slow-normal	normal	fast	very rapid
Ferrite %	40	60	25	10	0
Pearlite %	60	40	75	90	100
Graphite	I-3	I-4	I-3	I-4	I-3
Tensile, psi	68.000	81.900	97.000	112.900	172.000
Yield, psi	49.800	58.000	61.000	77.100	91.800
Elongation %	15.6	15.1	10.2	8.3	5.0
BHN	156	197	229	241	285
Annealed (keel blocks)					
Ferrite %	95	95	90	85	70
Pearlite %	5	5	10	15	30
Graphite	I-3	I-3	I-3	I-3	I-3
Tensile, psi	62.400	65.200	68.000	72.000	77.700
Yield, psi	47.200	47.400	49.000	54.000	54.400
Elongation %	—	24.6	23.2	19.1	10.2
BHN	146	143	152	163	174
Normalized (keel blocks)					
Ferrite %			20	25	5
Pearlite %			80	75	95
Graphite			I-3	I-4	I-3
Tensile, psi			96.200	95.900	115.500
Yield, psi			61.500	60.500	60.000
Elongation %			9.9	9.7	6.2
BHN			212	207	241

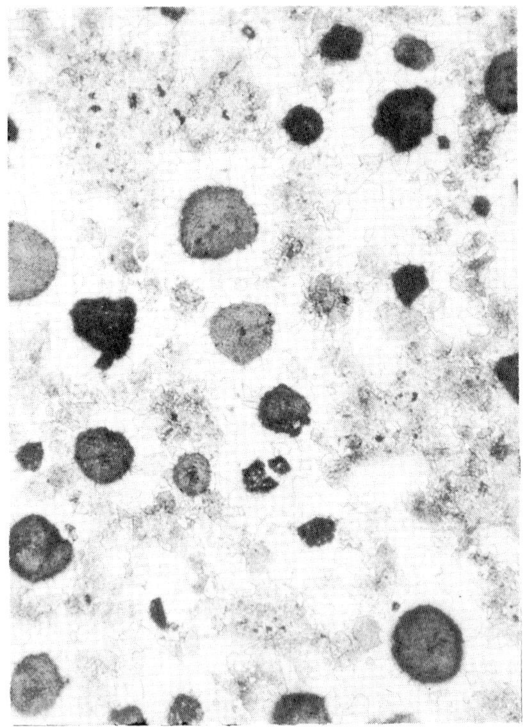

Fig. 13-19. K-5, annealed, nital etched; ×150.

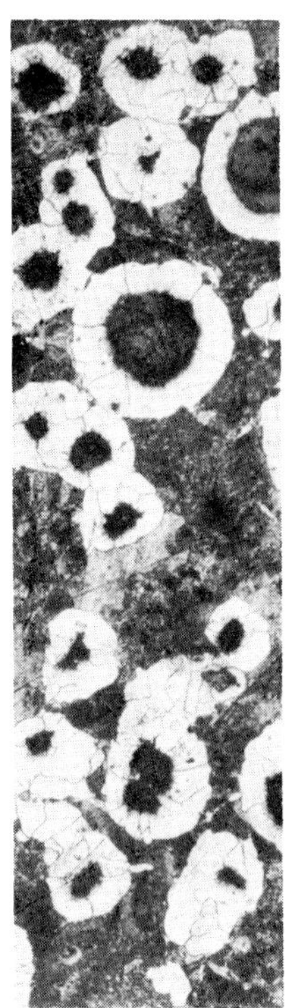

Fig. 13-20. K-3, normalized, etched nital; ×150.

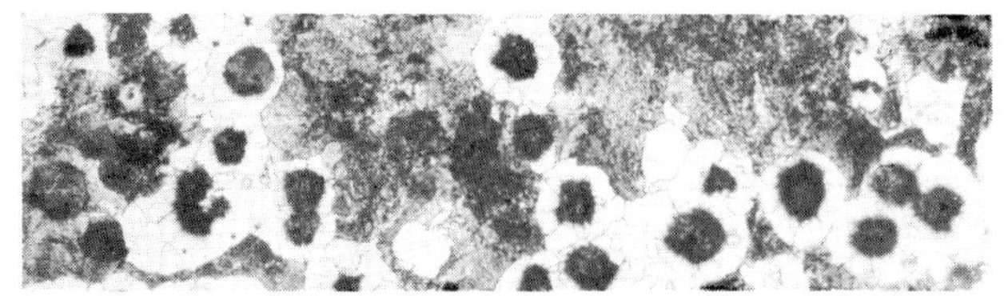

Fig. 13-21. K–4, normalized, nital etched; ×150.

Table 13-8. Summary of Heat Treating Ductile Irons

Heat Treatment	Temperature (°F) and Time	Microstructure Produced	Typical BHN	Type of DI*
Annealing				
First stage: carbide breakdown and homogenization	1550–1750F for 1–3 hr			
Second stage: pearlite breakdown and ferritization	1250–1300F for 5 hr, or 35°/hr from 1450–1200F	Fully ferritic	160–170 143–169	60-45-12 60-40-18
Stress relieving	1000–1250F for 1 hr; furnace cool to 700F, then cool in air	Little structural change if pealitic or ferritic at start	Reduces BHN 0–50 points	60-45-10 or 80-60-03
Sub-critical ferritizing	Hold for 5 hr at 1200–1300F, plus 1 hr per in./section	Ferritic	Reduces BHN	60-45-10
Normalizing and Temper	Austenitize at 1550–1600F for 1 hr. Cool in air. Temper at 800–1000F for 1 hr	Pearlitic or acicular depdending on alloy content, tempering time and temperature	240–350	80-60-03 or 100-70-03
Quench and Temper	Austenitize at 1550–1600F 1 hr. Quench in oil, salt, etc. Temper 800–1000F for 1 hr.	Acicular	270–350	120-90-02
Surface Hardening				
normalize				
induction	Varying	Acicular	50–62 Rc BHN	
flame	Varying	Acicular	50-62 BHN	
Normalize	Austenitize, 1600–1700F	Acicular or fine pearlite	223–302	100-70-03
Step Nomalize	Austenitize, 1600–1700F; hold at 1450F and 1375F	Pearlitic and some ferritic	180–241	80-55-06
Austempering	1550–1700F, 1 hr. plus 1 hr/section-inch, then quench as specified.			
Grade 1	Quench at 700–725F for 1/2–3 hr		269–341	125-80-10
Grade 2	Quench at 625–700F for 1–3 hr	Acicular (ausferrite)	302-363	150-100-07
Grade 3	Quench at 550–650F for 1-1/2–3 hr		341–444	175-125-04
Grade 4	Quench at 500–575F for 2–4 hr		388–477	200-155-01
Grade 5	Quench at 450–525F for 2–5 hr	Fine ausferrite	444–555	230-185-(no elongation specified)

*minimum values

14

Ductile Iron Casting Defects

George M. Goodrich

Taussig Associates, Inc.
Skokie, Illinois

■ INTRODUCTION

Most of the casting defects normally associated with gray cast iron also occur in ductile iron. In addition, a series of specialized casting defects peculiar only to ductile iron also occur. The causes of these defects and ways to reduce or eliminate them will be reviewed in this chapter. The causes for some ductile iron defects are not entirely clear and, in some instances, differences of opinion exist relative to the cause and cure. Effort has been made to objectively present the most recent available information.

Different techniques are used for the production of ductile iron. Each one has limitations and advantages. No single technique has universally been adapted for all types and casting sizes. In considering the causes of ductile iron casting defects, all the various ramifications and circumstances surrounding melting, magnesium treating, inoculating, pouring, gating, molding, and shakeout must be considered. This chapter is concerned with individual ductile iron casting defects. A description is presented and possible remedies have been suggested.

■ GRAPHITE FLOTATION

Graphite is significantly less dense than iron. Consequently, graphite that precipitates in the liquid iron has a tendency to float. When this flotation occurs prior to solidification in the mold, a concentration of graphite nodules occurs at the cope surface and immediately under cored surfaces. When these castings are broken, a region displaying a black fracture can be seen. Figure 14-1 shows an example of cope surface graphite concentration/flotation on Charpy impact bars.

Carbon concentrations in the black fractures are often as high as 5.0–6.0%. Mechanical properties, particularly tensile strength, percent elongation, and impact strength, can deteriorate significantly in the graphite flotation zone. Observers have reported a 25% reduction in tensile strength, over 80% reduction in percent elongation, and over 50% reduction in impact strength.[1]

The three most important factors associated with graphite flotation are: (1) carbon equivalent; (2) casting section size; and (3) solidification rate.

Fig. 14-1. Fractured surfaces of Charpy impact test bars showing graphite flotation. Note the distinct line of demarcation for the v-notched samples at the lower right. This demarcation line is

Graphite flotation may occur in all hypereutectic irons in which the carbon equivalent exceeds 4.5%. Carbon equivalent is defined by the expression:

$$CE = (\%C + \%Si/3) + (\%P\%/3)$$

The extent of flotation is small when irons are just hypereutectic (i.e., their carbon equivalent values just exceed 4.3%). Flotation effects are confined mainly to heavier sections at these carbon equivalent levels. When graphite flotation occurs, concentration of graphite nodules can be seen on cross sections through the flotation zone. Figure 14-2 shows flotation on a machined surface. Figure 14-3 shows the appearance of the graphite in the flotation zone.

Control of the base iron composition before magnesium treatment and inoculation is a practical and effective way to prevent flotation. The effect of carbon and silicon contents upon flotation are similar to those upon the liquidus arrest temperature (CEL). CEL is defined by the expression:

$$CEL = (\%C + \%Si/4) + (\%P/2)$$

To prevent flotation, it is recommended that the maximum CEL values for ductile iron castings of different section sizes shown in Table 14-1, at various casting temperatures, not be exceeded.[1] The relationship between carbon equivalent and section size has been expressed by

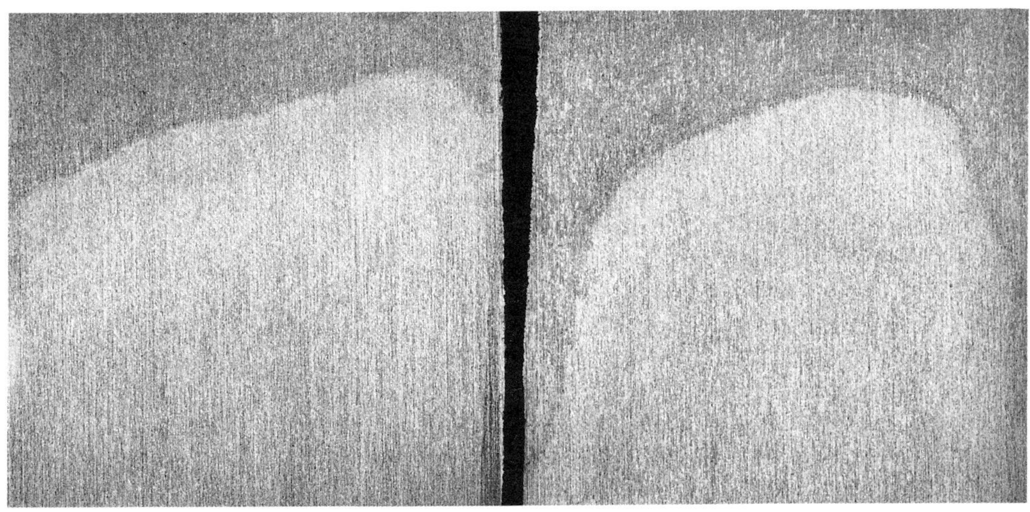

Fig. 14-2. Typical appearance of flotation on machined surfaces of blocks cast from ductile iron with 4.9% carbon equivalent.

223

TABLE 14-1. MAXIMUM CEL TO MINIMIZE RISK OF GRAPHITE FLOTATION

Pouring Temperature		Maximum CEL Value for Various Bar Section Sizes Cast in Green Sand Molds			
°F	°C	20 mm (0.787 in.)	30 mm (1.181 in.)	50 mm (1.968 in.)	80 mm (3.150 in.)
2400	1315	4.56	4.52	4.44	4.31
2450	1340	4.53	4.49	4.41	4.27
2500	1370	4.50	4.46	4.38	4.24
2550	1400	4.47	4.43	4.35	4.21
2600	1425	4.45	4.40	4.32	4.19
2650	1455	4.42	4.37	4.29	4.15

other investigators in practical forms. Tables 14-2, 14-3, and 14-4 are examples of the guidelines that have resulted.[2]

EXPLODED GRAPHITE

One of the more common forms of undesirable graphite is called exploded graphite. A photomicrograph of this graphite form is shown in Figure 14-4. Factors generally responsible for the occurrence of exploded graphite are the same as those associated with graphite flotation. These factors are carbon equivalent values in excess of 4.5%, overtreatment with magnesium in the presence of uncombined rare earth metals in excess of 0.02%, and extended solidification time, which often occurs in heavy-section castings. In certain cases, exploded graphite can be the forerunner of chunk graphite in heavy-section castings.

CARBIDES[2,3]

The term "carbides" covers a wide range of defects, from the presence of a few scattered carbides in the matrix to the complete absence of graphite with a carbidic matrix. The causes of carbide formation are discussed at length in Chapter 10. In general, heat treatment can salvage carbidic castings as described in Chapter 13. The relationships between carbides and composition were discussed in Chapter 4.

Carbides in ductile iron can occur in three broadly different forms: eutectic carbides, inverse chill and segregation carbides. Eutectic carbides account for the greatest proportion of carbides found in ductile iron. These carbides occur as a result of rapid cooling rate. Figure 14-5 shows eutectic carbides resulting from rapid solidification. Inverse chill is a carbide form that occurs in the interior of castings in the medium-size ranges. Figure 4-58 (p 105) shows inverse chill. Segregation carbides occur at eutectic cell boundaries,

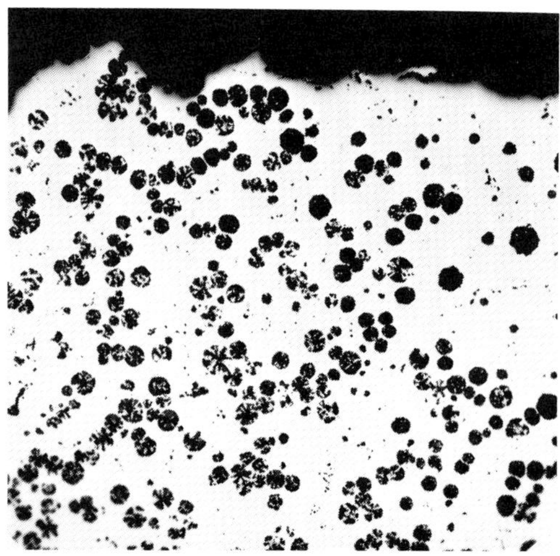

Fig. 14-3. Graphite in the flotation zone; ×35, unetched condition.

TABLE 14-2. GENERAL RECOMMENDATION FOR CARBON AND SILICON CONTENT OF UNALLOYED DUCTILE IRON

| Wall Thickness | | As-Cast Matrix | | | |
| | | Pearlitic* | | Ferritic** | |
Minimum	Maximum	%C	%Si	%C	%Si
1/8 in.	1/8 in.	3.85	2.85	4.25	2.85
	1/4 in.	3.80	2.85	4.00	2.85
	1/2 in.	3.75	2.75	3.90	2.85
	1 in.	3.70	2.75	3.80	2.85
	2 in.	3.60	2.75	2.65	2.85
	3 in.	3.40	2.75	2.50	2.85
	4 in. or more	3.25	2.75	3.45	2.85
1/4 in.	1/4 in.	3.85	2.80	3.95	2.85
	1/2 in.	3.80	2.80	3.90	2.85
	1 in.	3.75	2.80	3.85	2.85
	2 in.	3.70	2.80	3.80	2.85
	3 in .	3.65	2.70	3.70	2.85
	4 in. or more	3.60	2.70	3.60	2.85
1/2 in.	1/2 in.	3.80	2.75	3.85	2.80
	1 in.	3.75	2.75	3.80	2.80
	2 in.	3.65	2.75	3.75	2.80
	3 in.	3.60	2.65	3.70	2.80
	4 in. or more	3.55	2.65	3.65	2.80
1 in.	1 in.	3.60	2.55	3.75	2.70
	2 in.	3.55	2.55	3.65	2.70
	3 in.	3.50	2.55	3.60	2.65
	4 in. or more	3.40	2.40	3.55	2.65
2 in.	2 in.	3.55	2.25	3.55	2.45
	3 in.	3.40	2.20	3.50	2.45
	4 in. or more	3.40	2.20	3.45	2.45
3 in.	3 in.	3.40	2.15	3.40	2.45
	4 in. or more	3.40	2.15	3.40	2.45

*Mn = 0.45–0.65

**Mn = 0.25–0.15

TABLE 14-3. RECOMMENDED CARBON CONTENT FOR UNALLOYED DUCTILE IRON WITH A 2.5% Si PEARLITIC MATRIX

Maximum Wall Thickness in Inches	Total Carbon %
1/8	4.40
1/4	4.10
1/2	3.90
1	3.60
2	3.45
3	3.40
4 or more	3.35

TABLE 14-4. RECOMMENDED CARBON CONTENT FOR UNALLOYED DUCTILE IRON WITH A 2.5% Si FERRITIC MATRIX

Maximum Wall Thickness in Inches	Total Carbon %
1/8	4.35
1/4	4.05
1/2	3.85
1	3.55
2	3.40
3	3.35
4 or more	3.30

particularly in heavy-section castings, which cool slowly. Figure 14-6 shows an example of segregation carbides.

The occurrence of all three carbide forms is minimized with efficient inoculation, high nodule counts, and low levels of carbide-forming elements. These elements include chromium, boron, molybdenum, vanadium, manganese, and copper at more than about 1%, and are particularly important when considering the cause for cell boundary or segregation carbides. Chromium, vanadium and manganese at levels greater than 0.08%, 0.04% and 0.60%, respectively, can be especially harmful. These carbides are difficult to remove with heat treatment. Eutectic carbide and inverse chill can be minimized by maintaining a high silicon content. Silicon, however, can produce embrittling effects at levels in excess of 2.6%. In order to keep the overall tendency to form carbides to a minimum, the following measures are recommended:

1. Use high-purity charge materials, such as oxygen-blown or specially prepared pig iron, or carefully selected steel scrap.
2. Avoid low carbon content and, particularly, low final silicon content materials.
3. Add an efficient inoculant as late as possible in the treatment cycle. This inoculation step should definitely be performed after the magnesium addition has been completed.
4. Pour the castings as soon as possible after inoculation.
5. In critical cases, the inoculation procedure should include a late inoculant addition to the mold or pouring stream.
6. Avoid conditions likely to promote hydrogen pick-up, such as long runner systems, wet refractories, and high aluminum content.
7. For large castings, use low pouring temperatures in order to reduce the time between pouring the casting and the onset of solidification.

FLAKE GRAPHITE

Flake graphite is not a desirable graphite form in ductile cast iron. Relatively small quantities of flake graphite can reduce the overall mechanical properties of ductile iron castings and cause them to be defective. Flake and vermicular graphite are usually associated with insufficient retained magnesium. Magnesium levels of less than 0.03% can normally produce vermicular graphite. Figure 4-9 (p. 74) shows an example of vermicular graphite. Subversive trace elements such as titanium, tellurium, arsenic, lead, and antimony can cause this graphite form to occur. The use of rare earths can neutralize the effect of subversive trace elements. For this reason, most treatment alloys contain a small quantity of rare earths.

Possible methods for eliminating the formation of vermicular graphite are as follows:

1. Increase the magnesium content of the iron by increasing the amount of magnesium alloy or treating material. (See Chapter 4.)
2. Review treating practice and equipment. (See Chapter 8.)
3. Review melting and pretreating practices. (See Chapters 5, 6, and 7.)
4. Determine if subversive trace elements are present. If these elements are detected, take the

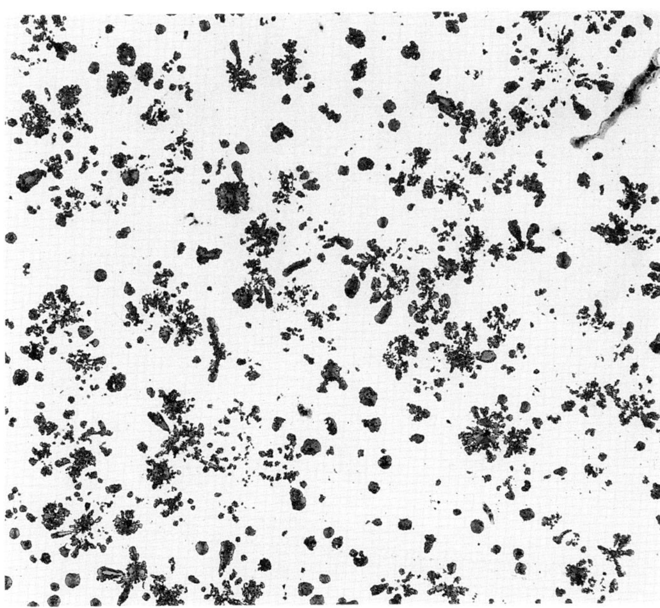

Fig. 14-4. Example of exploded graphite; ×50.

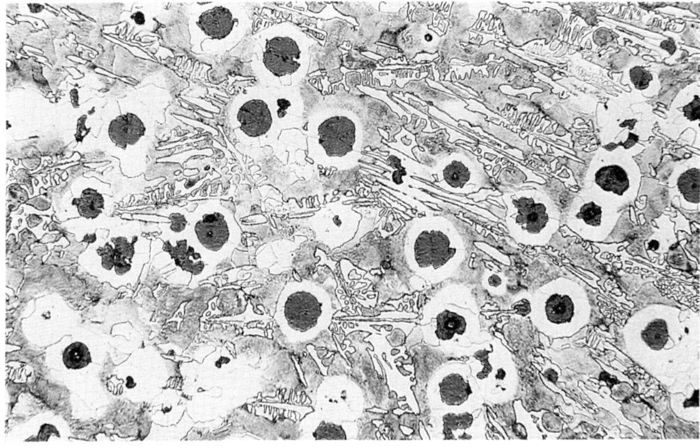

Fig. 14-5. Eutectic carbide (15% carbides); ×100, nital etched.

necessary action such as adding rare earths. (See Chapter 4.)

5. Control sulfur in molding sand to <0.10% to minimize the chance of unsatisfactory structures at the iron-sand interface.

CHUNK GRAPHITE [2]

Chunk graphite is most normally associated with heavy sections and slow solidification. Figure 14-7 shows a typical example. Chunk graphite is visible to the naked eye on fresh saw-cut surfaces through heavy sections. The mechanism responsible for the formation has not been definitely established. It is, however, believed to be in some way associated with the rate of solidification and the presence of subversive elements. Since this graphite form commonly occurs in thermal centers, one means of preventing the formation is to use chills on one side of the casting to drive the thermal center towards the surface. Over-inoculation is also a possible cause.

Recent investigations have indicated that low nodule count will minimize formation of chunk graphite. Low nodule counts can be promoted by superheating the base iron above 2750F (1510C), but the risk of inverse chill is increased. Holding the base iron for extended periods at temperatures above 2700F (1482C) also reduces nodule count. Inoculating with minimum quantities of ferrosilicon (0.1–0.25%) has also proven beneficial for reducing the occurrence of chunk graphite.

Since chunk graphite occurs in the last metal to solidify, low magnesium levels are generally present. Therefore, increasing the magnesium content can correct the condition. Additions of 0.001% antimony or 0.02% tin have also been reported to minimize the formation of chunk graphite.

In heavy sections, silicon content above 2.00% can aggravate chunk graphite formation. Therefore, silicon should be held between 1.00–2.00%; although, in practice levels below 1.85% are rare. At these low silicon levels, even with slow cooling, some carbides can form. Should this occur, a lower manganese level is suggested.

DROSS INCLUSIONS

Dross in ductile iron is a product of the magnesium treatment. It consists of a mixture of magnesium oxide, magnesium silicate, and/or magnesium sulfide. Since magnesium is continually being rejected by the iron and oxidizes on the surface of the metal, dross is probably the most common defect encountered in the production of ductile iron. The defect is more prevalent in heavy-section castings, but can be found in all casting sizes. Figures 14-8 and 14-9 show examples of the dross and elements associated with its occurrence; spectra for areas 1, 2, and 3 are shown in Figures 14-10–12.

Dross defects may appear to be cracks. Actually, these defects are folds or cold shuts caused by a thin film of dross. In extreme cases of dross-related defects, wrinkles

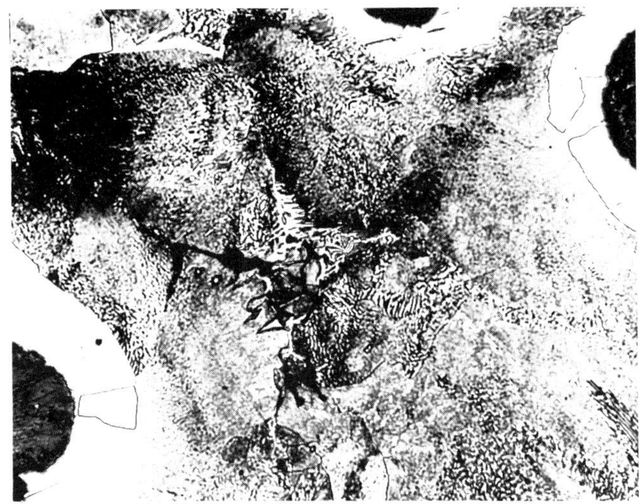

Fig. 14-6. Cell boundary carbides (white) caused by Mn segregation; ×1000, nital etched.

Fig. 14-7. Irregularly shaped graphite spheroids in a heavy ductile iron casting, unetched; original ×100.

are visible on the surface of the casting. This wrinkled surface is often referred to as elephant skin. All ductile iron castings generally are subject to some degree of dross related surface crazing or cold shuts. These defects most commonly appear on cope surfaces and can be detected with magnetic particle nondestructive testing techniques.

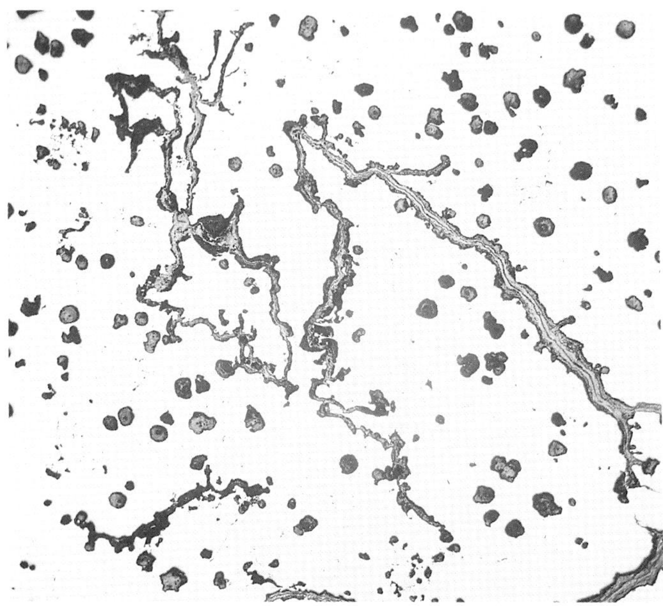

Fig. 14-8. Dross; ×50, unetched condition.

Dross defects can result from a number of different causes.

1. A high retained magnesium content. Therefore, reducing the quantity of magnesium added or supplementing the magnesium addition with rare earth will reduce the occurrence of dross.

2. Slow pouring, pouring cold metal or pouring metal that has an obvious slag skin. Maintaining clean ladles and proper skimming techniques are essential to pouring clean metal and dross-free castings.

3. A gating system that causes turbulent metal flow.

4. Hypereutectic irons with high carbon and silicon tend to be the worst dross producers. Therefore, a carbon equivalent that is suited to the section size and pouring temperature, as described in the section on graphite flotation, is also essential to producing dross free castings.

▬ PINHOLING[2]

The occurrence of pinholes in gray and ductile iron castings has long been a source of significant scrap losses in commercial foundries. But losses from pinholes in ductile iron are approximately 50% greater than in gray iron. Pinholes are normally classified according to the type of gas that caused them: magnesium vapor, hydrogen, nitrogen, carbon monoxide, and interactions involving both hydrogen and nitrogen.

Low pouring temperatures (2450F/1343C and lower) drastically increase the incidence of magnesium vapor pinholes in ductile iron castings. The area surrounding these pinholes is usually pearlitic with traces of carbides. This type of pinhole can be associated with cold pouring and the inability of entrapped gases and magnesium vapor to escape from the molten iron. The incidence of magnesium vapor pinhole defects can be reduced by increasing the pouring temperature and by adding up to 0.02% total rare earths to improve the fluidity of treated ductile iron. Increased fluidity allows the gases to escape more readily from the molten iron.

As a general rule, hydrogen and nitrogen are not introduced into ductile iron base iron during melting. However, if they are the magnesium treatment will remove them from the molten iron. Normal magnesium-treated ductile irons contain approximately 2 ppm hydrogen and up to 80 ppm nitrogen. When this is poured into 5% clay-bonded green sand with a moisture content of approximately 3.0%, only a very slight incidence of hydrogen pinholes is possible. However, if the moisture content of the green sand is increased to a level of 4–6%, severe pinholing will result. This increased hydrogen level not only results from wet sand, but also can be caused by wet ladle refractories.

The presence of increased quantities of strong oxide formers, such as magnesium and aluminum, promotes pinholing. These two elements drastically increase the reaction producing hydrogen. Since hydrogen activity promotes gas bubble growth, an increase in the severity of pinholing also results from reactions of these elements with steam in the mold:

$$Mg + H_2O \rightarrow MgO + 2H$$

$$2Al + 3H_2O \rightarrow Al_2O_3 + 6H$$

The stable oxides resulting from these reactions and the presence of magnesium silicate particles often act as sites for gas bubble nucleation. Therefore, hydrogen pinholing severity can be increased, due to the presence of aluminum and magnesium.

Aluminum is the most potent element for promoting hydrogen gas formation. For this reason, every effort should be made to use charge materials that do not contain aluminum. The influence of aluminum on pinholing in wet sand molds becomes more severe when aluminum is in the range of 0.045–0.060%. When aluminum is present in amounts approaching 0.10% and greater, this severity decreases.

Hydrogen pinholes are either round or egg-shaped and usually occur just below the surface of a casting. Hydrogen pinholes associated with wet sand usually have a smooth

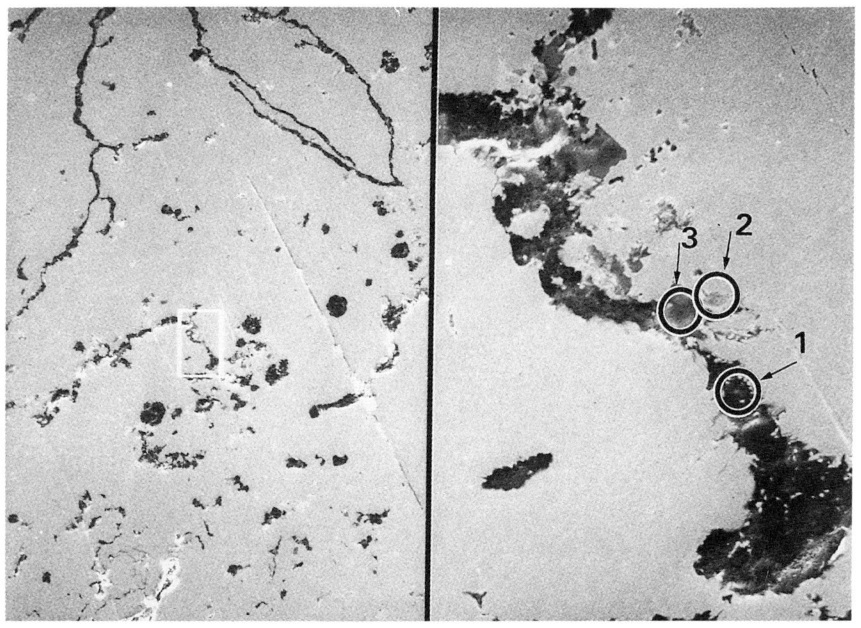

Fig. 14-9. Dross in ductile iron. Split-image scanning electron micrograph at ×75 and ×750.

229

interior covered with a crystalline graphite coating. The microstructure in the area of the pinholes often consists of a ferrite ring containing vermicular and flake graphite. Measures that can decrease the formation of hydrogen pinholes are:

1. Increase the seacoal content of molding sand to 6%.
2. Avoid exposing the liquid iron to moisture after magnesium treatment.
3. Use dried molds.
4. Increase pouring temperatures and venting to permit gas to escape from the metal.
5. Add 2–5 grams of tellurium per 1000 pounds of treated iron. A 30 ppm addition of boron is also effective.
6. Maintain aluminum levels below 0.04% in treated iron.
7. Add rare earth metals to react with hydrogen and nitrogen, minimizing pinhole formation.
8. Eliminate hydrogen-forming materials, such as cereal binders and wood flour, from the mold sand.

The other major source of gas porosity in ductile iron castings is nitrogen evolved from core and mold binders that contain nitrogen. These include:

1. Phenol/formaldehyde shell core binder.
2. Furan-type hotbox core binders.
3. Furan-type air set core and mold binders.

Nitrogen pinholes in ductile iron castings can take a spherical or irregular shape. The interior of these pinholes is usually dull gray in color. They occur in the area of the casting adjacent to the nitrogen-bearing bonded core or mold surface when the nitrogen level in ductile iron reaches 130 ppm.

The most effective way of eliminating nitrogen pinholes is to use nitrogen-free mold and core binders. Another approach is to introduce 0.02% rare earths or 0.10% aluminum. However, the introduction of aluminum can cause the nodule form of graphite to deteriorate.

Carbon monoxide is another source of pinholes. These pinholes form when the carbon present in the iron or carbonaceous materials present in the sand react with magnesium silicate and slag. This type of spherical pinhole usually occurs in patches on the cope side of the casting in an area where slag inclusions are present. It is usually the result of overtreatment with magnesium ferrosilicon alloys, resulting from high base sulfur and oxide levels.

Some researchers have reported that an addition of 2–5 grams of tellurium or bismuth per 100 pounds iron can reduce or eliminate pinholes. Both of these elements can cause the nodule form of graphite to deteriorate. Therefore, caution must be exercised when these elements are added for pinhole control. The use of rare earths in conjunction with these elements often will control the deleterious effects

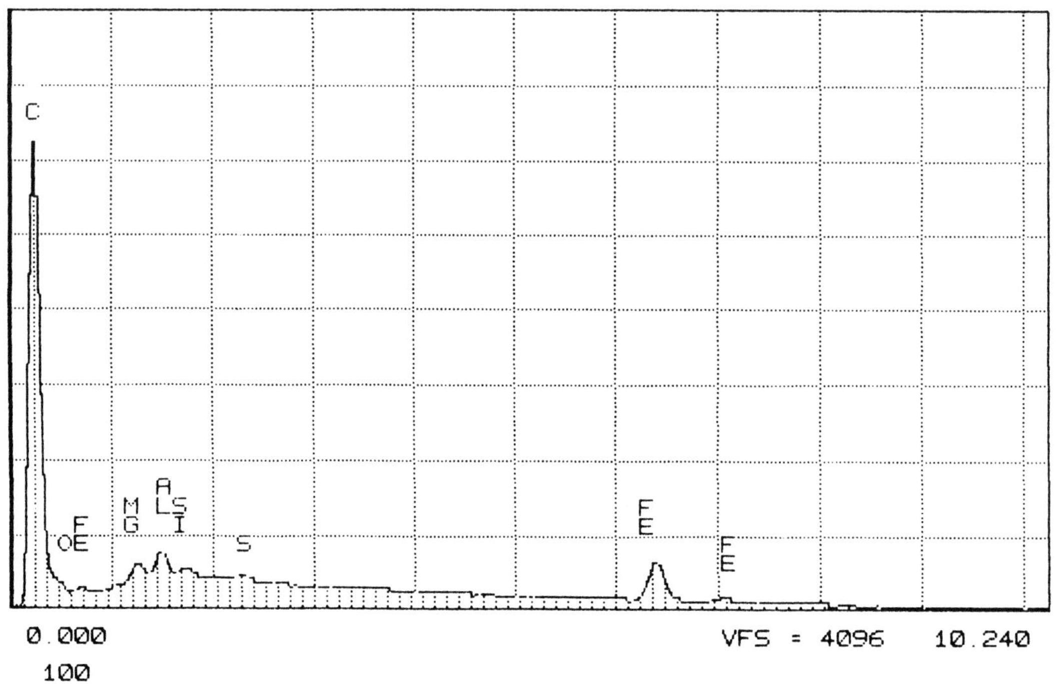

Fig. 14-10. Spectrum of elements for area 1 in Fig. 14-11. (Reference: Taussig Associates, Inc., Report No. 88085.)

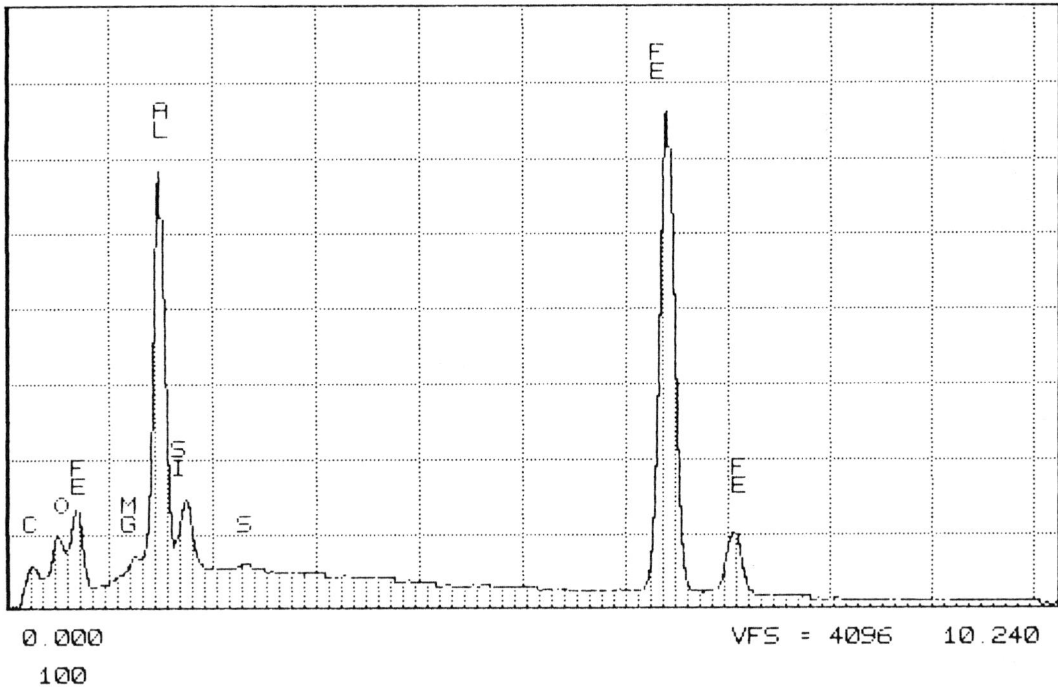

Fig. 14-11. Spectrum of elements for area 2 in Fig. 14-9. (Reference: Taussig Associates. Inc.. Report No. 88085.)

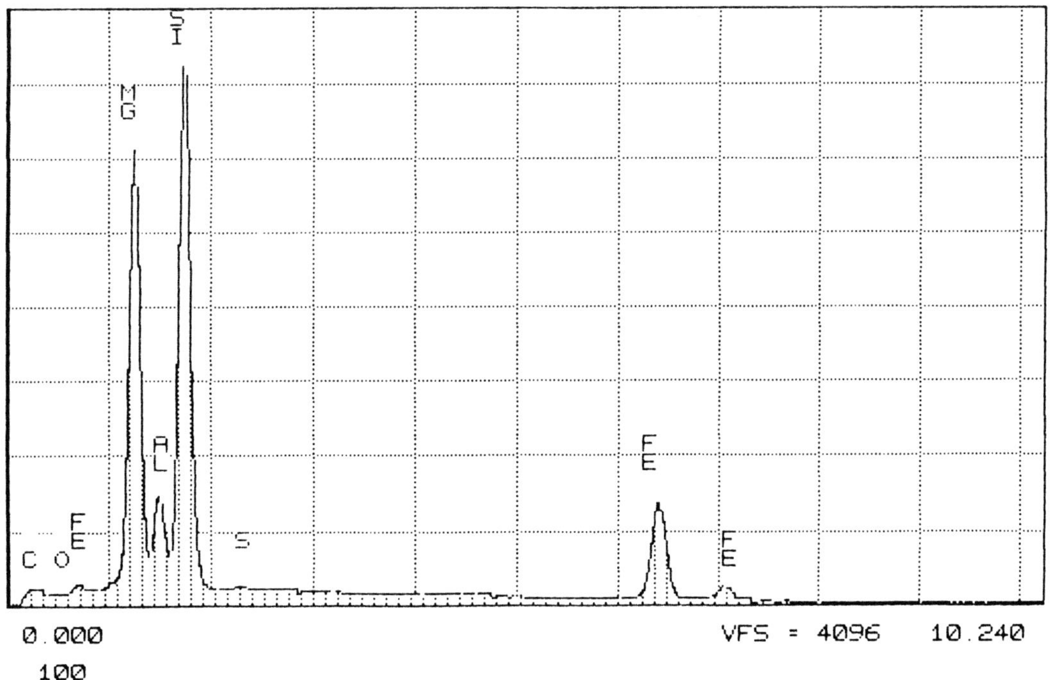

Fig. 14-12. Spectrum of elements for area 3 in Fig. 14-9. (Reference: Taussig Associates. Inc.. Report No. 88085.)

Fig. 14-13. Example of pinholes on the cope side of crankshaft castings.

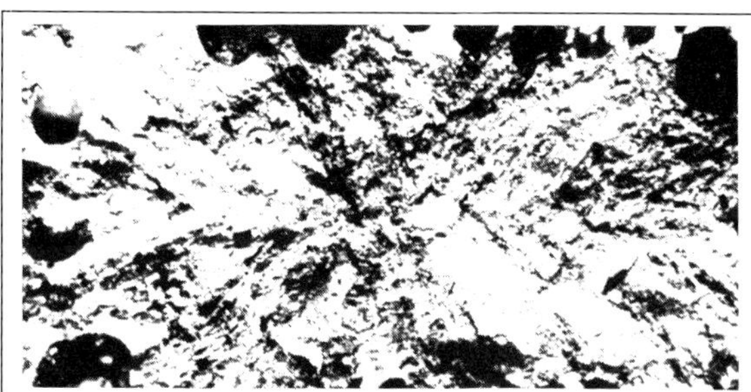

Fig. 14-14. Typical pinholes found in ductile iron cast in green sand in commercial foundries.

on graphite shape. Figures 14-13 and 14-14 are examples of pinholing.

SHRINKAGE[2]

Sponge or interdendritic shrinkage occurs during the last stages of solidification and can usually be eliminated when adequate quantities of graphite precipitate during solidification. This is very important in castings that have areas that are difficult to riser effectively. Figure 14-15 shows the importance of a suitable carbon and silicon relationship for promoting sufficient graphite precipitation. The formula

$$\%C + (\% \, Si/7) = 3.9$$

has been used as a guide to determine a safe range of chemistry that will produce sound castings. It should be noted that this formula is different than the one calculat-

ing carbon equivalent. This is very important in producing ductile iron castings in dry sand molds or hand rammed molds without risers.

Shrinkage will increase as pouring temperatures decrease. Shrinkage is more pronounced in light and medium sections, which have high surface-area-to-volume ratios and a high cooling rate as the metal fills the mold. When metal temperatures fall below 2300F (1260C), graphite precipitation begins. If the mold cavity is not full at this temperature, the effect of the volume increase due to graphite precipitation is lost, and excess shrinkage results. Very slow pouring or gating systems that cause excessive cooling can promote shrinkage.

It is important that the mold cavity be filled before the metal temperature falls to 2300F (1260C). The gating system should be designed to provide uniform metal distribution and temperature gradients. This uniformity is important to confine the expansion due to graphite precipitation to the mold cavity. Inconsistent results from day to day, or even between ladles, may mean that the pouring temperature is too close to 2300F (1260C) and should be raised.

Isolated bosses on heavy sections may require risering or chilling. Otherwise, these areas will act as a riser to light sections, particularly in green sand. Increasing the green sand hot strength and hard ramming the molds, along with faster pouring and higher metal temperatures (at least 2600F/1427C) will often eliminate shrinkage problems.

MISCELLANEOUS DEFECTS DUE TO MAGNESIUM OVERTREATMENT[2]

Adding magnesium in excess of 0.06% is one of the major causes of casting defects in ductile iron. It is also one of the most elusive and misunderstood causes of defects. The ideal magnesium range is 0.035–0.045%. Magnesium in excess of 0.05% tends to promote the formation of excessive

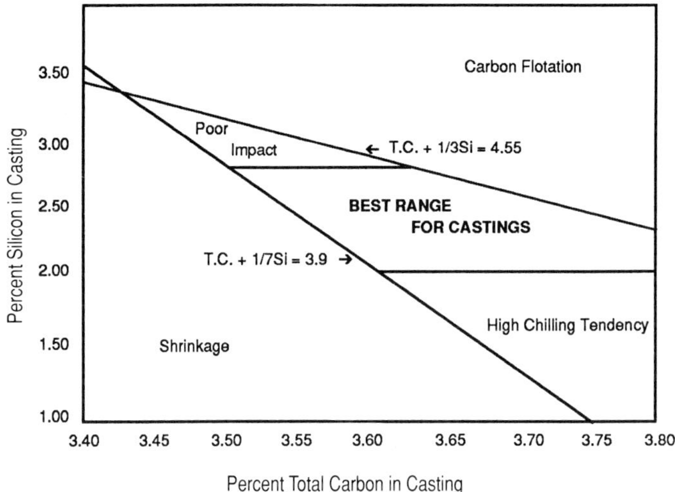

Fig. 14-15. Carbon and silicon ranges for ductile iron castings.

magnesium silicate and magnesium oxide dross or slag. Magnesium at this level also promotes the formation of carbon flotation. Magnesium content greater than 0.08% may cause graphite nodules to form crab-like or spiky shapes. As magnesium content increases, patches of sharp-ended graphite flakes appear. High magnesium also promotes carbide formation.

REFERENCES

1. American Foundrymen's Society, Inc; "Graphite Floatation in Ductile Iron Castings," AFS Special Report (1980).

2. A.F. Spangler, ed.; *The Ductile Iron Process*, Miller and Company (1985).

3. J.V. Dawson; "Causes and Prevention of Carbide Formation in Nodular (S.G.) Iron," *Foundry Trade Journal*, p 239 (Aug 14, 1989).

BIBLIOGRAPHY

GENERAL

Barton, R. "Nodular Iron: Possible Structural Defects and their Prevention." *Foundry Trade Journal,* vol 155, no. 3267 (Jul 14, 1983).

FLOTATION

Fuller, A.G. and T.N. Blackman. "Effects of Composition and Foundry Process Variables on Graphite Flotation in Hypereutectic Ductile Irons." *AFS Transactions,* vol 94, Des Plaines, IL: American Foundrymen's Society, Inc. (1986): 823.

Sun, G.X. and C.R. Loper, Jr. "Graphite Floatation in Cast Iron." *AFS Transactions,* vol 91, Des Plaines, IL: American Foundrymen's Society, Inc. (1983).

Zhang, J., C. Zhang, and Y. Xie. "Investigation of the Nodularization and Inoculation of Rare Earth-Treated Ductile Iron with Image Analyzing Computer." *AFS Transactions,* vol 91, Des Plaines, IL: American Foundrymen's Society, Inc. (1983).

CARBIDES

Dawson, J.V.; "Causes and Prevention of Carbide Formation in Nodular (S.G.) Iron," *Foundry Trade Journal* (Aug 14, 1989).

Evans, W.J., S.F. Carter and J.F. Wallace. "Factors Influencing the Occurrence of Carbides in Thin Sections of Ductile Iron." *AFS Transactions,* vol 89, Des Plaines, IL: American Foundrymen's Society, Inc. (1981).

Smolyakova, T.M. and M.V. Mozharov. "Formation of Carbides of Magnesium and Cerium in High-Strength Irons." Izv VUZ, *Chern. Met.* (60) (1978).

Tsutsumi, N. and Y. Matsukawa. "The Influence of Carbon Equivalent Value, Pouring Temperature and Cooling Rate on the Inverse Chill in Spheroidal Graphite Cast Iron." Report of the Casting Research Laboratory, Waseda University, no. 31 (1980): 43-51.

FLAKE GRAPHITE

Martin, F. and S.I. Karsay. "Localized Flake Graphite Structure as a Result of a Reaction Between Molten Ductile Iron and Some Components of the Mold." *AFS Transactions,* vol 87, Des Plaines, IL: American Foundrymen's Society, Inc. (1979).

Vorontosov et al.; "Anomalous Graphite in Castings of

Magnesium-Inoculated Iron." vol 1, *Izv. VUZ, Chern. Met.* (1979).

CHUNK GRAPHITE

Campomanes, E. "The Suppression of Graphite Deterioration in Heavy Ductile Iron Castings." *Giesserei 65* (20) (1978).

Church, N.L. and R.D. Schelling. "Detrimental Effect of Calcium on the Graphite Structure in Heavy-Section Ductile Iron." *AFS Transactions*, vol 78, Des Plaines, IL: American Foundrymen's Society, Inc. (1970).

Hoover, Jr., H.W. "A Literature Survey on Degenerate Graphite in Heavy Section Ductile Iron." *AFS Transactions*, vol 94, Des Plaines, IL: American Foundrymen's Society, Inc. (1986): 601.

Liu, P.C., C.L. Li, D.H. Wu and C.R. Loper, Jr. "SEM Study of Chunky Graphite in Heavy-Section Ductile Iron." *AFS Transactions*, vol 91, Des Plaines, IL: American Foundrymen's Society, Inc. (1983).

Lutsyak, V.G. et al. "Special Features of the Structure of the Spheroidal Graphite in Massive Castings." *Izv. VUZ, Chem. Met.* (1982).

Moore, W.H. "Graphite Decay in Nodular Iron." *Casting Engineering & Foundry World*, vol 14, no. 3 (Fall 1982).

Strizik, P. et al. "Contribution to the Mechanism of Formation of Chunky Graphite." *AFS International Cast Metals Journal* (Sep 1976).

Thury, W. "Formation and Avoidance of Graphite Degeneration in Thickwalled Castings of Spheroidal Graphite Iron." *AFS Cast Metals Research Journal* (Sep 1974).

Thury, W. "Prevention of Some Faults in Spheroidal Graphite Iron Castings." *Proceeding of the Quality Control of Engineering Alloys and the Role of Metals Science Symposium* (Delft, The Netherlands) N.p. (Mar 1977).

DROSS

BCIRA. "Prevention of Dross Defects in Nodular (SG) Iron Castings." BCIRA Broadsheet, no. 20, N.d.

"Identifying Ferroalloy Slag Inclusions in Ductile Iron." *Modern Casting*, vol 75, no. 10 (Oct 1985).

Levi, L.I. et al. "Amount and Composition of Oxide Inclusions of Nodular-Graphite Iron." *Russian Castings Production* (Nov 1969).

Murray, W.G. "Control of Dross Defects in Magnesium-Treated Cast Iron." *Foundry Trade Journal*, vol 124, no. 2672 (Feb 22, 1968).

Peng, X., Y-M Yang, N-X Ding, J.L. Mercer and J.F. Wallace. "Influence of In-the-Mold Chamber Design on Dross Formation in Ductile Iron Castings." *AFS Transactions*, vol 95, Des Plaines, IL: American Foundrymen's Society, Inc. (1987).

Subramanian, S.V. "Investigation of the Origin of Some Typical In-Mold Casting Defects." *AFS Transactions*, vol 91, Des Plaines, IL: American Foundrymen's Society, Inc. (1983).

Vashchenko, K.I. et al.; "Influence of Molten Metal Oxidation on Non-Metallic Inclusion Contents in Castings." *Russian Castings Production* (Mar 1972).

PINHOLE

Alfonsi, B. P. Granatiand and M. Petrucci. "Investigation of Pinhole Formation in Fe-C Alloys." Paper 16, Presented at the 45th International Foundry Congress (Budapest) (1978).

BCIRA. "Pinholes Formed by Hydrogen Gas During Solidification." BCIRA Broadsheet, no. 7, N.d.

Carter, S.F. et al. "Factors Influencing the Formation of Pinholes in Gray and Ductile Iron." *AFS Transactions*, vol 87, Des Plaines, IL: American Foundrymen's Society, Inc. (1979).

Crawford, J. "Blows and Pinholes." *Mini Talk #4—American Foundrymen's Society Current Awareness*, Des Plaines, IL: American Foundrymen's Society, Inc. #71-319, N.d.

Dawson, J.V. "Avoiding Pinholes in Ductile Iron." *Foundry* (Jun 1973).

Farquhar, J.D. "Nitrogen in Ductile Iron: A Literature Review." *AFS Transactions*, vol 87, Des Plaines, IL: American Foundrymen's Society, Inc. (1979).

Mazumdar, R.C.: "Causes of Blowholes and Pinholes in Cast and Ductile Iron Castings." *Molten Metal* (Jan/Feb 1975).

Naik, R.V. and J.F. Wallace. "Surface Tension–Nucleation Relations in Cast Iron Pinhole Formation." *AFS Transactions*, vol 88, Des Plaines, IL: American Foundrymen's Society, Inc. (1980).

Naro, F.L. "Variables Affecting the Formation of Porosity Defects in Iron Castings Prepared with Urethane Binder Systems." *AFS Transactions*, vol 82, Des Plaines, IL: American Foundrymen's Society, Inc. (1974).

Ryntz, Jr., E.F., R.E. Shroeder, W.W. Chaput and W.O. Rassenfoss. "The Formation of Blowholes in Nodular Iron Castings." *AFS Transactions*, vol 91, Des Plaines, IL: American Foundrymen's Society, Inc. (1983).

Subramanian, S.V. "Investigation of the Origin of Some Typical In-Mold Casting Defects." *AFS Transactions*, vol 91, Des Plaines, IL: American Foundrymen's Society, Inc. (1983).

Shrinkage

Anderson, U. "Shrinkage Defects and Feeding of SG Iron." *Stoberiet* (May 1976 and Jul 1976).

Fidos, H. "Volume Changes During the Solidification of Ductile Iron." 51st International Foundry Congress (Lisbon), Paper 8 (1984).

Gariepy, B. R. Goller. "Effect of Mold Material and Feeding on Volume Changes in Ductile Iron Castings." *AFS Transactions*, vol 89, Des Plaines, IL: American Foundrymen's Society, Inc. (1981).

Phelps, T.A., R.W. Heine and J.J. Uicker. "Analysis of Internal Unsoundness Casting Defects Using Artificial Intelligence Techniques." *AFS Transactions*, vol 97, Des Plaines, IL: American Foundrymen's Society, Inc. (1989).

Tafazzoli-Yadzi, M. "Volumetric Contraction and Shrinkage Cavities Behavior in Ductile Iron Castings." *AFS International Cast Metals Journal* (Dec 1977).

Wallace, J.F., P.K. Samal and J.D. Voss. "Factors Influencing a Shrinkage Cavity Formation in Ductile Iron." *AFS Transactions*, vol 92, Des Plaines, IL: American Foundrymen's Society, Inc. (1984).

Winter, B.P., T.R. Ostrom, D.J. Hartman, P.K. Trojan and R.D. Pehlke. "Mold Dilation and Volumetric Shrinkage of White, Gray, and Ductile Cast Irons." *AFS Transactions*, vol 92, Des Plaines, IL: American Foundrymen's Society, Inc. (1984).

Magnesium Overtreatment

Basutkar, P.K., C.S. Park, R.E. Miller and C.R. Loper, Jr. "Formation of Spiky Graphite in High Magnesium Ductile Iron Castings." *AFS Transactions*, vol 81, Des Plaines, IL: American Foundrymen's Society, Inc. (1973).

15

Ductile Iron Process Control

R. Alan Patrick

Buck Co., Inc.
Quarryville, Pennsylvania

INTRODUCTION

The ductile iron foundry of today—and of the future—must provide a means to continually improve quality, productivity and competitiveness.

There are three interdependent functions that must be fully developed and utilized in order to provide a Total Process Control (TPC) system:

- a dynamic *production operating system* that incorporates design, facilities, procedures and requirements;
- *statistical process control (SPC) techniques* to indicate where and when to make changes in the operating system;
- consistent training of personnel in procedures, requirements and SPC techniques for employees, managers and vendors, and developing teamwork among inter-company personnel, customers and vendors—the *human element*.

TOTAL PROCESS CONTROL SYSTEM

In the past, the foundry industry has concentrated its efforts primarily only on production operating systems, particularly on design, facilities and procedures. Now, it is imperative that process control be addressed in its entirety and that all functions affecting it be controlled.

The diagram in Figure 15-1 illustrates the interrelationship between those functions necessary to achieve a total process control system.

The objective of TPC is "quality productivity, economically." Following a discussion of the three previously mentioned primary functions necessary to achieve the TPC objective—changes in the production operating system, SPC techniques, and the human element—the various techniques used to accomplish TPC (those in the outer ring of 15-1) will be discussed.

PRODUCTION OPERATING SYSTEM

An effective production operating system must function proactively rather than reactively.

Any changes in the system must be made as the result of SPC monitoring or advanced quality planning, rather than through evaluation of scrap or rework material at final inspection. This provides a means for "prevention rather than detection" and "process control rather than product control," which are much less costly than waiting until final inspection, where all costs are already in the product.

Incorporation of corrective design changes, modification of facilities, implementation of effective procedures, and establishing sound foundry operating requirements are mandatory in the proper development of the production operating system.

SPC TECHNIQUES

SPC techniques are implemented by first collecting data and then charting data to pinpoint problems. Causes of problems are then indicated based on analysis of the charted data. Lastly, corrective action must be taken on all causes, by making changes to the production operating system in either design, facilities, procedures or requirements.

These techniques facilitate a process control system that is geared toward *continuous improvement.* They promote a system that, over time, will have the least amount of variation and will be running at the targeted optimum process level.

THE HUMAN ELEMENT

The human element has long been ignored as being an important function in process control. However, improvements in morale, cooperation, production, quality and cost control are all benefits that result when the human element is brought into the picture of process control.

Fundamental in the development of a TPC system is the improvement of: (1) management training to ensure the consistent implementation of effective quality policies that ensure customer satisfaction, (2) training employees to consistently produce quality work the first time around, and (3) training the vendor to provide a consistent quality product or service each and every time. Programs must be established within each of these groups that rely on requirements (specific statements that support good, sound foundry practices) to provide the necessary direction for improvement in consistency.

Teamwork development must be utilized in every aspect of human involvement in order to maximize performance. Management must initiate programs that produce a cooperative team effort between themselves and their employees, their vendors and their customers. There is a synergistic effect that takes place during a coordinated team effort, such that the combined effect through working together (teamwork) is greater than the sum of the individual effects.

OUTER RING

The outer ring of the TPC diagram depicts the specific techniques that may be used to implement statistical process

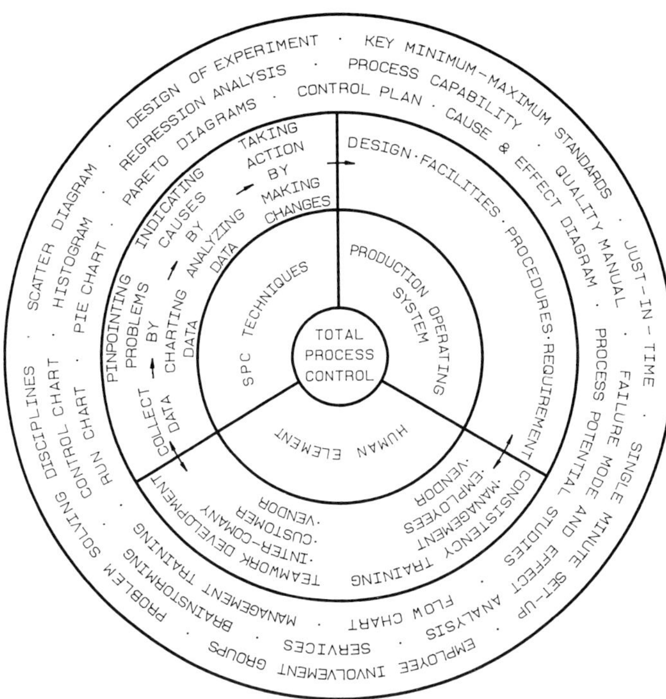

Fig. 15-1. A Total Process Control (TPC) system diagram.

control, develop the human element, and control the production operating system. The following section offers brief descriptions and examples of each of these techniques, with references for obtaining further detailed information.

■ TOTAL PROCESS CONTROL TECHNIQUES

MANAGEMENT TRAINING

It is the responsibility of management to implement quality policies that ensure a product that meets customer requirements. Management is responsible for the control of 85% of all process variation (quality problems) and must learn and practice those measures necessary to prevent variation.

Management must learn how to improve and maintain the quality consciousness of all employees. All employees must understand the costs involved in making the product right (and in making it wrong), and that making it right the first time eliminates the need for expensive repetition. Management must implement procedures for problem analysis and positive corrective action and must incorporate the use of SPC techniques throughout the production process.[1,3,5,8,13,15]

CHARACTERISTICS:

A. Base Iron
B. Transfer Ladle/Pour Ladle
C. Transfer Ladle
D. Heat Treat

Characteristic Affected	Detail	Evaluation Method	Frequency Sample	Analysis Method	Reaction to Out-of Control Condition
A	Chemical Analysis	Spectrometer Carbon & sulphur determinator	Every back-charge	Control chart	No Pour Correct & Retest
A	Temperature	Immersion Pyrometer	Every back-charge	Control Chart	No Pour Correct & Retest
B	Alloy Additions Chill Wedge Nodularity	Usual Inspection	Every Pour Ladle	Chill Wedge Micro Lug	Remove mold from production Lab Analysis
C	Pour Temperature	Immersion Pyrometer	Every Heat	Control Chart	No Pour Return to furnace
D	----------	Brinell Hardness Test	C=O per Mil. Std. 105D	Histogram	No shipment/ Lab Analysis

Submitted By:_____________________
    Metallurgist - XXXX Company, Inc.

Approved By:_____________________
    QA Representative

Fig. 15-2. An example of a control plan for grade 60-40-18 ductile iron.

CONTROL PLAN

A control plan is a comprehensive, written summary of the routine procedures necessary for controlling the ongoing quality of a process. Complete control plans should be written for each part being produced, detailing the specific requirements from the point of incoming inspection to final inspection.

The example given in Figure 15-2 is for grade 60-40-18 ductile iron only. It does not include the additional details necessary for the specific requirements of a particular part.[14,15]

JUST-IN-TIME (JIT)

The principle of Just-In-Time (JIT) is that an individual process must supply the succeeding process with only what is required, when it is required, and in the exact amount required.

A primary benefit of JIT is cost reduction, which should be a principal goal of every company. It is estimated that a reduction in cost of 10% is equal to a 100% increase in sales. The worst enemy to cost reduction is waste, and the greatest waste of all is producing defective products (investing materials, equipment and labor in something that cannot be sold).[13,15]

RUN CHARTS

Run charts (Fig. 15-3) are point plots of data in graphic form, used to analyze trends within a particular interval of time. The causes for all trends (six or more points, consecutively increasing or decreasing) or shifts in the average

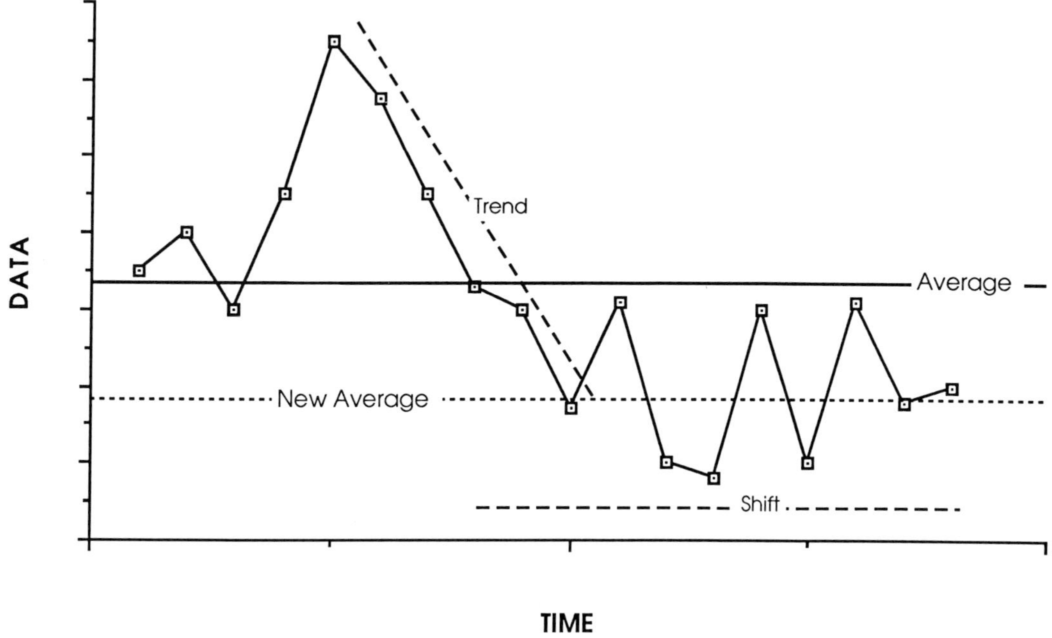

Fig. 15-3. Run chart showing trend and shift.

(eight consecutive points above or below the average) should be investigated. If any trend or shift is found to be advantageous to the process, they should be incorporated into it. If they are found to be detrimental, they should be excluded.[3-11]

CONTROL CHARTS

Run charts with statistically calculated upper control limits (UCL) and lower control limits (LCL) are control charts (Fig. 15-4). They are used to show whether the variability in a process is due to common cause (operating within the control limits) or special assignable cause (points outside the control limits).

A process is said to be "in statistical process control" when all special causes have been eliminated, or in other words, it is operating within the upper and lower control limits and is only being affected by normal variation.

Control charts are also used to show whether the process is capable of meeting the specification. If the upper and lower control limits fall within the upper specification limit (USL) and the lower specification limit (LSL), as in Figure 15-5, the process is said to be capable. However, it is not customary to show specification limits on control charts; therefore, Figures 15-5 and 15-6 are used for illustration only.

A process can be "in control" but still not meet the specification requirements, as is shown in Figure 15-6. If this condition exists, the process must be adjusted to satisfy the specification limits. If this cannot be done, a request must be made of the customer to change their specification limits so that the existing control limits can satisfy the new specification limits.[3-11]

PROCESS POTENTIAL STUDIES

A process potential study is a short-term examination of a process used to predict the long-term capability of producing products that meet customer specification.[14]

PROCESS CAPABILITY

Process capability is a measure of the ability of a specific process to meet specification limits.

Capability indices are calculated values that indicate the extent to which a process is or is not meeting its specification limits.

The C_p index (Fig. 15-7) is a numerical representation of the process variation, relative to the specification limits. When the C_p index is equal to one, the process is just meeting the specification limits (curve 1). With a C_p value greater

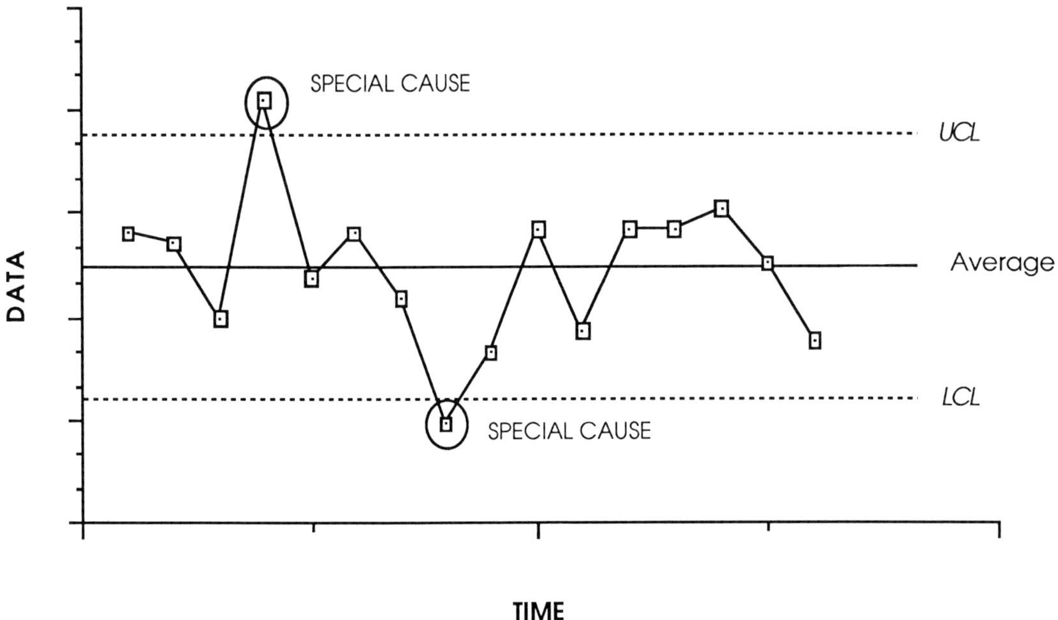

Fig. 15-4. Control chart indicating special assignable causes (points outside the control limits).

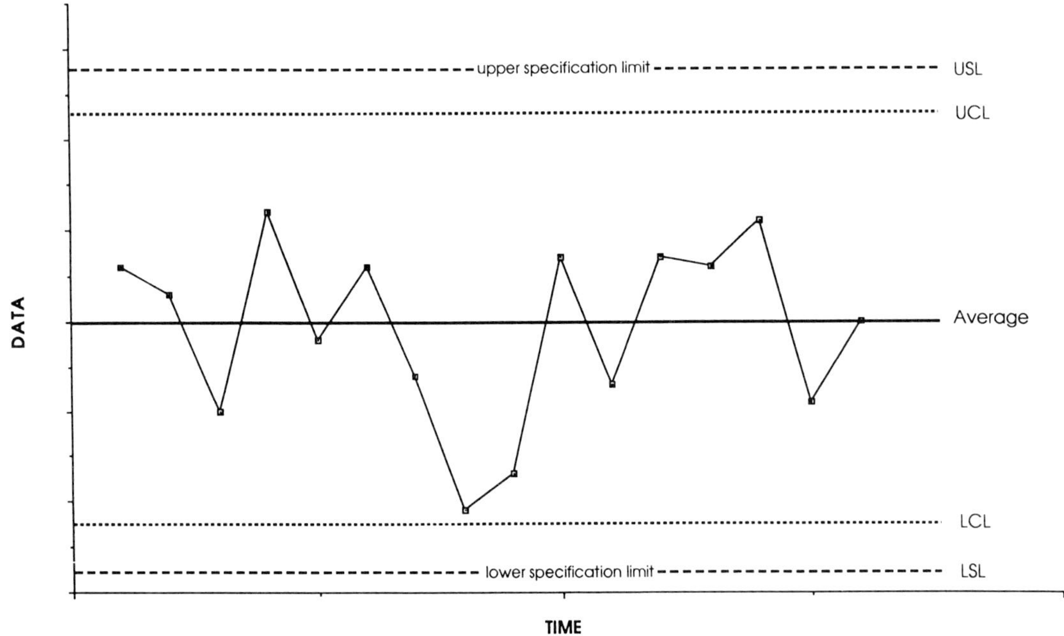

Fig. 15-5. A control chart indicating that process control limits fall within upper and lower specification limits.

than one, the process is within specification (curve 2), and the likelihood of producing defectives will be minimal; however, if the process is not on target, defectives can be produced (curve 3). If the C_p value is less than one, then defectives are being produced (curve 4).

Similarly, the C_{pk} index is a numerical representation of the process variation, relative to the specification limits, accounting also for how well the process average is centered on target.[3-6,8-11]

Pareto Diagrams

Vertical bar graphs used to display the frequency distributions of problems in the relative order of importance are called *pareto diagrams*. They are used to display attribute (countable) data only (Fig. 15-8).

The left hand side of the graph displays a vertical scale that represents the frequency at which various problems occur; the right hand side shows the cumulative percentage, representing the sum of the problems from left to right, which is illustrated in Figure 15-8 by the interconnecting point plot.

Discretion should be used in its interpretation; the most frequently occurring problems are not always the most important.[3,4,6,8,9,11]

Histogram

A vertical bar graph display of the frequency distribution of variable (measurable) data is a histogram. It offers a pictorial representation of the amount and type of variation that exists within a process (Fig. 15-9).[3-9,11]

Scatter Diagram

Scatter diagrams are point plots of data in graphic form used to study the possible correlation of one variable to another (Figs. 15-10a,b,c). The closer the plotted points are to a straight line, the stronger the correlation between the variables.[4-6,9,11]

Regression Analysis

Regression analysis provides a quantitative method of determining the relationship between variables. Examination of the variable data, in the form of a scatter diagram, should always be done prior to calculating a regression line.[3-6,9]

Design of Experiments

Design of experiments is a powerful technique in problem analysis used to study the effects of many variables at once, and determine their relative significance.[3,5,10]

Failure Mode and Effects Analysis (FMEA)

FMEAs are a formal identification and documentation of design or process activities, through team effort, which could eliminate or lessen the probability of a potential failure. A team of individuals knowledgeable in design, manufacturing, sales and quality should all participate in identifying potential failure modes.[3,10,14,16]

Single Minute Setup

This is a system that, through the use of specific techniques, makes it possible to drastically reduce setup times and make small-lot production more attractive. Inventory is reduced by permitting only those jobs that have been ordered to be produced.

The key to this system is a change in thinking from internal setups, which are carried out only when a machine is not operating, to external setups, which can be carried out while the machine is operating.[2]

Employee Involvement Groups

Probably one of the single most important techniques that can be utilized to improve quality and production is the employee involvement group. Ideally, these groups are from four to eight volunteer employees who meet regularly and use problem-solving techniques to analyze and recommend solutions to management for production- and quality-related problems. Sometimes referred to as quality circles, participation from these groups has shown that the employee work force is a vast wealth of untapped knowledge.[1,3,8,12,13,15]

Brainstorming

Brainstorming is a group problem-solving technique aimed at developing many ideas in a short time to resolve a specific problem. A team of concerned participants freely express their ideas without criticism, combining, changing, and adding to ideas until the best solution to the problem is determined.[1,6,8,10,11,15]

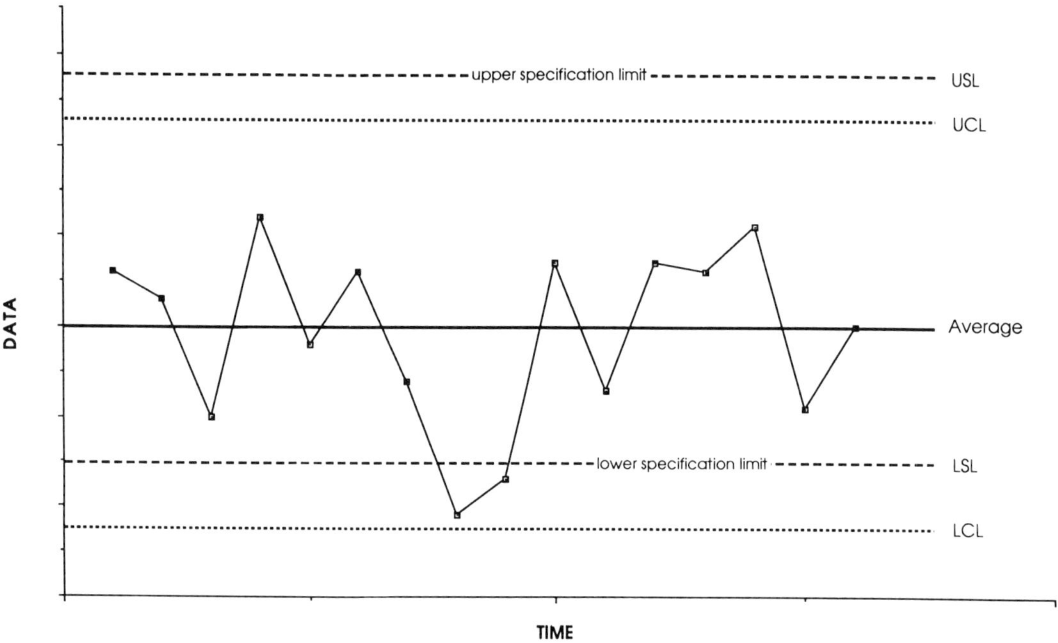

Fig. 15-6. A control chart showing a process "in control" but which does not meet the specification requirements.

CAUSE AND EFFECT DIAGRAM (FISH BONE DIAGRAM)

This diagram (Fig. 15-11) is a systematic method of graphically displaying the relationship between a specific problem or effect and all possible causes influencing it. The problem can relate to quality, economy or environment.

There are six main causal factors—material, method, manpower, movement, machine and measurement—that can be related to an effect or specific problem. Each of these main factors is represented in Figure 15-11 by a branch arrow pointing to the main arrow. The main branch items involved are further detailed by sub-branching down to the most detailed contributing factor.

This diagram is an effective tool that can be used for individual or group problem solving, such as brainstorming sessions.[3-6,8,11]

QUALITY MANUAL

A quality manual is a formal document that describes, in detail, the quality policies, procedures and requirements of a company, including its organizational structure. It exhibits evidence of the company's efforts to assure the quality of its product and to provide the methods for auditing and sustaining the quality system.[3]

PIE CHART

This chart (Fig. 15-12) is a circular graph where a complete circle is representative of 100% of the data being analyzed. The circle is subdivided into sections that represent the breakdown of all the data and their various percentages.[4,11]

KEY MINIMUM-MAXIMUM STANDARDS (KMS)

KMS is a system designed to promote optimum consistency in the daily performance of all personnel within an organization. Based on defect prevention rather than detection, this system addresses the need for conformance to process requirements as a means to monitor and control the ongoing quality of a process. All process requirements are established by a process group team, which has the options to enforce, modify or eliminate any requirements.

Example requirements are:

- all obviously defective materials and equipment shall be reported to the supervisor;
- for all jobs covered by a job instruction card, the card shall be posted at the work station before the first cycle of each production run;
- each element on all job instruction cards shall be performed as specified during every cycle of each production run.

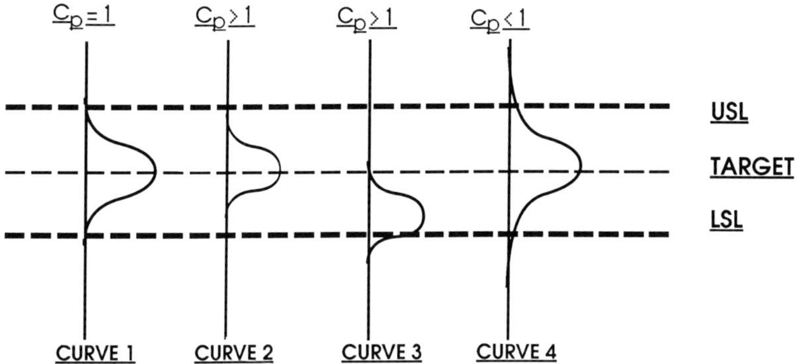

Fig. 15-7. Process variation is shown using the C_p index.

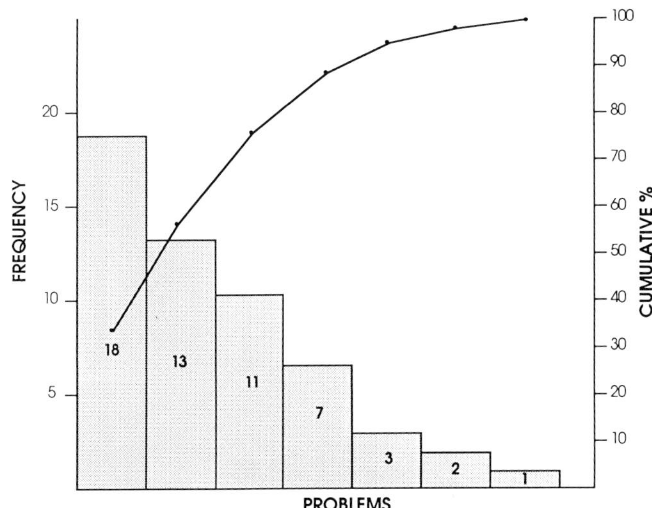

Fig. 15-8. One type of Pareto diagram.

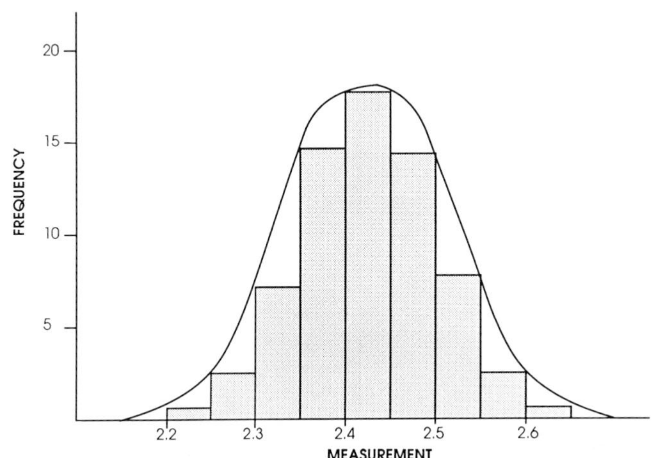

Fig. 15-9. Histogram showing the frequency distribution of variable (measurable) data.

243

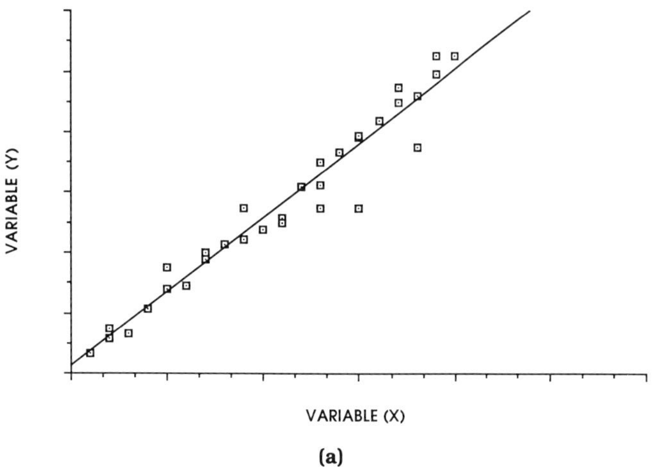

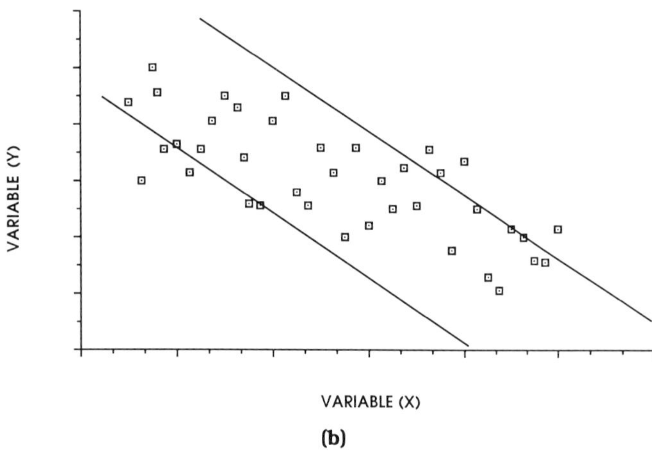

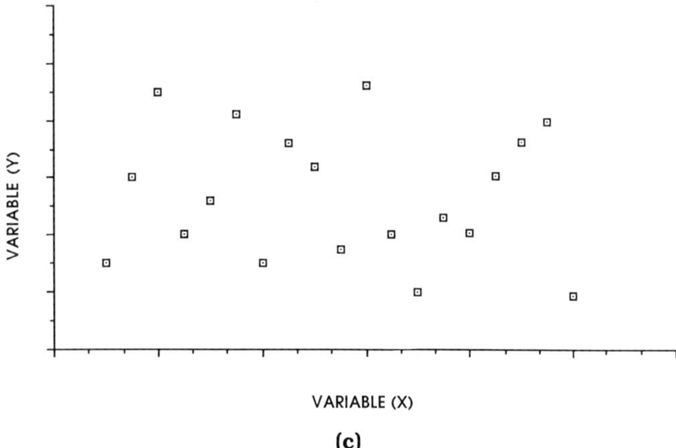

Fig. 15-10. Scatter diagrams showing: (a) strong positive correlation; (b) weak negative correlation; and (c) no correlation.

The key minimum-maximum standards system can be further used to control maintenance, safety and housekeeping.

This system fits nicely with the concepts of total process control in that the three principal functions used for control—production operating system, SPC and the human element—are integral to its success.[12]

PROBLEM SOLVING DISCIPLINES

Problem solving disciplines should be used to provide a comprehensive, step-by-step, team approach to problem solving:

    Step 1: Use team approach

    Step 2: Describe problem

    Step 3: Take interim action

    Step 4: Identify principal cause

    Step 5: Develop alternative solutions

    Step 6: Determine best solution

    Step 7: Take permanent corrective action

    Step 8: Compliment team.

This approach must be designed to provide immediate resolution to identified problems and positive corrective action to prevent recurrence.[1,3,5,6,8,10,13,15,17]

SERVICES

The incorporation of effective services, such as production control and scheduling, should provide key support in a TPC system. Scheduling of the product mix must be optimized to provide quality while maintaining maximum production. Similarly, cost accounting, engineering, maintenance, data processing and sales are other areas all of which must actively participate in total process control.[3,5,7,13,15]

FLOW CHART

This is a systematic method of diagraming the sequence of a process, detailing all operations and controls necessary for maintaining process capability and statistical control to meet customer requirements.

The flowchart shown in Figure 15-13a is an example of a batch process for the production of ductile iron.[3,10,11,15]

Each process control diamond could be broken down to provide complete detail of the entire process. A detailed process control example for melting is illustrated in Figure 15-13b.

PRODUCTION CONTROL: DUCTILE IRON FOUNDRY OPERATIONS

INCOMING MATERIALS AND RECEIVING INSPECTION

Well-defined requirements or specifications should be established for all material used in the production of ductile iron. All material suppliers should provide vendor certifications of composition, sizing and packaging. All materials should be visually examined for container damage, contamination and compliance to specifications.

Chemical and/or physical testing should be performed on those materials where vendor certification is considered impractical, and where normal variation in the material can result in inconsistency in the production process.

In foundries where materials other than ductile iron are produced, every effort should be made to keep all the different types of returns separated. Mixing gray or malleable iron sprue or scrap castings with ductile iron can lead to increased sulfur in the base iron and other serious problems. Return sprue or scrap castings should be considered incoming materials and treated in a manner that will maintain consistency in the base iron chemistry.

FURNACE CHARGES

All furnace charges must be weighed and charged on the basis of a calculated, predetermined base iron composition, if quality ductile iron is to be produced.

BASE IRON COMPOSITION

The composition of the base iron (the iron before the nodulization and inoculation treatments) requires careful control of carbon, silicon and sulfur. Rapid analysis of carbon and silicon is normally accomplished at the melting area with thermal analysis techniques for each tap from the furnace. Analysis of sulfur can be determined by a laboratory combustion routine (and carbon can be verified at the same time). Larger ductile iron producers also have vacuum spectrographic capabilities to keep an eye on tramp elements, as well as carbon, silicon and sulfur.

SPC charting techniques should be used to monitor and control all primary elements. Changes in the charge make-up should be made, as necessary, to maintain the required base iron chemistry.

BASE IRON TEMPERATURE CONTROL

Consistent base iron treatment temperatures are very important in producing a quality ductile iron. The treatment

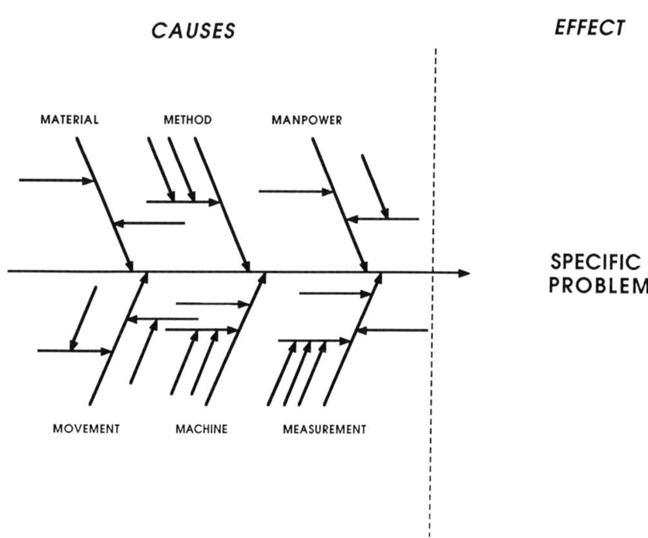

Fig. 15-11. A cause and effect diagram often used by problem-solving groups.

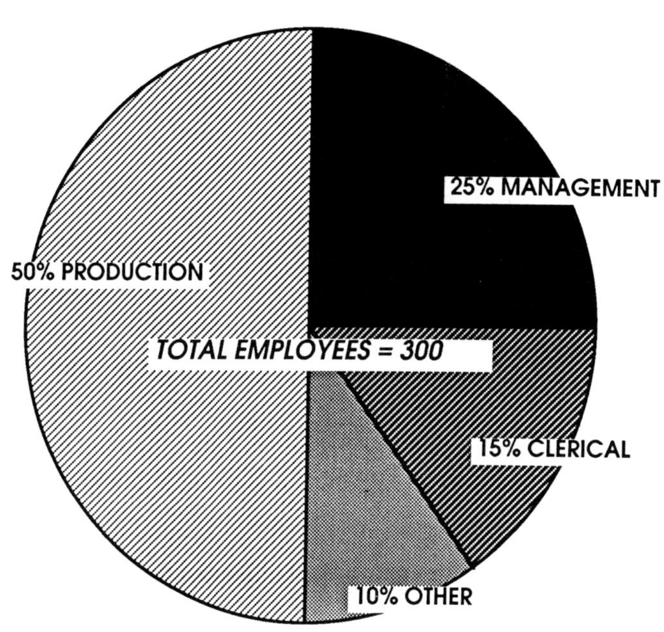

Fig. 15-12. A sample pie chart.

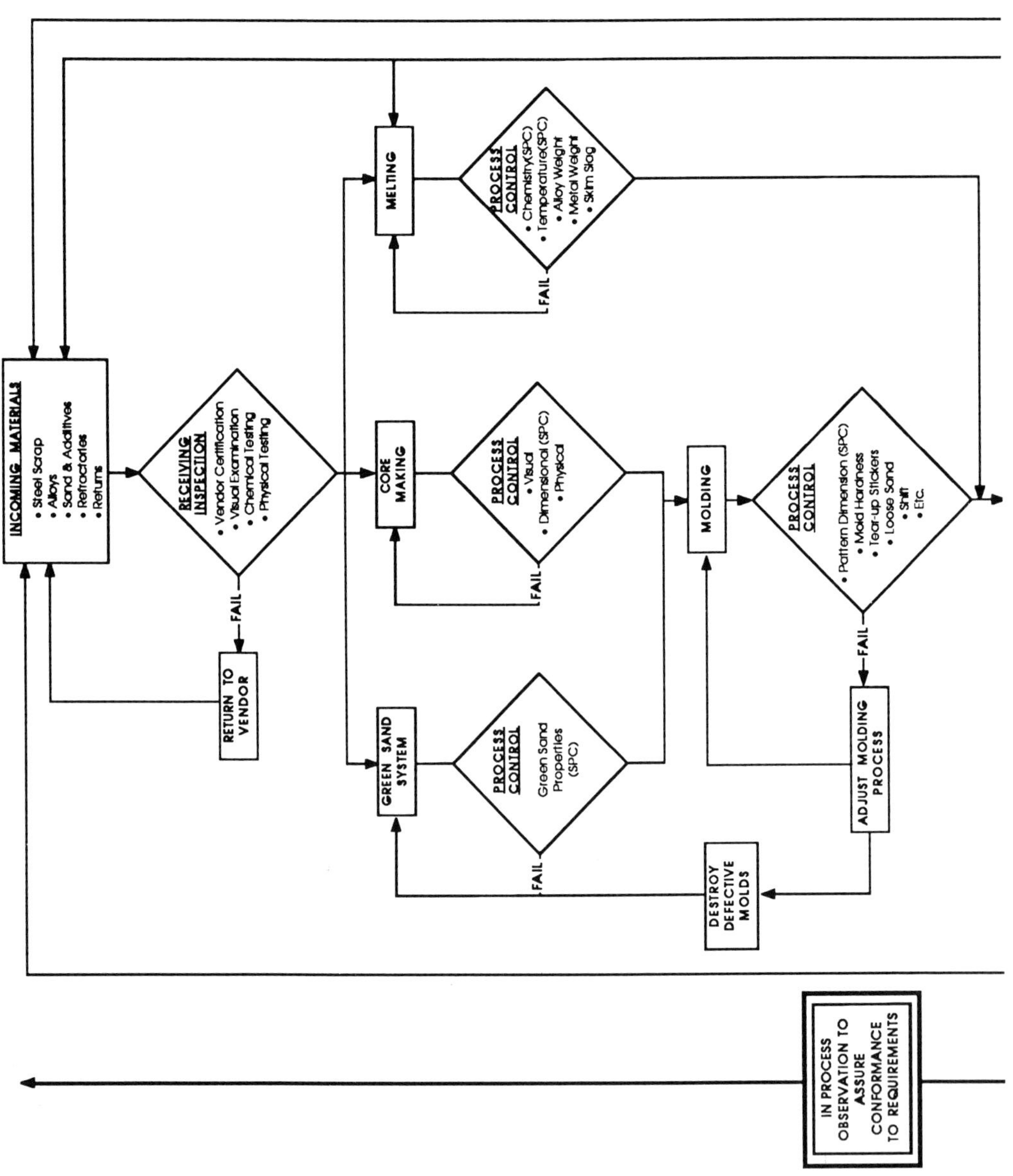
INCOMING MATERIALS
• Steel Scrap
• Alloys
• Sand & Additives
• Refractories
• Returns
RECEIVING INSPECTION
• Vendor Certification
• Visual Examination
• Chemical Testing
• Physical Testing
FAIL
RETURN TO VENDOR
MELTING
PROCESS CONTROL
• Chemistry(SPC)
• Temperature(SPC)
• Alloy Weight
• Metal Weight
• Skim Slag
FAIL
CORE MAKING
PROCESS CONTROL
• Visual
• Dimensional (SPC)
• Physical
FAIL
GREEN SAND SYSTEM
PROCESS CONTROL
Green Sand Properties (SPC)
FAIL
MOLDING
PROCESS CONTROL
• Pattern Dimension (SPC)
• Mold Hardness
• Tear-up Stickers
• Loose Sand
• Shift
• Etc.
FAIL
ADJUST MOLDING PROCESS
DESTROY DEFECTIVE MOLDS
IN PROCESS OBSERVATION TO ASSURE CONFORMANCE TO REQUIREMENTS

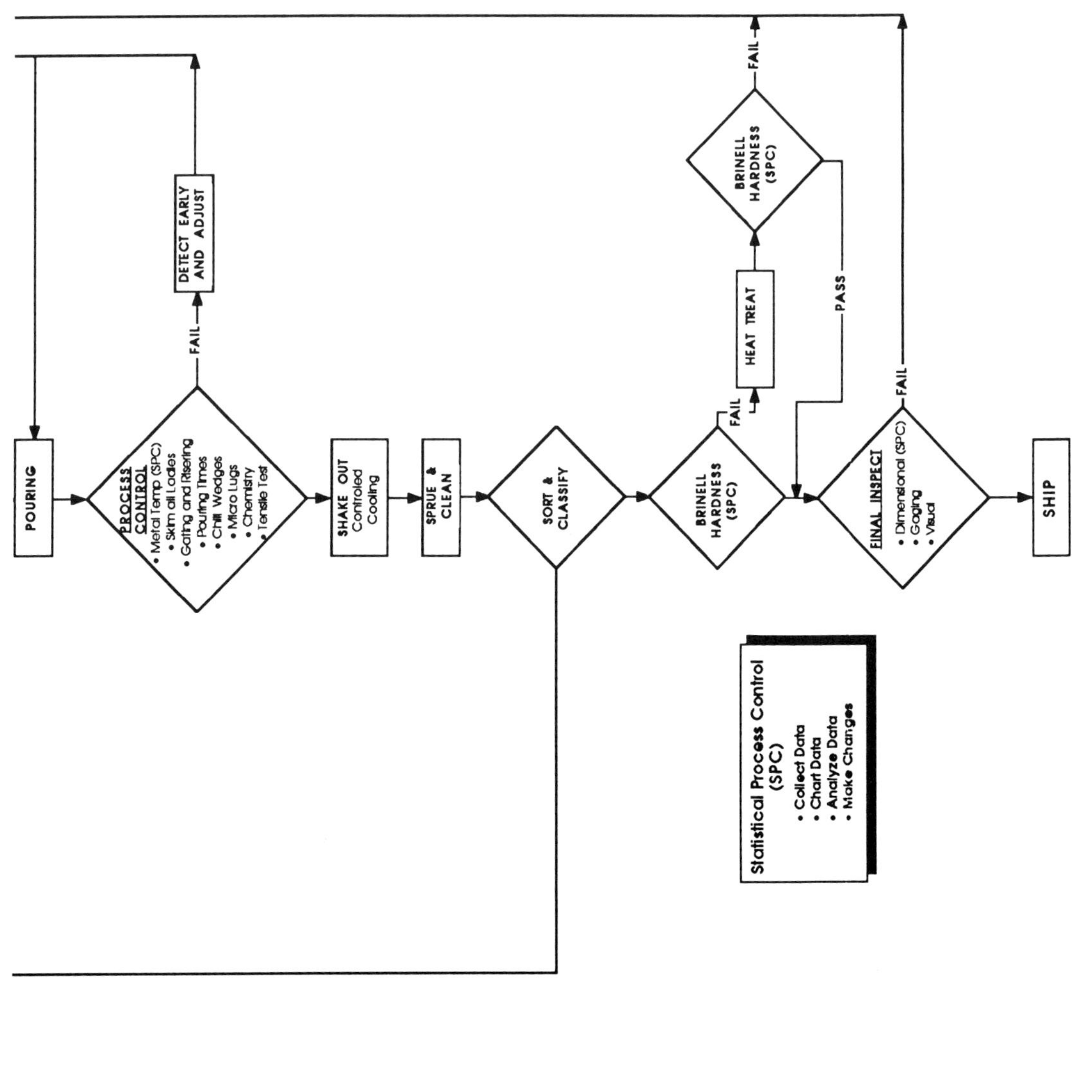

Fig. 15-13a. Flow chart: ductile iron production.

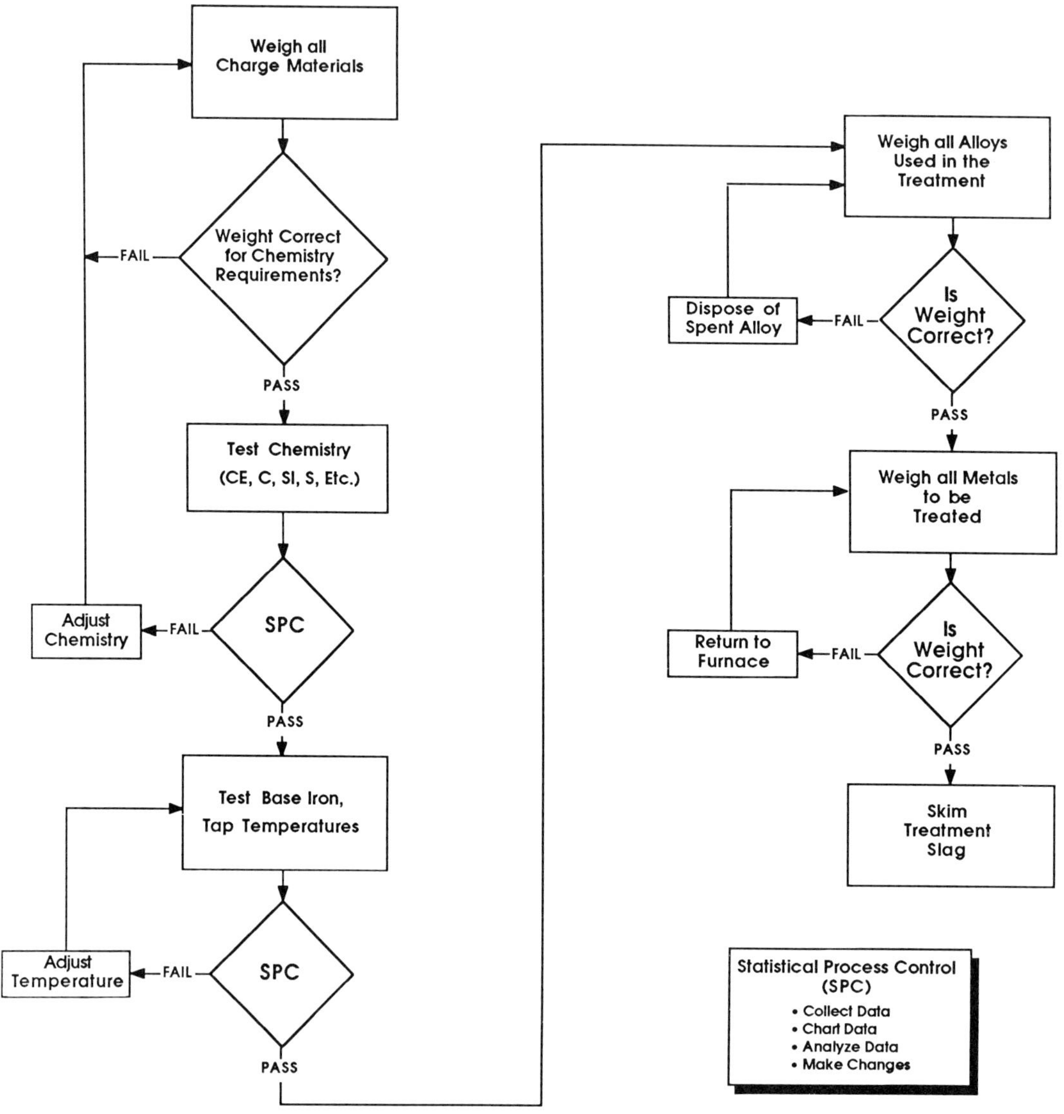

Fig. 15-13b. Flow chart: melting process control.

temperature range governs the pouring temperature range, which can have an influence on hardness and dimensional variation within the casting. High treatment temperatures are one of the principle causes of low treatment alloy recovery. However, low treating temperatures can cause excessively high treatment alloy recoveries, which promote the formation of carbides, pinholes, dross and misruns. A balance between all influencing variables must be developed to establish the optimum treatment temperature range.

WEIGHING

Weigh all metal to be treated. This is a very important step in the production of ductile iron. Large quantities of *scrap* ductile iron can be produced if inaccurately weighed amounts of base iron are used in treatment procedures. Undertreatment, due to an excess of base iron treated, can result in poor nodularity. Overtreatment, due to an insufficient amount of base iron relative to the treatment alloy, can result in carbides, pinholes, dross and misruns.

All alloys and inoculants used in the production of ductile iron should also be accurately weighed prior to use.

Metallographic Analysis

The microstructure of a standard test specimen, poured from the last metal of each treatment, should be evaluated to make sure that even the most faded treated iron produces acceptable castings. If the microsample is not acceptable, additional samples from the molds must be taken to determine whether or not the castings are acceptable. If the microsample was only marginally acceptable, going back to the last mold poured may be sufficient to determine whether unacceptable castings were produced. If, however, the microsample exhibited very poor nodularity, it will be necessary to go back further in the pouring sequence to determine which castings were affected. All castings having substandard nodularity must be immediately segregated from the normal production flow and scrapped.

The following steps are concerned with describing the equipment and typical procedures used for rapid metallographic analysis of ductile iron to determine the degree of graphite nodularity:

1. Photographic-quality metallographic samples are not required for control purposes. The typical polishing equipment might consist of a coarse abrasive (50–60 grit) sander or grinder, capable of quickly preparing a flat surface, and/or two or more high-speed disk grinders, which are used to progressively polish from 180–600 grit. A higher degree of polishing is not necessary, since the graphite particles will be well enough defined to read accurately at low magnification.

2. An inexpensive metallographic microscope capable of resolution of 50X to 100X is sufficient to determine if the graphite nodules are well-formed and no flake graphite is present. An AFS microexamination test coupon (see Figure 15-14) should be used for this rapid examination. The procedure is as follows:

   a) Pour sample from treated and inoculated iron.

   b) After solidification, cool the sample slowly in water.

   c) Fracture or cut off the sample.

   d) Rough grind a flat surface.

   e) Progressively polish the sample.

   f) Examine the graphite microscopically.

Chemical Analysis

A complete elemental analysis should be made on selected samples from each day's production. In particular, a com-

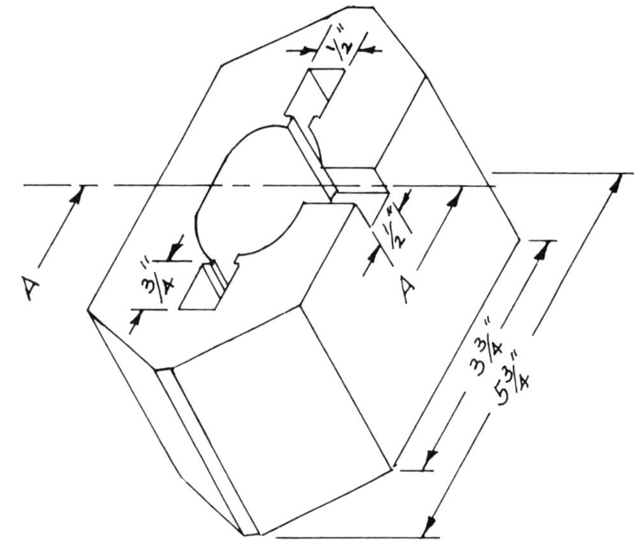

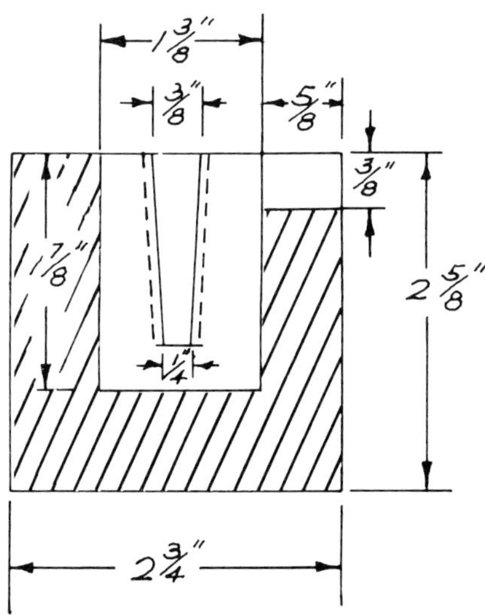

Fig. 15-14. Cores for DI microstructure samples.

plete analysis should be performed, for certification purposes, at the time of casting tensile test specimens.

The more important elements influencing the properties of ductile iron are:

- Carbon
- Silicon
- Magnesium
- Sulfur
- Manganese
- Phosphorous
- Nickel
- Copper
- Chromium
- Molybdenum

TENSILE TESTING

Y block or keel block samples should be poured at prescribed intervals of the production process to be able to provide certification of customer requirements. During initial heats, when standard practices are being established in foundries, tensile tests and yield strength tests should be made frequently in order to set up standard practice and correlate mechanical properties and microstructure.

Tensile test requirements are as follows:

1. Tensile bars should be machined from coupons from one-inch Y or keel blocks poured in core sand molds (see Figures 2-3, 2-4 and 2-5 and Table 2-7). Machined tensile bars should be 0.505 inches in diameter (see Fig. 2-6b).

2. Test bars should accompany castings through heat treating processes.

3. When castings are used in the as-cast condition, correlation of casting to test bar properties depends on the relative cooling rate of the castings versus the cooling rate of test bars. For this reason, a Brinell hardness audit, at a particular place on the casting, is a practice that is accepted by many customers for the release of their castings. Test bars can be pulled from castings to check the difference in properties. In thin-section castings, when rapid cooling occurs, hardness and tendency toward carbide formation is increased, tensile properties increase, and ductility is less than that in standard Y blocks. However, if desired, special test specimens can be cast to simulate the properties in specific sections.

HARDNESS TESTING

The hardness of the castings produced should be tested on a random basis at designated locations. Casting hardness must fall within a specified range according to customer specifications. The hardness of the castings can be correlated with the tensile strength by using a K-factor.

Calculating the ratio of tensile strength to Brinell hardness is one method of checking the quality of ductile iron. The K-factor, or tensile strength-Brinell hardness ratio, is obtained by dividing the tensile strength (in pounds per square inch) by the Brinell hardness number. An example is as follows:

K-factor = Tensile Strength / Brinell Hardness

K-factor = 86,000 / 200 = 430

Generally, in a good quality ductile iron, the number obtained (K-factor) is about 435 ± 15 for the ductile irons that are in the as-cast or annealed grades. For the normalized grades, a typical K-factor is about 470 to 480.

Variations from these ratios can come from a number of different sources. Some of these are massive carbides, vermicular graphite, flake graphite, carbon flotation, unsound test specimens, and incorrect tensile strength results or Brinell hardness readings.

VISUAL INSPECTION

Castings must be inspected for cracks, burned-in sand, cold shuts, shrinks, gas defects, etc. Scrap should be counted and categorized by the date and time produced, so that the causes of the scrap can be determined and corrected.

SPECIAL EQUIPMENT

As applications for ductile iron castings become more sophisticated and quality requirements become more exacting, a number of special types of inspection equipment have come into more general use. These include:

Fluoroscopic real-time x-ray imaging—a filmless radiographic unit with high resolution video imaging and recording capabilities for determination of internal casting defects.

Magnetic particle inspection equipment—used to detect surface and just-below-the-surface defects in castings.

Sonic and ultrasonic inspection equipment—for determining the percentage of nodular graphite present in individual castings and the presence of internal defects. This type of inspection equipment can be automated and used on a high-production basis.

Eddy current—when correlated with Brinell hardness, this technique can be used to provide a rapid measure of the relative hardness of ductile cast iron. It can also be used to inspect for surface and subsurface defects.

Coordinate measuring machine (CMM)—an automatic measuring system that can be used to inspect the dimensional and geometric accuracy of patterns, core boxes and castings to control quality during the manufacturing process.

Scanning electron microscope (SEM)—provides extremely high magnification analysis and semi-quantitative analysis for micropolished and three-dimensional fractured surfaces.

Eutectometric determination of graphite nodularity—involves the interpretation of eutectometer cooling curves for ductile iron to determine the degree of graphite nodularity and relative percentage of magnesium content.

■ REFERENCES

1. C.A. Aubrey, P.K. Felkins; *Teamwork: Involving People in Quality and Productivity Improvement*, UNIPUB/ Quality Resources, White Plains, NY (1989).

2. S. Shingo; *A Revolution in Manufacturing: The SMED System*, Rudra Press, Cambridge, MA (1985). Copies: Productivity Press P.O. Box 814, Cambridge, MA 02238, 617/497-5146.

3. J.M. Juran, F.M. Gryna, R.S. Bingham; *Quality Control Handbook*, 3rd ed., McGraw-Hill Book Co., USA (1979).

4. K. Ishikawa; *Guide to Quality Control*, UNIPUB/ Quality Resources, White Plains, NY (1988).

5. Western Electric Co., Inc.; *Statistical Quality Control Handbook*, 2nd ed., Delmar Printing Co., Charlotte, NC (1985). Copies: AT&T Technologies, Commercial Sales Clerk, Select Code 700-444, P.O. Box 19901, Indianapolis, Indiana 46219, 800/432-6600.

6. H. Kume; *Statistical Methods for Quality Improvement*, 3A Corporation, Chiyoda-ku, Tokyo (1988).

7. E.L. Grant, R.S. Leavenworth; *Statistical Quality Control*, 5th ed., McGraw-Hill Book Co., USA (1980).

8. R.T. Amsden, H.E. Butler, S.M. Amsden; *SPC Simplified: Practical Steps to Quality*, UNIPUB/Kraus International Publications, USA (1986).

9. E.R. Ott; *Process Quality Control*, McGraw-Hill Book Co., USA (1975).

10. J.L. Hradesky; *Productivity and Quality Improvement*, McGraw-Hill Book Co., USA (1988).

11. GOAL/QPC; *The Memory Jogger: A Pocket Guide of Tools for Continuous Improvement*, 2nd ed., GOAL/ QPC, Methuen, MA (1988). Copies: GOAL/qpc, 13 Branch St., Methuen, MA 01844, 508/685-3900.

12. K.M. Smith; *The Key Minimum-Maximum Standards process Control System*, American Foundrymen's Society, Inc., Des Plaines, IL (1983).

13. Japan Management Association, *KANBAN Just-In-Time at Toyota*, 2nd ed., Productivity Press, Cambridge, MA (1989). Copies: Productivity Press, P.O. Box 3007, Cambridge, MA 02140, 617/497-5146.

14. Ford Motor Co.; *Q-101 Quality System Standard*, Ford Motor Company, Quality Office, Dearborn, MI (1986).

15. General Motors; *Targets for Excellence*, General Motors Purchasing Activities, Detroit, MI (1987).

16. Ford Motor Co.; *Potential Failure Mode & Effects Analysis (FMEA)*, Ford Motor Company Office, Dearborn, MI (1988).

17. Ford New Holland; *8D Problem Solving*, Ford New Holland Office, New Holland, PA.

16

Welding

G. Garlough

Goulds Pumps, Inc.
Seneca Falls, New York

T. Stoecker

Sandy Hill Corp.
Hudson Falls, New York

■ INTRODUCTION

A wide variety of techniques and filler metals have been used to weld ductile iron. This chapter will deal with the two most common and successful methods to accomplish that end. These methods are (1) arc welding using nickel-base electrodes and (2) gas welding using cast ductile iron filler rods.

These two methods each have their advantages. The main advantage of arc welding using nickel-base electrodes is ease of use and versatility. The main advantage of gas welding using a cast ductile filler rod is its ability to produce a weld deposit similar in metallurgy to the base metal.

Each method has a different way of dealing with the high carbon (high hardenability) that tends to make ductile iron difficult to weld. Nickel's success as a filler metal is due to the fact that it does not combine with carbon to produce stable carbides. The solubility of carbon is very low in solid nickel, and it is rejected from solution to form graphite, reducing stresses in the weld zone.

Arc welding's success depends on lower heat inputs to minimize the temperature gradient in the heat-affected zone. Gas welding's success *depends* on the fact that it is a high heat input technique, which slows down the solidification rates, eliminating carbides from the weld zone and heat-affected zone. But in most cases, both techniques require preheat. Preheat will help prevent cracking by lowering thermal gradients and reducing residual stresses.

The equipment, materials, techniques and results of the two primary methods used to weld ductile iron will next be discussed.

■ ARC WELDING—NICKEL ELECTRODES

With the proper technique and in the right application, ductile iron can be welded with all the common arc welding processes. This section will cover only shielded metal arc welding "stick" (SMAW) and flux-cored arc welding

(FCAW)—the most widely used and most versatile techniques, which offer the widest selection of consumables.

EQUIPMENT

Shielded Metal Arc Welding

The power supply should supply the current and polarity required for the electrode used. For most electrodes, the power supply will be direct current (D.C.) with reverse polarity (i.e., the electrode is positive). It will also have a high, open-circuit voltage for arc starting, and have a dropping voltage amperage curve. Other required equipment includes grounding, leads, electrode holder and safety equipment.

Flux-Cored Arc Welding

The power supply will be a typical metal inert gas (MIG) power supply. The supply will have sufficient amperage and voltage to run the wire size used. Direct current, reverse or straight polarity, is used depending on the wire requirements. A wire feeder is required with a suitable MIG gun.* A shielding gas supply may be required, but most of the wires used to weld ductile iron are self-shielding. Individual manufacturers should be consulted on this matter.

***[Editors' Note: Water-cooled guns are also in use.]**

CONSUMABLES

There are three main categories of nickel-base electrodes available: high-nickel, medium-nickel and nickel-iron-manganese. The American Welding Society (AWS) has specifications covering two of these classes under ANSI/AWS A5.15-90 (see Table 16-1).

High-Nickel Electrodes

These are classified by A5.15-90 as ENi-CI and ENi-CI-A, and are chemically the same except ENi-CI-A provides better arc characteristics when used with irons having higher percentages of aluminum . These electrodes may be used where a high degree of machinability is required or when welding without preheat is appropriate. Their lower yield strength keeps the shrinkage stresses low.

Medium-Nickel Electrodes

These are classified by A5.15-90 as ENiFe-CI and ENiFe-CI-A, and cover the 55% nickel electrodes that account for the bulk of cast iron welding. The deposited composition of this electrode (diluted with iron) forms an alloy approaching the composition of the alloy group invar (70% Fe and 30% Ni).

The composition has a low coefficient of thermal expansion, making this electrode useful for heavy-section welding.

Nickel-Iron-Manganese Electrodes

A more recently developed, proprietary type of electrode for the welding of ductile iron, these electrodes have a deposited composition of approximately 44% nickel and 12% manganese. The manganese addition improves fusion to the base metal.

PREWELD HEAT TREATMENT

Arc welding should be used on ferritic grades of ductile iron. Nonferritic grades should be given a ferritizing heat treatment prior to welding. The main advantage of this is that it will produce a heat-affected zone that is less hard and brittle, due to the fact that the pearlite-free matrix is less saturated with carbon.

BASE METAL PREPARATION

Surface Cleaning

To produce sound welds, the weld area must be clean and free of dirt, oxides, moisture, oils, paints, etc. Cleaning may be done by wire brushing, grinding and abrasive blasting.

Defect Removal

This may be done by nonthermal methods, such as machining, grinding and chipping. Thermal methods include air-arc, plasma gouging and oxygen lance. Thermal methods need to be followed by grinding to remove heat-affected metal. Linear defects may tend to propagate and should be "pinned" prior to removal. This is done by drilling holes at the end of the defect, thus blunting the crack.

Preparing the Cavity or Joint

The weld cavity needs to be open and rounded. Steep side walls should be avoided. This allows for electrode manipulation and slag removal. Nickel-base electrodes produce a sluggish puddle with reduced penetration and wetting; hence, they require a more open cavity.

PREHEATING

Preheating has many advantages when welding ductile iron with a nickel-base electrode. Preheating (1) improves the fusion of the weld metal to the base metal; (2) prevents cracking by lowering the thermal gradient, reducing thermal stress; (3) reduces residual stresses and distortion; and

TABLE 16-1. CHEMICAL COMPOSITION OF FILLER METALS (ANSI/AWS A5.15-90*)

AWS Class[d]	UNS No.[e]	C	Mn	Si	P	S	Fe	Ni[f]	Mo	Cu[g]	Al	Other Elements Total
Shielded Metal Arc Welding Electrodes												
ENi-Cl	W82001	2.0	2.5	4.0	—	0.03	8.0	85 min.	—	2.5	1.0	1.0
ENi-Cl-A	W82003	2.0	2.5	4.0	—	0.03	8.0	85 min.	—	2.5	1.0–3.0	1.0
ENiFe-Cl	W82002	2.0	2.5	4.0	—	0.03	Rem.	45–60	—	2.5	1.0	1.0
ENiFe-Cl-A	W82004	2.0	2.5	4.0	—	0.03	Rem.	45–60	—	2.5	1.0–3.0	1.0
ENiFeMn-Cl	W82006	2.0	10–14	1.0	—	0.03	Rem.	35–45	—	2.5	1.0	1.0
ENiCu-A	W84001	0.35–0.55	2.3	0.75	—	0.025	3.0–6.0	50–60	—	35–45	—	1.0
ENiCu-B	W84002	0.35–0.55	2.3	0.75	—	0.025	3.0–6.0	60–70	—	25–35	—	1.0
Flux Cored Arc Welding Electrodes												
ENiFeT3-Cl[h]	W82032	2.0	3.0–5.0	1.0	—	0.03	Rem.	45–60	—	2.5	1.0	1.0

[a]The weld metal, core wire, or the filler metal, as specified, shall be analyzed for the specific elements for which values are shown in this table. If the presence of other elements is indicated, in the course of this work, the amount of those elements shall be determined to ensure that their total does not exceed the limit specified for "Other Elements Total" in the last column of the table.

[b]Single values shown are maximum, unless otherwise noted.

[c]"Rem." stands for Remainder.

[d]Copper-base filler metals frequently used in the braze welding of cast irons are no longer included in this specification.

[e]SAE/ASTM Unified Numbering System for Metals and Alloys.

[f]Nickel plus incidental cobalt.

[g]Copper plus incidental silver.

[h]No shielding gas shall be used for classification ENiFeT3-Cl.

*Reprinted from ANSI/AWS A5.15-90, "Specification for Welding Electrodes and Rods for Cast Iron," © 1990 American Welding Society, reprinted with permission.

(4) reduces the hardness of the heat-affected zone, thus improving machinability.

It is very important that preheating be done *in a uniform manner* (avoid localized preheating). Highly localized preheating will create even greater stresses in an area after welding, if the area is constrained by the surrounding casting.

Preheat Temperature Ranges

No preheat may be used for small cosmetic repairs, usually not requiring machining afterwards. Minimum heat inputs are used (small electrodes with low amperage) and, generally, high-nickel electrodes.

Medium preheats (300–600F/149–316C)* are used for most nickel-iron filler metals. This allows for higher heat inputs and higher deposition rates without producing cracks in the heat-affected zone.

High preheats (600–1100F/316–593C)* are generally used to maximize the chances for a successful weld. High preheats and high heat inputs will minimize carbides in the heat-affected zone.** The heat-affected zone will tend to be wider with high preheats, due to a lower thermal gradient, but with slow cooling, martensite will not be formed in the heat-affected zone. Rapid cooling can cause martensite to form, even though high preheat is used.

*[Ed. N.: In commercial practice, 600F (316C) preheat has been very successful.]

**[Ed. N.: Increased hardness in the weld may be due to: (1) carbides in the weld metal and in the zone of partial fusion; (2) martensite, as a result of rapid cooling from M_s to M_f temperatures; or (3) increased pearlite.]

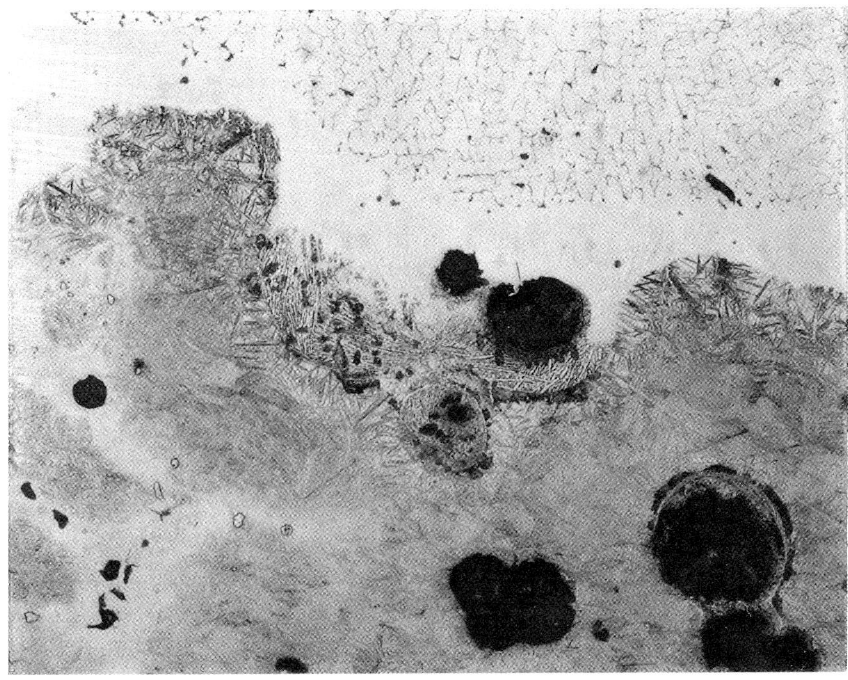

Fig. 16-1. Fifty-five percent nickel electrode hot welded on 4 in. riser of 80-60-03 ductile with medium preheat (×200).

Arc Welding Technique

Electrode current, electrode size and other parameters should be within the range recommended by the manufacturer.

The technique used in arc welding has a great influence on the thermal effects in the weld. Considerations are: 1) travel speed, 2) width of the weld bead, and 3) weld pattern. The greater the travel speed, the more rapid the cooling. The narrower the weld, the faster the cooling. The following are general techniques for a variety of welding methods:

- For general cosmetic repairs with no or low preheat, short stringer beads are used to minimize heat inputs. Peening the weld while it is still hot (above 1000F/538C) will help reduce stresses in the weld zone.
- For more critical repairs using medium or high preheats, greater heat inputs are allowed. Stringer beads with oscillation widths to three times the electrode diameter are used.

For larger cavities, the faces of the cavity (joint) are built up first, with the final passes tying in at the center. This will minimize stresses and distortion across a large cavity.

The final weld area is ground smooth and visually inspected for cracks, porosity, undercuts, etc.

Post Weld Heat Treatment

The primary requirement is that the casting be slowly cooled. The casting may be stress relieved. For critical castings, a high-temperature anneal (1650F/899C) will eliminate carbides produced in the heat-affected zone.

Arc Welding Results

Mechanical properties of the weld deposit using ENi-CI (90% nickel electrode) are reported in the range of 40,000 psi tensile strength, 38,000 psi yield strength, and 3–6% elongation, with a Brinell hardness of 200–300 BHN.

Physical properties of the weld deposit using ENiFe-CI (55% nickel electrode) are reported in the range of 60,000–80,000 psi tensile strength, 44,000–52,000 yield strength, 10–20% elongation, with a Brinell hardness of 200–300 BHN.

Figure 16-1 shows the various microstructures across a weld using ENiFe-CI (55% nickel electrode) on a pearlitic grade ductile (80-60-03) welded with medium preheat. The lower portion of the figure shows the base metal, the center portion shows the heat-affected zone, and the top portion is the filler metal. Figure 16-2 shows a weld made with the same electrode of the same pearlitic ductile, but with no

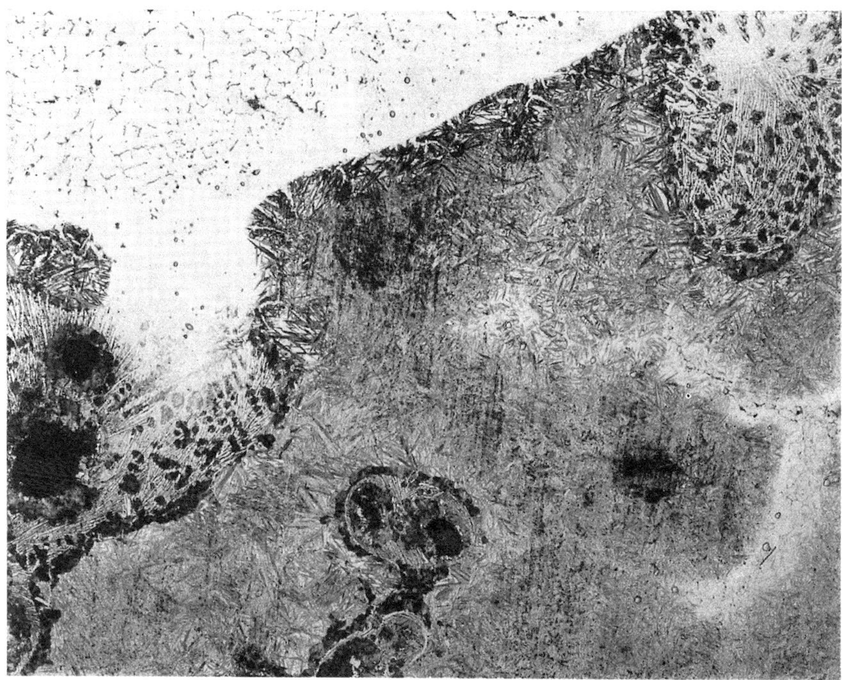

Fig. 16-2. Fifty-five percent nickel electrode cold welded on 4 in. riser of 80-60-03 ductile with no preheat (×200).

preheat. (A coarser structure in the heat-affected zone is evident.)

GAS WELDING— CAST DUCTILE IRON FILLER ROD

It is possible to use gas welding to salvage defective ductile iron castings before or after machining, producing weld deposits similar to the parent metal in chemical analysis, microstructure and physical properties. Ductile iron castings may be repaired in the field, since no heat treatment is required. Following is a description of the methods and techniques for making sound gas welded repairs of all sizes using cast filler rod. This process is able to produce a weld that is virtually indistinguishable from the base metal.

EQUIPMENT

Oxygen-acetylene gas welding equipment must be used with appropriate regulators and a powder-spray type torch with appropriate tip. Fuel gases should not be used.

Suitable preheat equipment usually consists of gas torches including custom torches for particular jobs. Natural gas- or butane-fueled burners may also be used for preheating. Insulating blankets should be available to help retain heat.

Grinding or scarfing devices for preparation of the area to be welded and for finishing the welded area are also required. Suitable operator protective equipment, for both grinding and welding, must be provided.

FILLER RODS AND FLUXES

The *filler rod* must be a cast ductile iron rod American Welding Society classification RCI-B. Typical rod sizing is 1/4 inch by 1/4 inch and 24 inches long. The rod's typical high magnesium content of 0.04–0.10% allows for losses of magnesium due to oxidation during welding.

A number of proprietary fluxes are available. Fluxes generally contain borax (sodium borate) or boric acid, sodium carbonate and iron oxide powder.

PREPARATION FOR WELDING

The area to be weld is prepared by the same methods and similarly as in arc welding, i.e., it must be clean and free from scale, rust or other impurities. Either chipping or grinding is appropriate.

PREHEATING

Gas welding is a high heat input method and requires high preheat temperature (800–1100 F/427–593C). Preheating is necessary for sound welds because it: (1) retards the cooling rate of the weld deposit preventing the formation of carbides (cementite or martensite) in the transition zone, and (2) it prevents or minimizes welding stress and/or strain.

Preheating with a gas torch in the area of the weld is usually sufficient for cosmetic repairs. For more complex shapes and extensive repairs, complete preheating of the casting is required. Cracking will result, however, from stress introduced by improper preheating and welding techniques.

WELDING PRELIMINARIES

Similarly sized welding torches and tips are used for welding ductile iron as those used for the same thickness of mild steel. For fusion welding, a neutral or slightly reducing (carburizing) flame should be used—excess acetylene. An oxidizing flame must be absolutely avoided.

Check your preheat. Temperature-indicating crayons or a contact pyrometer are recommended.

GAS WELDING TECHNIQUE

After preheating, the flux is sprinkled on the groove face, and the flame is directed to the bottom of the weld groove. When the bottom starts to melt, the torch flame is moved form side to side and on to the welding rod. While puddling may be accomplished by either backhand or forehand techniques, the latter is preferred in that the rod leads the progression of the weld.

The heated tip of the welding rod is dipped into the flux and then placed back into the weld pool. A slight stirring action may be used to help remove impurities as the weld metal accumulates. The amount of flux is adjusted so as to promote a fluid slag cover necessary for sound gas welds, as well as to ensure complete nodularity of the weld deposit.

POST WELD HEAT TREATMENT

Since the heat input is high with oxyacetylene welding, the casting must cool slowly, and to further retard the cooling

rate, the torch may be played over the weld deposit after solidification. It is also beneficial to cover the casting with insulating blankets after the torch is removed.

If a ferritic matrix is specified or desired, a ferritizing anneal is mandatory. Castings may be normalized or annealed so to provide a more matching matrix between weld deposit and casting. The welded area will respond to conventional heat treatments such as ferritizing, normalizing and quench-and-temper in a manner similar to the parent metal (assuming normal chemical composition). However, carbide-stabilizing elements will affect response to heat treatment. Stress relieving may be required for more complex castings requiring dimensional stability.

Post heat treatment is seldom required to improve the machinability of the repaired area.

GAS WELDING RESULTS

The weld deposit and basic material will have no visible indication of fusion lines, and the color match will be excellent. Depending on the cooling rate, the weld deposit will be basically Grade 80-60-03 ductile iron and be readily machinable (Fig. 16-3). Other grades of ductile iron can be achieved by heat treatment subsequent to welding.

Since the heat input is high with oxygen-acetylene welding, the tendency to form cementite (iron carbide) or untempered martensite in the weld deposit, transition zone, or heat-affected zone* is greatly reduced.

> ***[Ed. N.: Increased hardness in the weld may be due to (1) carbides in the weld metal and in the zone of partial fusion, (2) martensite as a result of rapid cooling from M_s to M_f temperatures, or (3) increased pearlite.]**

Microstructure of a proper weld deposit, compared to the parent metal, are normally smaller but have a greater number of graphite nodules in the pearlite or sorbitic matrix with no carbides present. The matrix will have a small bull's-eye of ferrite around the graphite nodule.

◼ GAS WELDING—AN ALTERNATIVE

A proprietary gas welding technique has been described as providing excellent metallurgical consistency between the weld and the casting. This is well known as the Duc-Wel Process.

Following is a list of basic equipment:
1) Suitable operator protective equipment;
2) Acetylene gas and oxygen with appropriate pressure regulators;

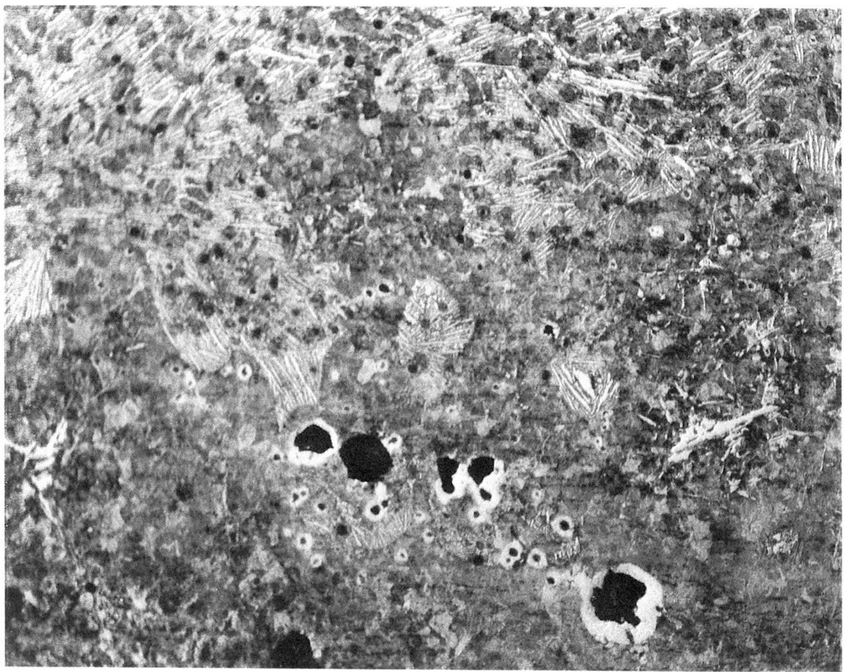

Fig. 16-3. Gas-welded ductile iron filler on 4 in. riser test piece of 80-60-03 ductile, high carbides in filler due to low silicon (2.30%) (×100).

3) Grinding or scarfing devices for preparation of the area to be welded and finishing of the welded area;

4) Burner (natural gas or butane) for preheat of part;

5) Powder spray type oxyacetylene torch fitted with appropriate tip.

The area to be welded *must* be clean and free from scale, rust or other impurities. This may be accomplished by any convenient means, such as chipping and/or grinding. Normal, good weld cleaning procedures are adequate. Conventional cleaning room equipment should prove satisfactory for this operation.

Preheating of the area to be welded serves two desirable purposes:

1) It retards post-welding cooling rate to avoid presence of cementite or martensite in the transition zone;

2) It avoids or minimizes excessive casting stress and/or strain.

Local preheating of the area to be welded with methane or LP gas torch to a temperature range of between 500 and 700F is sufficient to accomplish these goals. Complicated castings may require more elaborate preheating.

For fusion welding, a slight excess of acetylene is used to insure a slightly reducing flame. The flame is brought in contact with the area to be welded. When melting begins, the rod is introduced into the flame, and welding by puddling progresses in the usual manner. Puddling may be accomplished by either backhand or forehand techniques, with the latter preferred in that the rod leads the progression of the weld. Flux of special composition is added to the puddle through the flame in normal powder spray fashion.

The flux is so adjusted as to promote a fluid slag cover necessary to sound gas welds, as well as to insure complete nodularity of the weld deposit.

The cooling rate of the completed weld deposit may be further retarded by continuing to play the torch on the weld deposit after solidification. Post heat treatment is not required to accomplish a free machining welded area with excellent "color match." The welded area will respond to conventional heat treatments such as ferritizing, normalizing, and quench and temper in a manner like the parent metal (assuming normal chemical composition). Carbide-stabilizing elements will affect response to heat treatment.

Due to a more rapid cooling rate of the weld deposit than is probable with the parent metal, the nodule count of the weld deposit will be higher than that of the parent metal.

A properly executed weld will be essentially free from cementite (iron carbide) in the weld deposit or transition zone. It will posses a pearlitic or sorbitic matrix with a small bull's eye of ferrite around the graphite nodule.

SUMMARY

1. The process described is practical under normal operating conditions and is not a laboratory technique;

2. It is possible to salvage defective ductile iron castings, before or after machining, by oxyacetylene welding with resulting weld deposits similar to parent metal in chemical analysis, microstructure and physical properties;

3. Ductile iron parts that have been scrapped through machining error can be salvaged;

4. Ductile iron castings may be repaired in the field, since no heat treatment is required.

■■■ BIBLIOGRAPHY

American Welding Society. *Guide for Welding Iron Casting,* Miami, FL: American Welding Society (1989).

Anant-Narayan, S.N., A.J. Rickard and N. Stephenson. "Weldability of Spheroidal Graphite Type Ni-Resist Cast Iron," *Foundry Trade Journal* (Jul 1979).

Bishel, R.A. and H.R. Conway. "Flux-Cored Arc Welding for High Quality Joints in Ductile Iron," *Modern Casting* (Jan 1970).

Bowen, M. "The Welding of Ductile Irons," *British Foundryman,* vol 77, (1984).

Harding, R.A. "Progress in Joining Iron Castings," Institute of British Foundrymen, Report 1699 (Jul 1987).

Heine, H.J. "Welding Castings: A Review," *Foundry M & T* (Aug–Sep 1978).

Kelly, T.J., R.A. Bishel and R.W. Wilson. "Welding of Ductile Iron with Ni-Fe-Mn Filler Metal," *Welding Journal,* vol 64 (Mar 1985).

Roberts, D.T. "New Ductile Iron Welding Process...Saves Castings and $," *Modern Casting* (Feb 1969).

Voigt, R.C. and C.R. Loper, Jr. "Welding Metallurgy of Gray and Ductile Cast Irons," *AFS Transactions,* vol 94 (1986).

Wagner, A.J. "Perfect Welds of Gray and Ductile Iron," *Modern Casting* (Apr 1973).

GUIDELINES FOR FERROUS SCRAP: FS-90*

ISRI
CODE
NO.

200 **No. 1 heavy melting steel.**
Wrought iron and/or steel scrap 1/4 inch and over in thickness. Individual pieces not over 60 x 24 inches (charging box size) prepared in a manner to insure compact charging.

201 **No. 1 heavy melting steel 3 feet x 18 inches.**
Wrought iron and/or steel scrap 1/4 inch thickness and over. Individual pieces not over 36 inches x 18 inches (charging box size) prepared in a manner to insure compact charging.

202 **No. 1 heavy melting steel 5 feet x 18 inches.**
Wrought iron and/or steel scrap 1/4 inch and over in thickness. Individual pieces not over 60 inches x 18 inches (charging box size) prepared in a manner to insure compact charging.

203 **No. 2 heavy melting steel.****
Wrought iron and steel scrap, black and galvanized, 1/8 inch and over in thickness, charging box size to include material not suitable as No. 1 heavy melting steel. Prepared in a manner to insure compact charging.

204 **No. 2 heavy melting steel.****
Wrought iron and steel scrap, black and galvanized, maximum size 36 x 18 inches. *May include all automobile scrap properly prepared.*

205 **No. 2 heavy melting steel 3 feet x 18 inches.**
Wrought iron and steel scrap, black and galvanized, maximum size 36 x 18 inches. May include automobile scrap, properly prepared, however, to be free of sheet iron or thin gauged material.

206 **No. 2 heavy melting steel 5 feet x 18 inches.**
Wrought iron and steel scrap, black and galvanized, maximum size 60 x 18 inches. May include automobile scrap, properly prepared, however, to be free of sheet iron or thin gauged material.

207 **No. 1 busheling.**
Clean steel scrap, not exceeding 12 inches in any dimensions, including new factory busheling (for example, sheet clippings, stampings, etc.). May not include old auto body and fender stock. Free of metal coated, limed, vitreous enameled, and electrical sheet containing over 0.5 percent silicon.

207A **New black sheet clippings.**
For direct charging, maximum size 8 feet by 18 inches, free of old automobile body and fender stock, metal coated, lined, vitreous enameled and electrical sheet containing over 0.5 percent silicon, must lay reasonably flat in car.

208 **No. 1 bundles.**
New black steel sheet scrap, clippings or skeleton scrap, compressed or hand bundled, to charging box size, and weighing not less than 75 pounds per cubic foot. (Hand bundles are tightly secured for handling with a magnet.) May include Stanley balls or mandrel wound bundles or skeleton reels, tightly secured. May include chemically detinned material. May not include old auto body or fender stock. Free of metal coated, limed, vitreous enameled, and electrical sheet containing over 0.5 percent silicon.

---

*"Scrap Specifications Circular 1990," Institute of Scrap Recycling Industries, Inc., reprinted with permission. Specifications in force as of publication date.)

**The identical designations given for these two classifications are in accordance with established industry practices in specifying the materials desired.

209 No. 2 bundles.

Old black and galvanized steel sheet scrap, hydraulically compressed to charging box size and weighing not less than 75 pounds per cubic foot. May not include tin or lead-coated material or vitreous enameled material.

210 Shredded scrap.

Homogeneous iron and steel scrap, magnetically separated, originating from automobiles, unprepared No. 1 and No. 2 steel, miscellaneous baling and sheet scrap. Average density 50 pounds per cubic foot.

211 Shredded scrap.

Homogeneous iron and steel scrap magnetically separated, originating from automobiles, unprepared No. 1 and No. 2 steel, miscellaneous baling and sheet scrap. Average density 70 pounds per cubic foot.

212 Shredded clippings.

Shredded 1000 series carbon steel clippings or sheets. Material should have an average density of 60 pounds per cubic foot.

213 Shredded tin cans for remelting.

Shredded steel cans, tin coated or tin free, may include aluminum tops but must be free of aluminum cans, nonferrous metals except those used in can construction, and non-metallics of any kind.

214 No. 3 bundles.

Old sheet steel, compressed to charging box size and weighing not less than 75 pounds per cubic foot. May include all coated ferrous scrap not suitable for inclusion in No. 2 bundles.

215 Incinerator bundles.

Tin can scrap, compressed to charging box size and weighing not less than 75 pounds per cubic foot. Processed through a recognized garbage incinerator.

216 Terne plate bundles.

New terne plate sheet scrap, clippings or skeleton scrap, compressed or hand bundled, to charging box size, and weighing not less than 75 pounds per cubic foot. (Hand bundles are tightly secured for handling with a magnet.) May include Stanley balls or mandrel wound bundles or skeleton reels, tightly secured.

217 Bundled No. 1 steel.

Wrought iron and/or steel scrap 1/8 inch or over in thickness, compressed to charging box size and weighing not less than 75 pounds per cubic foot. Free of all metal coated material.

218 Bundled No. 2 steel.

Wrought iron or steel scrap, black or galvanized, 1/8 inch and over in thickness, compressed to charging box size and weighing not less than 75 pounds per cubic foot. Auto body and fender stock, burnt or hand stripped, may constitute a maximum of 60 percent by weight. (This percent based on makeup of auto body, chassis, driveshafts, and bumpers.) Free of all coated material, except as found on automobiles.

219 Machine shop turnings.

Clean steel or wrought iron turnings, free of iron borings, nonferrous metals in a free state, scale, or excessive oil. May not include badly rusted or corroded stock.

220 Machine shop turnings and iron borings.

Same as machine shop turnings but including iron borings.

221 Shoveling turnings.

Clean short steel or wrought iron turnings, drillings, or screw cuttings. May include any such material whether resulting from crushing, raking, or other processes. Free of springy, bushy, tangled or matted material, lumps, iron borings, nonferrous metals in a free state, grindings, or excessive oil.

222 Shoveling turnings and iron borings.

Same as shoveling turnings, but including iron borings.

223 Iron borings.

Clean cast iron or malleable iron borings and drillings, free of steel turnings, scale, lumps and excessive oil.

224 Auto slabs.

Clean automobile slabs, cut 3 feet x 18 inches and under.

225 Auto slabs.

Clean automobile slabs, cut 2 feet x 18 inches and under.

226 Briquetted iron borings.

Analysis and density to consumer's specifications.

227 Briquetted steel turnings.

Analysis and density to consumer's specifications.

228 Mill scale.

Dark colored, ranging from blue to black, ferro-magnetic iron oxide forming on the surface of steel articles during heating and working.

ELECTRIC FURNACE CASTING, AND FOUNDRY GRADES

229 Billet, bloom and forge crops.

Billet, bloom, axle, slab, heavy plate and heavy forge crops, containing not over 0.05 percent phosphorus or sulphur and not over 0.5 percent silicon, free from alloys. Dimensions not less than 2 inches in thickness, not over 18 inches in width, and not over 36 inches in length.

230 Bar crops and plate scrap.

Bar crops, plate scrap, forgings, bits, jars, and tool joints, containing not over 0.05 percent phosphorus or sulphur, not over 0.5 percent silicon, free from alloys. Dimensions not less than 1/2 inch in thickness, not over 18 inches in width, and not over 36 inches in length.

231 Plate and structural steel, 5 feet and under.

Cut structural and plate scrap, 5 feet and under. Clean open hearth steel plates, structural shapes, crop ends, shearings, or broken steel tires. Dimensions not less than 1/4 inch thickness, not over 5 feet in length and 18 inches in width. Phosphorus or sulphur not over 0.05 percent.

232 Plate and structural steel, 5 feet and under.

Cut structural and plate scrap, 5 feet and under. Clean open hearth steel plates, structural shapes, crop ends, shearings, or broken steel tires. Dimensions not less than 1/4 inch thickness, not over 5 feet in length and 24 inches in width. Phosphorus or sulphur not over 0.05 percent.

233 Cast steel.

Steel castings not over 48 inches long or 18 inches wide, and 1/4 inch and over in thickness, containing not over 0.05 percent phosphorus or sulphur, free from alloys and attachments. May include heads, gates, and risers.

234 Punchings and plate scrap.

Punchings or stampings, plate scrap, and bar crops containing not over 0.05 percent phosphorous or sulphur and not over 0.5 percent silicon, free from alloys. All materials cut 12 inches and under, and with the exception of punchings or stampings, at least 1/8 inch in thickness. Punchings or stampings under 6 inches in diameter may be any gauge.

235 Electric furnace bundles.

New black steel sheet scrap hydraulically compressed into bundles of size and weight as specified by consumer.

236 Cut structural and plate scrap, 3 feet and under.

Clean open hearth steel plates, structural shapes, crop ends, shearings, or broken steel tires. Dimensions not less than 1/4 inch in thickness, not over 3 feet in length and 18 inches in width. Phosphorus or sulphur not over 0.05 percent.

237 Cut structural and plate scrap, 2 feet and under.

Same as cut structural and plate scrap, 3 feet and under, except for length.

238 Cut structural and plate scrap, 1 foot and under.

Same as cut structural and plate scrap, 3 feet and under, except for length.

239 Silicon busheling.

Clean silicon bearing steel scrap, not exceeding 12 inches in any dimensions, including new factory busheling (for example, sheet clippings, stampings, etc.), having a silicon content of 0.05 percent to 5.0 percent.

240 Silicon clippings.

Clean steel scrap, including new factory busheling (for example, sheet clippings, stampings, etc.), may not include old auto body and fender stock. Free of metal coated, limed, vitreous enameled, and electrical sheet containing minimum one percent silicon.

241 Chargeable ingots and ingot butts.

Chargeable ingots and ingot butts for material to be suitable and acceptable to the consumer containing not over 0.05 percent phosphorus or sulphur and not over 0.05 percent silicon free of alloys.

242 Foundry steel, 2 feet and under.

Steel scrap 1/8 inch and over in thickness, not over 2 feet in length or 18 inches in width. Individual pieces free from attachments. May not include nonferrous metals, cast or malleable iron, cable, vitreous enameled, or metal coated material.

243 Foundry steel, 1 foot and under.

Same specifications as 2-foot material, except for length.

244 Springs and crankshafts.

Clean automotive springs and crankshafts, either new or used.

245 Alloy free turnings.

Clean shoveling steel turnings free from lumps, tangled or matted material, iron borings, or excessive oil containing not more than 0.05 percent phosphorus or sulphur, and free of alloys.

246 Alloy free short shoveling steel turnings.

Clean shoveling steel turnings, free of lumps, tangled or matted material, iron borings, or excessive oil, containing not more than 0.05 percent phosphorus or sulphur, and free of alloys.

247 Alloy free machine shop turnings.

Clean steel turnings, free of iron borings or excessive oil, containing not more than 0.05 percent phosphorus or sulphur and free of alloys. May not include badly rusted or corroded stock.

248 Hard steel cut 30 inches and under.

Automotive steel consisting of rear ends, crankshafts, driveshafts, front axles, springs, and gears prepared 30 inches and under. May not include miscellaneous small shoveling steel or any pieces too bulky for gray iron foundry use.

249 Chargeable slab crops.

Chargeable slab crops for material to be suitable and acceptable to the consumer containing not over 0.05 percent phosphorus and 0.05 percent sulphur and not over 0.05 percent silicon; free of alloys.

250 Silicon bundles.

Silicon sheet scrap, clippings or skeleton scrap, compressed or hand bundled, to charging box size, and weighing not less than 75 pounds per cubic foot, having a silicon content of 0.50 percent to 5.0 percent.

251 Heavy turnings.

Short, heavy steel turnings, containing not over 0.05 percent phosphorus or sulphur and free of alloys. May include rail Chips. May not include machine shop or other light turnings and must weigh not less than 75 pounds per cubic foot in the original state of production.

SPECIALLY PROCESSED GRADES TO MEET CONSUMER REQUIREMENTS*

Cast Iron Grades

252 Cupola cast.

Clean cast iron scrap such as columns, pipes, plates, and castings of a miscellaneous nature including automobile blocks and cast iron parts of agricultural and other machinery. Free from stove plate, burnt iron, brake shoes or foreign material. Cupola size, not over 24 inches x 30 inches, and no piece over 150 pounds in weight.

253 Charging box cast.

Clean cast iron scrap in sizes not over 60 inches in length or 30 inches in width, suitable for charging into an open hearth furnace without further preparation. Free from burnt iron, brake shoes, or stove plate.

254 Heavy breakable cast.

Cast iron scrap over charging box size or weighing more than 500 pounds. May include cylinders and driving wheel centers. May include steel which does not exceed 10 percent of the casting by weight.

255 Hammer block or bosses.

Cast iron hammer blocks or bases.

---

*Grades of scrap prepared especially to meet with steel mill or foundry requirements, individual specifications to be agreed on between consumer and supplier.

256 Burnt iron.

Burnt cast iron scrap, such as stove parts, grate bars, and miscellaneous burnt iron. May include sash weights or window weights.

257 Mixed cast.

May include all grades of cast iron except burnt iron. Dimensions not over 24 inches x 30 inches and no piece over 150 pounds in weight.

258 Stove plate, clean cast iron stove.

Free from malleable and steel parts, window weights, plow points, or burnt cast iron.

259 Clean auto cast.

Clean auto blocks; free of all steel parts except camshafts, valves, valve springs, and studs. Free of nonferrous and non-metallic parts.

260 Unstripped motor blocks.

Automobile or truck motors from which steel and nonferrous fittings may or may not have been removed. Free from driveshafts and all parts of frames.

261 Drop broken machinery cast.

Clean heavy cast iron machinery scrap that has been broken under a drop. All pieces must be of cupola size, not over 24 inches x 30 inches. and no piece over 150 pounds in weight.

262 Clean auto cast, broken, not degreased.

Clean auto blocks, free of all steel parts except camshafts, valves, valve springs and studs. Free of nonferrous and nonmetallic parts, and must be broken to cupola size, 150 pounds or less.

263 Clean auto cast, degreased.

Free of all steel parts except camshafts, valves, valve springs, and studs. Free of nonferrous and non-metallic parts, and must be broken into cupola size, 150 pounds or less.

264 Malleable.

Malleable parts of automobiles, railroad cans, locomotives, or miscellaneous malleable iron castings. Free from cast iron and steel parts and other foreign material.

265 Broken ingot molds and stools.

Broken ingot molds and stools, cast iron, maximum size 2 feet x 3 feet x 5 feet.

266 Unbroken ingot molds and stools.

Unbroken ingot molds and stools, cast iron.

Special Boring Grades

267 No. 1 chemical borings.

New clean cast or malleable iron borings and drillings containing not more than 1 percent oil, free from steel turnings, or chips, lumps, scale, corroded or rusty material.

268 Briquetted cast iron borings, hot process.

Cast iron borings, heated, briquetted, to a density of approximately 85 percent, oil and water content under one percent.

269 Briquetted cast iron borings, cold process.

Cast iron boring briquettes, free of steel and nonferrous material, hydraulically compressed into a cohesive solid, reasonably free of oil, and having a density of not less than 60 percent.

270 Malleable borings.

Clean malleable iron borings and drillings, free of steel turnings, scale, lumps and excessive oil.

271 No. 2 chemical borings.

New clean cast or malleable iron borings and drillings, containing not more than 1.5 percent oil, free from steel turnings, or chips, lumps, scale, corroded or rusty material.

Railroad Ferrous Scrap[*]

(2) Axles, steel.

Solid car and/or locomotive friction bearing, 8 inch diameter and under (free of axles with key-way between wheel seats, no axles of shorter lengths than distance between wheel seats to be included).

(2A) Axles, steel.

Solid car and/or locomotive friction bearing over 8 inch diameter (free of axles with key-

---

*Specifications of Association of American Railroads promulgated by its Purchases and Materials Management Division (revised 1973). Specifications in force as of publication date.

way between wheel seats, no axles of shorter length than distance between wheel seats to be included).

(3) Axles, steel.
Roller bearing 8 inch diameter and under (no axles of shorter lengths than distance between wheel seats to be included).

(3A) Axles, steel.
Roller beating over 8 inch diameter (no axles of shorter length than distance between wheel seats to be included).

(4) Spikes, track bolts and nuts, and lock washers, may include rail anchors.

(5) Tie plates.
Steel.

(6) Rail joints, angle and/or splice bars.
Steel.

(9) Bolsters and/or truck sides, frames: uncut.
Cast steel.

(11) Cast steel, No. 2.
Steel castings, over 18 inches wide and/or over 5 feet long.

(11A) Cast steel, No. 1.
Steel castings, 18 inches and under, not over 5 feet long, including cut truck side frames and bolsters.

(12) Cast iron, No. 1.
Cast iron scrap, such as columns, pipes, plates, and/or castings of miscellaneous nature, but free from stove plates, brake shoes, and burnt scrap. Must be cupola size, not over 24 inches x 30 inches in dimensions and no piece to weigh over 150 pounds. Must be free from foreign material.

(13) Cast iron, No. 2.
Pieces weighing over 150 pounds, but not more than 500 pounds. Free from burnt cast.

(14) Cast iron, No. 3.
Pieces weighing over 500 pounds; includes cylinders, driving wheel centers and/or all other castings. (Free from hammer blocks or bases.)

(15) Cast iron, No. 4.
Burnt cast iron scrap, such as grate bars, stove parts and/or miscellaneous burnt scrap.

(16) Cast iron brake shoes.
Brakes shoes of all types except composition-filled shoes.

(17) Couplers and/or knuckles.
Railroad car and/or locomotive steel couplers, knuckles and/or locks stripped clean of all other attachments.

(18) Frogs and/or switches, uncut.
Steel frogs and switches that have not been cut apart, exclusive of manganese.

(18A) Railbound manganese frogs and switch points with manganese inserts that have not been cut apart.

(23) Malleable.
Malleable parts of automobiles, railroad cars, locomotive and/or miscellaneous malleable castings.

(24) Melting steel, railroad, No. 1.
Clean wrought iron or steel scrap, 1/4 inch and over in thickness, not over 18 inches in width, and not over 5 feet in length. May include pipe ends and material 1/8 inch to 1/4 inch in thickness, not over 15 inches x 15 inches. Individual pieces cut so as to lie reasonably flat in charging box.

(27) Rail, steel No. 1.
Standard section tee rails, original weight 50 pounds per yard or heavier, 10 feet long and over. Suitable for rerolling into bars and shapes. Free from bent and twisted rails, frog, switch, and guard rails, or rails with split heads and broken flanges. Continuous welded rail may be included provided no weld is over 9 inches from the end of the piece of rail.

(28A) Rail, steel No. 2.
Cropped Rail Ends. Standard section, original weight of 50 pounds per yard and over 18 inches long and under.

(28B) Rail, steel No. 2 cropped rail ends.
Standard section, original weight of 50 pounds per yard and over, 2 feet long and under.

(28C) Rail, steel, No. 2, cropped rail ends.
Standard section, original weight 50 pounds per yard and over, 3 feet long and under.

(29) Rail, steel, No. 3.
Standard section tee, girder and/or guard

rails, to be free from frog and switch rails not cut apart, and contain no manganese, cast, welds, or attachments of any kind except angle bars. Free from concrete, dirt, and foreign material of any kind.

(30)   Sheet scrap, No. 1.

Under 3/16 inch thick, may include hoops, band iron and/or steel, scoops and/or shovels (free of wood). Must be free from burnt or metal coated material, cushion, or other similar springs.

(31)   Sheet scrap, No. 2.

Galvanized or tinned material and/or gas retorts, and/or any other iron or steel material not otherwise classified.

(32)   Steel, tool. (Specify kind in offering.)

(33)   Steel, manganese.

All kinds of manganese, rail, guard rails, frogs and/or switch points, cut or uncut.

(34)   Steel, spring.

Coil and/or elliptical, minimum thickness 1/4 inch may be assembled or cut apart.

(34A)   Steel, spring.

Coil only.

(35)   Structural, wrought iron and/or steel uncut.

All steel or steel mixed with iron from bridges, structures and/or equipment that has not been cut apart, may include uncut bolsters, brakebeams, steel trucks, underframes, channel bars, steel bridge plates, frog and/or crossing plates and/or other steel of similar character.

(36)   Tires.

All locomotive, not cut to specified lengths.

(38)   Turnings. No. 1.

Heavy turnings from wrought iron and/or steel railroad axles or heavy forgings and/or rail chips, to weigh not less than 75 pounds per cubic foot. Free from dirt or other foreign material of any kind. Alloy steel scrap may be excluded from these specifications by mutual agreement between buyer and seller.

(38A)   Turnings, drillings and/or borings.

No. 2 cast, wrought steel and/or malleable iron borings, turnings and/or drillings mixed with other metals.

(40)   Wheels, No. 1.

Cast iron car wheels.

(42)   Wheels, No. 3.

Solid cast steel, forged, pressed and/or rolled steel car and/or locomotive wheels, not over 42 inches diameter. (Specify kind in offering.)

(45)   Destroyed steel cars.

Bodies of steel cars cut apart sufficiently to load. (Specify kind.)

(45A)   Destroyed steel car sides and box car roofs.

Cut to a maximum length of...and a maximum width of...suitable for use in super presses and shears without additional preparation.

Conversions for units—Imperial and metric—to SI, classified

Acceleration

1 ft/s² = 0·304 8 m/s²

Angular momentum

1 lb ft²/s = 0·042 140 1 kg m/s²

Area

1 acre = 4 046·86 m² = 0·404 686 ha
1 ft² = 929·03 cm²
1 in² = 6·451 6 cm²
1 mile² = 2·589 99 km²

Calorific value, volume basis

1 therm/gal = 23·208 GJ/m³
1 Btu/ft³ = 37·258 9 kJ/m³

Concentration

1 gr/100 ft³ = 0·022 883 5 g/m³
1 gr/gal = 14·253 8 mg/l
1 lb/gal = 0·099 78 kg/l
1 oz/gal = 6·236 02 g/l
1 p p m (in water) = 1 mg/l

Density

1 lb/ft³ = 16·018 5 kg/m³
1 lb/in³ = 27·679 9 g/cm³
1 ton/yd³ = 1 328·94 kg/m³

Energy, heat, work

1 Btu = 1·055 06 kJ
1 cal$_{IT}$ = 4·186 8 J
1 ft lbf = 1·355 82 J
1 hph = 2·684 52 MJ
1 kWh = 3·6 MJ
1 therm = 105·506 MJ

Force

1 kgf = 9·806 65 N
1 lbf = 4·448 22 N
1 pdl (poundal) = 0·138 255 N
1 tonf = 9·964 02 kN

Force (weight)/unit length

1 lbf/ft = 14·593 9 N/m
1 lbf/in = 175·127 N/m

Fuel consumption

1 mile/gal = 0·354 006 km/l
1 mile/US gal = 0·425 1 km/l

Heat flow rate

1 Btu/h = 0·293 071 W

Heat-transfer coefficient

1 Btu/ft²h°F = 5·678 26 W/m²K

Illumination

1 lm/ft² = 1 foot-candle
     = 10·763 9 lx (lm/m²)

Length

1 ft = 0·304 8 m
1 in = 25·4 mm
1 μin (0·000 001 in) = 0·025 4 μm
1 mile = 1·609 34 km
1 thou = 25·4 μm

Luminance

1 cd/ft² = 10·763 9 cd/m²
1 cd/in² = 1 550·00 cd/m²
1 foot-lambert = 3·426 26 cd/m²

Mass

1 gr (grain) = 64·798 mg
1 lb = 0·453 592 37 kg
1 oz = 28·349 5 g

Mass per unit length

1 lb/ft = 1·488 16 kg/m
1 lb/in = 17·858 kg/m

Moment of inertia

1 lb·ft² = 0·042 140 1 kg m²
1 lb in² = 2·926 4 kg cm²

Momentum

1 lb ft/s = 0·138 255 kg m/s

Power

1 hp = 745·70 W
1 ft lbf/s = 1·355 82 W

Pressure, stress

1 ftH₂O = 2·989 07 kPa (N/m²)
1 inHg = 3·386 39 kPa (kN/m²)
1 inH₂O (1 in w g) = 249·089 Pa (N/m²)
1 kgf/cm² = 98·066 5 kPa (kN/m²)
1 lbf/ft² = 47 880 3 Pa (N/m²)
1 lbf/in² = 6·894 76 kPa (kN/m²)
1 mmHg (1 torr) = 133·322 Pa (N/m²)
1 mmH₂O = 9·806 65 Pa = 1 kgf/m²
1 pdl/ft² = 1·488 16 Pa (N/m²)
1 std atmos = 101·325 kPa (kN/m²)
1 tonf/ft² = 107·252 kPa (kN/m²)
1 tonf/in² = 15·444 3 MPa (MN/m², N/mm²)
1 torr = 133·322 Pa (N/m²)

Specific heat-capacity

1 kcal/kg °C = 4 186·8 J/kg K

Specific surface

1 in²/lb = 14·223 3 cm²/kg

Specific volume

1 ft³/lb = 62·428 0 dm³/kg
1 ft³/ton = 27·869 6 cm³/kg
1 gal/lb = 10·022 4 dm³/kg
1 in³/lb = 36·127 3 cm³/kg

Thermal capacity per unit volume

1 Btu/ft³ °F = 67·066 1 kJ/m³ K

Thermal conductivity

1 Btu in/ft² h °F = 0·144 228 W/mK

Torque

1 lbf ft = 1·355 82 N m
1 lbf in = 0·112 985 N m

Velocity

1 ft/s = 0·304 8 m/s
1 mile/h = 0·044 704 m/s
1 mile/h = 1·609 34 km/h

Viscosity, dynamic

1 lb/ft s = 1·488 16 kg/m s
1 lbf h/ft² = 0·172 369 MN s/m²
1 lbf s/ft² = 47·880 3 N s/m²
1 slug/ft s = 47·880 3 N s/m²

Viscosity, kinematic

1 ft²/s = 0·092 903 m²/s

Volume

1 fl oz = 28·413 1 cm³
1 ft³ = 28·316 8 dm³
1 gal = 4·546 09 dm³
1 gal US = 3·785 41 dm³
1 in³ = 16·387 1 cm³
1 L or l (litre) = 1 dm³
1 pt (pint) = 0·568 261 dm³
1 qt (quart) = 1·136 52 dm³
1 yd³ = 0·764 555 m³

Volume rate of flow

1 ft³/s (1 cusec) = 0·028 316 8 m³ s
     = 28·316 8 L/s

Useful approximations to remember:

1 Btu ≈ 1 kJ, 1 ft ≈ 30 cm, 1 gal ≈ 4½ L, 1 ha ≈ 2½ acre, 1 hp ≈ ¾ kW, 1 in ≈ 25 mm,
1 kg ≈ 2¼ lb, 1 kgf ≈ 10 N, 1 kgf/cm² ≈ 14½ lbf/in² ≈ 1 bar, 1 km ≈ ⅝ mile, 1 l ≈ 1¾ pt,
1 lbf ≈ 4½ N, 1 std atmosphere ≈ 0·1 MPa ≈ 1 bar, 1 therm ≈ 100 MJ, 1 yd ≈ 0·9 m

SI, Metric Non-SI and Non-Metric Conversions

This table of commonly used units is in alphabetical order (note: μ = micro), with SI units and those to be used with SI in heavier type. Mass units such as the lb, ton and kg when used as (obsolescent) force units are correctly shown with an f appended: lbf, tonf, kgf; and the kgf or 'kilogram weight' is sometimes termed the kilopond, kp, as a force unit. In SI, mass is expressed in kg, g etc., and the Mg or t; force in newtons, N; pressure or stress as N/m^2 or Pa; fluid pressure as Pa, or sometimes as bar for the convenience of its size. Sometimes the MN/m^2 (or MPa) is expressed as N/mm^2 for stress or strength, but normally the multiplying prefix is attached only to the first term: MN/m^2, not N/mm^2 or daN/mm^2.

For quantities of gas the volume, temperature and pressure have to be stated. A volume at 'normal' or 'standard' temperature and pressure is not known precisely because these terms have varied in meaning, and Nm^3 for a 'normal' cubic metre is an obsolete Continental term. It is necessary to use m^3 (s.t.p.) and to specify s.t.p. = 0 °C, 101·325 kPa.

Also 1 m^3 (s.t.p.) = 1 m^3 (0 °C, 760 mmHg) = 35·31 ft^3 (0 °C, 760 mmHg) = 37·32 ft^3 (60 °F, 760 mmHg).

Temperature conversion: A temperature of −40 °F is −40 °C, and from there each 9 °F = 5 °C. Ex. A temperature of 59 °F is 99 °F above −40 (°F or °C); each 99 °F = 55 °C; 55 °C above −40° is a temperature of 15 °C. And 0 °C is 273·15 K; each 1 °C = 1 K. Ex. 15 °C = (273·15 + 15) K = 288·15 K.

1 **a** (are) = **100 m²**
1 Å (angström) = 0·1 nm
1 acre = **4 046·86 m²**
1 atm = **101·325 kPa** = 760 mmHg

1 **bar = 100 kPa** = 14·503 8 lbf/in²
1 board foot = **2·359 74 dm³**
1 Btu = **1·055 06 kJ**
1 Btu/ft² h °F = **5·678 26 W/m²·K**
1 Btu/ft³ = **37·258 9 kJ/m³**
1 Btu/ft³ °F = **67·066 1 kJ/m³·K**
1 Btu/h = **0·293 071 W**
1 Btu in/ft² h °F = **0·144 228 W/m·K**
1 Btu/lb = **2 326 J/kg**
1 Btu/lb °F = **4 186·8 J/kg·K**
1 bu (bushel) = **36·368 7 L**

°C = **K** (temperature interval)
1 cal$_{IT}$* = **4·186 8 J** (*Int'n'l Table)
1 cal$_{15}$* = **4·185 5 J** (*ref. to H₂O, 15 °C)
1 cal$_{th}$* = **4·184 J** (*thermochemical)
1 cd/ft² = **10·763 9 cd/m²**
1 cd/in² = **1 550·00 cd/m²**
1 chain = **20·116 8 m**
1 chain (engr's) = **30·48 m**
1 chaldron = **1·309 27 m³**
1 Ci (curie) = **3·7 × 10¹⁰ Bq** (becquerel)
1 **cm = 0·393 701 in**
1 **cm² = 0·155 000 in²**
1 **cm³ = 0·061 023 7 in³**
1 cord = **3·624 56 m³**
1 cP (centipoise) = **1 mPa·s**
1 cSt (centistokes) = **1 mm²/s**
1 ctl (cental) = **45·359 2 kg** = 100 lb
1 cwt = **50·802 3 kg**
1 cycle/s = **1 Hz**

1 **d** (day) = **86 400 s**
1 **daN/mm² = 10 N/mm²** = 0·647 5 tonf/in²
1 **dm³ = 1 L** = 0·035 314 7 ft³
1 drachm (mass) = **3·887 93 g**
1 dram = **1·771 85 g**
1 dwt (pennyweight) = **1·555 17 g**
1 **dyn = 10 μN**

1 erg = **0·1 μJ**

1 **g** = 0·002 204 62 lb
1 gal (UK) = **4·546 09 L**
1 gal/lb = **10·022 4 L/kg**
1 gauss = **0·1 mT** (millitesla)
1 gill = **0·142 L**
1 **GN/m² = 1 GPa** = 64·749 tonf/in²
1 gr (grain) = **64·798 mg**
1 gr/100 ft³ = **0·022 883 5 g/m³**
1 gr/gal = **14·253 8 mg/L**

1 **h = 60 min = 3 600 s**
1 **ha = 10 000 m²** = 2·471 05 acre
1 hbar = **10 N/mm²** = 0·6475 tonf/in²
1 hp = **745·70 W**
1 hp h = **2·684 52 MJ**
1 **Hz** (hertz) = 1/s (frequency)

1 **in = 25·4 mm**
1 **in² = 645·16 mm²**
1 **in³ = 16·387 1 cm³**
1 **in⁴ = 41·623 1 cm⁴**
1 inHg = **3·386 39 kPa**
1 inH₂O = **249·089 Pa**
1 in rope = **8·092 mm dia.**
1 in w.g. (inH₂O) = **249·089 Pa**
1 in²/lb = **14·223 3 cm²/kg**
1 in²/s = **645·16 cSt**
1 in³/lb = **36·127 3 cm³/kg**

1 **J = 1 N·m** = 0·737 562 ft lbf

K = °C (temperature interval)
1 **kg** = 2·204 62 lb
1 **kg/m³** = 0·062 428 lb/ft³
1 **kgf = 9·806 65 N**
1 kgf = **2·204 62 lbf = 1 kp**
1 **kgf/cm² = 98·066 5 kPa = 98·066 5 kN/m²**
1 **kgf/mm² = 9·806 65 N/mm²** = 0·634 97 tonf/in²
1 **kJ** = 0·277 778 W·h
1 **kJ** = 0·947 8 Btu
1 km = 0·621 371 mile
1 **km² = 247·105 acre**
1 **kN/m² = 1 kPa** = 0·145 038 lbf/in²
1 knot (UK) = **1·853 18 km/h**
1 kp (kilopond) = **1 kgf**
1 k.s.i. (1000 p.s.i.) = **6·894 76 MPa**
1 kW·h = **3·6 MJ** = 3 412 Btu

1 **m** = 1·093 61 yd = 3·280 84 ft
1 **m² = 1·195 99 yd²** = 10·763 91 ft²
1 **m³ = 1·307 95 yd³** = 35·314 67 ft³
1 **μm** = 39·370 1 μin
1 mil (0·001 in) = **0·025 4 mm**
1 mile = **1·609 34 km**
1 minim = **59·193 9 mm³**
1 **mm** = 0·039 370 1 in
1 **mm² = 0·001 55 in²**
1 mmHg = **133·322 Pa** = 1 torr
1 mmH₂O = **9·806 65 Pa** = 1 kgf/m²
1 **MN/m² = 1 MPa = 1 N/mm²** = 145·038 lbf/in²

1 N = 0·224 809 lbf
1 N = 7·233 01 pdl
1 N = 0·101 971 6 kgf
1 **N/m² = 1 Pa**
1 N/mm² = **1 MPa** = 145·038 lbf/in²
    = 0·064 75 tonf/in²

1 Oe (oersted) = **79·577 5 A/m**
1 oz = **28·349 5 g**
1 oz apoth. (or troy) = **31·103 5 g**
1 oz/gal = **6·236 g/L**

1 P (poise) = **0·1 Pa·s**
1 **Pa = 1 N/m²**
1 **Pa** = 0·000 145 038 lbf/in²
1 **Pa** = 0·020 885 4 lbf/ft²
1 **Pa** = 0·000 295 300 inHg
1 pdl = **0·138 255 N**
1 pdl/ft² = **1·488 16 Pa**
1 perch = **5·029 2 m**
1 p.p.m. (in water) = **1 mg/L**
1 pole = **5·029 2 m**
1 p.s.i. (lbf/in²) = **6·894 76 kN/m² (kPa)**
1 pt (pint) = **0·568 261 L**

1 qt (quart) = **1·136 52 L**

°R = **5/9 K** (temperature interval)
1 **rad** = 57·295 779 5°
1 rod = **5·029 2 m**

1 slug = **14·593 9 kg** = 32·174 lb
1 slug ft² = **1·355 82 kg·m²**
1 slug/ft s = **47·880 3 Pa·s**
1 std atmosphere = **101·325 kPa**
1 std atmosphere = 760 mmHg

$°F = °R = 5/9\ K$ (temp. interval)
1 fathom $= 1·8288$ m
1 fl oz $= 28·4131$ cm^3
1 foot-candle $= 1$ lm/ft^2
1 foot-lambert $= 3·42626$ cd/m^2
1 ft $= 0·3048$ m
1 ft$^2 = 929·030$ cm$^2 = 0·092\,903$ m^2
1 ft$^3 = 0·028\,3168$ m$^3 = 6·228\,83$ gal
1 ft$^4 = 83·3097$ dm^4
1 ftH$_2$O $= 2·989\,07$ kPa
1 ft lbf $= 1·355\,82$ J
1 ft lbf/lb $°F = 5·380\,32$ J/kg·K
1 ft lbf/s $= 1·355\,82$ W
1 ft pdl $= 0·042\,140\,1$ J
1 ft/s $= 0·3048$ m/s
1 ft/s$^2 = 0·3048$ m/s^2
1 ft^2 h $°F$/Btu in $= 6·933\,47$ m·K/W
1 ft^2/h $= 0·092\,903$ m^2/h
1 ft^2/s $= 0·092\,903$ m^2/s
1 ft^3/lb $= 62·4280$ dm^3/kg
1 ft^3/s (1 cusec) $= 28·3168$ L/s
1 ft^3/ton $= 0·027\,8696$ L/kg
1 furlong $= 0·201\,168$ km

1 L or l (litre) $= 1$ dm$^3 = 0·001$ m$^3 =$
  $0·219\,97$ gal
1 lb $= 0·453\,592\,37$ kg
1 lb ft/s $= 0·138\,255$ kg·m/s
1 lb ft$^2 = 42·140\,1$ g·m^2
1 lb/ft $= 1·488\,16$ kg/m
1 lb/ft$^3 = 16·018\,5$ kg/m^3
1 lb/ft s $= 1·488\,16$ kg/m·s
1 lb/in $= 17·858\,0$ kg/m
1 lb in$^2 = 2·926\,40$ kg·cm^2
1 lb/in$^3 = 27·679\,9$ Mg/m^3
1 lbf $= 4·448\,22$ N
1 lbf ft $= 1·355\,82$ N·m
1 lbf/ft $= 14·593\,9$ N/m
1 lbf/ft$^2 = 47·880\,3$ Pa
1 lbf h/ft$^2 = 0·172\,369$ MPa·s
1 lbf in $= 0·112\,985$ N·m
1 lbf/in $= 175·127$ N/m
1 lbf/in^2 (p.s.i.) $= 6·894\,76$ kN/m^2 (kPa)
1 lbf s/ft$^2 = 47·880\,3$ Pa·s
1 link $= 0·201\,168$ m
1 lm/ft$^2 = 10·763\,9$ lx $= 1$ ft candle
1 lm/m$^2 = 1$ lx; (lumen, lux)

1 St (stokes) $= 0·000\,1$ m^2/s
1 super foot $= 0·092\,9$ m^2

1 t (tonne) $= 1$ Mg
1 t $= 0·984\,204$ ton $= 2204·62$ lb
1 therm $= 105·506$ MJ
1 therm $= 100\,000$ Btu
1 therm/gal $= 23·208\,0$ GJ/m^3
1 thou $= 25·4\ \mu$m
1 ton $= 1·016\,05$ t $= 1016·05$ kg
1 ton (US) $= 907·184\,74$ kg
1 ton (refrig.) $= 3·516\,85$ kW
1 ton/yd$^3 = 1·328\,94$ t/m^3
1 tonf $= 9·964\,02$ kN
1 tonf/ft$^2 = 107·252$ kPa
1 tonf/in$^2 = 15·444\,3$ MPa $= 15·444\,3$ N/mm^2
1 torr $= 133·322$ Pa

1 US gal $= 3·785\,41$ L

1 W $= 1$ J/s

1 yd $= 0·9144$ m
1 yd$^2 = 0·836\,127$ m^2
1 yd$^3 = 0·763\,555$ m^3

T (tera) $= 10^{12}$; G (giga) $= 10^9$; M (mega) $= 10^6$; k (kilo) $= 10^3$; h (hecto) $= 10^2$; da (deca) $= 10$; d (deci) $= 10^{-1}$; c (centi) $= 10^{-2}$; m (milli) $= 10^{-3}$
μ (micro) $= 10^{-6}$; n (nano) $= 10^{-9}$; p (pico) $= 10^{-12}$

Index

—J

Just-In-Time (JIT),   238

—K

Key Minimum-Maximum Standards (KMS),   242
Kinetic undercooling,   15

—L

Ladle,   178
   inoculating ductile iron,   171–176
   pressure ladle,   162–164
   shaking,   143
Lanthanum,   86
Late metal stream inoculation,   171–172
Lead,   97, 226
Line-frequency melters,   120–122
Liquid iron-carbon alloys, structure of,   1
Liquidus arrest temperature,   223
Low-frequency induction furnaces,   120–122

—M

Machinability of Ni-resist austenitic ductile irons,   64
Magnesium,   95, 102, 186
   additions, elemental,   150
   additions using the plunging method,   161–162
   alloyed additions of,   150–154
   in ductile iron,   83–84
   fading,   7, 87
   levels associated with flake graphite,   226
   loss,   155
   nodulizer,   148–149
   overtreatment,   224, 232
   porous plug method for adding,   158–159
   pressure ladle additions of,   162-164
   recovery,   155–156
   residuals,   203
   sulfide,   77
Magnesium treatment, 223, 229
   dross inclusions resulting from,   227
   methods,   155–166
   of Ni-resists,   54
   open-ladle, pour-over process of,   156
   sandwich method of,   156–157
   selection of,   165–166
   tundish cover ladle used for,   157–158

Magnetic particle inspection,   250
Magnetic properties of ductile iron,   27–28
Management training,   237
Manganese,   211
   charge material,   138
   in ductile iron,   79–83
MAP Process, 162
Martensite,   254, 257
Mechanical properties,   199–203
   definitions of,   23–27
   influence of section size on,   197–203
Medium-frequency melters,   122
Medium-nickel electrodes,   253
Melt
   desulfurization,   77–78
   surface tension,   10
Melting
   direct-arc electric furnace,   124–128
   efficiencies of induction furnaces,   118–120
   practices, 110–131
   temperature (Liquidus) of ductile iron,   27–31
   using coreless induction furnaces,   118–124
   vertical channel furnace,   116–118
Metal cleanliness and handling,   178-179
Metal Inert Gas (MIG),   253
Metallic charge materials,   133–134
Metallographic analysis of sample specimens,   249
Microstructure, influence of section size on,   197–203
Models for growth of compacted and degenerated graphite,   15–16
Modulus of elasticity,   24–25
Modulus of rigidity of ductile irons,   25
Mold cooling,   204–207, 216
Molding,   179
Molds, hard ramming,   232
Molybdenum,   88–93, 211

—N

Ni-resist austenitic ductile cast iron,   47
   annealing,   64
   austenite stability of,   63–64
   base charge for,   53
   basic melting practice for,   54
   carbon equivalent calculations for,   55
   corrosion resistance of,   42
   electrical and magnetic properties of,   49
   erosion resistance of,   41–42
   gating and risering of,   55
   heat treatment of,   64, 64–65, 207

—U

—V

—W